Principles of Pharmacogenetics and Pharmacogenomics

The study of pharmacogenetics and pharmacogenomics focuses on how our genes and complex gene systems influence our response to drugs. Recent progress in the science of clinical therapeutics has led to the discovery of new biomarkers that make it technically easier to identify groups of patients that are more or less likely to respond to individual therapies. The aim is to improve personalized medicine – not simply to prescribe the right medicine, but to deliver the right drug at the right dose at the right time. This textbook brings together contributions from leading experts to discuss the latest information on how human genetics has an impact on drug response phenotypes. It presents not only the basic principles of pharmacogenetics, but also clinically valuable examples that cover a broad range of specialties and therapeutic areas. The first section of the book outlines critical concepts in pharmacogenetics and pharmacogenomics, including genetic testing, genotyping technologies, and adverse drug effects. The next section discusses the legal, ethical, and social implications of pharmacogenomics. The second half of the book details many of the main therapeutic areas, including oncologic drugs, cardiovascular drugs, statins, drug-induced long-QT syndrome, diabetes drugs, respiratory drugs, gastrointestinal drugs, rheumatoid arthritis drugs, obstetric drugs, psychiatric drugs, pain and anesthesia drugs, HIV and antiretroviral drugs, pediatrics, and fetal and neonatal medicine. This textbook provides an introduction to pharmacogenetics and pharmacogenomics for health care professionals, medical students, pharmacy students, graduate students, and researchers in the biosciences.

Online resources for this book can be found at www.cambridge.org/altman.

Resources include:

- Link to the Pharmacogenomics Knowledgebase
- Study guides
- Images from the book
- Discussion questions
- Content updates

RUSS B. ALTMAN is Chairman of the Bioengineering Department and Professor of Bioengineering, Genetics, and Medicine at Stanford University. His primary research interests are in the application of computing technology to basic molecular biological problems of relevance to medicine. His group builds the PharmGKB (www.pharmgkb.org).

DAVID FLOCKHART is the Harry and Edith Gladstein Chair in Cancer Genomics and Professor of Medicine, Medical Genetics and Pharmacology at the Indiana University School of Medicine in Indianapolis. He is also the Director of the Division of Clinical Pharmacology. His research is focused on clinically relevant applications of pharmacogenetics and drug interactions.

DAVID B. GOLDSTEIN is the Richard and Pat Johnson Distinguished University Professor and Director of the Center for Human Genome Variation at Duke University. He is also Professor of Molecular Genetics and Microbiology at Duke. His principal interests include human genetic diversity, the genetics of disease, and pharmacogenetics.

Principles of Pharmacogenetics and Pharmacogenomics

Edited by

Russ B. Altman

Stanford University

David Flockhart

Indiana University

David B. Goldstein

Duke University

CAMBRIDGE UNIVERSITY PRESS
Cambridge, New York, Melbourne, Madrid, Cape Town,
Singapore, São Paulo, Delhi, Tokyo, Mexico City

Cambridge University Press
32 Avenue of the Americas, New York, NY 10013-2473, USA

www.cambridge.org
Information on this title: www.cambridge.org/9780521885379

First published 2012

Printed in China by Golden Cup

A catalog record for this publication is available from the British Library.

Library of Congress Cataloging in Publication data

Principles of pharmacogenetics and pharmacogenomics / [edited by] Russ B. Altman, David Flockhart,
David B. Goldstein.
p. ; cm.
Includes bibliographical references and index.
ISBN 978-0-521-88537-9 (hardback)
1. Pharmacogenetics. I. Altman, Russ. II. Flockhart, David A. III. Goldstein, David B.
[DNLM: 1. Pharmacogenetics. 2. Pharmacological Phenomena – genetics. QV 38]

RM301.3.G45P76 2012
615′.7–dc23 2011018883

ISBN 978-0-521-88537-9 Hardback

Additional resources for this publication at www.cambridge.org/altman

Contents

Contributors — *page* vii

Introduction — xi
Russ B. Altman, David Flockhart, and David B. Goldstein

Part I: Critical Concepts

1. Introduction to Population Diversity and Genetic Testing — 3
   Michael D. Caldwell, Ingrid Glurich, Kimberly Pillsbury, and James K. Burmester
2. Genotyping Technologies — 12
   Cristi R. King and Sharon Marsh
3. Pharmacokinetics: Absorption, Distribution, Metabolism, Excretion Overview Chapter — 21
   Terrence Blaschke
4. Overview: Adverse Drug Reactions — 27
   Matthew R. Nelson
5. PharmGKB, a Centralized Resource for Pharmacogenomic Knowledge and Discovery — 38
   Li Gong and Teri E. Klein
6. DrugBank — 55
   David S. Wishart
7. Ethical Considerations for Pharmacogenomics: Privacy and Confidentiality — 66
   Sandra Soo-Jin Lee
8. Informed Consent in Pharmacogenomic Research and Treatment — 74
   Mark A. Rothstein
9. Legal Trends Driving the Clinical Translation of Pharmacogenomics — 81
   Barbara J. Evans

Part II: Therapeutic Areas

10. Oncologic Drugs — 97
    Uchenna O. Njiaju and M. Eileen Dolan

11. Pharmacogenetics and Pharmacogenomics of Cardiovascular Disease 115
Daniel Kurnik and C. Michael Stein

12. Statin-Induced Muscle Toxicity 125
Russell A. Wilke, Melissa Antonik, Elenita I. Kanin, QiPing Feng, and Ronald M. Krauss

13. Genomics of the Drug-Induced Long-QT Syndrome 136
Dan M. Roden, Prince J. Kannankeril, Stefan Kääb, and Dawood Darbar

14. Pharmacogenetics of Diabetes 145
Mark C. H. de Groot and Olaf H. Klungel

15. Pharmacogenetics – Therapeutic Area – Respiratory 154
Kelan Tantisira and Scott Weiss

16. Pharmacogenomics Associated with Therapy for Acid-Related Disorders 175
Takahisa Furuta

17. Pharmacogenetics of Rheumatology: Focus on Rheumatoid Arthritis 188
Robert M. Plenge, Yvonne C. Lee, Soumya Raychaudhuri, and Daniel H. Solomon

18. Pharmacogenetics of Obstetric Therapeutics 202
David Haas and Jamie Renbarger

19. Pharmacogenomics of Psychiatric Drugs 217
David Mrazek

20. Pain and Anesthesia 224
Konrad Meissner and Evan D. Kharasch

21. HIV and Antiretroviral Therapy 238
Amalio Telenti

22. Application of Pharmacogenetics and Pharmacogenomics in Pediatrics: What Makes Children Different? 249
Jennifer A. Lowry and J. Steven Leeder

23. Fetal and Neonatal Pharmacogenomics 263
Yair Blumenfeld

Contributors

Russ B. Altman
Chairman
Department of Bioengineering
Stanford University
Stanford, CA

Melissa Antonik
Division of Endocrinology and Metabolism
Department of Medicine
Medical College of Wisconsin
Milwaukee, WI

Terrence Blaschke
Professor Emeritus
Stanford School of Medicine
Stanford University
Stanford, CA

Yair Blumenfeld
Division of Maternal-Fetal Medicine
Department of Obstetrics and Gynecology
Stanford University
Stanford, CA

James K. Burmester
Marshfield Clinic, Research Foundation
Marshfield, WI

Michael D. Caldwell
Principal Investigator
Marshfield Clinic, Research Foundation
Marshfield, WI

Dawood Darbar
Division of Cardiovascular Medicine
Vanderbilt University School of Medicine
Nashville, TN

Mark C. H. de Groot
Heart and Lung Institute
University Hospital
Utrecht
The Netherlands

M. Eileen Dolan
Department of Medicine
Comprehensive Cancer Research Center
The University of Chicago
Chicago, IL

Barbara J. Evans
Health Law and Policy Institute
University of Houston Law Center
Houston, TX

QiPing Feng
Division of Clinical Pharmacology
Department of Medicine
Vanderbilt University Medical Center
Nashville, TN

David Flockhart
Harry and Edith Gladstein Chair in Cancer Genomics
Division of Clinical Pharmacology
Indiana University School of Medicine
Indianapolis, IN

Takahisa Furuta
Center for Clinical Research
Hamamatsu University School of Medicine
Hamamatsu
Japan

Ingrid Glurich
Marshfield Clinic, Research Foundation
Marshfield, WI

David B. Goldstein
The Richard and Pat Johnson Distinguished University Professor
Director, Center for Human Genome Variation
Duke University Medical Center
Durham, NC

Li Gong
Department of Genetics
Stanford University Medical Center
Stanford University
Stanford, CA

David Haas
Indiana University School of Medicine
Division of Clinical Pharmacology
Wishard Memorial Hospital
Indianapolis, IN

Stefan Kääb
Department of Medicine
Vanderbilt University School of Medicine
Nashville, TN

Elenita I. Kanin
University of Wisconsin Hospital and Clinics
Madison, WI

Prince J. Kannankeril
Department of Pediatrics
Vanderbilt University School of Medicine
Nashville, TN

Evan D. Kharasch
Russell D. and Mary B. Shelden Professor of Anesthesiology
Department of Anesthesiology
Washington University in St. Louis
St. Louis, MO

Cristi R. King
Faculty of Pharmacy and Pharmaceutical Sciences
University of Alberta
Edmonton, Alberta
Canada

Teri E. Klein
Department of Genetics
Stanford University Medical Center
Stanford University
Stanford, CA

Olaf H. Klungel
Division of Pharmacoepidemiology
Utrecht Institute for Pharmaceutical Sciences
Utrecht University
Utrecht
The Netherlands

Ronald M. Krauss
Senior Scientist and Director, Atherosclerosis Research
Children's Hospital Oakland Research Institute
Oakland, CA

Daniel Kurnik
Departments of Medicine and Pharmacology
Division of Clinical Pharmacology
Vanderbilt University School of Medicine
Nashville, TN

Sandra Soo-Jin Lee
Stanford Center for Biomedical Ethics
Stanford University
Stanford, CA

Yvonne C. Lee
Division of Rheumatology, Immunology, and Allergy
Brigham and Women's Hospital
Boston, MA

J. Steven Leeder
Division of Clinical Pharmacology and Medical Toxicology
Department of Pediatrics
Children's Mercy Hospitals and Clinics
Kansas City, MO

Jennifer A. Lowry
Division of Clinical Pharmacology and Medical Toxicology
Department of Pediatrics
Children's Mercy Hospitals and Clinics
Kansas City, MO

Sharon Marsh
Department of Internal Medicine
Washington University in St. Louis
St. Louis, MO

Konrad Meissner
Department of Anesthesiology and Intensive Care Medicine
Universitätsklinikum Greifswald
der Ernst-Moritz-Arndt-Universität Greifswald AöR
Greifswald
Germany

David Mrazek
Chair of Department of Psychiatry and Psychology and Director of the SC Johnson Genomics of Addictions Program
Mayo Clinic
Rochester, MN

Matthew R. Nelson
Pharmacogenetics
GlaxoSmithKline
Research Triangle Park, NC

Uchenna O. Njiaju
Department of Medicine
Comprehensive Cancer Research Center
The University of Chicago
Chicago, IL

Kimberly Pillsbury
Marshfield Clinic, Research Foundation
Marshfield, WI

Robert M. Plenge
Brigham and Women's Hospital
Harvard Medical School
Boston, MA

Soumya Raychaudhuri
Division of Rheumatology, Immunology, and Allergy
Brigham and Women's Hospital
Boston, MA

Jamie Renbarger
Indiana University School of Medicine
Division of Pediatrics
Wishard Memorial Hospital
Indianapolis, IN

Dan M. Roden
Department of Medicine
Vanderbilt University School of Medicine
Nashville, TN

Mark A. Rothstein
Herbert F. Boehl Chair of Law and Medicine
Institute for Bioethics, Health Policy, and Law
University of Louisville School of Medicine
Louisville, KY

Daniel H. Solomon
Brigham and Women's Hospital
Harvard Medical School
Boston, MA

C. Michael Stein
Departments of Medicine and Pharmacology
Division of Clinical Pharmacology
Vanderbilt University School of Medicine
Nashville, TN

Kelan Tantisira
Channing Laboratory
Brigham and Women's Hospital
Harvard Medical School
Cambridge, MA

Amalio Telenti
Institute of Microbiology
University Hospital Center
University of Lausanne
Lausanne
Switzerland

Scott Weiss
Channing Laboratory
Brigham and Women's Hospital
Harvard Medical School
Cambridge, MA

Russell A. Wilke
Division of Clinical Pharmacology
Department of Medicine
Vanderbilt University Medical Center
Nashville, TN

David S. Wishart
Departments of Computing Science and Biological Sciences
University of Alberta
Edmonton, Alberta
Canada

Introduction

Health care is moving toward a more individualized approach that has been termed "personalized medicine." The underlying causes for this transition are many; they include the ability to genotype and sequence DNA, the increasing emphasis on consumerism among patients, and changes in the pharmaceutical industry worldwide and particularly at the U.S. Food and Drug Administration (FDA) and its sister regulatory agencies around the world. Pharmacogenetics and pharmacogenomics both involve the study of how genetics exerts an impact on drug response phenotype. For our purposes, the term "pharmacogenetics" connotes single genes that dominate the effects on a drug response, whereas "pharmacogenomics" connotes systems of many genes that create complex drug response phenotypes. It is clear that pharmacogenetics and pharmacogenomics are the core elements of the future of personalized medicine.

The emergence of robust and effective patient advocacy groups over the past thirty years has led to organized demands by patients for medicines that are more effective and that have fewer side effects. This was fueled by the Institute of Medicine "To Err is Human" report of 1999, which estimated that more than 50,000 Americans die each year because of medical errors, in particular, involving prescription drugs. Health care organizations have registered the clinical and financial dangers inherent in medication errors, and more precise prescribing is now a central part of quality control and even part of the marketing campaigns for hospitals in the United States. The pharmaceutical industry is experiencing the death of the "blockbuster" model of drug development in which one dose of a single medication can be used to treat everyone, including men and women, people of all races, infants, adolescents, adults, and the elderly. The success of therapies that treat a carefully selected subset of the population, such as Herceptin™ in the treatment of breast cancer, demonstrates that focusing a therapy on a population with a better benefit-to-risk ratio need not incur economic calamity. Last, the inexorable progress of science within clinical therapeutics has led to the discovery of new biomarkers for therapeutic effect that make it technically easier to identify groups of patients who are more or less likely to respond to individual therapies. Measures of DNA sequence (both focused genotyping and full sequencing) are the biomarkers whose cost has dropped most precipitously, with an accuracy approaching perfection.

The revolution occurring in the use of biomarkers to assess the risks and benefits of drugs is not confined to new prescription medicines, but includes the entire therapeutic armamentarium. Both the FDA in the United States, through its efforts on age-old medicines such as warfarin and tamoxifen, and the National Institutes of Health, through its funding of basic research (e.g., the Pharmacogenetics Research Network), have shown that they expect all existing therapies to be evaluated for the potential of more targeted use. As a result, health care providers and administrators will rapidly need to understand the optimal selection and use of these new biomarkers to provide the best care possible. Consistent with this need, research and reimbursement agencies are stressing the importance of data on "comparative effectiveness" between existing medications – an emphasis that will inevitably involve the use of validated biomarkers that will soon be integrated into routine medical care.

Although the current focus of pharmacogenomic research is on biomarkers derived from inherited (germline) DNA, there is increasing interest in somatic biomarkers from tumors, proteomics, and metabolomics. The initial focus on DNA is natural: it is relatively stable and easier to handle than other more degradable biologic materials like RNA and protein. In addition, we have a detailed map of the human genome, and sentinel examples of the use of DNA are already available as role models. These advantages are not enough to move this science into the clinic, however.

For genetic testing to realize its full potential to improve drug selection and dosing, we must integrate the science and communicate its clinical value within the curricula of pharmacy and medical schools. As part of this effort, we recognized the need for a book that presents not only the basic principles of pharmacogenetics, but also the clinically valuable examples that cover a broad range of clinical specialties and therapeutic areas. Our intended audience is medical and pharmacy students, as well as practicing professionals. This book represents our first attempt to create such a text. It represents the work of many scientists and practicing physicians in a wide range of medical specialties, and is designed to provide not only a broad overview of the science underlying this testing, but also a strong, practical element for clinical practice. We are grateful to all these contributors not only for the many hours of toil involved in creating this work, but also for their continued efforts to educate a new generation of health care professionals, not simply to prescribe the right medicine, but to deliver "the right drug at the right dose at the right time."

Russ B. Altman, Dave Flockhart, and David B. Goldstein

PART I

Critical Concepts

Introduction to Population Diversity and Genetic Testing

1

Michael D. Caldwell, Ingrid Glurich, Kimberly Pillsbury, and James K. Burmester

OVERVIEW OF HOW GENETIC DIVERSITY ARISES

Genetic diversity arises from differences in the genome of humans. Mapping of the human genome has revealed that humans are approximately 99.9 percent identical relative to their DNA sequences, with differences occurring at the rate of one change in every 100 to 300 bases along the sequence of 3 billion bases that comprise the human genome (1, 2). These differences in coding sequences at these sites are termed *single-nucleotide polymorphisms* (SNPs) and are considered significant when they occur with a minimum frequency of 1 percent in a given population. Whereas genetic differences of 0.1 percent among individual humans appear negligible, occurrences of ~10 million SNPs have been cataloged to date. These SNPs may occur in coding or noncoding regions of the genome and may affect gene expression or disease susceptibility. Whereas SNPs are estimated to account for 90 percent of all genetic variability, other genetic differences have been detected and may stem from errors during DNA replication, including copy number variations, insertions, deletions, inversions of bases, or other mutational events caused by environmental factors (3). This genetic variability contributes to genetic diversity among populations as well as its individual members. Genetic diversity has an impact on all manner of human traits from external appearance to disease susceptibility and response to pharmacological agents.

To gain a greater appreciation of the magnitude of the impact that genetic diversity has on human phenotypes, it is best to begin with an exploration of its historical development in modern humans. In general, race and ethnicity, which are largely defined culturally by phenotypic traits or defined by man-made designations of geographic origin, become much less distinct at the genomic level. Mutations can occur within *alleles* that occupy a distinct site within a given gene or genomic region and thereby help to define traits. It is thought that population genetic diversity is largely due to a combination of allelic mutation at specific sites along the genome and the selective pressure from population segregation that occurred during the migration of humans out of Africa over a period of about 200,000 years (4). Each offspring inherits two alleles, one from each parent. Thus, if either parent is not a carrier of the wild-type allele originally encoded in the genome due to local mutations within the allele, the offspring may inherit the mutant allele and be heterozygous for a given trait. The mutation may be dominant (i.e., expressed), recessive (carried silently), or, in some cases, coexpressed. When a new mutation is associated with a beneficial trait, it is thought that *positive selection* occurs, allowing those carrying the beneficial allele to survive, reproduce, and pass on the trait to their offspring at a frequency dictated by its pattern of inheritance. However, the major contribution to genetic diversity occurred because of the geographical isolation that resulted from tribal resettlement following migration that produced colonies of individuals with a reduced genetic pool representative of the founders of each new colony.

Figure 1.1 shows the migration pattern across Europe and Asia. As the oldest "modern humans," Africans have had the most time to accumulate changes in their DNA, making them the most genetically diverse race and the ancestral foundation for the evolution of all other races as they are presently defined. As tribes of humans left Africa and settled in a new location, these migrants distributed a subset of the total gene pool in the locations where they resettled. Figure 1.2 shows the genetic relatedness of Africans with Europeans and Asians. The traceable effect on the evolving gene pool at the site of resettlement exerted by the small number of individuals newly migrated into an area is known as the *founder effect.* The founders reflect their own limited genetic variation relative to the larger population from which they originated. At the same time, any new mutation that arose in the colony was not dispersed back into the ancestral

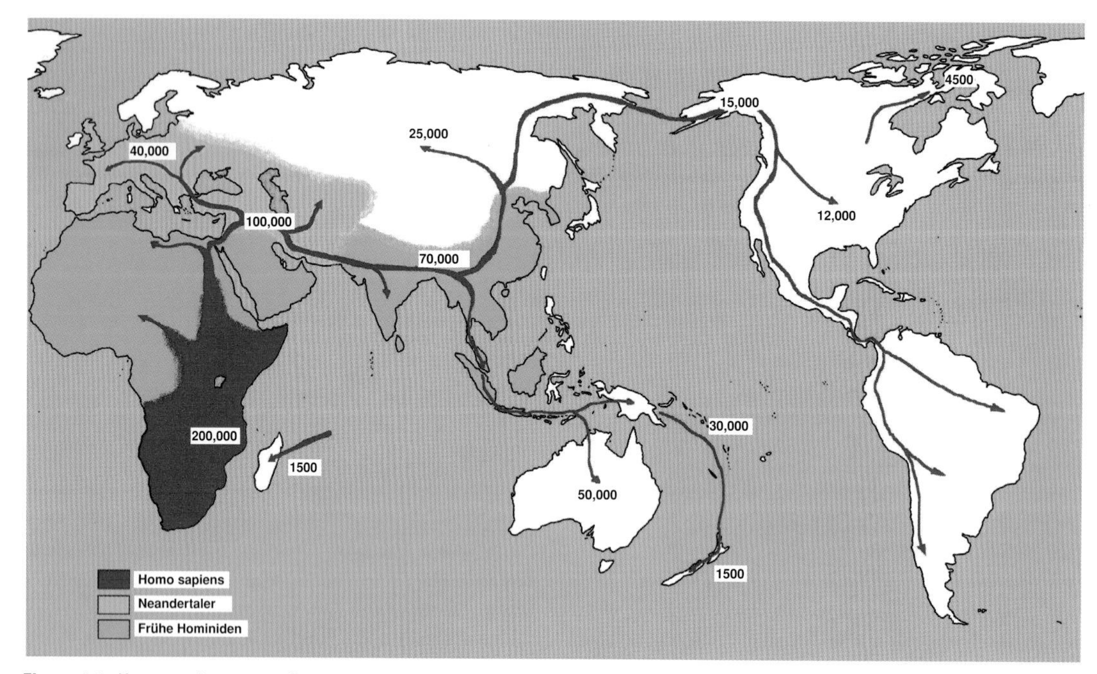

Figure 1.1. Homo sapiens spreading over the world.

foundation population in Africa. This resulted in isolated local offspring that carried the new mutation. When the mutation was associated with disease, the frequency of that disease increased in relative proportion to its allele frequency and penetrance within the population.

Within Africa, genetic diversity correlates with cultural variation and the origin of tribal language (5). Africans are descended from 14 ancestral populations that settled across the continent, establishing cultural and linguistic boundaries. The voluntary migrations out

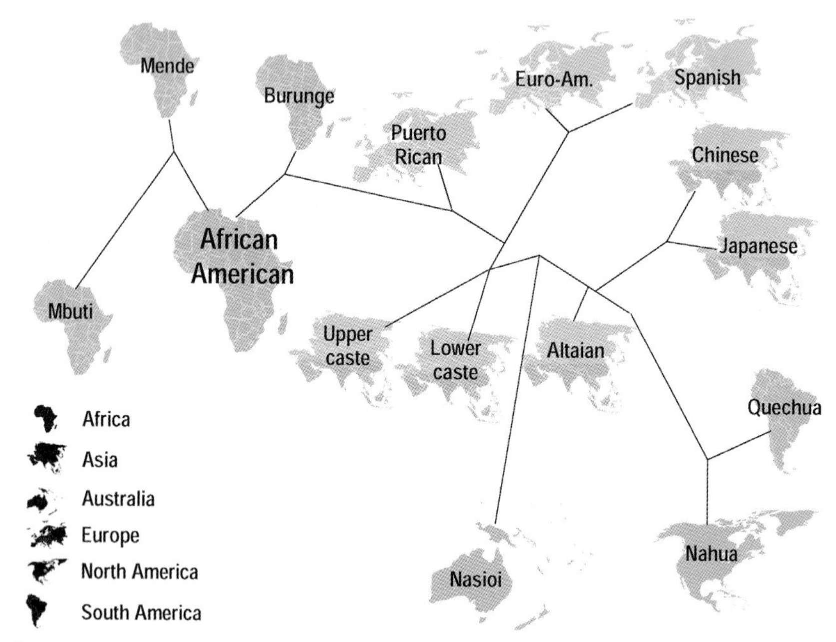

Figure 1.2. A display of population differences (adapted from Race, Genetics, and Healthcare. National Coalition for Health Professional Education in Genetics [NCHPEG]. Located at http://www.nchpeg.org/raceandgenetics/index.asp).

of Africa occurred from East Africa, whereas the involuntary migrations that occurred with the slave trade were from West Africa. As a result, African Americans have mixed ancestry from the expanse of West Africa. The ratios of genetic origins reflected among African Americans is, on average, 71 percent West African, 8 percent from other parts of Africa, and 13 percent European. The identification of genetic admixture underlying genetic diversity has introduced challenges in integrating genetics into best medical practice. Whereas mapping of the human genome and improved understanding of genetic diversity have given an appreciation of where genetic differences and similarities lie, advancement in our understanding of optimal application of genetics to the management of human health depends on a deeper understanding of how to expand and bridge our application of genetic diversity. Approaches that have been undertaken to advance our understanding include:

- further definition of the human genome through the HapMap Project, and
- association studies that are designed to specifically define clinical manifestations consequential to genetic diversity.

ROLE OF THE HapMap PROJECT

The objective of the International HapMap Project was to establish common patterns of variations in human DNA sequences that could help investigators discover genetic factors influencing vulnerability to disease and variability in drug response (6). The HapMap Project was central to cataloging the identity, position, and frequency of polymorphisms that occur in the human genome across races and ethnicities. This project has developed genome-wide maps for European, African, and Asian populations. The project compiled data on 270 DNA samples from four distinct populations. Samples from two populations represented trios that included two parents and one adult child: these were a Utah (U.S.) population of largely Northern and Western European ancestry ($n = 30$ trios) and a cohort of Yoruba people in Ibadan, Nigeria ($n = 30$ trios). The other two populations consisted of unrelated Japanese from Tokyo, Japan ($n = 45$), and Han Chinese from Beijing, China ($n = 45$). The HapMap Project made it possible to identify 100,736 SNPs that appeared uniquely in each ethnic group (7).

Furthermore, the HapMap Project delineated relationships among SNPs. Ten million SNP sites are estimated within the human genome (6). The HapMap Project capitalized on the observation that SNP alleles in close proximity to each other are often strongly associated and inherited as a "block" (8). When two SNPs are thus associated, they are said to be in linkage disequilibrium (LD). When SNPs exhibit a high degree of LD, their respective alleles are almost always inherited together. This allows detection of the presence of one SNP in an individual's genome to be highly predictive of the presence of the other SNPs in its haplotype block (8, 9). Those SNPs that are used to identify haplotype blocks are called "tag SNPs." The HapMap Project took advantage of the presence of tag SNPs to create the haplotype map of chromosomes (10, 11). Genotyping for tag SNPs as an indicator to detect coinheritance of SNPs in haplotype blocks obviates the need to genotype all 10 million common SNPs within the genome to detect SNPs associated with genetic traits (8, 12, 13).

An example of a haplotype block is shown in Figure 1.3. This pharmacogenetic example highlights the gene VKORC1, which encodes an enzyme involved in the reduction of vitamin K1 into an active cofactor for the synthesis of γ-glutamyl carboxylated clotting factors. The anticoagulant warfarin blocks the activity of this enzyme, thus reducing the levels of proteins essential for clot formation. Variant alleles have been detected within VKORC1 genes, and these define the amount of the enzyme that is made. In turn, the level of activity of the enzyme, along with several other factors defined to date, determines the dose of warfarin required to achieve stable levels of anticoagulation. Haplotype blocks H1 and H2 are present in individuals that require less warfarin to achieve optimal anticoagulation, and haplotype blocks H7, H8, and H9 are present in individuals that require higher doses of warfarin (14). These blocks can be identified by use of the tag SNPs that are highlighted in Figure 1.3, and, as shown, genetic testing of these tag SNPs can be used to modulate the warfarin dose.

In phase I of the HapMap Project, 1 million SNPs were genotyped (7) and another 2.1 million SNPs were genotyped in phase II (15). Park et al. (16) compared the genomic profiles of the four populations in the HapMap Project database and identified the ethnically variant single-nucleotide polymorphisms (ESNPs) using the nearest shrunken centroid method (NSCM). From the top eighty-two ESNPs initially selected to classify the populations, Zhou and Wang (17) established a set of sixty-four SNPs that classified the HapMap ethnic populations more efficiently.

Several studies have emphasized the importance of population diversity within the context of the HapMap database, so that the haplotype maps for different populations would help in the identification of similarities and differences among ethnicities (18). For example, Lin, Hwang, and Tzeng (19) analyzed the practicality of using SNP data from the HapMap database to represent the overall Taiwanese (TWN) population in association studies. Results of the study showed that only the Han Chinese in the Beijing population of the HapMap database characterized the TWN in terms of allele and haplotype frequencies and could be used in association studies to represent the TWN. A study by Ouyang and

	Haplotypes	Frequency		Dose (mg/d)
H1	***C***---C---G---***A***---T---C---***T***---C---***T***---G	12%	2.9	**Low Dose**
H2	***C***---C---G---***A***---G---C---***T***---C---***T***---G	24%	3.0	
H7	***T***---C---G---***G***---T---C---***C***---G---***C***---A	35%	6.0	
H8	***T***---A---G---***G***---T---C---***C***---G---***C***---A	8%	4.8	**High Dose**
H9	***T***---A---C---***G***---T---T---***C***---G---***C***---G	21%	5.5	

Figure 1.3. Polymorphisms are listed in order for positions 381, 861, 2653, 3673, 5808, 6009, 6484, 6853, 7566, and 9041 of reference sequence AY587020 in GenBank (adapted from Rieder et al. *N Eng J Med.* 2005;352:2285–93).

Krontiris (18) found that a highly conserved haplotype structure could be established among ethnic populations having African ancestry and non-African populations. The study also found evidence of common ancestry in all the population groups.

There are plans to extend the HapMap Project to include seven other populations to provide information on less common variants that have the potential to affect individuals (15). These populations are a Luhya cohort in Webuye, Kenya; the Maasai in Kinyawa, Kenya; the Tuscans in Italy; the Gujarati Indians in Houston; the Chinese in Denver; a cohort with Mexican ancestry in Los Angeles; and a cohort with African ancestry in the southwestern region of the United States (15). Whereas it is expected that most common haplotypes are found in all human populations, the frequency of a specific haplotype may differ among populations. Another important future goal of the HapMap Project will be to create molecular phenotypes for the DNA samples and to combine the SNP information with associated structural variations (15).

OVERVIEW OF ASSOCIATION STUDY DESIGN AND CONSIDERATIONS

The availability of the HapMap database has greatly facilitated the study of the relative association of disease traits with specific SNPs or groups of SNPs defined by a haplotype. In these studies, DNA variants from a cohort of individuals affected with the disease of interest are compared with the sequence of a cohort of individuals not affected by the disease. Statistical methods are used to determine whether differences in allele frequency are significant after adjustment for multiple comparisons. Then it is important to test any newly identified genetic association in additional cohorts to replicate the finding and to determine whether the association relates to different racial groups.

The initial approach to association studies was limited to sequencing potential candidate genes for variants that were present at higher frequency in affected individuals in comparison with unaffected controls (20). Some candidate gene studies were based on the detection of a mutation in a gene of an individual affected with the disease. Often these studies were done within families by comparing affected and unaffected family members. Whereas candidate gene analysis is useful in some instances, an inherent limitation to this approach is that families represent a very narrow example of genetic diversity because of their high degree of relatedness. As a result, association may be much more limited or absent in the broader population, which will exhibit higher levels of genetic diversity.

Availability of the HapMap database allowed an alternative approach to association studies in which SNPs are used as genetic markers to discover associations between disease and genomic regions. This indirect approach for detecting sequence variation and disease gene identification is more effective and efficient than the use of the candidate gene approach, because it allows a small set of tag SNPs to identify the common patterns of variation in the genome with a high probability of detection for a disease-gene association (6). Candidate genomic regions thus identified are then more densely mapped by genotyping additional SNPs occurring within the regions that are in close proximity to each other. Detection of high levels of LD among the SNPs helps to narrow the region and allows identification of genes or other genomic regions that underlie the detected association (21).

The newest approach to association studies is the *genome-wide association study* (GWAS) or *whole genome association study*. This approach involves a more global examination of genetic variation across the entire genome for the purpose of detecting genetic associations and linking them to observable traits. These studies use a case-control design in which both cases and controls are genotyped. Data are analyzed using bioinformatics approaches to identify regions exhibiting genetic variability between case and control cohorts. If a higher frequency of an SNP is detected in people with the disease, a level of association is defined. This approach allows identification of genomic regions associated with disease in a non–hypothesis-driven manner (22). A list of genetic associations identified by GWAS is summarized at http://www.genome.gov/gwastudies/. In many GWAS studies, the variant that is identified as associated with the disease is unlikely to be the variant that actually causes the disease. In many cases, the variant is in LD with the causal variant, and additional functional studies are required to understand the physiological role of the variants.

Several genotyping platforms are available to conduct GWAS and other association studies. The Affymetrix genotyping array is designed to use randomly selected, evenly spaced SNPs across the whole genome. The Illumina platform uses fewer SNPs, but uses HapMap tag SNPs. Because the use of tag SNPs is more efficient than using the same number of randomly selected SNPs, the difference in power can be as large as 20 percent, with coverage in Africans being most difficult. The use of tag SNPs for association studies results in a 5 percent to 10 percent reduction of power compared with using all HapMap SNPs (23).

LINKAGE DISEQUILIBRIUM STRUCTURE IN POPULATIONS OF DIVERSE GEOGRAPHICAL ORIGIN

The history of the population being studied affects LD patterns within the population. As the ancestral population from which all others evolved, it follows that the most diverse haplotype structure is found in Africa. As distance from Africa increases, diversity in haplotype structure decreases, reflecting the origin of humans in Africa and migration and segregation patterns around the world. This is consistent with each group of migrants drawing a sample, but not a complete set, of haplotypes from the population at the original site they left. Linkage disequilibrium patterns describe population history, human migration, natural selection, and localization of preferred recombination sites.

Recombination hot spots are loci prone to exhibit variability in sequence and are similar across different racial groups. However, environmental pressures have heavily influenced resulting genetic expression. "Bottlenecks" are one example of how environment can have an impact on genetic expression. Genetic bottlenecks are brought about by cataclysmic events that markedly reduce the number of humans, consequentially causing the extinction of many genetic lineages within the population, and thereby decreasing genetic diversity. One example of a bottleneck occurred about 70,000 years ago in Africa (24). It is proposed that the supereruption of the volcano Toba spewed volcanic ash into the atmosphere, resulting in drought and famine. As a result, it is estimated that only 5,000 females survived in Africa. When a bottleneck reduces a population that later expands, the result can be *genetic drift* in which changes in allele frequency occur independently of selection pressure, sometimes at the cost of elimination of beneficial adaptations. Bottlenecks and genetic drift are associated with decreased LD.

Population stratification due to *population admixture*, in which a cohort reflects a mixture of several subpopulations, may also have an impact on LD. In this scenario, a subpopulation with altered frequencies for expression of a given allele will cause the perception of an increased LD. For example, population admixture can occur in the presence of founder effect where a rare allele in the originating generation has penetrated one of the subpopulations in the admixture resulting in allele frequencies that are incongruent with natural selection. This may result in the misleading appearance of a higher frequency of a genetic disease within the combined population. To control for type 1 error due to the presence of population admixture in association studies, data are subjected to Hardy-Weinberg equilibrium testing (25).

There is low portability of HapMap tag SNPs in Africans, because the LD length in these individuals is smaller than the length in whites and Asians. In contrast to European and Asian data, where tag SNPs are valuable for identifying a haplotype block, these same tag SNPs do not efficiently identify haplotype blocks in Africans. By contrast, tag SNPs transfer well among white populations living in distant regions of the world, but they do not transfer nearly as well between more distant populations such as African Americans and Africans (26–28). It is also difficult to estimate how well tag SNPs will cover rare SNPs.

POPULATION STRATIFICATION ISSUES IN STUDY DESIGN

In association studies using SNPs, a subpopulation of people with an index condition are compared with otherwise-matched individuals derived from the same population. However, the same limitations that apply to familial candidate gene analysis may apply when this association is tested in populations that exhibit higher genetic diversity in comparison with the index population. An understanding of population structure is critical to avoiding false associations and allows assessment of the extent to which population stratification may affect the results of an association study (29). Nonrandom distribution of individuals in an association study results in *population stratification.* In the presence of population admixture, an association may be inferred if the trait distribution and allele prevalence differ between the subpopulations, even though no biological influence on the trait is present and the locus is not linked to any gene that influences the trait. In this case, population stratification can lead to the false association of a DNA variant with a disease.

Another form of population stratification leading to false association in a case-control design occurs when the population being studied has subpopulations, and the case and control groups have different proportions of each subpopulation represented in each study arm (30). Allele and genotype frequencies between the case and control groups should be the same. Several current methods for assessing population stratification involve SNP frequency analyses (31, 32). Whereas association studies are generally performed using large sample sizes to increase the strength of the association and the significance of the findings, it is important to determine whether population stratification is present to avoid false-positive results. For example, a study by Yamaguchi-Kabata et al. (29) showed that, within the same Japanese population, genetic differentiation created subpopulations according to geographic region. Different stratifications of the Japanese subpopulations in the case and control groups caused more false-positive results with increasing sample size (29).

The choice of the optimal statistical method for a population-based association study should be a function of study population sampling. Four population-based statistical methods were evaluated in different population stratification levels for their ability to reduce the influence of the stratification in an association study (33). Data from the HapMap Project were used to imitate a stratified population, and the four statistical methods were applied, including the traditional case-control test (TCCT), structured association (SA), genomic control (GC), and principal components analysis (PCA) (26, 33–35). PCA had a low rate of false-positive results with a high level of accuracy, making it the best choice to correct for population stratifications in association studies (33). SA and PCA had comparable results as long as adequate ancestral informative markers were included in the SA analysis (33). The ability of GC to correct for population stratification depended on the level of stratification; the GC analysis was effective only in studies with low levels of population stratification (33).

Most GWAS attempt to correct for population stratification during analysis of the genetic data. Several methods have been developed to correct for admixture in association studies (34–37). Other issues in study design that have an impact on the analysis of association include small sample size, small effect of alleles on disease causation, parent of origin effects, interaction between environment and genes, and copy number changes throughout the genome at genes of interest. Because cases are more likely than controls to be related in diseases with a genetic basis, the assumption of independence of observations is violated. This may lead to overestimation of the size of the association (38).

POPULATION STRATIFICATION ISSUES RESULTING FROM EXTRAPOLATING RESULTS FROM ONE POPULATION TO ANOTHER

The effect of population structure on common SNP variation is considerable and emphasizes the need for understanding both ancestry and stratification in association studies (39). Some variants linked to common disease risk in Americans of European descent have significantly different frequencies from those in other American ancestral populations, making it difficult to study the effects of the variant on disease risk (28, 40). This also limits the ability to extrapolate findings for a single population to the entire human genome.

Admixture among races is common and increasing, making it difficult to apply personalized medicine to patient treatment based solely on the frequency of polymorphisms of medical relevance in a racial population. Personalized treatment of disease and response to drug therapy rely on inherited genetic individuality. This individuality necessitates genetic testing of each patient for the polymorphisms prevalent in the racial background. However, because of admixture, it becomes necessary to expand the testing panel to cover disease-causing polymorphisms frequent in other racial groups to ensure that patients of mixed background receive complete genetic testing appropriate to them or to concomitantly use panels that define ethnic diversity. The application of genetic testing to pharmacogenetics requires that all potentially important alleles are tested on the basis of the race of the patient. Fortunately, multiplex genetic testing platforms are available, making it possible to test all of the known relevant functional alleles in genes that modify drug response.

Characterization of common SNPs among ancestral populations helps test the influence of common variants on disease risk (41). The degree to which common variants may account for disease risk across populations depends in part on whether alleles are common, or at least shared, among different populations. Even if SNPs are common among differing populations, the frequency of variants may differ between groups (27). Common alleles that influence the risk of common disease vary greatly in frequency across populations (27). Population-specific natural selection may have perpetuated variations in frequency of the genetic variants that play a role in the common disease risk for different populations (28). A study comparing the differentiation in frequencies of disease-associated SNPs and random SNPs in the genomes of Europeans and Africans found that ethnicity was not a good predictor of either disease-associated or random variants (28). Thus, ethnicity cannot easily be tied to risk of genetically based common diseases, because frequencies of the risk alleles do not vary significantly among ancestral populations.

COMMON VARIANTS VERSUS RARE VARIANTS

The "common disease, common variant" hypothesis states that the genetic impact on common diseases is due to a limited number of DNA variants with a frequency greater than 1 percent within the disease-causing gene. At the outset of the HapMap Project, there were few data supporting the identification of genes for common disease. As a result, the common disease, common variant hypothesis remained controversial. The alternate hypothesis, the "rare variant hypothesis," states that inherited predisposition to common diseases is the result of additive affects of multiple low-frequency dominant alleles, each independently contributing variants of multiple genes, each of which confers a moderate but detectable increase in risk (42). Taken to the extreme, the rare variant hypothesis would predict that each disease-causing mutation might only be found once in the population. In many cases, rare variants are most pronounced in specific ethnic groups because of a founder effect. Both the common disease/common variant and common disease/rare variant causes of disease are correct, depending on the gene and disease examined. In the case where the risk-associated allele is very rare, by definition, multiple genes must be involved to manifest a common disease. By contrast, the common disease/common variant hypothesis holds that genetic variants that cause disease should be present at least at 1 percent to 5 percent across a population.

Rare variants are important in inherited disease, especially where the mutation results in a highly penetrant manifestation of disease. In these cases, the mutation causes disease in almost all individuals that inherit the mutation. Approximately 5 percent of colorectal cancers are inherited in a familial Mendelian manner. These cancers are caused by highly penetrant deleterious mutations in genes such as HNPCC, mismatch repair genes (MLH1 and MSH2), Wnt signaling genes (APC, AXIN1, and CTNNB1), and others. Similarly, approximately 5 percent of breast cancer results from rare inherited mutations in the BrCa1, BrCa2, or FANCF genes.

In many cases, multiple genes have been identified for the common diseases, with each new variant contributing a small amount (1.0–1.5) to the odds ratio (OR) attributed to development of disease. Because of the low contribution to OR, it has been important to do very large studies and then replicate the observation in several additional cohorts. In addition, because of the small ORs, studies have had to account for confounding effects of population structure, population admixture, and the testing of a very large number of SNPs.

POSITIVE SELECTION AND DISEASE INCIDENCE

Genetic diversity among populations is thought to support survival of the population through positive selection. Positive selection predicts an increase in protective gene alleles that support fitness and is thereby considered associated with the emergence of new phenotypes. Population segregation also clearly adds to this selection. Selection is also thought to be driven by environmental factors present in a given geographical location that increases the risk of a population for certain diseases. The increase in frequency of a protective gene allele is thought to counterbalance the environmental factor resulting in increased fitness of individuals with the protective allele, whereas those with disease susceptibility exhibit lower survival. Because environmental factors are not evenly distributed across all geographic portions of the world, the protective allele would then increase in frequency in localized regions but not in other distant populations. Genes affected by positive selection could cover a diverse set of phenotypic characteristics including host-pathogen interactions, reproduction, dietary influences, and facial and body appearances and attributes.

An example of positive selection that occurred in Africa at the level of the β-globin gene was the emergence of red blood cell sickling based on protection against malarial infection of the red blood cells by *Plasmodium falciparum* and *Plasmodium vivax* organisms. Individuals that carried one mutant allele had greater resistance to malaria than homozygous wild-type individuals and thus were more protected from malarial disease. As a result, the allele frequency has risen under selective pressure in malaria-infested regions of Africa. However, this trait can also be deleterious, because individuals that are homozygous for the mutant β-globin allele develop sickle-cell anemia.

The identification of genes underlying common diseases holds substantial potential impact on human health. However, genetic diversity will challenge whether genetic observations made in one population will be applicable to another population. Over time, the allelic frequencies at specific sites may have undergone changes across distinct populations, and alleles that are important in the development of complex diseases in one population may not be relevant in another population because of the genetic divergence of the populations. It is therefore imperative that the impact of genetics and the environment be compared across populations to accurately assess disease susceptibility, response to therapy, and environmental triggers for disease.

GENETIC DIVERSITY AND PERSONALIZED MEDICINE

A thorough understanding of genetic diversity brings with it the promise of personalized medicine. Broadly defined, *personalized medicine* is health management informed by knowledge of the underlying genetics of each individual. For example, medical management of patients by the use of a personalized medicine approach would assess genetically encoded capacity for drug metabolism and disease susceptibility. Personalized medicine represents a substantial shift in the current population-based paradigm for medical practice (evidence-based medicine) that extrapolates algorithms based on the average observed global experience of the broader population and applies them to its individual members. For example, scientists have classically conducted drug trials using randomized case-control experimental designs within a given population and administer only those drugs that achieve a defined standard of safety and efficacy to the population at large. However, it is well known that there is considerable interindividual variation in patient response to pharmacological agents attributable to genetically based capacity for drug metabolism (43). In many cases, the polymorphisms that determine drug *A*bsorption, *D*istribution, *M*etabolism, and *E*limination (ADME) vary markedly in frequency across different individuals and racial heritages.

An understanding of population diversity is central to developing personalized medicine. For example, to develop relevant genetic tests that predict response to medications, it is imperative to understand the genetic variation prevalent among diverse races that define drug responsiveness (44). Furthermore, it is important to recognize how population diversity confounds the definition of the genetic impact on disease susceptibility and how this must be taken into account when designing studies that test disease association with genetic factors.

As our understanding of genetic diversity expands, it grows increasingly clearer that the "one size fits all" approach to medicine is no longer relevant in most circumstances. An example of the growing acceptance of the concept of personalized medicine can be found in the approval of the drug Bidil in 2005 by the Food and Drug Administration specifically for African Americans with heart failure. This was possible only after strong drug efficacy was noted in this subpopulation, but not in the broader population (44). A deeper understanding of genetic diversity among populations has concomitantly blurred the artificial distinctions of race and ethnicity that have risen historically based on phenotypic characteristics and social organization. Ideally, to realize personalized medicine, as genetic diversity among populations and their individual members becomes more defined and the optimization of study design increases, a clearer understanding of disease processes relevant to specific subpopulations will emerge. These developments will allow for identification of relevant risk factors, for preventive medicine strategies, and for development of effective postdiagnostic pharmacological treatment and medical care tailored in a population- and patient-specific manner.

REFERENCES

1. Guttmacher AE & Cullins FS. Genomic medicine – a primer. *N Engl J Med.* 2002;**347**:1512–20.
2. Altshuler D, Daly MJ, & Lander ES. Genetic mapping in human disease. *Science.* 2008;**322**:881–8.
3. Kidd JM, Cooper GM, Donahue WF, Hayden HS, Sampas N, Graves T, Hansen N, Teague B, Alkan C, Antonacci F, Haugen E, Zerr T, Yamada NA, Tsang P, Newman TL, Tüzün E, Cheng Z, Ebling HM, Tusneem N, David R, Gillett W, Phelps KA, Weaver M, Saranga D, Brand A, Tao W, Gustafson E, McKernan K, Chen L, Malig M, Smith JD, Korn JM, McCarroll SA, Altshuler DA, Peiffer DA, Dorschner M, Stamatoyannopoulos J, Schwartz D, Nickerson DA, Mullikin JC, Wilson RK, Bruhn L, Olson MV, Kaul R, Smith DR, & Eichler EE. Mapping and sequencing of structural variation from eight human genomes. *Nature.* 2008;**453**:56–64.
4. Relethford JH. Genetic evidence and the modern human origins debate. *Heredity.* 2008;**100**:555–63.
5. Tishkoff SA, Reed FA, Friedlaender FR, Ehret C, Ranciaro A, Froment A, Hirbo JB, Awomoyi AA, Bodo JM, Doumbo O, Ibrahim M, Juma AT, Kotze MJ, Lema G, Moore JH, Mortensen H, Nyambo TB, Omar SA, Powell K, Pretorius GS, Smith MW, Thera MA, Wambebe C, Weber JL, & Williams SM. The genetic structure and history of Africans and African Americans. *Science.* 2009;**324**:1035–44.
6. International HapMap Consortium. The International HapMap Project. *Nature.* 2003;**426**:789–94.
7. International HapMap Consortium. A haplotype map of the human genome. *Nature.* 2005;**437**:1299–1320.
8. Gabriel SB, Schaffner SF, Nguyen H, Moore JM, Roy J, Blumenstiel B, Higgins J, DeFelice M, Lochner A, Faggart M, Liu-Cordero SN, Rotimi C, Adeyemo A, Cooper R, Ward R, Lander ES, Daly MJ, & Altshuler D. The structure of haplotype blocks in the human genome. *Science.* 2002;**296**:2225–9.
9. McVean GA, Myers SR, Hunt S, Deloukas P, Bentley DR, & Donnelly P. The fine-scale structure of recombination rate variation in the human genome. *Science.* 2004;**304**:581–4.
10. Daly MJ, Rioux JD, Schaffner SF, Hudson TJ, & Lander ES. High-resolution haplotype structure in the human genome. *Nature Genet.* 2001;**29**:229–32.
11. Johnson GC, Esposito L, Barratt BJ, Smith AN, Heward J, Di Genova G, Ueda H, Cordell HJ, Eaves IA, Dudbridge F, Twells RC, Payne F, Hughes W, Nutland S, Stevens H, Carr P, Tuomilehto-Wolf E, Tuomilehto J, Gough SC, Clayton DG, & Todd JA. Haplotype tagging for the identification of common disease genes. *Nature Genet.* 2001;**29**: 233–7.

12. Carlson CS, Eberle MA, Rieder MJ, Smith JD, Kruglyak L, & Nickerson DA. Additional SNPs and linkage-disequilibrium analyses are necessary for whole-genome association studies in humans. *Nature Genet.* 2003;**33**:518–21.
13. Goldstein DB, Ahmadi KR, Weale ME, & Wood NW. Genome scans and candidate gene approaches in the study of common diseases and variable drug responses. *Trends Genet.* 2003;**19**:615–22.
14. Rieder MJ, Reiner AP, Gage BF, Nickerson DA, Eby CS, McLeod HL, Blough DK, Thummel KE, Veenstra DL, & Rettie AE. Effect of VKORC1 haplotypes on transcriptional regulation and warfarin dose. *N Engl J Med.* 2005;**352**:2285–93.
15. International HapMap Consortium. A second generation human haplotype map of over 3.1 million SNPs. *Nature.* 2007;**449**:851–61.
16. Park J, Sohyun H, Lee YS, Kim S, & Lee D. SNP@Ethnos: a database of ethnically variant single-nucleotide polymorphisms. *Nucleic Acids Res.* 2007;**35**:D711–15.
17. Zhou N, & Wang L. Effective selection of informative SNPs and classification on the HapMap genotype data. *BMC Bioinformatics.* 2007;**8**:484.
18. Ouyang C & Krontiris T. Identification and functional significance of SNPs underlying conserved haplotype frameworks across ethnic populations. *Pharmacogenet Genomics.* 2006;**16**:667–82.
19. Lin E, Hwang Y, & Tzeng C. A case study of the utility of the HapMap Database for Pharmacogenomic haplotype analysis in the Taiwanese population. *Mol Diag Ther.* 2006;**10**:367–70.
20. Botstein D & Risch N. Discovering genotypes underlying human phenotypes: past successes for Mendelian disease, future approaches for complex disease. *Nature Genet.* 2003;**33**:228–37.
21. Risch N & Merikangas K. The future of genetic studies of complex human diseases. *Science.* 1996;**273**:1516–17.
22. Pearson TA & Manolio TA. How to interpret a genome-wide association study. *JAMA.* 2008;**299**:1335–44.
23. Ke X, Taylor MS, & Cardon LR. Singleton SNPs in the human genome and implications for genome-wide association studies. *Eur J Hum Genet.* 2008;**16**:506–15.
24. Ambrose SH. Late Pleistocene human population bottlenecks, volcanic winter, and differentiation of modern humans. *J Hum Evol.* 1998;**34**:623–51.
25. Deng HW, Chen WM, & Recker RR. Population admixture: detection by Hardy-Weinberg test and its quantitative effects on linkage-disequilibrium methods for localizing genes underlying complex traits. *Genetics.* 2001;**157**:885–97.
26. Conrad DF, Jakobsson M, Coop G, Wen X, Wall JD, Riosenberg NA, & Pritchard JK. A worldwide survey of haplotype variation and linkage disequilibrium in the human genome. *Nat Genet.* 2006;**38**:1251–60.
27. deBakker PI, Burtt NP, Graham RR, Guiducci C, Yelensky R, Drake JA, Bersaglieri T, Penney KL, Butler J, Young S, Onofrio RC, Lyon HN, Stram DO, Haiman CA, Freedman ML, Zhu X, Cooper R, Groop L, Kolonel LN, Henderson BE, Daly MJ, Hirschhorn JN, & Altshuler D. Transferability of tag SNPs to genetic association studies in multiple populations. *Nat Genet.* 2006;**38**:1298–1303.
28. Ding K & Kullo IJ. Methods for the selection of tag SNPs: a comparison of tagging efficiency and performance. *Eur J Hum Genet.* 2007;**15**:228–36.
29. Yamaguchi-Kabata Y, Nakazono K, Takahashi A, Saito S, Hosono N, Kubo M, Nakamura Y, & Kamatani N. Japanese population structure, based on SNP genotypes from 7003 individuals compared to other ethnic groups: effects on population-based association studies. *Am J Hum Genet.* 2008;**83**:445–56.
30. Pritchard JK & Rosenberg NA. Use of unlinked genetic markers to detect population stratification in association studies. *Am J Hum Genet.* 1999;**65**:220–8.
31. Price AL, Patterson NJ, Plenge RM, Weinblatt ME, Shadick NA, & Reich D. Principal components analysis corrects for stratification in genome-wide association studies. *Nat Genet.* 2006;**38**:904–9.
32. Purcell S, Neale B, Todd-Brown K, Thomas L, Ferreira MA, Bender D, Maller J, Sklar P, de Bakker PI, Daly MJ, & Sham PC. PLINK: a tool set for whole-genome association and population-based linkage analyses. *Am J Hum Genet.* 2007;**81**:559–75.
33. Zhang F, Wang Y, & Deng H. Comparison of population-based association study methods correcting for population stratification. *PLoS One.* 2008;**3**:e3392.
34. Pritchard JK, Stephens M, Rosenberg NA, & Donnelly P. Association mapping in structured populations. *Am J Hum Genet.* 2000;**67**:170–81.
35. Devlin B & Roeder K. Genomic control for association studies. *Biometrics.* 1999;**55**:997–1004.
36. McKeigue PM, Carpenter JR, Parra EJ, & Shriver MD. Estimation of admixture and detection of linkage in admixed populations by a Bayesian approach: application to African-American populations. *Ann Hum Genet.* 2000;**64**:171–86.
37. Tang H, Peng J, Wang P, & Risch NJ. Estimation of individual admixture: analytical and study design considerations. *Genet Epidemiol.* 2005;**28**:289–301.
38. Kathiresan S, Newton-Cheh C, & Gerszten RE. On the interpretation of genetic association studies. *Eur Heart J.* 2004;**25**:1378–81.
39. Guthery SL, Salisbury BA, Pungliya MS, Stephens JC, & Bamshad M. The structure of common genetic variation in United States populations. *Am J Hum Genet.* 2007;**81**:1221–31.
40. Lohmueller KE, Mauney MM, Reich D, & Braverman JM. Variants associated with common disease are not unusually differentiated in frequency across populations. *Am J Hum Genet.* 2006;**78**:130–6.
41. Yang Q, Rabinowitz D, Isasi C, & Shea S. Adjusting for confounding due to population admixture when estimating the effect of candidate genes on quantitative traits. *Hum Hered.* 2000;**50**:227–33.
42. Pritchard JK. Are rare variants responsible for susceptibility to complex diseases? *Am J Hum Genet.* 2001;**69**:124–37.
43. Altman RB. Genetic sequence data for pharmacogenomics. *Curr Opin Drug Discov Devel.* 2003;**6**:297–303.
44. Daar AS & Singer PA. Pharmacogenetics and geographical ancestry: implications for drug development and global health. *Nat Rev Genet.* 2005;**6**:241–6.

2 Genotyping Technologies

Cristi R. King and Sharon Marsh

With the release of the Human Genome Mapping Project data (1) and the subsequent International HapMap Project (2), a wealth of genotype information is now publically available to researchers. Pharmacogenomics can utilize this information to its advantage by screening patient samples for known functional or tagging polymorphisms and deriving associations with drug outcome and toxicity. In addition, where no known functional polymorphisms have been identified in genes involved in drug pathways, technologies have emerged to perform whole-genome screens to find novel genome regions for further study.

Often considered the "gold standard" of genotyping, Sanger sequencing performed on the same DNA region in multiple individuals (resequencing) can be used to identify both new and previously reported polymorphisms. However, this process is not cost-effective, and analysis time can be slow. Consequently, it is often used as a quality-control step to confirm genotypes reported through the various technologies discussed later in this chapter.

There is no ideal genotyping technology for every project. The laboratory environment, sample number, number of polymorphisms to be screened, and cost are all factors involved in selecting the appropriate system. Therefore, technologies have developed to fit into low-, medium-, high-, and super-high-throughput laboratories. In addition, many systems have now moved beyond simple single-nucleotide polymorphism (SNP) screening and can incorporate screening for insertions and deletions (indels), methylation, allelic imbalance, and copy number variations.

LOW THROUGHPUT

For projects where a small number of samples and a small number of polymorphisms need to be assessed, it is usually not cost-effective to invest in a specific genotyping technology. Basic polymerase chain reaction (PCR) and agarose gel electrophoresis can be used to identify large polymorphic repeats. For example, the number of tandem 28-bp repeats in the thymidylate synthase enhancer region can be assessed by straightforward PCR and gel electrophoresis (3). For SNPs and single-base indels, PCR-based methods such as PCR-restriction fragment length polymorphism (RFLP) and allele-specific PCR are ideal in low-throughput situations, because no specialized equipment is needed.

PCR-RFLP

One of the simplest genotyping systems makes use of highly conserved enzyme restriction sites that are altered in the presence or absence of the polymorphism – RFLPs. A region containing the polymorphism of interest is amplified by standard PCR. The PCR product is then incubated with the appropriate restriction enzyme at a temperature optimal for cleavage of the restriction site. The restriction sites will be cut depending on the allele, which can easily be visualized using agarose gel electrophoresis and ethidium bromide staining. If the fragments caused by the cutting are short, either the agarose gel percentage can be increased, or low-melting-point agarose can be used for better resolution. This method does not require specialized equipment beyond what is standard for molecular biology laboratories (PCR block, electrophoresis equipment, ultraviolet box, or other gel documentation system), and it requires minimal training.

Not every allele to be tested will lie within a restriction site; therefore, this is not an ideal system for every research project. Allele-specific PCR requires the same equipment and similar training and can be a useful substitute in a low-throughput setting.

Allele-Specific PCR

Allele-specific PCR requires that PCRs contain a common primer plus a primer that terminates with a base specific to one allele. This results in PCR product only when the specific allele is present. If the sample is heterozygous, both sets of primers will yield product that can be visualized by the use of agarose gel electrophoresis (4). A similar premise can also be used to detect gene deletions, where the primers span the region of the deletion, leading to no PCR product if the gene is present or product when the gene is absent. This technique has been used to analyze whole-gene deletions such as GSTM1 and GSTT1. Typically, the genetic region of known copy number is amplified within the same reaction to rule out PCR failure as a cause for the lack of gene product (5).

This technique does not require the presence of a restriction site containing the polymorphism. However, not all regions of the genome are amenable to primer design, and it is not always possible to force a primer into the region terminating with the SNP. In addition, stringent PCR conditions are required to ensure that the primers do not nonspecifically bind to the incorrect allele. The concept of allele-specific PCR is the basis of several higher-throughput technologies.

MEDIUM THROUGHPUT

Medium-throughput technologies are ideal for the majority of retrospective studies, and a plethora of commercial instruments are available. Most require specialized training and an initial outlay of equipment cost to perform assays. In addition to standard polymorphism or phenotype association studies, several of these instruments can be adapted to clinical settings and/or clinical laboratory improvement amendment laboratories (CLIAs) where genotyping of patients for clinical decision making can be performed. The availability of prevalidated commercial kits and tests for many of such systems allows for greater quality control, and reduces the time usually required to design and optimize assays.

Taqman® (Applied Biosystems)

The Applied Biosystems (ABI) Taqman® system is based on real-time allele-specific PCR technology. PCR is performed along with the addition of a probe specific to one of the alleles. This probe contains a fluorescent reporter dye at one end and a quencher at the other. When the reporter and quencher are in close proximity, no fluorescence is emitted. During PCR, if the probe perfectly matches the SNP sequence and annealing occurs, the PCR extension from the primers displaces the probe by cleaving the 5′ end. This releases the reporter dye, allowing distance between it and the quencher. The resulting fluorescence can be detected on a real-time PCR machine. If the probe does not anneal, cleavage does not occur. Thus, the reporter dye and quencher remain in close proximity, and no fluorescence is recorded (6). This is a fast 384-well system that requires only a real-time PCR machine, which can also be used to analyze copy number variation and RNA expression. Many predesigned, ready-to-use assays are available through ABI (https://products.appliedbiosystems.com/ab/en/US/adirect/ab?cmd=catNavigate2&catID=600765), and new assays can be custom ordered.

Infiniti™ (Autogenomics)

The automated Infiniti™ microarray system can analyze multiplexed PCR product for a variety of assays for up to 24 samples at a time. Upon completion, the PCR product is then transferred into the Infiniti™ instrument for further processing, allowing for minimal technical error or contamination and minimal hands-on time by the technician. Detection primer extension occurs when fluorescence-labeled dCTP is added to the PCR product, as well as detection primers, with its 5′ ends containing an "anti-zip code." These anti-zip codes are complementary to the capture probes (zip codes) that are located on the BioFilmChip® array (7, 8). To enhance efficiency, the 3′ ends of the detection primers contain a complementary sequence to the SNP site and target sequence. Extension only occurs if there is an exact match between the 3′ end of the detection primer and the target sequence. After primer extension, there is a hybridization step between the zip codes and anti-zip codes on the BioFilmChip®. Only primers with the incorporated fluorescent label are detected, and genotypes are determined.

Ready-to-use kits are available for purchase to assess panels of polymorphisms relevant to drug response (http://www.autogenomics.com/1/products.php). For example, the warfarin assay kit will analyze CYP2C9 and VKORC1 polymorphisms. It is not always possible to be cost-effective by batching samples to fill a 96-well or 384-well plate for other genotyping systems. Consequently, the relatively low sample volume (24 samples maximum) makes this a useful technology for the clinical setting where results are needed within a 24-hour time frame.

Tag-It™ Mutation Detection (Luminex Molecular Diagnostics)

The Tag-It™ microsphere-based approach by Luminex Molecular Diagnostics (formerly Tm Bioscience) allows for multiplexing up to 100 tests in a single reaction tube with their unique 100 xMAP microsphere sets. While up to 24 samples are analyzed at a time, multiplex

PCR is performed. Upon completion, unique "tagged" allele-specific primers at the SNP of interest allow for extension while incorporating biotinylated cytidine triphosphate (CTP). Shrimp alkaline phosphatase (SAP) inactivates any additional nucleotides, and exonuclease I degrades any additional primers to increase efficiency. "Anti-tagged" microspheres are then bound to the 5′-tagged allele-specific primer extension (ASPE) product (7, 9–11). The anti-tagged microspheres contain a unique fluorescent bead identification specific to a particular bioassay (7). Streptavidin-conjugated R-phycoerythrin dye binds to the biotinylated CTP, allowing for signal amplification and fluorescence to determine SNP genotyping (7, 9, 12). Flow cytometry is used to detect the fluorescence ratio that allows for single-tube multiplexing (6). Genotypes are read out on the Luminex xMAP flow cytometer instrument with custom software.

The xMAP technology offers many genotyping applications, including pharmacogenomics, DNA sample profiling, and population genetics. This technology allows for the additional analysis of a wide range of assays. Research assays include various cardiac and cancer markers, in addition to isotyping assays. Allergy testing, cystic fibrosis screening, and infectious disease analysis are included for diagnostic screening. Thus, a variety of assays can be performed by the use of the xMAP technology.

Pyrosequencing® (QIAGEN)

Pyrosequencing® is an automated sequencing-by-synthesis technique used to detect nucleotide incorporation by chemiluminescence (6, 9, 13–15). Pyrosequencing has an advantage over other genotyping systems in that it reports sequences surrounding the actual SNP site, which also allows for added quality control.

Based on a 96-well format, and more recently on a 24-well format, a regular PCR is performed. Upon PCR completion, with the use of SNP-specific PCR primers (one of which is biotinylated), PCR product is then bound to streptavidin beads (6, 7). A sequencing primer is annealed to the PCR product mixture containing the target DNA sequence (16). The reaction begins when a nucleotide is incorporated in a predetermined order that is complementary to the nucleotide being extended. Pyrophospate is released in proportion to the amount of deoxynucleotide triphosphates (dNTPs) incorporated (16). ATP sulfurylase uses the released pyrophosphate to generate ATP, which is then used up by the luciferase enzyme to generate light proportionate to the amount of ATP generated (6, 7, 16, 17). Apyrase degrades the excess dNTPs, and the light generated is detected by camera and is displayed as Pyrograms® (Figure 2.1).

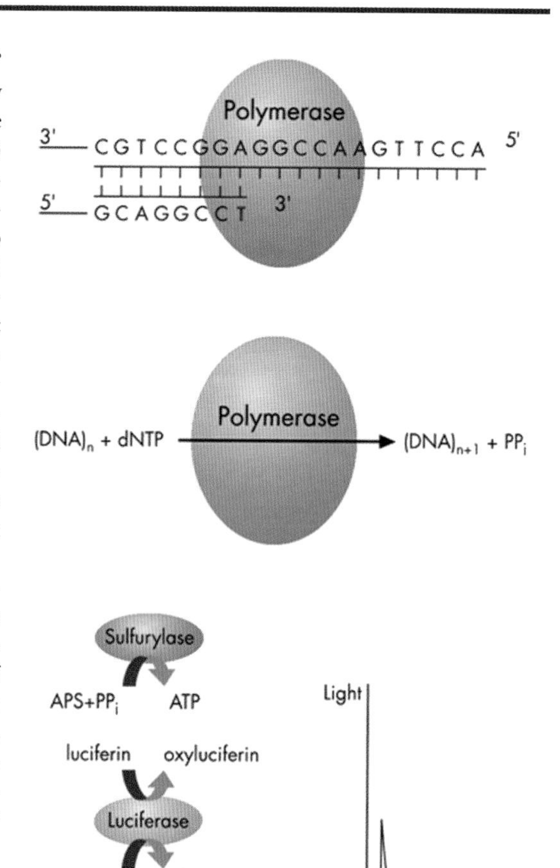

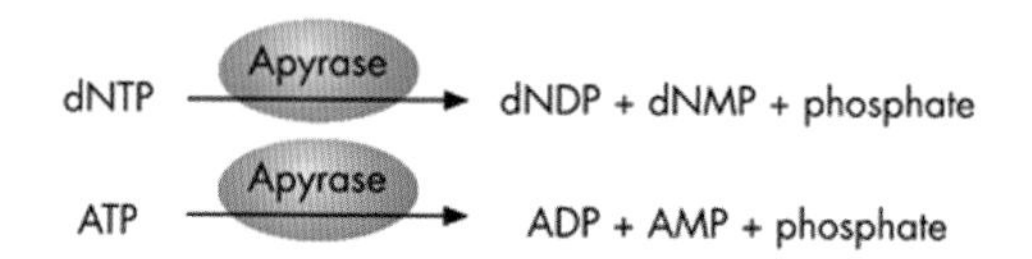

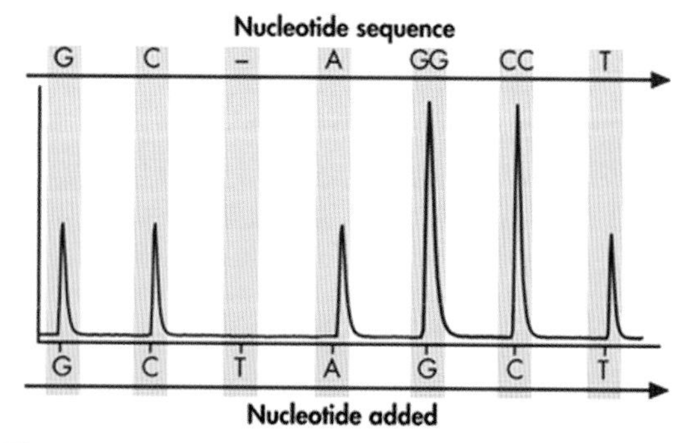

Figure 2.1. Example of a medium-throughput genotyping system. Pyrosequencing® technology is sequencing by synthesis. Published with permission from QIAGEN.

Pyrosequencing® can be used in a variety of analyses. Besides SNP and indel analysis, Pyrosequencing® has allele quantification capabilities and whole-genome sequencing potential (see "Massively Parallel Sequencing"). CpG methylation sites can also be used for methylation analysis (18). A range of premade kits are available for purchase (http://www.pyrosequencing.com/DynPage.aspx?id=14600&mn1=1409&mn2=1528). This platform thus has the potential and flexibility to serve a wide range of research needs.

Invader® Assay (Hologic, formerly Third Wave Technologies)

Hologic's Invader® technology is based on a flap endonuclease enzyme's structure-specific cleavage (9, 19, 20). Using two allele-specific probes and one Invader probe, Invader technology includes two basic reactions: a primary ASPE reaction and a secondary signal amplification and detection (19, 21, 22). The primary ASPE probe, which matches the variant allele on the target strand, also binds at the SNP site to form a 5′ flap. This 5′ flap is recognized by the Cleavase enzyme, which cleaves the overhang at the SNP site. Thus, differences at the SNP site on the target DNA may easily be differentiated (20). This cleavage allows for the release of the 5′ flap, which serves as an Invader for the secondary reaction (9). Amplification occurs as the probes cycle on and off the target strand, constantly releasing 5′ flaps. The secondary reaction entails the 5′ flap released in the primary reaction serving as an Invader probe on a labeled fluorescence resonance energy transfer (FRET) probe. Upon cleavage by Cleavase, the resulting reaction and amplification allow for signal probe cleavage and SNP detection. Samples are then read on a fluorescence plate reader with custom software.

This simple Invader® system allows for versatility in analyzing numerous assays while running up to 32 samples, including controls. Pharmacogenetics, infectious disease, and genetic assays are available for analysis. In addition, Hologic provides analyte-specific reagents for drug metabolism, cardiovascular risk analysis, and mutation analysis. Food and Drug Administration–approved in vitro diagnostic pharmacogenetic testing of UGT1A1 is also available (http://www.twt.com/clinical/ivd/ugt1a1.html).

HIGH THROUGHPUT

For screening large sample sets with large numbers of polymorphisms, a multiplex system is preferable for both speed and cost. Multiplex systems are ideal for studies where multiple candidate genes need to be assessed. Both specialized training and outlay of equipment cost are also required for these systems.

iPLEX® (Sequenom)

The MassARRAY® System is used to analyze the widely used MassEXTEND® assays, which use matrix-assisted laser desorption/ionization time-of-flight mass spectrometry to discriminate alleles (9, 16, 23). Genotypes are determined by the differentiation in mass between alleles based on a mixture of regular nucleotides and mass-modified 2′, 3′-dideoxynucleoside-5′-triphosphates used to increase accuracy (24, 25).

iPLEX® assays use 36-plex PCRs on a 384-well plate where the target DNA is amplified with the mass-modified terminators and extension is based on the SNP site. SAP cleaves a phosphate from unincorporated dNTPs to deactivate them for further reactions. The iPLEX® assay is a modified version of the MassEXTEND® assay in that all primer extension reactions are terminated after a single, mass-modified base extension (26, 27). The iPLEX® primer extension occurs by the use of the iPLEX® cocktail mix. With mass-modified terminators, the iPLEX® assay does not need to terminate 2–3 bases after the SNP for a larger mass separation, which is typical of the MassEXTEND® assays (27). The MassARRAY® System then analyzes the primer extension product of varying masses to determine the genotype (Figure 2.2).

The iPLEX® System is useful for SNP validation and fine mapping projects where customized genotyping panels are required.

SUPER-HIGH THROUGHPUT

When no clear candidate genes and/or polymorphisms are available, chip-and-bead arrays offer a fast scanning approach to narrow down genomic regions that can be further analyzed using more targeted approaches. These arrays allow either a very large amount of specific genes and SNPs to be assessed or a genome-wide approach where SNPs spanning the entire genome can be assessed in one assay. The wealth of data derived from these arrays requires complex statistical analysis, and the systems are typically used to identify novel genome regions previously not known to be associated with drug response.

BeadArray™ System (Illumina)

Illumina's BeadArray™ System is capable of super-high-throughput multiplexing and is based on allele-specific primer extension reaction and hybridization of individual beads separated into optical bundles for genotyping detection (6, 9, 28, 29). Two popular BeadArray™ assays by Illumina are the GoldenGate and Infinium assays. The GoldenGate assay can analyze from 96 to 1,536 SNPs simultaneously by the use of a discriminatory DNA polymerase allowing a technician to generate more than

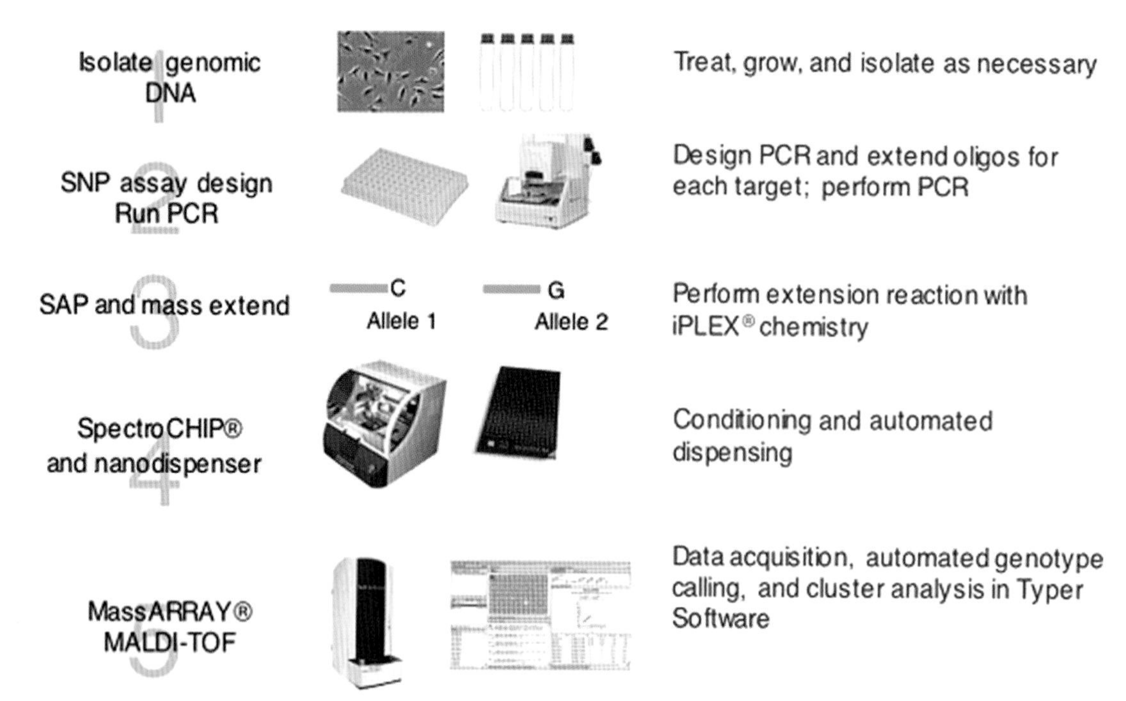

Figure 2.2. Example of a high-throughput genotyping system. The iPLEX® can be used for custom-built multiplexed genotyping panels. Published with permission from Sequenom®.

300,000 genotypes in approximately six hands-on hours. In addition, this platform has the flexibility to run either 16 or 96 samples in parallel.

ASPE is based on the use of two allele-specific probes that hybridize at the 3′ end of the SNP site and one locus-specific probe that binds downstream of the SNP (9). A typical assay entails the hybridization of genomic DNA that is immobilized on a solid support to the matching primers, and unbound probes are washed away (20). Enzymatic extension and ligation occur, followed by PCR amplification with a fluorescent tag for allele calling (9). Another hybridization step allows for the binding of the product to the array matrix, followed by fluorescent SNP detection.

The Infinium assay allows for unlimited levels of multiplexing, and the BeadChip format allows for single-tube preparation genotyping without PCR or ligation steps. It also provides higher genomic coverage and high-throughput genotyping. These BeadArray™ assays by Illumina can be good for a variety of analyses including linkage analysis, methylation profiling, sequencing, and whole-genome SNP genotyping.

Genome-Wide Human SNP Array 6.0 (Affymetrix)

Affymetrix provides a variety of platforms for genetic analysis. One example is the Genome-Wide Human SNP Array 6.0, which contains more than 1.8 million genetic markers for analysis, including 946,000 nonpolymorphic copy number probes. With such an enormous amount of SNPs available, this array is ideal for large-scale, high-powered association studies that are based on allele discrimination by hybridization and fluorescence detection (9). Because the SNP 6.0 array contains a combination of nonpolymorphic (CN) and polymorphic (SNP) probes, the researcher can conduct copy number variation (CNV), UPD, consanguinity, parent of origin, and mosaicism, all in a single-array experiment.

The SNP 6.0 assay has five main stages and takes three days to complete. First, the genomic DNA is digested with the restriction enzymes *Nsp*I and *Sty*I. Second, the PCR product is ligated to a specific adaptor upon recognition. Third, PCR amplification and purification occur and are based on the recognition of the adaptor sequence. (For a detailed graphic, please see: http://www.affymetrix.com/support/technical/datasheets/genomewide_snp6_datasheet.pdf.) Fourth, the PCR amplicon is fragmented and labeled with a fluorescent tag. Last, it is hybridized to the array for SNP and CNV detection (Figure 2.3) (30). The assay can be run with 48 or 96 samples at any one time, including internal quality controls to ensure accuracy.

Besides high-powered association studies, Affymetrix provides assays for targeted genotyping using the GeneChip® Scanner Targeted Genotyping System. This system allows for genotyping 3,000–20,000 SNPs at once

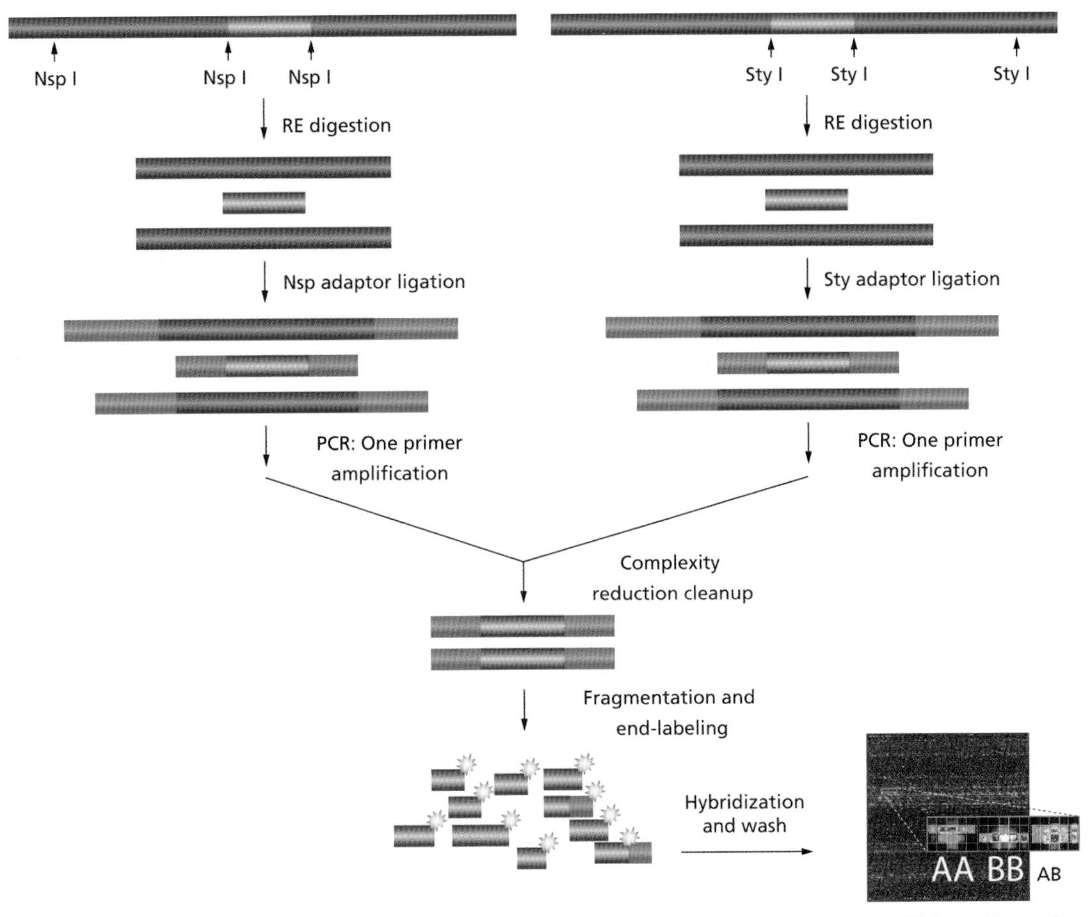

Figure 2.3. Example of a super-high-throughput genotyping system. The Affymetrix Genome-Wide Human SNP Array 6.0 contains more than 906,600 SNPs. Image courtesy of Affymetrix.

for more focused areas of research. In addition, Affymetrix markets a DMET® Panel that analyzes 1,936 pharmacogenetic markers for in-depth drug metabolism and transport analysis (http://www.affymetrix.com/products_services/arrays/specific/dmet.affx). CNV can be assessed using the SNP Array 6.0, and a more comprehensive analysis of chromosome rearrangements for cytogenetics can be performed using the Cytogenetics Whole-Genome 2.7M Array, which contains 2.7 million copy number markers.

MASSIVELY PARALLEL SEQUENCING

Although sequencing has long been considered the gold standard of genotyping technologies, it is simply not cost-effective for large projects involving many genes or amplicons to be sequenced. The development of massively parallel sequencing (sometimes called next-generation sequencing) eliminates this problem by allowing up to 100 million bases of DNA sequence to be produced within two days (31). These systems can be used to identify known variants in candidate genes or to screen genes in depth for novel variants. In addition, entire genomes can be sequenced. Future technologies will allow single-molecule sequencing (32). Although still at an early stage, massively parallel sequencing is now at the forefront of pharmacogenomics research.

454 Sequencing™ Technology (Roche)

454 Sequencing™ technology is based on the Pyrosequencing® technology described in "Pyrosequencing® (QIAGEN)" (32). However, unlike Pyrosequencing®, which is performed in liquid phase, 454 Sequencing™ uses a solid-phase system. With the liquid-phase system, a limited stretch of sequence can be synthesized as the nucleotide dispensation eventually dilutes out the necessary enzymes for the reaction to occur. The solid-phase system eliminates this problem, allowing for longer stretches of sequence to be captured.

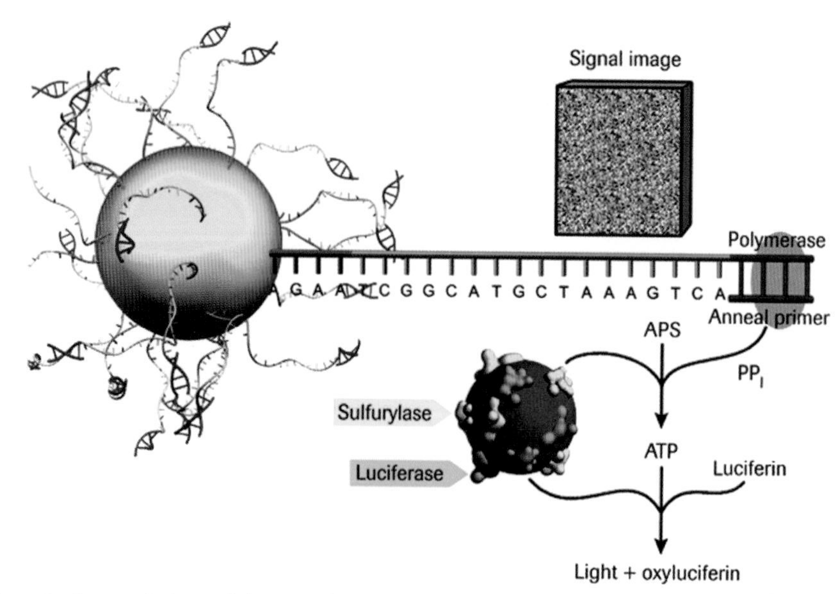

Figure 2.4. Example of a massively parallel sequencing system. This is a sequencing reaction of the 454™ Genome Sequencer System. Millions of copies of a single clonal fragment are contained on each DNA Capture Bead. Published with permission from 454 Sequencing ©Roche Diagnostics GmbH.

The process uses emulsion-based clonal amplification with primers that are both target specific and contain a tag allowing the fragments to bind to beads. Each bead contains an individual single-stranded DNA copy and is amplified using emulsified PCR (em-PCR™), resulting in millions of copies of the amplicon per bead. Each bead occupies a single well on a plate, and a typical Pyrosequencing® reaction ensues (Figure 2.4). The resulting signal is recorded using a charge-coupled device (CCD) camera, visualized in the form of a Flowgram and analyzed using Genome Sequencer FLX™ software. The background is reduced by flushing the system between each nucleotide addition. The current system allows up to 500 bp of sequence at a rate of approximately 1 billion bases per day. The system is highly accurate in comparison with Sanger sequencing (33), and the throughput and efficiency are obviously greatly enhanced.

In addition to sequencing up to 500-bp stretches of DNA, this system can be used for transcriptome analysis to identify transcription factor binding sites, methylation analysis (after bisulfate transformation of the sample), and SNP and indel identification. 454 Sequencing™ technology has been used successfully to sequence the entire human genome (34) and the woolly mammoth genome (35) and to determine the percentage of a patient's tumor cells containing novel mutations (36).

SOLiD™ (Applied Biosystems)

The Supported Oligonucleotide Ligation and Detection (SOLiD™) system from ABI can produce up to 4 Gb of sequence per run (six days) and generates sequencing reads of up to 35 bp in length. The recently released SOLiD™ 3 system can generate up to 20 Gb per run and can run two flow cells in parallel. DNA fragments are ligated to oligonucleotides that are attached to beads. Subsequent amplification is performed using emulsion PCR. Beads are immobilized onto a glass flow cell. Extension is performed by using primers that anneal to the oligonucleotides ligated to the beads. Nucleotide incorporation is measured as fluorescence, the fluorophore is washed off, and more nucleotides are added. After seven cycles, the primer is removed and a new extension primer offset by one base is annealed. This is repeated five times, and the adapter sequence is used to align the resulting fluorescence nucleotide reads (32).

While providing greater flexibility, this system can be successfully used for targeted resequencing to identify new polymorphisms in patient samples. In addition, SOLiD™ can be used for whole-genome resequencing, transcriptome analysis, and methylation analysis (http://www3.appliedbiosystems.com/AB_Home/applicationstechnologies/SOLiDSystemSequencing/index.htm).

Genome Analyzer (Illumina, formerly Solexa)

The Genome Analyzer uses flow cells with eight lanes. Oligonucleotide anchors are bound to the lanes. The DNA to be sequenced is fragmented, the 5′ ends repaired, and dATP incorporated onto the 3′ end. The adenine on the 3′ end triggers the ligation of the fragments to specific oligonucleotide anchors. The anchored DNA fragments are amplified by bridge PCR. This causes the amplicons to fold and anneal to spare oligonucleotide adaptors in the flow cell. Subsequent rounds of PCR generate clusters of identical amplicons. Sequencing is performed by annealing a primer to the anchor and using dye terminator sequencing extension (32). This instrument can generate up to 36 bp of sequencing reads, and a 2.5-day run yields up to 1.5 Gb of data. Up to twelve samples can be multiplexed per channel.

The Genome Analyzer has a wide range of applications. Novel sequencing, targeted resequencing, transcriptome analysis, copy number variation analysis, and methylation analysis can all be performed using this system (http://www.illumina.com/pages.ilmn?ID=250). Thus, the Genome Analyzer allows for great flexibility while meeting many research needs.

Second- and Third-Generation Sequencing Technologies

At the time this volume was going to press, several new high-throughput sequencing technologies were emerging. These platforms include Helicos, Pacific Bioscience, and Complete Genomics. Helicos promises True Single Molecule Sequencing (tSMS™) technology. This uses a sequencing-by-synthesis method but bypasses the need for DNA amplification by analyzing single molecules of DNA, thus significantly reducing the time required for each assay (37). Pacific Biosciences works on the principle of Single Molecule Real Time (SMRT™) Biology using uninterrupted template-directed synthesis to achieve sequencing on single DNA molecules for routine medical screening (38). Complete Genomics is described as a third-generation sequencing platform and offers a service that promises to be a complete solution for human genome screening. Sequencing is performed using DNA Nanoarrays and advantage can be taken of their Data Center, where a comparison of thousands of sequenced genomes can be performed (39). Although they are in their infancy stage, these technologies demonstrate the rapid growth of targeted medical resequencing and suggest that it will play a large role in pharmacogenomics in the near future.

CONCLUSIONS

A wide range of technologies is available for genotyping and sequencing, covering all needs from low throughput to super-high throughput. The decision on which platform to use will depend on the individual needs of each laboratory, in terms of both the imminent and projected future projects. Specialized equipment and training are required for most systems, and no single technology is ideal for all applications. Genotyping technologies currently can generate vast amounts of data in short periods of time. This has led to the need for improving data storage and handling, with dedicated servers needed for the data generated by massively parallel sequencing, and an increased need for complex statistical methodologies to extract value from the data. Processing the information generated by these systems and converting it into clinical applications is the next challenge for pharmacogenomics research.

DISCUSSION QUESTIONS

1. Other than SNPs, what types of variation can be assessed by using different genotyping platforms?
2. What studies would be suitable for low-, medium-, or high-throughput genotyping systems?
3. What would determine whether a genome-wide versus a candidate gene study should be performed?
4. What are the challenges involved once large amounts of genotyping or sequencing data have been generated?

ACKNOWLEDGMENTS

The authors wish to thank Katie Montgomery and Roche Diagnostics; Lennart Suckau and QIAGEN; Jodee Steinberg and Sequenom; and Marcia Bock, Mary Napier, and Affymetrix for their help. C.R.K. is supported by R01 HL074724, R01 HL57951, R01 HL58036, and the American Heart Association. S.M. is supported by the University of Alberta.

REFERENCES

1. Sachidanandam R, Weissman D, Schmidt SC, et al. A map of human genome sequence variation containing 1.42 million single nucleotide polymorphisms. *Nature.* 2001;**409**:928–33.
2. Altshuler D, Brooks LD, Chakravarti A, et al. A haplotype map of the human genome. *Nature.* 2005;**437**:1299–320.
3. Horie N, Aiba H, Oguro K, et al. Functional analysis and DNA polymorphism of the tandemly repeated sequences in the 5′-terminal regulatory region of the human gene for thymidylate synthase. *Cell Struct Funct.* 1995;**20**:191–7.
4. Rose CM, Marsh S, Ameyaw MM, et al. Pharmacogenetic analysis of clinically relevant genetic polymorphisms. *Methods Mol Med.* 2003;**85**:225–37.
5. Voso MT, D'Alo F, Putzulu R, et al. Negative prognostic value of glutathione S-transferase (GSTM1 and GSTT1) deletions in adult acute myeloid leukemia. *Blood.* 2002;**100**:2703–7.
6. Freimuth RR, Ameyaw M-M, Pritchard SC, et al. High-throughput genotyping methods for pharmacogenomic studies. *Curr Pharmacogenomics.* 2004;**2**:21–33.

7. King CR, Porche-Sorbet RM, Gage BF, et al. Performance of commercial platforms for rapid genotyping of polymorphisms affecting warfarin dose. *Am J Clin Pathol.* 2008;**129**:876–83.
8. Vairavan R. AutoGenomics, Inc. *Pharmacogenomics.* 2004;**5**:585–8.
9. Kim S & Misra A. SNP genotyping: technologies and biomedical applications. *Annu Rev Biomed Eng.* 2007;**9**:289–320.
10. Bortolin S, Black M, Modi H, et al. Analytical validation of the tag-it high-throughput microsphere-based universal array genotyping platform: application to the multiplex detection of a panel of thrombophilia-associated single-nucleotide polymorphisms. *Clin Chem.* 2004;**50**:2028–36.
11. Strom CM, Janeczko RA, Anderson B, et al. Technical validation of a multiplex platform to detect thirty mutations in eight genetic diseases prevalent in individuals of Ashkenazi Jewish descent. *Genet Med.* 2005;**7**:633–9.
12. Ugozzoli L, Wahlqvist JM, Ehsani A, et al. Detection of specific alleles by using allele-specific primer extension followed by capture on solid support. *Genet Anal Tech Appl.* 1992;**9**:107–12.
13. King CR & Scott-Horton T. Pyrosequencing(R): a simple method for accurate genotyping. *Methods Mol Biol.* 2006;**373**:39–56.
14. Marsh S, King CR, Garsa AA, et al. Pyrosequencing of clinically relevant polymorphisms. *Methods Mol Biol.* 2005;**311**:97–114.
15. Ronaghi M. Pyrosequencing sheds light on DNA sequencing. *Genome Res.* 2001;**11**:3–11.
16. Isler JA, Vesterqvist OE, & Burczynski ME. Analytical validation of genotyping assays in the biomarker laboratory. *Pharmacogenomics.* 2007;**8**:353–68.
17. Ronaghi M, Uhlen M, & Nyren P. A sequencing method based on real-time pyrophosphate. *Science.* 1998;**281**:363, 365.
18. Marsh S. Pyrosequencing applications. *Methods Mol Biol.* 2007;**373**:15–24.
19. Lyamichev V, Mast AL, Hall JG, et al. Polymorphism identification and quantitative detection of genomic DNA by invasive cleavage of oligonucleotide probes. *Nat Biotechnol.* 1999;**17**:292–6.
20. Hall JG, Eis PS, Law SM, et al. Sensitive detection of DNA polymorphisms by the serial invasive signal amplification reaction. *Proc Natl Acad Sci U S A.* 2000;**97**:8272–7.
21. de Arruda M, Lyamichev VI, Eis PS, et al. Invader technology for DNA and RNA analysis: principles and applications. *Expert Rev Mol Diagn.* 2002;**2**:487–96.
22. Wang DG, Fan JB, Siao CJ, et al. Large-scale identification, mapping, and genotyping of single-nucleotide polymorphisms in the human genome. *Science.* 1998;**280**:1077–82.
23. Pusch W, Wurmbach JH, Thiele H, et al. MALDI-TOF mass spectrometry-based SNP genotyping. *Pharmacogenomics.* 2002;**3**:537–48.
24. Fei Z, Ono T, & Smith LM. MALDI-TOF mass spectrometric typing of single nucleotide polymorphisms with mass-tagged ddNTPs. *Nucleic Acids Res.* 1998;**26**:2827–8.
25. Braun A, Little DP, & Koster H. Detecting CFTR gene mutations by using primer oligo base extension and mass spectrometry. *Clin Chem.* 1997;**43**:1151–8.
26. Gabriel S, Ziaugra L, & Tabbaa D. SNP genotyping using the Sequenom MassARRAY iPLEX platform. *Curr Protoc Hum Genet.* 2009; Chapter 2:Unit 2.12.
27. Oeth P, Beaulieu M, Park C, et al. iPLEX™ Assay: Increased Plexing Efficiency and Flexibility for MassARRAYSystem Through Single Base Primer Extension with Mass-Modified Terminators. http://www.agrf.org.au/docstore/snp/iPlex .pdf.
28. Michael KL, Taylor LC, Schultz SL, et al. Randomly ordered addressable high-density optical sensor arrays. *Anal Chem.* 1998;**70**:1242–8.
29. Steemers FJ, Ferguson JA, & Walt DR. Screening unlabeled DNA targets with randomly ordered fiber-optic gene arrays. *Nat Biotechnol.* 2000;**18**:91–4.
30. Kennedy GC, Matsuzaki H, Dong S, et al. Large-scale genotyping of complex DNA. *Nat Biotechnol.* 2003;**21**:1233–7.
31. Bushman FD, Hoffmann C, Ronen K, et al. Massively parallel pyrosequencing in HIV research. *AIDS.* 2008;**22**:1411–15.
32. Voelkerding KV, Dames SA, & Durtschi JD. Next-generation sequencing: from basic research to diagnostics. *Clin Chem.* 2009;**55**:641–58.
33. Huse SM, Huber JA, Morrison HG, et al. Accuracy and quality of massively parallel DNA pyrosequencing. *Genome Biol.* 2007;**8**:R143.
34. Wheeler DA, Srinivasan M, Egholm M, et al. The complete genome of an individual by massively parallel DNA sequencing. *Nature.* 2008;**452**:872–6.
35. Miller W, Drautz DI, Ratan A, et al. Sequencing the nuclear genome of the extinct woolly mammoth. *Nature.* 2008;**456**:387–90.
36. Ley TJ, Mardis ER, Ding L, et al. DNA sequencing of a cytogenetically normal acute myeloid leukaemia genome. *Nature.* 2008;**456**:66–72.
37. Milos P. Emergence of single-molecule sequencing and potential for molecular diagnostic applications. *Exp Rev Mol Diag.* 2009;**9**:659–66.
38. Eid J, Fehr A, Gray J, et al. Real-Time dna sequencing from single polymerase molecules. *Science.* 2009;**323**:133–8.
39. Drmanac R, Sparks AB, Callow MJ, et al. Human genome sequencing using unchained base reads on self-assembling DNA nanoarrays. *Science.* 2010;**327**:78–81.

Pharmacokinetics: Absorption, Distribution, Metabolism, Excretion Overview Chapter

3

Terrence Blaschke

VARIABILITY IN DRUG RESPONSE AND PHARMACOKINETICS

The variability in the response of individual patients to drugs was recognized long before the advent of our earliest understanding of the influence of genetics on drug response more than fifty years ago (1–3). Several reports have reviewed the topic of variability in some detail, attempting to identify and quantify the sources of variability (4, 5). Harter and Peck (4) estimated the relative contribution of different sources of variability to the cumulative total variability, and thought that pharmacokinetic variability accounted for about half of the total variability (4). Genetic factors contribute to the variability in pharmacokinetics and, in particular, they affect the enzymes and transporters involved in the absorption, distribution, metabolism, and elimination of drugs. This chapter will outline the basic principles of pharmacokinetics as they apply to the disposition of drugs in humans.

Pharmacokinetics is defined as the study of the time course of drug concentrations in the body, and can be separated into components describing the absorption, distribution, metabolism, and excretion of a drug, often abbreviated as ADME. The term pharmacodynamics refers to the study of the relationship between drug dose or concentration and the intensity and time course of pharmacological, clinical, or toxicological responses. In its simplest concept, pharmacokinetics can be thought of as "what the body does to the drug" and pharmacodynamics is "what the drug does to the body."

ABSORPTION

Because most drugs are administered orally, absorption of the drug from the gastrointestinal tract is necessary, and can be an important source of variability. The rate and extent to which a drug reaches the systemic circulation in unchanged form is known as its "bioavailability." Many factors have an impact on bioavailability, including formulation, lipid solubility, dissolution, and intestinal and hepatic enzymes and transporters. The latter have been reviewed in detail elsewhere (6). Bioavailability is often quantified as the fraction or percentage of the administered dose that reaches the systemic circulation, using the notation "F" for a drug's bioavailability. As an orally administered drug traverses the enzymes and transporters in the intestine and liver, a substantial fraction of the dose may not be absorbed or it may be lost (7). This phenomenon of reduced systemic drug availability is known as the first-pass effect, and such drugs typically have values of F that are less than 0.1 or 10 percent.

Figure 3.1 illustrates the blood or plasma concentration versus the time profile after a single dose of an orally administered drug. Features to note are the maximum plasma concentration, or C_{max}, the time after dosing that the C_{max} is reached, or T_{max}, and the area-under-the-curve, AUC, a measure of the systemic exposure to the drug.

DISTRIBUTION

Once a drug enters the circulation it generally does not remain confined to the blood, but is distributed into various tissues. This process is called "distribution" and describes the transfer of the drug from the circulation into other places within the body, in particular, from plasma to tissue. A characteristic of the distribution process is that it is *reversible*. The extent to which a drug is distributed again depends on a number of factors, including its lipid solubility, the extent to which it is bound to plasma proteins or cellular components in the blood, and interactions with a variety of transporters on the surface of cells, both within the blood and in tissues. Thus, as with bioavailability, both drug characteristics and host factors can affect the distribution process.

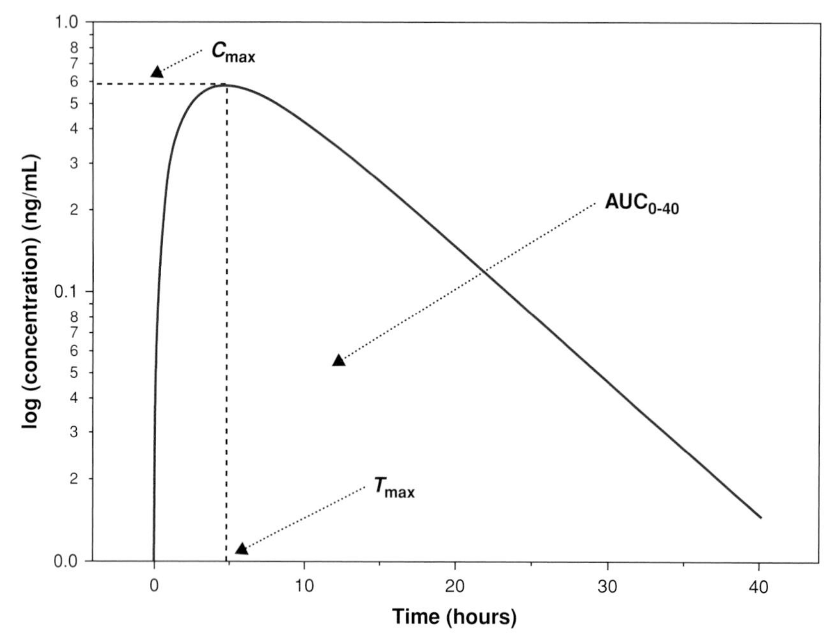

Figure 3.1. Concentration vs. time plot, oral dosing. This is a typical profile of a drug given as a single oral dose, plotted on a semilogarithmic scale. For an immediate release formulation, concentrations rise to a maximum concentration (C_{max}) attained at a particular time (T_{max}). At C_{max}, the absorption and the elimination of the drug are equal. Thereafter, concentrations fall in a log-linear fashion. The AUC is a measure of the systemic exposure to the drug.

A useful term that is used to quantify the distribution of a drug in the body is the "volume of distribution," usually designated by the notation "VD." The volume of distribution of a drug is the volume into which the *total amount of drug in the body* (usually designated as "Ab") would have to be diluted to be equal to the plasma concentration, or "[Cp]." For the vast majority of drugs there is *no physiological space or compartment that corresponds with the volume of distribution.*

Under certain circumstances, especially when a drug is given intravenously, the distribution of the drug from its site of administration may occur at different rates in different tissues. The distribution of a drug into certain tissues, in particular those that are small in total mass and highly perfused by the circulation, occurs very quickly, whereas distribution into tissues with larger masses and slower perfusion rates occurs more slowly. Examples of highly perfused tissues include the brain, heart, liver, and kidneys. Examples of slowly perfused tissues include fat, skin, and muscle. In this so-called two-compartment model distribution into the highly perfused tissues (compartment 1 or central compartment, V_C) is rapid, and occurs within a few minutes after drug administration, whereas the distribution into the slowly perfused tissues occurs over a much longer time. Figure 3.2 illustrates the plasma concentration versus time profile of a prototypical drug with first-order kinetics (discussed later in this chapter) given by intravenous bolus injection. When the plasma concentration is plotted on a semilogarithmic scale, the drug disappears from plasma in two phases, the first exponential fall, designated $T_{1/2\alpha}$, representing the distribution phase, in which the drug enters the slowly perfused tissues from the blood or plasma (compartment 2 or tissue compartment, V_T). As shown in the inset, this process is governed by a first-order rate constant, k_{12}, and the return of drug from compartment 2 to the blood or plasma is governed by the constant k_{21}. Because distribution processes are reversible, distribution equilibrium is reached when the instantaneous mass transfer of drug to and from each compartment is equal. The second phase represents the elimination of the drug from the body. This monoexponential process, designated $T_{1/2\beta}$, is the half-life of elimination. The concept of the half-life of a drug is discussed in "Half-Life."

Thus far, this chapter has defined and described the processes by which an orally administered drug gets into the systemic circulation and is distributed throughout the body. For drugs given parenterally (intramuscularly or subcutaneously), but not intravenously, bioavailability is a pharmacokinetic parameter that has meaning and must be determined. For drugs that are administered intravenously, the bioavailability, or *F*, is 1 or 100 percent, but distribution from the blood into the tissues still occurs, and may occur at different rates in different tissues or "compartments."

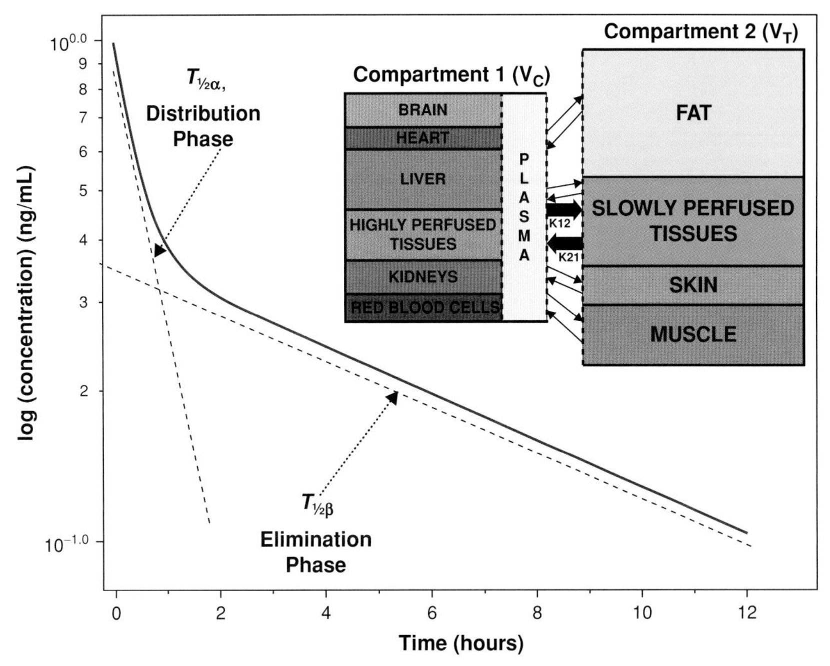

Figure 3.2. Concentration vs. time plot, intravenous dosing. This graph illustrates the concentration vs. time profile of a drug given as a single intravenous bolus dose, disappearing from blood or plasma as the sum of two monoexponential components. The initial exponential, $T_{1/2\alpha}$, represents the distribution phase, and the second exponential, $T_{1/2\beta}$, is the elimination phase. The inset illustrates the bidirectional movement of drug between plasma or blood and tissues that equilibrate rapidly (compartment 1, V_C) and those that equilibrate more slowly (compartment 2, V_T).

ELIMINATION

Once a drug is in the body, it must be eliminated, or removed, from the body. Elimination processes are *irreversible* and can be divided into two general categories: (1) metabolism and (2) excretion. Metabolism refers to *the chemical alteration* of drugs within the body. The site of chemical alteration is often the liver, but other organs (the gastrointestinal tract, kidneys, lung, skin, etc.) may metabolize drugs as well. It is important to recognize that the metabolites produced by metabolic processes may have pharmacological activity. In fact, for some drugs, only the metabolite(s) are pharmacologically active, and such drugs are known as "prodrugs." Excretion refers to the removal of drug from the body *without chemical alteration*. This usually occurs via the kidneys, but the lungs, the biliary tract, the gastrointestinal tract, and/or sweat may be involved as well.

The metabolism of drugs is frequently a two-stage process, often referred to as phase 1 and phase 2 reactions. In phase 1 reactions, enzymes located in the liver, gut, kidney, or other organs perform oxidation, reduction, or hydrolytic reactions. In phase 2, enzymes form a conjugate of the phase 1 product. The enzymes involved in phase 1 reactions are cytochrome P450s (P450 or CYP), flavin-containing monooxygenases, and epoxide hydrolases. The enzymes involved in phase 2 reactions are transferases, such as sulfotransferases, UDP-glucuronosyltransferases, glutathione-*S*-transferases, and *N*-acetyltransferases. The net result of these reactions, in general, is to convert a lipid-soluble drug into more water-soluble metabolites that are more easily eliminated from the body through the bile or kidneys. For a detailed review of drug metabolism, see Gonzalez and Tukey, chapter 3 in *Goodman and Gilman's The Pharmacological Basis of Therapeutics* (8). As will be discussed elsewhere in this volume, virtually all of the enzymes in each of these families contain genetic polymorphisms that can affect the activity of these reactions (9). In addition, the access to these enzymes, which are intracellular, is often modulated by transporters in the cell membrane that may also contain polymorphisms affecting the influx or efflux of drugs. For recent updates of this interplay, see Hagenbuch (10) and Benet, Cummins, and Wu (11).

Excretion of the unchanged drug is performed mainly in the kidneys, and occurs by filtration of the unbound drug (the drug in plasma that is not bound to plasma protein) through the glomerular membrane, transport of the drug from plasma and interstitial fluid into the lumen of the nephron, and reabsorption of the drug from the lumen, dependent on the lipid solubility and concentration gradient that is established as water and other solutes are reabsorbed in the distal nephron. Not only is the unchanged drug eliminated by these processes in the kidney, but conjugated and unconjugated metabolites of a drug are also often excreted into the urine by the kidneys (12).

CLEARANCE

Because a drug may be a substrate for many isoforms of phase 1 and 2 enzymes and of various transporters, there is often substantial variability in elimination, and elimination is difficult to predict in a given individual. However, elimination can be quantified by using the concept of clearance (13, 14). Underpinning this concept is that drugs circulating in the blood are brought to organs of elimination (usually the liver or kidneys) by the blood flowing to each organ of elimination, and as long as metabolism by enzymes or binding to transporters in those organs is not saturated (i.e., elimination is "first order," true for most drugs), a constant fraction or proportion of the drug delivered to each organ will be removed from the circulating blood during each pass through the organ. This fraction is called the extraction ratio, or ER.

Because a constant fraction of drug is eliminated per unit time, mathematically, the rate of change of drug in the body (Ab) is equal to:

$$D(\mathrm{Ab})/dt = -K(\mathrm{Ab}), \tag{3.1}$$

where K has units of reciprocal time (e.g., per hour).

A very small number of drugs that are used clinically demonstrate zero-order elimination, indicating that a constant amount of drug is eliminated per unit time. Mathematically, the rate of change of drug in the body (Ab) is expressed by:

$$D(\mathrm{Ab})/dt = -K, \tag{3.2}$$

where K has units of mass/time (e.g., milligrams per hour).

Some drugs have mixed first-order and zero-order elimination, and are said to have nonlinear elimination processes.

As defined earlier, the amount of drug in the body is the volume of distribution times the plasma concentration, or:

$$\mathrm{Ab} = \mathrm{VD} \times [\mathrm{Cp}]. \tag{3.3}$$

Substituting Eqn. 3.3 into Eqn. 3.1 and dividing both sides by [Cp],

$$D(\mathrm{VD})/dt = -K \times (\mathrm{VD}). \tag{3.4}$$

Figure 3.3 illustrates the relationship between the volume of distribution and the removal of the drug by

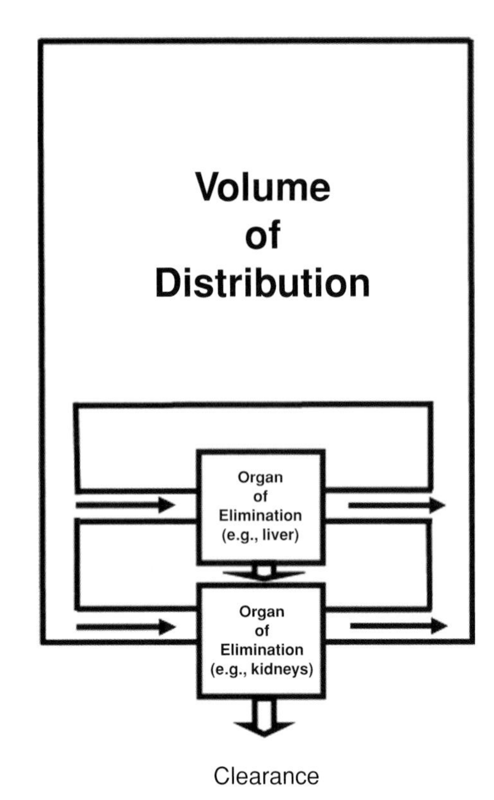

Figure 3.3. Physiological basis of clearance. The blood or plasma concentration of a drug is equal to Ab/VD, and in first-order elimination a constant fraction of the blood or plasma flowing to an organ of elimination (represented by the arrows) is "cleared" of drug per unit time. As a result of the parallel, mammillary nature of the circulatory system, in which organs are connected to one another through the blood, total clearance (CL_T) is the sum of the clearances by each organ of elimination.

the organs of elimination for drugs that have first-order kinetics as shown in Eqn. 3.1.

For such drugs, a constant fraction of the drug reaching any organ of elimination is eliminated during a single pass through that organ. For the entire body, then, the total body clearance is the volume of blood (or plasma) from which the drug is irreversibly removed per unit time. It has units of volume/time (e.g., milliliters per minute or liters per hour). Clearance, therefore, is a quantitative, physiological parameter that describes the elimination of an exogenous or endogenous compound by an organ in terms of the blood flow to that organ and the irreversible removal of the compound during a single pass through the organ. Mathematically,

$$\text{Clearance} = \text{blood flow} \times \text{extraction ratio or}$$
$$\mathrm{CL} = \mathrm{Q} \times \mathrm{ER}, \tag{3.5}$$

where Q is the mean flow and ER is the extraction ratio.

The important characteristics of clearance are that it is independent of concentration for linear (first-order)

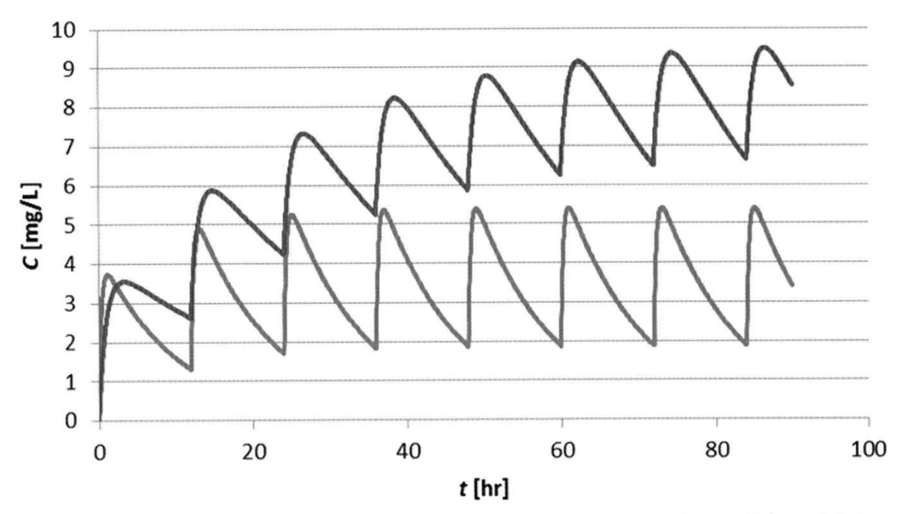

Figure 3.4. Pharmacokinetic profiles during multiple dose regimens. Most drugs are given in a multiple oral dosing regimen for varying durations of time. This graph indicates plasma concentration versus time profiles for two regimens given at the same dose at twelve-hour dosing intervals. The upper, red profile is for a drug with a half-life of seventeen hours, while the lower, blue profile is for a drug with a half-life of seven hours. Note that the drug with the short half-life (due to higher clearance) reaches steady-state conditions within twenty-four hours, but has a lower average steady-state concentration. The drug with the seventeen-hour half-life (due to slower clearance) takes about three days to attain steady state, with average concentrations that are about 2.5 times higher than the short half-life, high-clearance drug. Note also that the peak-to-trough concentration ratios are greater for the drug with the short half-life compared to the drug with the long half-life. This observation would hold for all dosing intervals.

systems and is additive for individual organs of elimination:

$$CL_{Total} = CL_{Hepatic} + CL_{Renal} + CL_{Other\ organs}. \quad (3.6)$$

Using the relationships in Eqns. 3.1, 3.3, and 3.4, total body clearance can be calculated as follows:

$$CL_{Total} = dose/AUC_{0\text{-}\infty}, \quad (3.7)$$

where dose is expressed as a mass (e.g., mg or g), and $AUC_{0\text{-}\infty}$ is the area-under-the-curve of the blood or plasma concentration from time = 0 extrapolated to infinity.

For drugs administered by the oral route, CL_{Oral} is equal to:

$$CL_{Total} = (F \times dose)/AUC_{0\text{-}\infty}, \quad (3.8)$$

where F is the fractional bioavailability between 0.0 and 1.0.

HALF-LIFE

The most commonly used and understood pharmacokinetic parameter is half-life, usually abbreviated as $T_{1/2}$. Half-life is the time required to go from any given concentration of a drug in blood or plasma to one-half of that concentration and, again, for drugs with first-order pharmacokinetics, half-life is *independent of concentration.* Half-life is, however, a hybrid pharmacokinetic parameter, which, as illustrated in Figures 3.1 and 3.2 and derived from the equations in this chapter, is directly proportional to the volume of distribution and inversely proportional to the total body clearance (CL_T) from the body. Mathematically, it is represented by:

$$T_{1/2} = \frac{VD}{CL_{Total}} \times (\ln 2). \quad (3.9)$$

Factors responsible for variability in VD have been discussed previously, as have factors that can affect metabolism and elimination and thus the extraction ratio in an organ.

CLEARANCE, HALF-LIFE, AND DOSING REGIMENS

When a drug is administered in a multiple-dose regimen for a few days, weeks, or even a lifetime, there is another pharmacokinetic issue to consider, and that is the pharmacokinetics and exposure to a drug in such settings. With repeated doses of a drug given for a sufficient time, the amount of a drug in the body (Ab) and the concentrations of a drug in the blood or plasma will reach a condition referred to as "steady state" that, from a pharmacokinetic point of view, is attained when there is essentially no fluctuation in drug concentrations (i.e., the peak and trough concentrations) over a dosing interval. This is illustrated in Figure 3.4

At steady state, the concentration of drug in the blood or plasma is described by the following relationship:

$$\text{Rate of drug administration} = \text{rate of drug elimination} \quad (3.10)$$

or

$$\text{Bioavailability } (F) \times \text{dosing rate} = \text{plasma concentration } ([\text{Cp}]) \times \text{total clearance } (\text{CL}_{\text{Total}}). \quad (3.11)$$

A final consideration is the time needed to reach steady state and the degree of fluctuation between the highest and lowest plasma concentration during a dosing interval, and these are also illustrated in the Figure 1.4 When the same dose is given at the same dosing interval for a period of time, it takes approximately three to five half-lives for the plasma concentrations to attain steady-state conditions. The fluctuation between the highest and lowest concentration largely depends on the half-life, but is also influenced by the rate of absorption of the drug.

SUMMARY

Knowledge of the terminology and definitions of a few basic parameters of pharmacokinetics and a basic understanding of the principles of drug disposition should equip the reader to appreciate the impact of pharmacogenomics of the host on many of the processes governing the absorption, distribution, metabolism, and excretion of a drug, to be able to understand and critically review the pharmacokinetic literature, and to be aware of the other sources, including nongenetic factors that modify pharmacokinetics that are responsible for the often considerable variability in drug responses (15). For a more advanced textbook on pharmacokinetics and pharmacodynamics, the reader is referred to the book by Rowland and Tozer (16).

REFERENCES

1. Kalow W. Unusual responses to drugs in some hereditary conditions. *Can Anaesth Soc J.* 1961;**8**:43–52.
2. Motulsky AG. Drug reactions enzymes, and biochemical genetics. *JAMA.* 1957;**165**:835–7.
3. Kalow W. *Pharmacogenetics: Heredity and the Response to Drugs.* Philadelphia: W.B. Saunders Co.; 1962.
4. Harter JG & Peck CC. Chronobiology. Suggestions for integrating it into drug development. *Ann N Y Acad Sci.* 1991;**618**:563–71.
5. Rowland M, Sheiner LB, Steimer J-L, & Sandoz AG. *Variability in Drug Therapy: Description, Estimation, and Control: a Sandoz Workshop.* New York: Raven Press; 1985.
6. Shugarts S & Benet LZ. The role of transporters in the pharmacokinetics of orally administered drugs. *Pharm Res.* 2009;**26**:2039–54.
7. Benet LZ, Izumi T, Zhang Y, Silverman JA, & Wacher VJ. Intestinal MDR transport proteins and P-450 enzymes as barriers to oral drug delivery. *J Control Release.* 1999;**62**:25–31.
8. Gonzalez FJ & Tukey RH. Drug metabolism. In: Goodman LS, Gilman A, Brunton LL, Lazo JS, Parker KL, eds. *Goodman & Gilman's The Pharmacological Basis of Therapeutics.* 11th ed. New York: McGraw-Hill; 2006:71–92.
9. Meyer UA. Pharmacogenetics – five decades of therapeutic lessons from genetic diversity. *Nat Rev Genet.* 2004;**5**:669–76.
10. Hagenbuch B. Drug uptake systems in liver and kidney: a historic perspective. *Clin Pharmacol Ther.* 2010;**87**:39–47.
11. Benet LZ, Cummins CL, & Wu CY. Unmasking the dynamic interplay between efflux transporters and metabolic enzymes. *Int J Pharm.* 2004;**277**:3–9.
12. Remmer H. The role of the liver in drug metabolism. *Am J Med.* 1970;**49**:617–29.
13. Rowland M, Benet LZ, & Graham GG. Clearance concepts in pharmacokinetics. *J Pharmacokinet Biopharm.* 1973;**1**:123–36.
14. Wilkinson GR & Shand DG. Commentary: a physiological approach to hepatic drug clearance. *Clin Pharmacol Ther.* 1975;**18**:377–90.
15. Tucker GT. Variability in human drug response. Presented at the Proceedings of the Esteve Foundation Symposium VIII, Sitges, Spain, 7–10 October 1998. 1st ed. New York: Excerpta Medica; 1999.
16. Rowland M, Tozer TN. *Clinical Pharmacokinetics and Pharmacodynamics: Concepts and Applications* (Fourth Edition). Philadelphia: Lippincott Williams & Wilkins; 2011.

4

Overview: Adverse Drug Reactions

Matthew R. Nelson

Adverse drug reactions occur during drug development and in clinical practice with approved medicines. They are responsible for the termination of approximately 20 percent of investigational drugs in the pharmaceutical pipeline. Approximately 1 percent of marketed drugs are withdrawn or restricted postmarketing because of safety-related issues. Adverse drug reactions affect millions of people every year, are responsible for a significant fraction of hospitalizations, and are a leading cause of death in developed countries. Thus, patients, the medical community, health care providers, regulatory agencies, and the pharmaceutical industry have a compelling interest to understand these adverse reactions and identify factors that influence them.

In this chapter we define adverse drug reactions and several related and commonly used terms; evaluate their impact on drug development, public health, and individual patient well-being; provide an overview of the contribution of known genetic variants to adverse drug reaction risk; and discuss efforts to identify genetic adverse drug reaction risk factors and incorporate them into development and clinical practice.

Several terms are in active use to describe an untoward outcome that a patient experiences while taking a drug, including adverse event, adverse effect, and adverse drug reaction. Although these terms are often used interchangeably, they have different definitions and meanings, and their use should be deliberate and appropriate. The World Health Organization has defined an adverse event as "any untoward medical occurrence that may present during treatment with a pharmaceutical product but which does not necessarily have a causal relationship with treatment." In contrast, an adverse drug reaction is defined as "a response to a drug which is noxious and unintended, and which occurs at doses normally used in man" (1). The difference between these definitions is one of causation. An adverse effect describes a co-occurrence of drug and harmful outcome, whereas an adverse drug reaction assumes or has demonstrated a causal relationship between drug and outcome. An adverse effect is the noxious and unintended effect of the drug, differing from adverse drug reaction only in its drug-centric rather than patient-centric view of the outcome. There are other definitions for these terms (2), and in some research settings it is important to clarify which one is in use. This definition of adverse drug reactions excludes therapeutic failures, intentional or accidental overdose, errors in administration, or noncompliance. It is inclusive of prescription drugs, over-the-counter medications, and other forms of dietary supplements.

An event or reaction is described as "serious" (i.e., serious adverse event or serious adverse drug reaction) if it has a significant impact on the well-being of the patient, including being fatal, life-threatening, significantly disabling, requiring or prolonging hospitalization, resulting in long-term incapacity, or causing congenital anomaly (3).

The pharmacological classification of adverse drug reactions into its subtypes has a long history and continues to be revisited and revised (4). The modern classification system is built on the proposal by Rawlins and Thompson (5). Their original proposal consisted of two types of reactions. Type A reactions are those that are dose related and predictable based on the pharmacological action of the drug, and they are often reproducible in animal models. One prominent example of this type of reaction is the serious bleeding caused by warfarin, which is prescribed to inhibit blood clotting. Another example is acetaminophen-induced liver injury due to accidental or intentional overdose, which is the leading cause of acute liver failure in the United States, Great Britain, and most of Europe (6).

Type B reactions are not related to dose or to drug pharmacology, commonly referred to as idiosyncratic, or due to a combination of factors that are unique to an individual or group (7). They tend to be rarer

Table 4.1. Adverse Drug Reaction Classification

Type of Reaction	*Mnemonic*	*Features*	*Examples*	*Management*
A: Dose-related	Augmented	Common; Related to pharmacological action of the drug; Predictable; Low mortality	Toxic effects; Warfarin bleeding, digoxin toxicity; Side effects; Anticholinergic effects of tricyclic antidepressants	Reduce dose or withhold; Consider effects of concomitant therapy
B: Non-dose-related	Bizarre	Uncommon; Not related to pharmacological action of drug; Unpredictable; High mortality	Immunological reactions; Penicillin hypersensitivity; Idiosyncratic reactions; Reye's syndrome; Pseudoallergy (e.g., ampicillin rash)	Withhold and avoid in future
C: Dose-related and time-related	Chronic	Uncommon; Related to the cumulative dose	Hypothalamic-pituitary-adrenal axis suppression by corticosteroids	Reduce dose or withhold; withdrawal may have to be prolonged
D: Time-related	Delayed	Uncommon; Usually dose-related; Occurs or becomes apparent some time after the use of the drug	Teratogenesis (e.g., adenocarcinoma with diethylstilbestrol); Carcinogenesis Tardive dyskinesia	Often intractable
E: Withdrawal	End of use	Uncommon; Occurs soon after withdrawal of the drug	Opiate withdrawal syndrome; Myocardial ischemia (β-blocker withdrawal)	Reintroduce and withdraw slowly
F: Unexpected failure of therapy	Failure	Common; Dose-related; Often caused by drug interactions	Inadequate dosage of oral contraceptive; clopidogrel thrombosis	Increase dosage; Alternative drug; Consider effects of concomitant therapy

Adapted from (2).

and more serious than type A reactions and, in general, are unpredictable. Type B reactions can be further categorized as immunological or nonimmunological in nature. Immunological reactions can involve immediate immunoglobulin E-mediated allergic reactions, such as penicillin hypersensitivity, as well as delayed reactions involving major histocompatibility complex-dependent CD4+ and CD8+ T-cell drug recognition through T-cell receptors, such as carbamazepine severe skin reactions (SSRs) (8, 9). Nonimmunological type B reactions are sometimes referred to as metabolic. These may be characterized as nonimmunological based on the absence of immune-related symptoms, such as rash, fever, eosinophilia, or the rapid recurrence of symptoms upon rechallenge with the drug. One example of a metabolic idiosyncratic reaction is Reye's syndrome, a rare but serious reaction to aspirin resulting in fatty liver and encephalopathy. This unpredictable reaction occurs almost exclusively in children with a concomitant viral infection (10).

Since the original classification scheme of Rawlins and Thompson was proposed, four additional categories have come into use and are summarized in Table 4.1 adapted from the review by Edwards and Aronson (2).

Many adverse events can be caused by a wide variety of possible factors. Take liver injury, or hepatotoxicity, as an example. In a study of 308 consecutive patients with acute liver failure admitted over a forty-one-month period from 1998 to 2001, liver failure was attributed to acetaminophen overdose (39 percent), idiosyncratic drug reactions (13 percent), hepatitis B (7 percent), hepatitis A (4 percent), ischemic hepatitis (6 percent), autoimmune hepatitis (4 percent), Wilson disease (3 percent). and Budd-Chiari syndrome (2 percent) (11). Several methods and tools have been developed and proposed for assessing drug causality in the occurrence of adverse events. One of the most widely used tools is the Roussel Uclaf causality assessment method (RUCAM), developed in a series of consensus meetings in the 1980s (12). The RUCAM includes seven criteria by which the suspected causal drug is scored, including time to onset, course of the reaction, risk factors, concomitant drugs, the presence of possible non-drug-related causes, information about the drug causing the observed event,

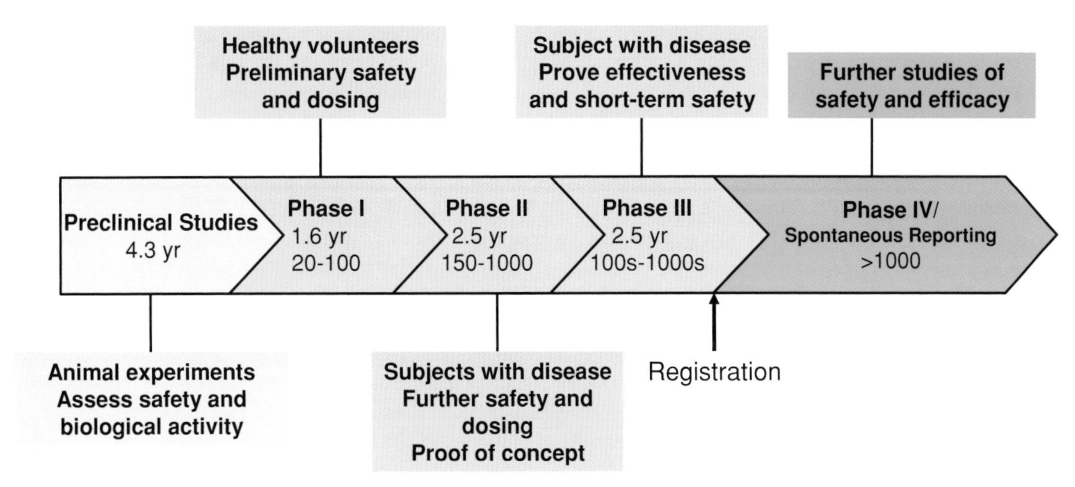

Figure 4.1. Clinical development process.

response to readministration, or availability of laboratory tests to confirm drug involvement. Responses can add to or subtract from the overall score, which can take on values between -9 and 15. The sum of the component scores classifies the causality as excluded (score ≤ 0), unlikely (1–2), possible (3–5), probable (6–8), and highly probable (>8). As with the adverse drug reaction classification scheme, alternative approaches continue to be proposed and evaluated.

IMPACT OF ADVERSE DRUG REACTIONS

Adverse drug reactions incur significant costs for patients, our public health, and drug developers. A common refrain from drug safety experts is that there is no drug that is free of risk. All drugs have the potential to cause harm under the wrong set of circumstances. The potential for adverse reactions is tremendous when the prevalence of drug usage is considered. It was recently estimated that 52 percent of the adult population of the United States is taking at least one prescription drug – 12 percent are taking five or more (polypharmacy) – at any given time. Broadening this to all forms of medication – including prescription, over-the-counter drugs, or dietary supplements – those percentages increase to 82 percent and 29 percent, respectively. The rates of drug usage and polypharmacy increase substantially with aging, as do the risks of adverse reactions (13).

Several studies have investigated the incidence of adverse drug reactions. These primarily assessed adverse drug reactions in a hospital setting, either investigating hospital visits and/or admissions due to drug reactions or the occurrence of reactions while hospitalized. Although estimates vary, the conclusions are consistent: a significant fraction of morbidity and mortality in developed countries with Western medical systems are due to adverse reactions to drugs. These reactions are responsible for approximately 2.5 percent of emergency department visits for unintentional injuries, for between 4 percent and 9 percent of hospital admissions, and for deaths of approximately 0.15 percent of hospitalized patients (14–16). Adverse drug reactions are a leading cause of death. The most common reactions that lead to emergency department visits include dermatologic, gastrointestinal, neurological, metabolic/endocrine, and bleeding problems. A large-scale surveillance study of emergency department visits in the United States found the top five drugs to be insulins (8.0 percent), warfarin (6.2 percent), amoxicillin (4.3 percent), aspirin (2.5 percent), and trimethoprim-sulfamethoxazole (2.2 percent). Although a large fraction of reactions that led to hospital visits were immunological in nature (33 percent), the majority of those requiring hospitalization are type A (Table 4.1), related to drug dose, and avoidable in half of the instances. The average hospital stay due to an adverse drug reaction is four days and accounts for a sizeable portion of health care costs.

Studies of adverse drug reactions treated or occurring in hospital settings have found that most reactions could be avoided with greater care in prescribing the right drug, determining appropriate doses, ensuring proper administration, providing proper monitoring, and avoiding interactions with other drugs or dietary supplements. However, a significant proportion of reactions are idiosyncratic, without well-known and avoidable risk factors.

The path from the discovery of a potential new therapeutic drug to its approval for marketing is clinical development (Figure 4.1). This is a detailed process that begins with animal experiments to assess several aspects of safety – including carcinogenicity and mutagenicity – dose dependence, metabolism, and kinetics. Given favorable preclinical indications of safety and efficacy, the new chemical entity undergoes a phased introduction into humans to assess safety and disposition, identify tolerable and effective doses, and determine relative patient benefit. Only one in five drugs that begin clinical development is subsequently approved for use in humans. Approximately 20 percent of that attrition is attributed to problems with safety (17). This process can take from twelve to fifteen years and cost upward of $800 million (18).

During the entire course of development, the number of subjects that are treated with an investigational drug is usually fewer than a few thousand. Such numbers are too small, and often the treatment periods are too short, to identify uncommon or rare idiosyncratic adverse drug reactions before drug approval. Consequently, many serious safety concerns are identified only after a drug is on the market. Safety concerns may be identified in phase IV clinical trials or through other forms of post-marketing surveillance, or pharmacovigilance. During a large industry-sponsored trial (APPROVe), the safety-monitoring board identified an increase in the risk of cardiovascular events due to the cyclooxygenase-2 inhibitor rofecoxib (Vioxx). This resulted in the withdrawal of rofecoxib – an effective treatment for osteoarthritis and acute pain – from the market (19). Many other drug-related adverse reactions are identified through spontaneous reporting mechanisms in the United States and elsewhere (20, 21).

Approximately 8.2 percent of drugs that are approved in the United States subsequently acquire one or more black-box warnings not present at approval (22). Approximately 1 percent of marketed drugs are withdrawn or restricted because of safety-related issues, and approximately half of those occur within the first two years on the market (23). Thus, patients, the medical community, health care providers, regulatory agencies, and the pharmaceutical industry have a compelling interest to identify factors that influence adverse drug reaction risks and improve our ability to manage them. Some of the risk factors known to contribute to adverse drug reactions include drug dose (depending on reaction type), environmental history and exposures, general health, diet, concomitant medications, age, sex, ethnicity, and genetic variants (10). The factors that best explain variation in risk are generally drug specific and may be population specific.

GENETICS OF ADVERSE DRUG REACTIONS

The potential of genetic risk factors has received increasing attention in recent years. There are several reasons for this. Most importantly, several adverse drug reactions have been shown to be strongly influenced and reasonably predictable by genetic variants. Second, genotype-based tests are very appealing in their ease, accuracy, and time independence of measurement. Assuming there is an appropriate assay, genetic risk factors can be reproducibly measured at any point during a patient's life regardless of age, disease, or environmental status, with very few exceptions. Third, genotyping technologies have made dramatic advances in recent years, making genetic investigations increasingly cost-effective. Finally, we cannot overlook the elevated status that genetics research has held within both scientific and popular culture during the past fifty years.

Many examples exist of the role that common genetic variation can play in adverse drug reaction risk, and they are increasing almost on a monthly basis. The Food and Drug Administration has listed several drugs with risks of adverse reactions that are subject to genetic modification (24). A selection of some of the most compelling or interesting examples available at the time of the preparation of this chapter are presented in Table 4.2, ordered by genetic effect estimates. Clopidogrel is at the low end of the range with a modest threefold increased risk of cardiovascular events due to therapeutic failure in patients with variants in the *CYP2C19*-metabolizing enzyme that do not efficiently convert the inactive prodrug to the active form efficiently. Several drugs are at the high end of the effect range, wherein the adverse reactions occur almost exclusively within carriers of the genetic risk factors and at relatively high rates. These include carbamazepine, abacavir, and allopurinol.

It is surprising that only three forms of genetic mechanisms underlie the risk of the adverse reactions listed in Table 4.2: drug transport, metabolism, and immune response. Drug transport and metabolism both contribute to drug disposition through absorption, distribution, metabolism, and excretion (referred to as ADME, sometimes also referred to as drug-metabolizing enzymes and transport, or DMET). ADME variation is involved in seven of the fifteen examples, with the largest effects associated with *UGT1A1*, a phase II enzyme that modifies the drug or its metabolites (glucouronidation) to facilitate elimination via biliary or urinary tracts (see Chapter 10 for additional detail). Given this mechanism, these adverse drug reactions are all dose related (type A or E), and, in some instances, risk may be modified through genotype-based dose modifications.

A notable absence from this list of dose-dependent adverse drug reactions is bleeding events due to warfarin anticoagulation, a related adverse reaction strongly influenced by the metabolizing enzyme *CYP2C9* and drug target *VKORC1*. This was not included in the listed examples because, under current dose management schemes that do not include genetic testing, the genotypes that influence warfarin dose are generally not associated with severe bleeding (25).

The remaining eight examples in Table 4.2 are adverse drug reactions affected by genes encoding the highly polymorphic human leukocyte antigens (HLAs) involved

Table 4.2. Examples of Adverse Drug Reactions with Major Genetic Risk Factors

	Adverse Drug Reaction			Genetic Risk Factor				
Drug	*Type*[a]	*Reaction*	*Prevalence*	*Risk Allele*	*Freq.*[b]	*Effect*[c]	*Population*[d]	*Ref.*
Clopidogrel	F	Cardiovascular events	0.13	*CYP2C19*2/3/4/5*	0.03	3	European	40
Ximelagatran	B	Hepatotoxicity	0.08	*HLA-DRB1*0701*	0.08	4	European	31
Gefitinib	A	Diarrhea	0.28	*ABCG2* Q141K	0.07	5	European	41
Isoniazid	A	Hepatotoxicity	0.15	*CYP2E1*1* & *NAT2*[d]	0.13	7	European	42
Amoxicillin-clavulinate	B	Hepatotoxicity	<0.001	*HLA-DRB1*1501*	0.15	10	European	43
Lumiracoxib	B	Hepatotoxicity	0.013	*HLA-DRB1*1501*	0.15	13	European	44
Simvastatin (high dose)	A/B	Myopathy	0.02	*SLCO1B1*	0.15	17	European	34
Irinotecan	A	Neutropenia	0.20	*UGT1A1*28*	0.32	28	European	45, 46
Ticlopidine	B	Hepatotoxicity (cholestatic)	<0.001	*HLA-A*3303*	0.14	36	Japanese	47
Tranilast	A	Hyperbilirubinemia	0.12	*UGT1A1*28*	0.30	48	European	36
Mercaptopurine	A	Neutropenia, other toxicity	0.12	*TPMT*2/3A/3B/3C*[f]	0.05	49	European	48
Flucloxacillin	B	Hepatotoxicity	<0.001	*HLA-B*5701*	0.04	81	European	32
Allopurinol	B	Severe cutaneous reaction	<0.001	*HLA-B*5801*	0.15	678	Han Chinese	49
Abacavir	B	Hypersensitivity reaction	0.08	*HLA-B*5701*	0.04	>1000	European	50
Carbamazepine	B	Stevens-Johnson syndrome	0.003	*HLA-B*1502*	0.04	>1000	Han Chinese	51

[a] Reaction classification (see Table 4.1).
[b] Allele frequency of the susceptibility variant.
[c] Genetic effect is the estimate of the genotype relative risk for those homozygous for the susceptible genotype in comparison with the low-risk homozygotes.
[d] Population effect was discovered in and described in the table.
[e] Frequency of the *CYP2E1*1* and *NAT2* slow acetylator homozygous genotype.
[f] Estimated cumulative frequency of TPMT-deficient alleles.

in adaptive immune response. The four strongest genetic risk factors are alleles of the *HLA-B* gene that are common within the populations in which they were discovered. Two of them are the same allele, *HLA-B*5701*, associated with the common hypersensitivity reaction due to abacavir and the hepatotoxicity due to flucloxacillin. The five largest HLA effects on risk are all class I genes, whereas the three smallest are due to the class II gene *HLA-DRB1* (or at least haplotypes characterized by *HLA-DRB1* alleles).

The prevalence estimates of the adverse drug reactions listed in Table 4.2 are highly variable, ranging from very rare (<1/10,000) to very common (0.28). However, there are two characteristics that they all have in common. The first is that the known genetic risk factors are all common in the population in which they were identified, with frequencies greater than 3 percent. The second is that they all have relatively large effects. We should not conclude from these observations that all genetic risk factors will have a common and large effect. It is probably no accident that it is the common and large-effect variants that we have the greatest statistical power to identify, and, in many instances, the only kinds that we could identify given the limited sample sizes and genotyping methods used.

FINDING GENETIC RISK FACTORS

Genetic studies of adverse drug reactions differ from common disease studies in two important ways. First, they differ in the nature of genetic effects of potential interest. In common disease research, even small genetic effects are considered highly relevant to shed light on disease etiology. However, in adverse drug reaction research, it is assumed that clinically useful genetic risk factors must have large genetic effects, capable of identifying those at greatest risk. Second, they differ in the sample sizes available for research. Disease genetic studies are often large, with case and control samples in the hundreds to thousands. (See reference 26 for a review of research methods in disease genetics.) Most sample sizes available to study adverse reactions are small (see Figure 4.1). Consequently, even for relatively common reactions, the number of subjects that experience the adverse reaction during development will be small. For rare reactions – often recognized only after drug approval – identification and recruitment of affected individuals into research studies is often costly, time consuming, and difficult. Therefore, most pharmacogenetic studies of adverse drug reactions are limited to identify variations that increase risk manyfold (27).

A variety of study designs and genotyping strategies may be used to identify genetic risk factors. The same study designs that are applicable to the study of common disease and other complex traits can and have been used in the context of adverse drug reactions. Most studies use a case-control design, identifying subjects that satisfy criteria for having experienced an adverse drug reaction as cases and matching them to suitable drug-treated or population controls. The choice of controls will be influenced by the frequency of the adverse reaction, frequency of the disease for which the drug is being tested or prescribed, as well as cost and expedience. Investigations into rare adverse reactions, in general, require large government or academic networks to identify and ascertain cases. In most such studies, controls are ascertained in a different manner, and recruiting drug-treated controls can be challenging. Population controls are usually a convenient and cost-effective option as long as important sources of confounding are considered. The use of well-characterized and previously genotyped controls can be superior to the use of drug-matched controls by reduction in study costs, the provision of large numbers, and the opportunity for avoiding genetic confounding by genotype-based matching. This topic has been previously reviewed in this context (27, 28).

Common adverse reactions studied during the course of clinical development will usually have well-matched drug-treated controls readily available that are typically used for study. Studies involving clinical trials may also study quantitative intermediate traits related to adverse reactions, such as alanine aminotransferase or bilirubin levels for hepatotoxicity, QT interval length from electrocardiograms for prolonged QT or Torsades de Pointes, or absolute neutrophil counts for neutropenia. Association analysis of the intermediate trait in the entire collection of drug-treated subjects will generally have greater statistical power than a comparison of subjects divided into case and control groups with thresholds of the same trait that are often arbitrary in nature (29).

There are three general genotyping strategies to consider when designing an association study: candidate variants, candidate genes, and genome-wide screening. In some instances, a single or small number of genetic variants or genes may have a suspected role in the reaction. Most of the effects listed in Table 4.2 were identified with such a candidate gene approach. Candidates for genotyping are selected on the basis of prior knowledge of enzymes and transporters that may play a role in drug disposition, therapeutic action of the drug (e.g., *VKORC1*, a vitamin K receptor, is the target for warfarin), or the nature of the adverse reaction. The heavy involvement of HLA genes in hepatotoxicity, hypersensitivity, and severe cutaneous reaction in the list of genes known to influence class B adverse reactions make this region a high priority for most studies. Similarly, *UGT1A1* is a prime candidate for adverse reactions involving hepatotoxicity, hyperbilirubinemia, and neutropenia. The candidate gene approach can be very effective when there is confidence in the set of selected candidates: genotyping can be conducted quickly, thoroughly, relatively inexpensively, and with greater statistical power because of a lower burden of multiple testing.

In many instances, there may not be adequate preexisting evidence for the selection of candidate genes, or there may be concerns that the list of candidates is not sufficiently inclusive. Contemporary genome-wide genotyping panels provide a very good measure of common genetic variation for most non-African human populations (see Chapter 10). Three genetic risk factors reported in Table 4.2 – for the drugs ximelagatran, simvastatin, and flucloxacillin – were discovered as part of genome-wide association studies (GWAS). Genome-wide panels have the clear advantage of providing a survey of common variation over the entire genome, including nearly all known gene coding and noncoding regions.

Potential downsides to a GWAS approach include the statistical burden of multiple testing, the impact on study power, the general restriction to associations due to common variations, and the challenges in localizing functional genetic variants. When searching for very large effects of common variants, even modestly sized collections of cases and controls can yield high statistical power (27). The effect of *HLA-B*1502* on carbamazepine-induced Stevens-Johnson syndrome could be identified with 80 percent power with as few as nine cases, adjusting for 500,000 tests.

Follow-up of associated regions to localize and identify causal genetic variants has proven to be challenging in common disease research for both candidate gene and GWAS (30). In contrast, for the confirmed adverse drug reaction genetic risk factors listed in Table 4.2, most of the causal variants are known. This is the result of the extensive research into the function of metabolizing enzymes and human histocompatibility and their mechanistic relationships with the drug and the adverse reaction.

The list of known risk factors of large effect listed in Table 4.2 is limited to a very restricted set of gene classes: ADME and HLA. Furthermore, to date, four reasonably powered GWAS related to adverse drug reactions have failed to identify other kinds of major genetic risk factors (25, 31–34). There are several possible explanations for this observation. First, these examples represent a small number of known adverse drug reactions. Further examples may ultimately reveal several additional genetic mechanisms. Second, it may be that drug disposition and adaptive immune response are the primary contributors to genetic risk. If additional genome-wide research in this area fails to identify the alternative genes/gene classes involved, future investigations may be safely limited to careful study of ADME and HLA genes.

Finally, large-scale studies of the genetics of common diseases and several other complex traits have uniformly failed to explain more than a small portion of their heritable variation. This has left the genetics community puzzling over where this additional genetic variability may be found (35). Several case studies have pointed toward the presence of low-frequency and rare variants with relatively large effects – including protein-disrupting single-nucleotide polymorphisms and large copy number variants – as one possible source. Unfortunately, the heritability of adverse drug reactions is unknown because of the rare co-occurrence of reactions in multiple family members. If rare, highly penetrant variants are involved, current GWAS methods will generally fail to find them. Well-conducted studies of ADME and HLA genes do investigate low-frequency variants of known function. Rapid advances in high-throughput sequencing and bioinformatics will eventually make it feasible to search for rare functional variants on a genome-wide scale. However, there are several study design and statistical analysis challenges to overcome. A vanguard of successful early studies will be needed to demonstrate the effectiveness of such a strategy.

CONDUCTING GENETIC STUDIES

Adverse drug reactions are a serious concern to public health and to pharmaceutical developers. As a result, genetic studies in this field are conducted by publicly funded academic groups and by drug companies. When adverse drug reactions are common, they will generally lead to the termination of drug development. However, when the potential clinical benefit of such a drug is high enough, genetic studies may be explored in an effort to improve the overall benefit-risk ratio. These studies are almost always conducted by the drug developers with clinical trial data. Tranilast-induced hyperbilirubinemia is a successful example. During a phase III clinical trial, significantly elevated rates of hyperbilirubinemia (12 percent) were observed in the tranilast-treated arm. Given the known role of a polymorphic TA repeat in the promoter of *UGT1A1* on population levels of bilirubin (Gilbert's syndrome), this variant was typed and shown to be primarily responsible for the elevations observed in response to tranilast treatment. This finding suggested that the bilirubin elevations were not likely to be indicators of liver damage (36). (The development of tranilast for the treatment of restenosis after percutaneous transluminal coronary revascularization was ultimately terminated because of the lack of efficacy, though not safety.) More commonly, adverse drug reactions with moderate frequencies (>1 percent, e.g., clopidogrel, lumiracoxib, and abacavir) are studied in one or more phase IV trials.

Rare adverse reactions are studied only after a drug is on the market, the safety risk is recognized, and sufficient numbers of cases accrue. Identifying and collecting cases can be a challenging task and may require a variety of approaches. There are several retrospective case collection approaches. Some countries have adverse event-reporting registries that can identify cases that may be approached retrospectively for study participation through cooperation with patients' physicians. Large patient databases built on top of electronic medical records, such as those maintained by public or private health networks, are increasingly being mined for patients with drug prescriptions and medical codes, laboratory results, or pathology reports matching adverse event criteria. These patient records can be further scrutinized for evidence of drug causality and selected patients approached to participate. Hospital discharge records are another source. Prospective collection of cases is also possible, particularly when the condition a drug is prescribed for is treated by specialists (e.g., abacavir hypersensitivity cases through HIV clinics), or when the adverse reaction must be treated in specialized wards (e.g., SSRs in burn units).

Fewer than half of the examples in Table 4.2 were conducted or sponsored by the companies developing or marketing the drug. Many genetic studies are conducted by independent, publicly funded clinical or academic networks. Recently, another model has emerged in this area. The International Serious Adverse Events Consortium (SAEC), formed in 2007, is a joint partnership of more than ten pharmaceutical companies, the Wellcome Trust, several academic networks and participants, and the Food and Drug Administration. (37) The SAEC was formed to sponsor and perform research directed toward the discovery of genetic risk factors for important serious adverse drug reactions that may help understand their causes and predict patients who are at the greatest risk. The consortium charter requires that it make all research data – including demographic, clinical, and genotype data – publicly available to qualified researchers within one year of data collection. The SAEC is also actively involved in improving research methods, promoting education, and encouraging needed infrastructure developments to facilitate future research in this area for both academic and commercial use.

Table 4.3. Characteristics of Genetic Tests

	Sensitivity	*Specificity*	*PPV*[a]	*NPV*	*Avoid One Case*[b]	*Excluded, No Reaction*[c]	*Treated, Reaction*[d]
Carbamazepine SSR	1.0	0.97	0.092	1.0	333	10	0
Abacavir hypersensitivity (clinically suspected)	0.46	0.98	0.61	0.95	28	1	1
Abacavir hypersensitivity (immunologically confirmed)	1.0	0.97	0.48	1.0	37	1	0
Allopurinol SSR	1.0	0.852	0.0007	1.0	10,000	1481	0

[a] PPV and NPV were calculated assuming the adverse reaction prevalence reported in Table 4.2.
[b] The number of subjects needed to screen to avoid one adverse reaction.
[c] The number of marker-positive subjects, of those screened[b], excluded from treatment that would not have developed the adverse reaction.
[d] The number of marker-negative subjects, of those screened[b], that would develop the adverse reaction.

The initial research of the SAEC has focused on the study of SSRS (including Stevens-Johnson syndrome and toxic epidermal necrolysis) and drug-induced liver injury (DILI). With genome-wide genotyping of more than 1 million single-nucleotide polymorphisms, the consortium quickly identified *HLA-B**5701* as the major common genetic risk factor for flucloxacillin liver injury. It also verified previous reports that haplotypes involving *HLA-DRB1**1501* increase risk of hepatotoxicity due to amoxicillin-clavulanate. Implicit in their research reported to date on more than 200 cases with DILI is that the common genetic risk factors are drug specific and that there do not appear to be any major risk factors similar to those at the bottom of Table 4.2 outside of the HLA region. Future work of the SAEC will include increasing case sample sizes for SSR and DILI – as well as expansion into other important adverse reactions – through contributions of case collections from participating companies and academic networks. The pooled resources of the consortium are also exploring emerging genetic technologies to determine the role of rare variants in adverse reaction risk.

CLINICAL APPLICATION

The examples of genetic risk factors for adverse drug reactions discussed in this chapter all have very large effects. To some it would seem obvious that most of these would have immediate applications to clinical care. However, of the examples listed in Table 4.2, only irinotecan, mercaptopurine, abacavir, and carbamazepine are recommended for genetic testing. The value of genetic testing comes down to how well a test can predict those who will and will not develop the reaction and how much that will improve patient care (38).

Positive and negative predictive values are two important measures when assessing the ability of a genetic test to predict which patients will and will not experience an adverse drug reaction. Positive predictive value (PPV) is the probability of experiencing the adverse drug reaction given that a person is positive for the test (in most cases listed, meaning that they carry at least one copy of the risk allele). Conversely, negative predictive value (NPV) is the probability that a person will not experience the adverse reaction given that they are negative for the test. Unlike the common diagnostic measures of sensitivity (probability of a positive test given a reaction) and specificity (probability of a negative test given no reaction), PPV and NPV are closely tied to the prevalence of the reaction. A test may have relatively high values of sensitivity and specificity; however, if the frequency of the reaction is very rare, a test may result in a wrong decision most of the time (38, 39).

It is instructive to contrast the most compelling examples of carbamazepine, abacavir, and allopurinol. The test characteristics for each drug are shown in Table 4.3. Carbamazepine is prescribed for the treatment of epilepsy and bipolar disorder. Severe skin reactions are reported to have an intermediate prevalence in treated Han Chinese, approximately 0.3 percent. A retrospective study of 44 Taiwanese cases and 101 tolerant controls found that all cases were carriers of the *HLA-B**1502* allele, compared with only three controls. This allele is both highly sensitive and specific, and, given the moderate prevalence, has a PPV of almost 10 percent (i.e., one in ten *HLA-B**1502* carriers treated with carbamazepine is expected to develop SSR). On the basis of these data, if carbamazepine was only prescribed to *HLA-B**1502*-negative patients, we can estimate that, for every 333 patients screened, 322 would be treated and 11 would not. Of those 11 not treated with carbamazepine, only one would be predicted to experience the adverse reaction and ten would not. Given the severity of this adverse reaction and the cost to treat it, several regulatory agencies have advised that genetic testing should be conducted in patients of East Asian ancestry. It turns out that *HLA-B*1502* is very rare (<0.1 percent) in other populations, such that screening for this allele is not cost-effective for non-East Asians.

Abacavir is a nucleoside reverse transcriptase inhibitor for treatment of HIV infection that was found to cause a hypersensitivity reaction – including fever, rash, and gastrointestinal and constitutional symptoms – in approximately 8 percent of patients. Approximately 3 percent of patients were shown by a skin patch test to have an immunological response to abacavir. A prospective trial of 1956 HIV patients who had not previously been treated with abacavir demonstrated that the sensitivity and specificity for *HLA-B*5701* in immunologically confirmed hypersensitivity cases is very high. The PPV is almost 50 percent and the NPV is 100 percent, indicating that half of *HLA-B*5701* carriers to receive abacavir experience an immunologically confirmed hypersensitivity reaction, whereas none of the *HLA-B*5701*-negative patients do. The high PPV and NPV values are possible in this case because the frequency of the risk allele and the outcome are similar to one another. Only 37 patients need to be screened for the risk allele to avoid one adverse reaction, and only one person would be excluded from treatment who would not have had a reaction. Based on these data, screening for *HLA-B*5701* has become standard practice to determine which HIV patients would benefit most from abacavir treatment.

A counter to the examples of carbamazepine and abacavir – where clinical utility has been demonstrated – is allopurinol. Allopurinol causes rare (prevalence < 0.01 percent) SSR. A study of 51 Han Chinese cases and 135 treated controls showed that all SSR cases were *HLA-B*5801* carriers compared with only 15 percent of the controls. Subsequent studies have confirmed this association, and, unlike *HLA-B*1502, HLA-B*5801* is present at frequencies greater than 1 percent in most populations and has been subsequently shown to be associated with SSR in several other Asian and European populations. However, the very low prevalence of the adverse reaction results in a PPV estimate of approximately 0.07 percent. Assuming a prevalence of 1 in 10,000 in the Han Chinese population, 10,000 patients would have to be screened to avoid one SSR occurrence, and 15 percent of those screened would be denied treatment. Estimates of the number needed to screen to avoid one reaction are higher in Europeans. Based on the available evidence, screening for *HLA-B*5801* is not cost-effective and is not regularly used in clinical practice.

In the instance of allopurinol and most of the other known examples, the cost of genetic screening before prescribing a medication with known genetic risk factors is not justified by the expected clinical benefit. However, there are two trends that may change this cost-benefit ratio. First, technological advances are making it easier to measure more genetic variants at a lower cost. Second, genetic research in disease and drug response continues to identify variants with potential clinical value. Because germline genetic variants do not change over the life span, it will become increasingly cost-effective to measure all relevant variants early in each person's life such that the genetic information is readily available to inform all subsequent diagnostic and treatment decisions. Because new genetic discoveries will continue to be made into the foreseeable future, knowing which genetic variants will be the most valuable to measure is not possible. (Perhaps the case could be made for complete HLA genotyping.) This conundrum may be solved when complete genome resequencing becomes routine. The availability of such complete population-level genetic information – combined with rich and integrated electronic medical records – will also facilitate the kind of research necessary to validate genetic hypotheses and make decisions about their applications in medical practice.

REFERENCES

1. World Health Organization. *The Importance of Pharmacovigilance: Safety Monitoring of Medicinal Products.* Geneva: World Health Organization Uppsala Monitoring Centre; 2002.
2. Edwards IR & Aronson JK. Adverse drug reactions: definitions, diagnosis, and management. *Lancet.* 2000;**356**:1255–9.
3. International Conference on Harmonization Good Clinical Practice (GCP) ICH E6 (R1) Guidelines. 1996. http://www.craconnection.com/resources/ich-gcp-resources/53-international-conference-on-harmonisation-good-clinical-practice-guidelines-craconnectioncom.html (June 29, 2011).
4. Aronson JK & Ferner RE. Joining the DoTS: new approach to classifying adverse drug reactions. *BMJ.* 2003;**327**:1222–5.
5. Rawlins MD & Thompson JW. Pathogenesis of adverse drug reactions. In: Davies DM, ed. *Textbook of Adverse Drug Reactions.* Oxford: Oxford University Press; 1977:10.
6. Lee WM. Acetaminophen-related acute liver failure in the United States. *Hepatol Res.* 2008;**38**:S3–8.
7. Uetrecht J. Idiosyncratic drug reactions: past, present, and future. *Chem Res Toxicol.* 2008;**21**:84–92.
8. Posadas SJ & Pichler WJ. Delayed drug hypersensitivity reactions – new concepts. *Clin Exp Allergy.* 2007;**37**:989–99.
9. Uetrecht J. Immune-mediated adverse drug reactions. *Chem Res Toxicol.* 2009;**22**:24–34.
10. Ulrich RG. Idiosyncratic toxicity: a convergence of risk factors. *Annu Rev Med.* 2007;**58**:17–34.
11. Ostapowicz G, Fontana RJ, Schiodt FV, Larson A, Davern TJ, Han SH, et al. Results of a prospective study of acute

liver failure at 17 tertiary care centers in the United States. *Ann Intern Med.* 2002;**137**:947–54.
12. Danan G & Benichou C. Causality assessment of adverse reactions to drugs. I. A novel method based on the conclusions of international consensus meetings: application to drug-induced liver injuries. *J Clin Epidemiol.* 1993;**46**:1323–30.
13. *Patterns of Medication Use in the United States*, 2006: a *Report from the Slone Survey.* Boston, MA: Slone Epidemiology Center; 2006.
14. Lazarou J, Pomeranz BH, & Corey PN. Incidence of adverse drug reactions in hospitalized patients: a meta-analysis of prospective studies. *JAMA.* 1998;**279**:1200–5.
15. Pirmohamed M, James S, Meakin S, Green C, Scott AK, Walley TJ, et al. Adverse drug reactions as cause of admission to hospital: prospective analysis of 18 820 patients. *BMJ.* 2004;**329**:15–19.
16. Budnitz DS, Pollock DA, Weidenbach KN, Mendelsohn AB, Schroeder TJ, & Annest JL. National surveillance of emergency department visits for outpatient adverse drug events. *JAMA.* 2006;**296**:1858–66.
17. Dimasi JA. Risks in new drug development: approval success rates for investigational drugs. *Clin Pharmacol Ther.* 2001;**69**:297–307.
18. Dimasi JA, Hansen RW, & Grabowski HG. The price of innovation: new estimates of drug development costs. *J Health Econ.* 2003;**22**:151–85.
19. Krumholz HM, Ross JS, Presler AH, & Egilman DS. What have we learnt from Vioxx? *BMJ.* 2007;**334**:120–3.
20. Bate A, Lindquist M, & Edwards IR. The application of knowledge discovery in databases to post-marketing drug safety: example of the WHO database. *Fundam Clin Pharmacol.* 2008;**22**:127–40.
21. Harmark L & van Grootheest AC. Pharmacovigilance: methods, recent developments and future perspectives. *Eur J Clin Pharmacol.* 2008;**64**:743–52.
22. Lasser KE, Allen PD, Woolhandler SJ, Himmelstein DU, Wolfe SM, & Bor DH. Timing of new black box warnings and withdrawals for prescription medications. *JAMA.* 2002;**287**:2215–20.
23. Wysowski DK & Swartz L. Adverse drug event surveillance and drug withdrawals in the United States, 1969–2002: the importance of reporting suspected reactions. *Arch Intern Med.* 2005;**165**:1363–9.
24. *U.S. Food and Drug Administration. Table of Valid Genomic Biomarkers in the Context of Approved Drug Labels.* June 23, 2009. Available from: http://www.fda.gov/Drugs/ScienceResearch/ResearchAreas/Pharmacogenetics/ucm083378.htm
25. Wadelius M, Chen LY, Lindh JD, Eriksson N, Ghori MJ, Bumpstead S, et al. The largest prospective warfarin-treated cohort supports genetic forecasting. *Blood.* 2009;**113**:784–92.
26. McCarthy MI, Abecasis GR, Cardon LR, Goldstein DB, Little J, Ioannidis JP, et al. Genome-wide association studies for complex traits: consensus, uncertainty and challenges. *Nat Rev Genet.* 2008;**9**:356–69.
27. Nelson MR, Bacanu SA, Mosteller M, Li L, Bowman CE, Roses AD, et al. Genome-wide approaches to identify pharmacogenetic contributions to adverse drug reactions. *Pharmacogenomics J.* 2009;**9**:23–33.
28. Nelson MR, Bryc K, King KS, Indap A, Boyko AR, Novembre J, et al. The Population Reference Sample, POPRES: a resource for population, disease, and pharmacological genetics research. *Am J Hum Genet.* 2008;**83**:347–68.
29. Schork NJ, Nath SK, Fallin D, & Chakravarti A. Linkage disequilibrium analysis of biallelic DNA markers, human quantitative trait loci, and threshold-defined case and control subjects. *Am J Hum Genet.* 2000;**67**:1208–18.
30. Ioannidis JP, Thomas G, & Daly MJ. Validating, augmenting and refining genome-wide association signals. *Nat Rev Genet.* 2009;**10**:318–29.
31. Kindmark A, Jawaid A, Harbron CG, Barratt BJ, Bengtsson OF, Andersson TB, et al. Genome-wide pharmacogenetic investigation of a hepatic adverse event without clinical signs of immunopathology suggests an underlying immune pathogenesis. *Pharmacogenomics J.* 2008;**8**:186–95.
32. Daly AK, Donaldson PT, Bhatnagar P, Shen Y, Pe'er I, Floratos A, et al. HLA-B*5701 genotype is a major determinant of drug-induced liver injury due to flucloxacillin. *Nat Genet.* 2009;**41**:816–19.
33. Takeuchi F, McGinnis R, Bourgeois S, Barnes C, Eriksson N, Soranzo N, et al. A genome-wide association study confirms VKORC1, CYP2C9, and CYP4F2 as principal genetic determinants of warfarin dose. *PLoS Genet.* 2009;**5**:e1000433.
34. Link E, Parish S, Armitage J, Bowman L, Heath S, Matsuda F, et al. SLCO1B1 variants and statin-induced myopathy – a genomewide study. *N Engl J Med.* 2008;**359**:789–99.
35. Maher B. Personal genomes: the case of the missing heritability. *Nature.* 2008;**456**:18–21.
36. Danoff TM, Campbell DA, McCarthy LC, Lewis KF, Repasch MH, Saunders AM, et al. A Gilbert's syndrome UGT1A1 variant confers susceptibility to tranilast-induced hyperbilirubinemia. *Pharmacogenomics J.* 2004;**4**:49–53.
37. Holden AL. The innovative use of a large-scale industry biomedical consortium to research the genetic basis of drug induced serious adverse events. *Drug Discov Today Technol.* 2007;**4**:75–87.
38. Constable S, Johnson MR, & Pirmohamed M. Pharmacogenetics in clinical practice: considerations for testing. *Expert Rev Mol Diagn.* 2006;**6**:193–205.
39. Visser H & Hazes JM. How to assess musculoskeletal conditions. Prognostics. *Best Pract Res Clin Rheumatol.* 2003;**17**:403–14.
40. Mega JL, Close SL, Wiviott SD, Shen L, Hockett RD, Brandt JT, et al. Cytochrome p-450 polymorphisms and response to clopidogrel. *N Engl J Med.* 2009;**360**:354–62.
41. Cusatis G, Gregorc V, Li J, Spreafico A, Ingersoll RG, Verweij J, et al. Pharmacogenetics of ABCG2 and adverse reactions to gefitinib. *J Natl Cancer Inst.* 2006;**98**:1739–42.
42. Huang YS, Chern HD, Su WJ, Wu JC, Chang SC, Chiang CH, et al. Cytochrome P450 2E1 genotype and the susceptibility to antituberculosis drug-induced hepatitis. *Hepatology.* 2003;**37**:924–30.
43. O'Donohue J, Oien KA, Donaldson P, Underhill J, Clare M, MacSween RN, et al. Co-amoxiclav jaundice: clinical and histological features and HLA class II association. *Gut.* 2000;**47**:717–20.
44. Wright TM. MHC II haplotype marker for lumiracoxib injury. Available at: http://www aasld org/conferences/

educationtraining/Documents/Hepatoxicity%20Slides/Wright.pdf (April 8, 2009).

45. Innocenti F, Undevia SD, Iyer L, Chen PX, Das S, Kocherginsky M, et al. Genetic variants in the UDP-glucuronosyltransferase 1A1 gene predict the risk of severe neutropenia of irinotecan. *J Clin Oncol.* 2004;**22**:1382–8.
46. Rouits E, Boisdron-Celle M, Dumont A, Guerin O, Morel A, & Gamelin E. Relevance of different UGT1A1 polymorphisms in irinotecan-induced toxicity: a molecular and clinical study of 75 patients. *Clin Cancer Res.* 2004;**10**:5151–9.
47. Hirata K, Takagi H, Yamamoto M, Matsumoto T, Nishiya T, Mori K, et al. Ticlopidine-induced hepatotoxicity is associated with specific human leukocyte antigen genomic subtypes in Japanese patients: a preliminary case-control study. *Pharmacogenomics J.* 2008;**8**:29–33.
48. Relling MV, Hancock ML, Rivera GK, Sandlund JT, Ribeiro RC, Krynetski EY, et al. Mercaptopurine therapy intolerance and heterozygosity at the thiopurine S-methyltransferase gene locus. *J Natl Cancer Inst.* 1999; **91**:2001–8.
49. Hung SI, Chung WH, Liou LB, Chu CC, Lin M, Huang HP, et al. HLA-B*5801 allele as a genetic marker for severe cutaneous adverse reactions caused by allopurinol. *Proc Natl Acad Sci U S A.* 2005;**102**:4134–9.
50. Mallal S, Phillips E, Carosi G, Molina JM, Workman C, Tomazic J, et al. HLA-B*5701 screening for hypersensitivity to abacavir. *N Engl J Med.* 2008;**358**:568–79.
51. Chung WH, Hung SI, Hong HS, Hsih MS, Yang LC, Ho HC, et al. Medical genetics: a marker for Stevens-Johnson syndrome. *Nature.* 2004;**428**:486.

5 PharmGKB, a Centralized Resource for Pharmacogenomic Knowledge and Discovery

Li Gong and Teri E. Klein

Differential response to the standard dose of drug therapy in patients is commonly seen in clinical practice. Environmental factors, diet, age, gender, disease severity, interacting drugs, and genetic variation all contribute to the variability in drug response. The importance of genetic variation in drug response has become more prominent with the emergence of pharmacogenomics and personalized medicine. Pharmacogenomics is the study of how genetic differences affect responses to drugs. By knowing more about a person's genetic makeup, clinicians will be better able to assess the risks and benefits associated with medications to maximize treatment success. Pharmacogenomics has the potential to have an impact on many steps of medical care, from diagnosis to tailored drug prescription, and from basic drug discovery to clinical trial design.

In the past two decades, substantial knowledge has accumulated about genetic events contributing to differences in drug responses, some resulting in drug-labeling changes and clinical practices (1, 2). U.S. drug labels for 6-mercaptopurine, warfarin, and irinotecan contain dosage adjustments based on *TPMT, CYP2C9* and *VKORC1*, and *UGT1A1* genotypes (3–5). As the U.S. Food and Drug Administration (FDA) requires more pharmacogenomic information to be included on drug labels, pharmacogenomics will be increasingly accepted and integrated into mainstream clinical practice. We are already observing a steady shift from the "trial and error" approach to a more knowledge-guided personalized approach toward drug therapy. To realize the full potential of pharmacogenomics, many formidable challenges still need to be overcome. These challenges include the ethical, economical, and legal issues associated with the ever expanding field of genetic testing, as well as the strong need to increase genetic literacy among patients and health care providers. Easier access to well-characterized clinical outcome and biometric data on patients under treatment are also crucial to identify and track genotype-phenotype relationships. With the combined efforts of researchers and health care providers to use the knowledge of pharmacogenomics in drug treatment and diagnosis, the barriers between scientific discovery and the clinical application of pharmacogenomics will diminish over time to realize the full benefits of the field.

The completion of the human genome sequence and the availability of affordable high-throughput genotyping technology have enabled pharmacogenomic researchers to study drug response in the context of the entire genome. The amount of data accumulated from various genome-wide studies grows exponentially each year. With data scattered throughout the literature and more than a thousand biological databases, as well (6), it is a significant challenge for researchers to find relevant data and to turn those data into distilled knowledge. Establishing centralized domain-specific knowledge bases is a more efficient way to manage and query information within specific biomedical fields. The Pharmacogenomics Knowledge Base (PharmGKB, http://www.pharmgkb.org) is a centralized Web-based resource for pharmacogenomic data and knowledge. Originated in 2000 as a National Institutes of Health (NIH)-sponsored pharmacogenetics knowledge base for the scientific community (7), PharmGKB has become a preeminent resource that disseminates comprehensive and up-to-date knowledge and data in pharmacogenomics (Figure 5.1). The knowledge domain includes extensively reviewed information on genes and important variants that are implicated in drug response, and it presents condensed information in the form of drug-centered pathways, very important pharmacogene summaries (VIPs), and annotation of functionally important variants and relationships among gene, drug, and disease. The primary data domain comprises both genotype and phenotype data submitted by the scientific community. As of June 2009, PharmGKB contained information on more than 800 genes with genotype information (20,970 subjects that have been genotyped with 3,401,867

Knowledge domain:

Functional annotation for variants that impact drug-response phenotypes (2,374 concise summaries)

- Major repository for pathways of drug action and metabolism (60 manually curated pathways)
- Publication-quality in-depth summaries for "very important pharmacogenes" (VIPs) (39 in-depth summaries)
- Gene-drug-disease interactions from the literature, tagged with relevance to pharmacokinetics (PK) and pharmacodynamics (PD) (3,626 manually curated articles)

Data domain:

- Data repository for Pharmacogenetics Research Network (PGRN) (810 genes with genotyping information, 20,970 subjects that have been genotyped with 3,401,867 variants reported)
- Data repository for international data-sharing consortia (International Warfarin Pharmacogenetics Consortium [IWPC] and International Tamoxifen Pharmacogenomics Consortium [ITPC])
- Searchable summary and links to high-impact datasets of relevance to pharmacogenomics

Figure 5.1. The PharmGKB content overview. PharmGKB is composed of two complementary domains: a knowledge domain and a primary data domain. The knowledge domain includes very important pharmacogene (VIP) summary, annotation of variants of pharmacogenomic interest, drug response pathways, and relationships among genes, drugs, and diseases. The primary data domain contains both genotype and phenotype data. Establishing correlations between genotypes and phenotypes will generate novel pharmacogenomics relationships, and the knowledge gained can in turn be used to catalyze scientific research and discovery.

variants reported), 2,374 variants with functional annotation, 594 drugs with supporting data, 60 drug pathways, 39 in-depth pharmacogene summaries, and more than 3,626 curated articles. Whether searching for candidate genes for a specific drug treatment or disease susceptibility, or a specific genetic variation and its functional consequence regarding drug response, PharmGKB's highly structured knowledge content and flexible query interface provide the infrastructure to easily retrieve information of interest and advanced tools to support hypothesis generation.

This chapter describes PharmGKB's knowledge and data contents, basic navigation and search of the database, and Web site contents. Acute lymphoblastic leukemia (ALL) is used as an example to demonstrate how clinicians can use PharmGKB to find pharmacogenomic information related to a specific disease. A study design workflow is also presented on how PharmGKB can be used to facilitate study design and downstream analysis to stimulate scientific discovery.

PharmGKB HOME PAGE AND BASIC QUERY

PharmGKB's home page (www.pharmgkb.org) is the entry point for most users (Figure 5.2). It was designed to highlight the interests of its users and to ensure that the primary contents of the knowledge base are easily accessible to them. The menu tabs at the top of the page provide access to the top-level section of the PharmGKB site. **Home** is the front page where we highlight the knowledge and data content, mission, contact information, and registration; **Search** is the main search page where the user can search by either free text or canned queries, or browse information by domain; **Submit** describes how a user can submit genotype, phenotype, pathway, or literature data; **Help** includes an extensive list of background information, downloads, and educational and technical references; **PGRN** lists all members involved in the NIH Pharmacogenetics Research Network, their research interests, submissions to PharmGKB, and information in support of the PGRN; **Contributors** is the section where people are listed who have contributed data to PharmGKB; **Clinical PGx** catalogs drugs with pharmacogenomic information in the context of FDA-approved drug labels; **My PharmGKB** is the section for our registered users to view and edit their profile, submission, and Web site statistics and only appears after a user has logged in. A user can ask questions of the PharmGKB Team regarding any aspect of PharmGKB by clicking the **Feedback** button, which is located at the top right corner of every PharmGKB Web page. To orient our users to the most important

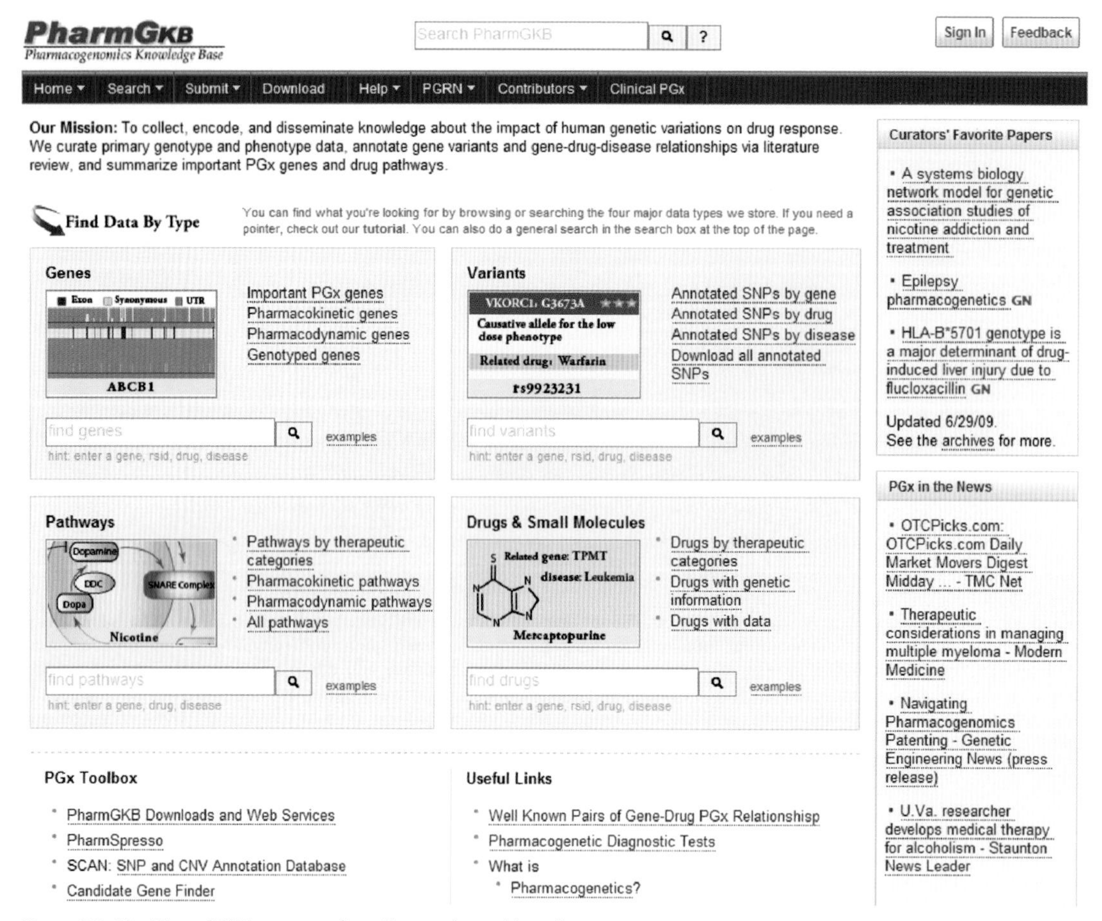

Figure 5.2. The PharmGKB home page (http://www.pharmgkb.org).

knowledge content pertinent to pharmacogenomics, we present four featured types of data in the center of the home page: **Genes** with pharmacogenetic significance; genetic **variants** associated with drug response; **pathways** for drug transport, metabolism, and action; and **drugs** for which there is pharmacogenomic knowledge. The current "favorite" papers selected biweekly by our curators and a dynamically generated news feature are presented on the right side of the home page.

Users may query and browse PharmGKB in many ways. The general search box above the menu bar allows free text search across all domains in the PharmGKB database, whereas the focused search embedded in the four main data types narrows the search results by data categories. For example, a search in the genes search box (e.g., drug names, variants, and gene names) will return a list of genes related to that search text. Similarly, a search under the variant search box will only return a list of variants whose documentation contains the input query. If a user is looking for information on "which genetic variants play a role in response to gefitinib?" she can enter "gefitinib" in the **variant** search box, which will return the list of variants in genes *EGFR*, *ABCG2*, and *CYP2D6* that are associated with gefitinib response (Figure 5.3). The highly integrated nature of our database and focused search strategy provide users with flexibility in how they find information and allow them to ask mechanistic questions about the relationships between drugs and genes and genetic variation.

PharmGKB gene, drug, and disease pages are structured similarly with a tab system. Figure 5.4 demonstrates the general layout of a PharmGKB gene page for *VKORC1*, which encodes a key enzyme in the vitamin K cycle with a significant contribution to the variable response of warfarin therapy. The **Overview** tab for the gene lists the official and alternative names and symbol for the gene, its genomic boundaries, and associated OMIM phenotypes. The **VIP** tab includes links to

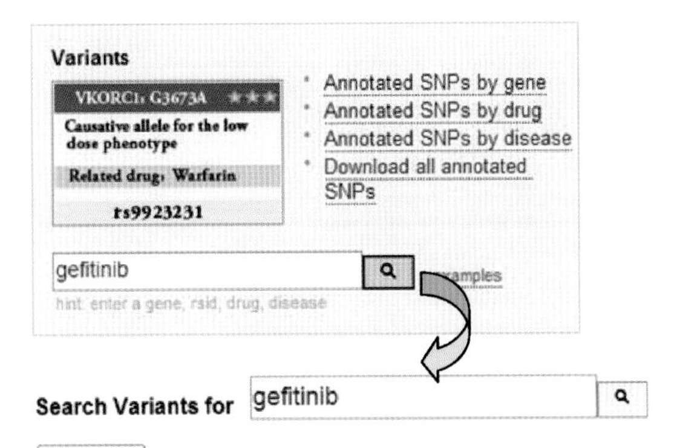

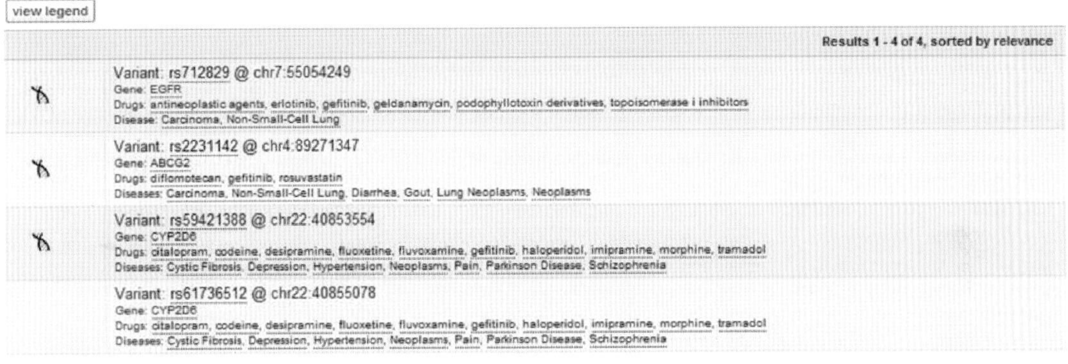

Figure 5.3. Example of focused search for variants associated with response to gefitinib.

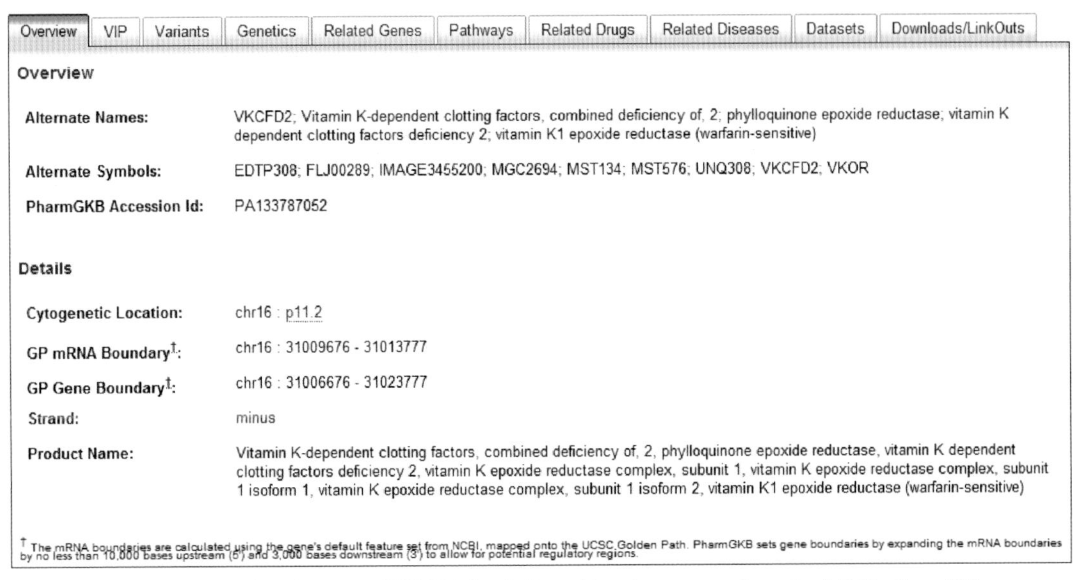

Figure 5.4. Example of PharmGKB Gene page (*VKORC1*, vitamin K epoxide reductase complex, subunit 1). The PharmGKB gene page is organized by a tab system and provides detailed information on synonyms, phenotype, genomic location, variant information, associated pathways, related drugs and diseases from the literature, as well as download/cross-references to other complementary genomic databases.

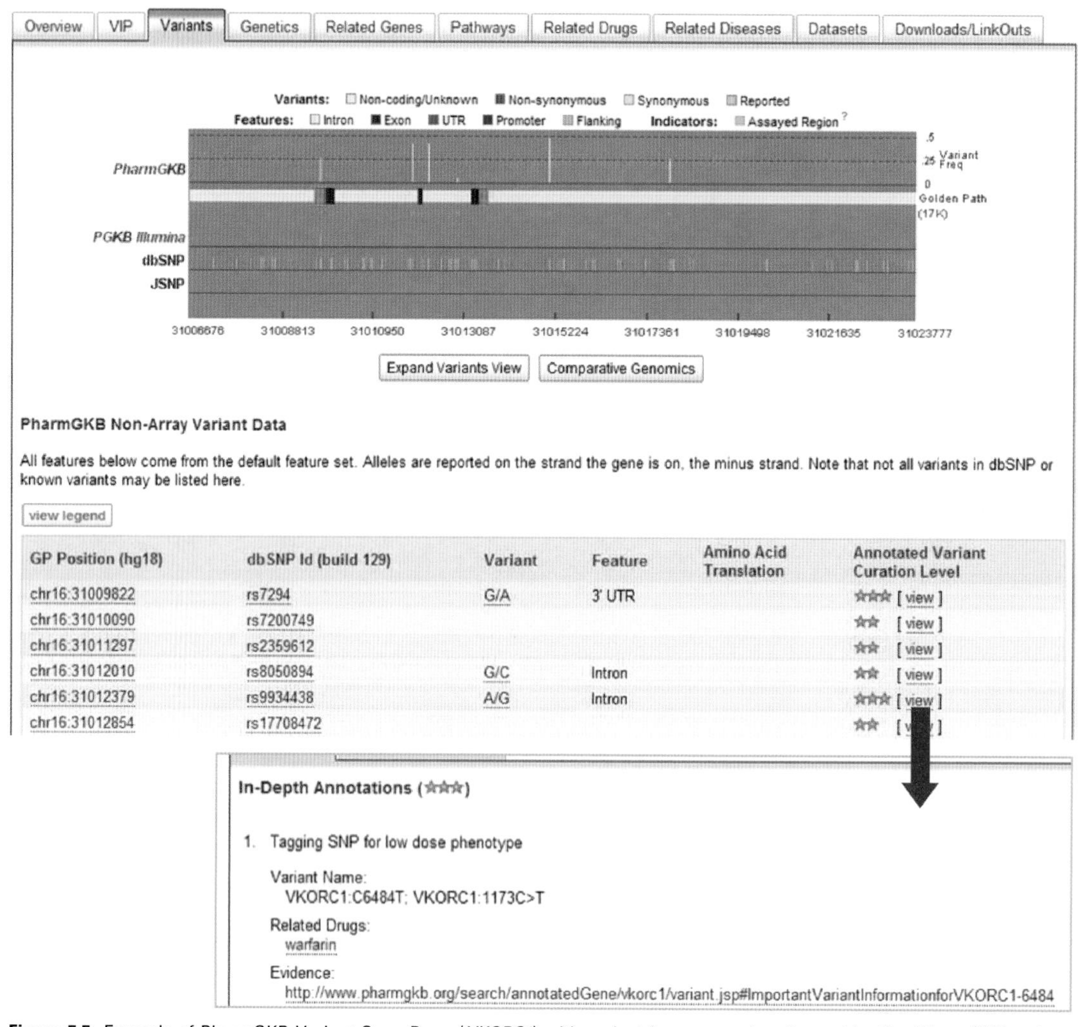

Figure 5.5. Example of PharmGKB Variant Gene Page (*VKORC1*) with variant browser and variant table. The PharmGKB variant page contains a graphic browser on the top and a variant table below. The variant browser displays all the polymorphisms across the span of the gene of interest from various resources such as PharmGKB primary data, SNP arrays (Illumina and Affymetrix), dbSNP, and jSNP. Each tick on the browser represents a variant from the respective resource. Gene features are color coded to differentiate exons, introns, promoters, and UTRs. The variant table lists the details for nonarray genotype data in PharmGKB, such as their genomic positions, functional roles, frequencies, and assay types.

the in-depth pharmacogene summary with introductory information describing the gene, important variants/haplotypes, and their impact in drug response. The **Variant** tab includes a graphic browser on the top and variant table below (Figure 5.5). The **Datasets** tab contains the phenotype data stored in PharmGKB for the gene of interest. The **Pathway** tab lists all related pathways for the gene. Gene-drug-disease relationships can be found under the **Related Drugs** and **Related Diseases** tabs. Links for the download of experimental data for the gene and cross-references to other complementary genomic databases are under the **Download/LinkOuts** tab.

KNOWLEDGE DOMAIN: SIGNIFICANT PHARMACOGENES/VARIANTS, DRUG PATHWAYS, AND GENE-DRUG-DISEASE RELATIONSHIPS

PharmGKB collects and summarizes knowledge on genes and genetic variations related to the safety and efficacy of drug treatments, drug metabolism or site of action pathways, and complex relationships among gene, drug, and diseases supported by evidence from the primary literature (8, 9). The diverse knowledge content at PharmGKB can be differentiated not only by types, but also by levels of curation. The three levels of curation are clearly labeled throughout our knowledge base by the use of a "star" system: (1) noncurated (noted as one star on the Web site) is knowledge that has been programmatically generated yielding a higher volume with wider breadth but lower quality (less specificity); (2) curated (noted as two stars on the Web site) is knowledge that has been reviewed manually by a scientific curator and yields a moderate volume of knowledge and a higher level of quality than the noncurated facts; and (3) in-depth (noted as three stars on the Web site) detailed descriptions such as the pathways or VIP pages that are extremely high in quality, but are very time consuming to prepare.

Significant Pharmacogenes/Variants

Variability in drug response is often related to functional changes in genes/proteins that are involved in the pharmacokinetic (PK) or pharmacodynamic (PD) processing of the drugs. The PK genes generally include drug metabolism enzymes such as cytochrome P450s (or CYPs) and sulfotransferases, and drug-uptake or efflux transporters such as members of the *OATP* and *p*-glycoprotein families. The PD genes usually involve the drug target, its downstream signaling molecules, and other molecules that might modulate the biological context where the drug-target interaction happens. Variation in any element that controls the PK or PD process may lead to variability in drug efficacy or safety. PharmGKB presents knowledge for key genes and important variants in the forms of (1) in-depth summary for **VIPs** (three stars) and (2) the abbreviated summary for variants of pharmacogenomic interests (**Annotations**, two stars) such as functional variants, tagging SNPs, and variants in association studies. The VIPs are structured summaries that include detailed information about a given gene, including its important polymorphisms, haplotypes, phenotypes, interacting drugs and complete mapping information, and supporting literature references. An allele frequency table may also be included if the specific variant has been studied extensively in different populations. To significantly increase the breadth of coverage for important variants, PharmGKB has begun to catalog and highlight variants of pharmacogenomic significance with the use of concise summaries. In contrast to the VIPs, which are encyclopedia-like summaries that require large amounts of manual curation, the two-star annotated variants are minimally curated, yet still bring key phenotypic consequences of individual variants to our users. Two variants (*VKORC1*:rs9923231 and *CYP4F2*:rs2108622) that are important for warfarin dosing illustrate these two different levels of curation (Figure 5.6). Because nomenclature inconsistencies for genes and variants present a major challenge when searching for relevant pharmacogenomic information, PharmGKB includes lists of current as well as historical terms used for genes and their variants in both our VIPs and variants of interest efforts. For example, rs9923231, a common promoter variant for *VKORC1*, has been mentioned in the literature as "G3673A" and "-1639G>A" (Figure 5.4). By including the alternative names that have been used in the literature as references, our users can quickly reconcile the multiple naming issues when conducting comprehensive searches in biomedical literature for genetic variations.

Drug Pathways

Many of the early successful examples in pharmacogenetics focused on one gene and its candidate SNPs. Although studies of individual genes are valuable for understanding their function in drug response, it is well accepted that genes and their proteins do not act in isolation, but rather as components of larger pathways or networks. Similarly, most drug affects are determined by the interplay of multiple gene products in biological and pharmacological pathways that modulate the pharmacokinetics and pharmacodynamics of the drug (10, 11). PharmGKB drug-centered pathways aim to provide an overview of how genes are involved in the pharmacokinetics and pharmacodynamics of drugs. The pathway diagrams use standard shapes and colors to represent genes, drugs, metabolites, and interactions. Users can click on each of these objects to go directly to an individual drug and gene page to access more specific information. Each pathway has a textual summary description and a downloadable evidence spreadsheet that provides literature support for each interaction depicted on the pathway diagram. Unlike pathway resources that primarily focus on common biological and physiological processes (e.g., KEGG, Reactome, Biocarta, GenMAPP), PharmGKB is the only resource with a primary focus on drug-centered pathways, in particular, PK pathways (12–15). The genes included in the PK and PD pathways can serve as candidate genes for a pharmacogenetic study of the drug and can be used to help interpret findings from a genome-wide association study. At present, PharmGKB has sixty pathways that cover drugs used in a wide variety of disease classes such as asthma, cancer, cardiovascular disease, diabetes, depression, HIV, and inflammatory diseases. PharmGKB pathways and VIPs are now published monthly in *Pharmacogenetics and Genomics* (16). Figure 5.7 shows an example of a published drug response pathway for the chemotherapy agent etoposide. This pathway illustrates the complex gene-drug interaction network

There are **Three Important Variants** for VKORC1.

1. VKORC1: G3673A (rs9923231)
2. VKORC1: C6484T (rs9934438)
3. VKORC1: G9041A (rs7294)

1. VKORC1: G3673A (-1639 G>A) on AY587020 (rs9923231) 

Gene HGNC Name: VKORC1

Indicate if this variant should be linked to a haplotype worksheet: VKORC1*2

Variant Summary:

G3673A, or -1639 G>A as it is commonly called in the literature, is a polymorphism in the promoter region of VKORC1 that is believed to be the causative SNP for the low dose phenotype. Luciferase assays show that the activity of the G allele was increased by 44% over the activity of the A allele [15888487]. Additionally, analysis of VKORC1 mRNA isolated from human liver samples showed that carriers of the A allele at position 3673 had reduced amounts of VKORC1 mRNA [15930419]. Both of these studies support the contention that the G3673A SNP likely disrupts the binding of a transcription factor in the promoter region of VKORC1 which in turn leads to a lower amount of VKORC1 mRNA transcript, and presumably fewer functional copies of the mature VKORC1 protein.

The G3673A or -1639 G>A variant has been genotyped in a number of different populations (see Table). This polymorphism has pronounced differences in its frequency by ethnic group as it is actually the majority allele in Asian populations. This variant is also quite common in Caucasians, with an allele frequency typically around 40% in predominantly Caucasian populations.

Population	N	Allele Frequency of "A"	PMID
Japanese	93	93%	17049586
Swedish	181	39%	17048007
Japanese (anticoagulated)	260	89%	16890578
Japanese (healthy)	228	94%	16890578
Spanish (anticoagulated)	105	52%	16611310
Florida VA hospital	356	34%	16580898
German	200	42%	16270629
English	297	47%	15947090

Key PubMed IDs: 15888487 15930419 17049586 17048007 16890578 16611310 16580898 16270629 15947090 15883587 15790782 16432637 17161452 16141794

Genomic Variant & GenBank ID: G3673A (-1639 G>A) on AY587020

mRNA Variant & GenBank ID: N/A

Protein Variant & GenBank ID: N/A

dbSNP rs#: rs9923231 sometimes also appears in the literature as rs17878363

GoldenPath Position: chr16:31015190;(hg18)

Drugs/Substrates: Warfarin, coumarin, acenocoumarol

Phenotypes/Diseases: Atrial Fibrillation, Coagulation Protein Disorders, Hemorrhage, Vascular Diseases

Please list any phenotype datasets that are of particular relevance to this variant: WUSTL warfarin dosing data, group A

B.

Variant rs2108622 at chr19:15851431 in CYP4F2

Data | Annotations

Curated Annotations (★★)

1. This variant is found to have clinical impact on stable warfarin dose. Patients with 2 TT alleles were reported to require approximately 1 mg/day more warfarin than patients with 2 CC alleles. This variant affects enzyme acitivity and was shown to decrease 20-HETE production in a reconstituted recombinant protein system to approximately 60% of the wild-type enzyme.

   Variant Name:
   CYP4F2:V433M
   Related Drugs:
   warfarin
   Evidence:
   PMID:17341693
   PMID:18250228

Figure 5.6. Example of PharmGKB variant annotation (*VKORC1*:rs9923231 in-depth annotation and *CYP4F2*:rs2108622 brief annotation). PharmGKB collects and annotates variants of pharmacogenomic interest at two curation levels: in-depth VIP annotation (three stars) and brief annotation (two stars). (**A**) rs9923231, a *VKORC1* variant important for warfarin dosing, is annotated as an in-depth VIP variant. The VIP variant annotation includes a detailed functional summary, associated haplotypes, allele frequency, interacting drugs, complete mapping, and supporting literature references. (**B**) rs2108622, a *CYP4F2* variant that has recently been implicated in warfarin dosing, is annotated with a brief summary, key phenotypic consequences, alternative names, and supporting evidence.

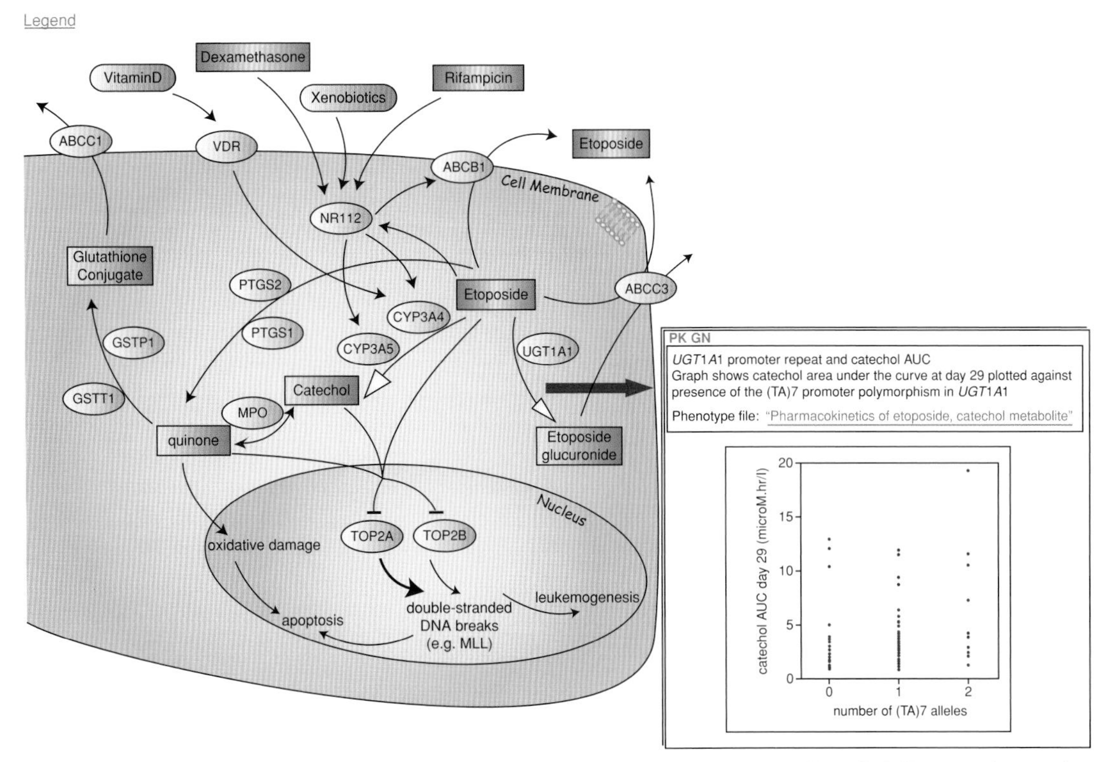

Figure 5.7. Example of PharmGKB pathway (etoposide pathway). The etoposide pathway on PharmGKB illustrates the complex gene-drug interaction network that is involved in the metabolism, cellular disposition, mechanism of action, and delayed toxicity of etoposide. Etoposide is a commonly used chemotherapy agent with a broad range of antitumor activity. It is metabolized by CYP3A4 and CYP3A5, whose expressions are regulated by the vitamin D receptor (VDR) and PXR (NR1I2). Polymorphisms in these genes have been shown to influence the clearance rate of etoposide and may be used as predictors of etoposide PK variability. Etoposide exerts its antitumor activity through inhibition of topoisomerase II. Along with its *O*-demethylated metabolites, catechol and quinone, etoposide stabilizes the double-stranded DNA breaks and induces apoptosis. One of the most serious side effects of etoposide is its ability to induce secondary malignancies, which has recently been attributed to its inhibition of topoisomerase II beta, whereas the antitumor activity largely depends on inhibition of the α-isoform. The PharmGKB pathway diagrams use standard shapes and colors to represent genes, metabolites, drugs, and interactions. Users can also click on each of these objects to go directly to the individual drug and gene page to access more specific information.

that is involved in the metabolism, cellular disposition, mechanism of action, and delayed toxicity of etoposide (17).

Gene-Drug-Disease Relationships

Capturing the complex interactions between genes, drugs, and diseases is a continuing focus in pharmacogenetic and pharmacogenomic research. Most of the accumulated knowledge is only found in journal articles and books. PharmGKB curators routinely scan the primary literature to annotate gene-drug-disease relationships and present the knowledge at PharmGKB in the **Related Genes, Related Drugs**, and **Related Diseases** sections. Figure 5.8 shows an example of gene-drug-disease relationships for the antidiabetic drug rosiglitazone (Avandia®). Relationships between the drug-gene and drug-disease are tagged with Categories of Pharmacogenetic Knowledge (PK, PD, GN [Genotype], FA [Molecular & Cellular Functional Assays], CO [Clinical Outcome]) for easy classification and retrieval, and are linked to the original journal article for detailed information. Recently, we have implemented an automated literature pipeline that identifies papers through text-mining approaches (18) and then highlights potentially relevant terms for quick orientation to pertinent information (19). The automatic scan of applicable literature can be found in the **Non-curated Information** section, along with the metabolizing enzymes and drug target information automatically retrieved from DrugBank (20).

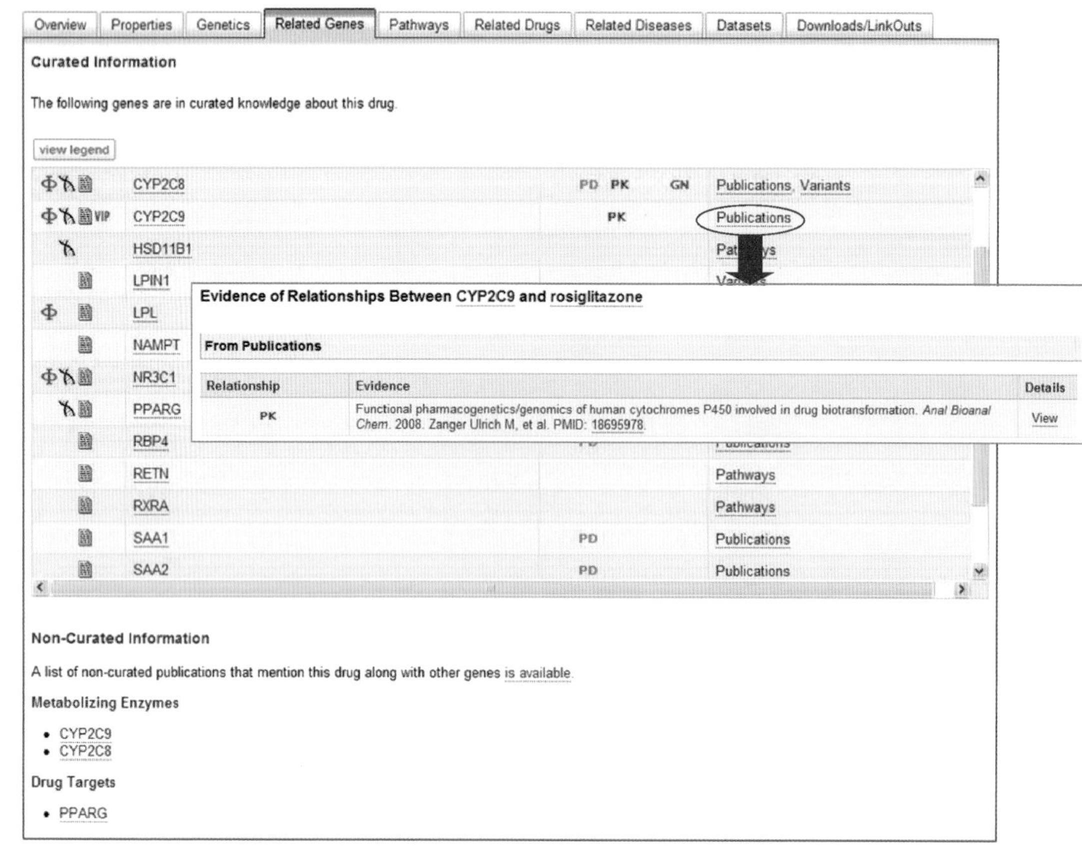

Figure 5.8. Example of PharmGKB curated publications (rosiglitazone). PharmGKB annotates gene-drug-disease relationship and presents the knowledge in the Curated Publications section. Relationships between the drug-gene and drug-disease are tagged with categories of pharmacogenetic knowledge (PK, PD, GN, FA, CO) for easy classification and retrieval, and linked to the original journal article for detailed information.

PRIMARY DATA DOMAIN

In addition to our knowledge mission, PharmGKB is the central primary data repository for pharmacogenetic and pharmacogenomic studies. Initially, the data repository was seeded by studies from the Pharmacogenetics Research Network (PGRN) and focused on a handful of genes. As whole-genome scans become more widely used, PharmGKB has expanded its capacity to accommodate large-scale high-throughput data that may involve a large number of samples assayed across the entire genome. For example, SNP array experiments on pharmacogenomics of statin therapy can now be viewed and downloaded from PharmGKB (https://www.pharmgkb.org/search/browseAlpha.action?browseKey=snpArray). As of June 2009, more than 20,972 subjects had been genotyped with 3,461,867 variants reported in PharmGKB. In addition, there were 461,324 distinct phenotype measurements. The large volume of integrated primary data in PharmGKB offers potential for meta-analysis between related pharmacogenomic phenotypes. More recently, PharmGKB has moved into a leadership role in creating data-sharing consortia for high-impact pharmacogenomics collaborations. The goal of the consortia is to create merged international datasets from a large multicenter, multinational cohort of patients to develop the best strategy for predicting drug efficacy and toxicity. Two example consortia are the International Warfarin Pharmacogenetics Consortium (IWPC) (http://www.pharmgkb.org/views/project.jsp?pId=56) (21) and the International Tamoxifen Pharmacogenomics Consortium (ITPC) (https://www.pharmgkb.org/views/project.jsp?pId=63). The IWPC pooled data from more than 5,000 patients and derived a globally relevant warfarin pharmacogenetic dosing equation (22). The IWPC dataset is carefully curated, stored, and managed at PharmGKB. The detailed demographic and phenotype information (e.g., ethnicity, age, height, weight, dose, and target International Normalized Ratio or INR) for all subjects in the study as well as associated genotypes is available for download to researchers worldwide (Figure 5.9A). Besides in-house primary data, PharmGKB regularly surveys literature and external archival databases to identify important datasets of relevance to pharmacogenomics, and to provide access to these datasets for our users through a searchable summary of the study tagged with pertinent genes, drugs, and diseases by the use of standardized vocabularies. Figure 5.9B illustrates an example dataset summary for the genome-wide association study of addiction housed at the National Center for Biotechnology Information database of Genotypes and Phenotypes (dbGaP). The experimental results of the study from dbGaP can be accessed through the link under **External Data Links** tab.

CLINICAL PHARMACOGENOMICS (PGx)

In April 2009, PharmGKB introduced the **Clinical PGx** domain to bring accurate and up-to-date pharmacogenomic information in the context of FDA-approved drug labels to clinicians and the general public (Figure 5.10). This page includes information for all the drugs for which the FDA drug label has been revised to include specific pharmacogenomics recommendations, as well as drugs with mounting pharmacogenomic evidence. Clicking on a drug on this page opens a new window displaying pharmacogenomic drug-labeling information, FDA-approved/cleared diagnostic tests for the drug, and links to relevant PharmGKB resources. A summary table of pharmacogenomic diagnostic tests for many of the most-studied genetic variant-drug interactions is also available from the useful link section on the home page (https://www.pharmgkb.org/resources/forScientificUsers/pharmacogenomic_diagnostic_tests.jsp).

DISEASE CASE STUDY: HOW TO FIND IMPORTANT PHARMACOGENOMIC INFORMATION ASSOCIATED WITH A DISEASE AND ITS THERAPEUTICS (THE ALL EXAMPLE)

Cancer has a strong genetic component, making it an ideal field for pharmacogenomic research. Anticancer drugs, in general, have a narrow therapeutic index and exhibit significant interindividual variability in their efficacy and toxicity. With multiple treatments available for many cancer types, there is a pressing need for tools to aid decision making on drug selection. Methods to identify genetic variants that contribute to variable clinical outcomes associated with chemotherapy drugs have been an area of active research for many decades. Variants in the drug-metabolizing enzymes (e.g., *TPMT*, *UGT1A1*, and *CYP2D6*) have been identified to be associated with response to chemotherapy drugs (23–25). Gene expression profiling has also revealed gene signatures for classifying disease and predicting treatment outcome (26, 27). The great strides made in the field of cancer pharmacogenomics generated deeper insights into the pathogenesis of cancer and the mechanisms of drug sensitivity, as well, to provide clinicians the opportunities to personalize chemotherapy to maximize efficacy and reduce toxicity for each patient.

Research in childhood ALL is a classic example of how pharmacogenomic information can be used to benefit clinical practices. Without the introduction of any new chemotherapy drugs in the past 30 years, the overall survival of patients with ALL has improved from less than 10 percent in the 1960s to more than 80 percent at present (28). One of the key reasons behind the advance in cure rate is the improved ability to identify patients early on who are at high risk of treatment failure or severe toxicity. A well-characterized pharmacogenetic marker for chemotherapy agents used to treat ALL is human thiopurine methyltransferase (*TPMT*). *TPMT* is a phase II metabolizing enzyme catalyzing the metabolism and intracellular inactivation of thiopurine drugs such as 6-mercaptopurine and azathiopurine. *TPMT* activity is highly variable, with approximately 10 percent of the patients having intermediate enzymatic activity and 0.3 percent being deficient for *TPMT* activity. Numerous studies have shown that patients with inherited *TPMT* deficiency are exposed to a higher concentration of the active drug, and are at risk for severe or fatal hematopoietic toxicity when given the normal dose of the drug. *TPMT* deficiency has also been linked to a higher risk of second malignancies among patients with ALL. Several genetic polymorphisms in *TPMT* (notably *TPMT*2*, *TPMT*3A*, and *TPMT*3C*) have been discovered that lead to the reduced activity of the enzyme. A genetic test on *TPMT* genotype has been developed to identify patients who are at higher risk of severe bone marrow toxicity and thus require significant dose reductions to avoid life-threatening toxicity. This example demonstrates the clear utility of using pharmacogenomic knowledge for dose selection and optimizing drug therapy. The early groundbreaking research in ALL pharmacogenomics was conducted by PGRN scientists; the relevant knowledge and significant datasets are archived at PharmGKB; and knowledge is represented in the form of drug pathways, significant variant annotation, and literature annotation.

Searching for Specific Information on the PharmGKB

To search for information relevant to a disease in PharmGKB, simply type the disease name in the general search box, for example, "ALL," and then open the

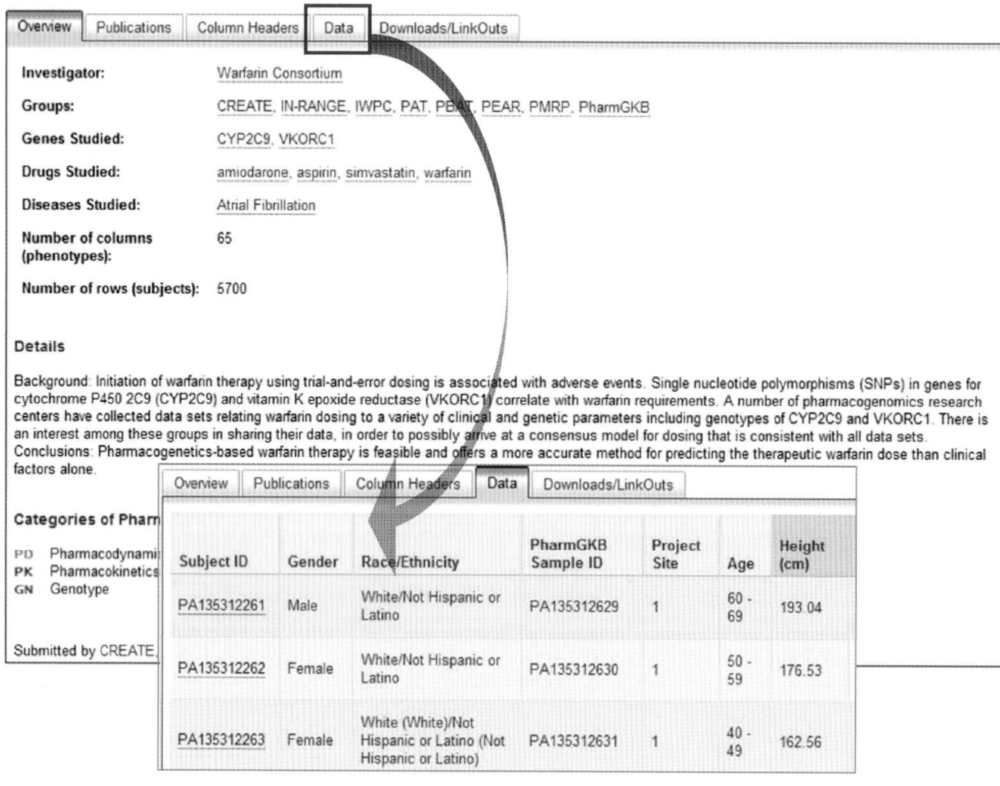

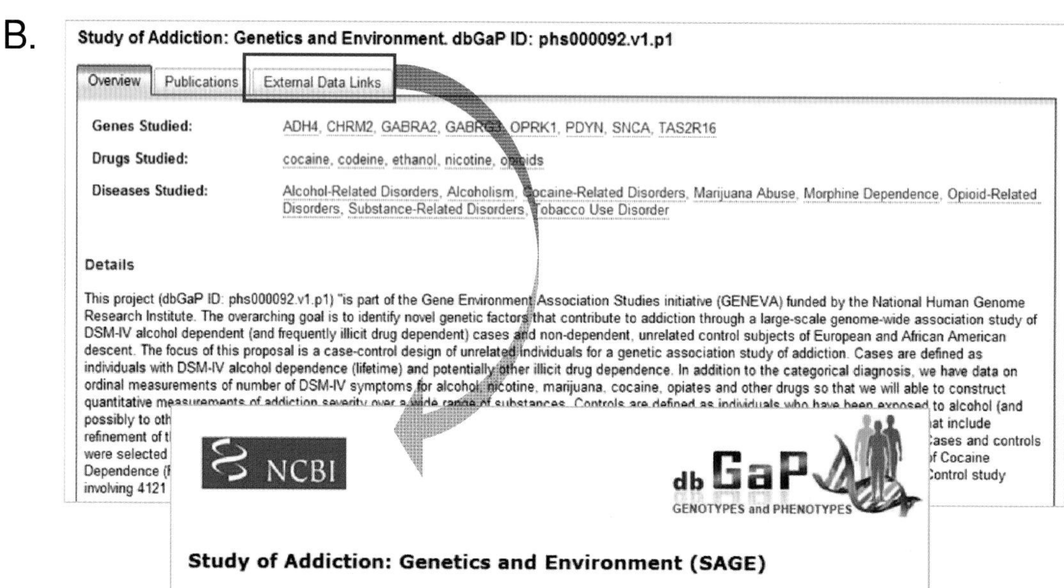

Figure 5.9. Example of high-impact datasets with relevance to pharmacogenomics, housed at PharmGKB or at external archival databases. PharmGKB identifies and curates high-impact datasets from pharmacogenomic studies. (**A**) Example of a carefully curated dataset housed at PharmGKB (IWPC dataset). (**B**) Example dataset summary for important pharmacogenomic study housed at external archival databases (*study of addiction: genetics and environment* from dbGAP, http://www.ncbi.nlm.nih.gov/projects/gap/cgi-bin/study.cgi?study_id=phs000092.v1.p1).

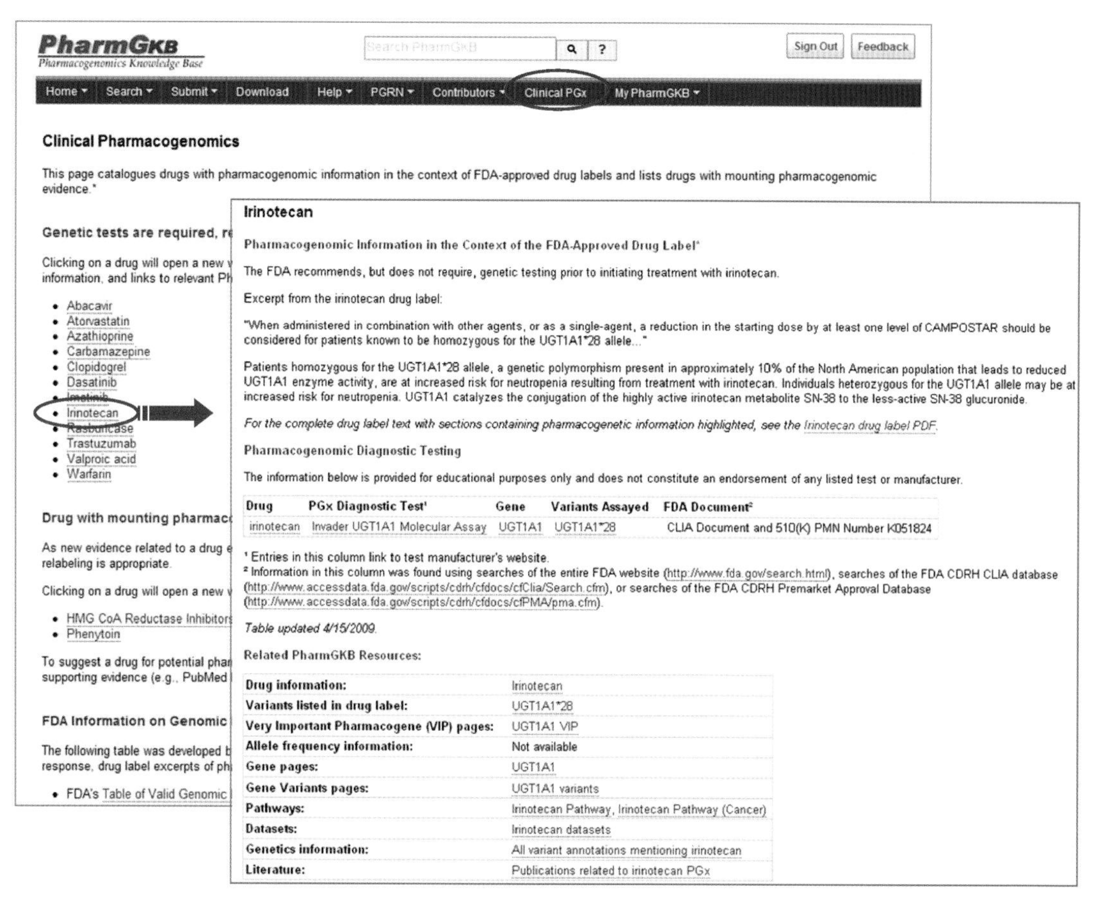

Figure 5.10. Clinical pharmacogenomics (PGx) page. The PharmGKB clinical PGx page catalogs drugs with pharmacogenomic information in the context of FDA-approved drug labels and lists drugs with mounting pharmacogenomic evidence.

PharmGKB disease page for ALL (precursor cell lymphoblastic leukemia-lymphoma) (Figure 5.11). PharmGKB Disease names are imported from Medical Subject Headings (MeSH, http://www.nlm.nih.gov/mesh/). Precursor cell lymphoblastic leukemia-lymphoma is the official disease name in MeSH for ALL. On the ALL disease page, the primary phenotype data files associated with ALL can be found under the **Datasets** tab. These include PK studies on drugs used in ALL treatments and gene expression studies done to correlate drug resistance in children with ALL. The pharmacogenomic knowledge accumulated from ALL studies can be found under the **Genetics, Pathways, Related Genes,** and **Related Drugs** tabs. Under the **Genetics** tab, users can find lists of genetic variations that have been associated with treatment response in ALL, including ones associated with end-of-induction minimal residual disease in ALL. In the **Pathways** section, there are five pathways on pharmacokinetics and/or pharmacodynamics of the drugs used to treat ALL (doxorubicin, etoposide, methotrexate, thiopurine, and vinca alkaloids). Clicking on the pathway name opens the pathway diagram that provides a quick overview of genes that are involved in the metabolism, transport, and action of the drug of interest. The genes and drugs on the pathways are all hyperlinked to the specific gene and drug page for more detailed information. For example, *TPMT* is shown on the thiopurine pathway diagram to convert mercaptopurine into an inactive metabolite called methylmercaptopurine (meTGMP), thereby decreasing the formation of the active thioguanine nucleotides. To learn more about polymorphisms for *TPMT*, clicking on the *TPMT* gene from the pathway image will display detailed variant annotations from the *TPMT* variant page and VIP page. Variants that lead to decreased enzymatic activity of *TPMT* can be easily identified by viewing the star annotations column in the variant table. Under the **Related Genes** and **Related Drugs** tabs, related genes and drugs from the literature for ALL are presented and tagged with our categories of knowledge. Clicking on View under the Details column in related genes/drugs from the literature section will lead to links to the original paper documenting the relationship between ALL and respective genes and drugs.

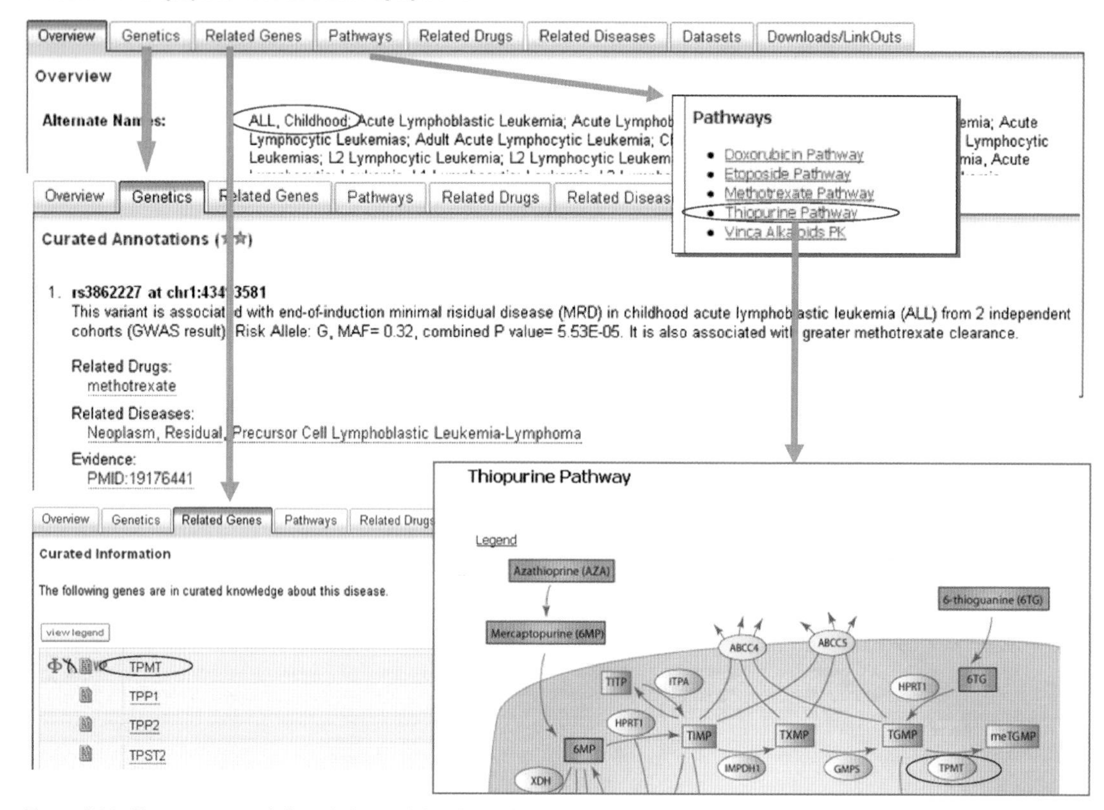

Figure 5.11. Pharmacogenomic knowledge and significant datasets related to acute lymphoblastic leukemia (ALL) on PharmGKB.

KNOWLEDGE-GUIDED STUDY DESIGN AND DATA ANALYSIS USING PharmGKB (ADVERSE DRUG REACTIONS ASSOCIATED WITH STATIN TREATMENT EXAMPLE)

Adverse drug reactions (ADRs) are a major concern in current drug therapy across all diseases. Each year, millions of people in the United States experience an ADR using marketed drugs, and ADR is ranked as one of the leading causes of illness and death. ADRs are also the top reason for drug withdrawal from the market and are responsible for the termination of approximately 20 percent of investigational drugs in the drug development pipeline (29). Given the significant impact, lowering ADR rates is a major goal for both the medical field and the pharmaceutical industry. Although ADRs can result from a variety of factors (e.g., age, organ function, drug interactions), genetic factors play a significant role in the incidence and severity of ADRs (2). Identifying genetic markers to predict which individuals are at greater risk for ADRs has been a high-priority area of pharmacogenomic research (30). Once identified and validated, the genetic variant information can be incorporated into a diagnostic test that will help predict a patient's response to a specific drug and guide treatment and dosage selections. Currently, the two principle approaches to study ADRs are the candidate gene approach and the genome-wide approach. The candidate gene approach selects genes for study based on knowledge of the target or metabolic pathways of the drugs used. Alternatively, the genome-wide approach uses a broad discovery-based method by examining hundreds of thousands of SNPs simultaneously. These two approaches are complementary, each with its advantages and disadvantages. The rational selection of genes in the candidate gene approach normally leads to biologically meaningful results with a low-cost, moderate sample size and a low false-discovery rate. However, novel genes might be missed. The genome-wide approach allows identification of new candidate genes that were previously unknown to be important in response to a drug. This later type of study, however, normally requires increased expense, has a large sample size, and has significantly higher risk of false discovery because of multiple comparisons. Each of the novel genes or variants identified from the genome scan will need to be evaluated individually, and the positive findings need to be replicated in independent datasets. The public SNP databases now contain more than 18 million validated SNPs, and the number is continuing to grow (31, 32). These SNPs are of various degrees of data quality, with only a small portion of them characterized in terms of population frequencies, and an even smaller fraction that have been functionally characterized. It becomes an increasing challenge to sort through a vast number of SNPs to select appropriate ones for pharmacogenetic studies. The extensive knowledge content stored at PharmGKB can help scientists quickly identify sets of genes and important variants that might reasonably be expected to modulate pharmacogenomic phenotypes. Scientists conducting studies using the candidate gene approach can use PharmGKB knowledge to front-load a study with relevant PGx candidate genes and variants. Scientists pursuing the genome-wide approach can use knowledge extracted from PharmGKB to filter and prioritize results by looking for genes and variants with a pharmacogenomic rationale. We will use ADRs associated with statin treatments to demonstrate how PharmGKB can be used to guide the knowledge-based study design with the use of the candidate gene approach.

The β-hydroxy-β-methylglutaryl-coenzyme A (HMG-CoA) reductase inhibitors, collectively known as statins, are currently one of the most effective medications for managing elevated concentrations of low-density lipoprotein cholesterol. Statins reduce the frequency of coronary heart diseases by as much as 21 percent to 43 percent and are the most prescribed class of drugs in America and worldwide. Six statins are currently marketed in the United States, including atorvastatin, simvastatin, rosuvastatin, pravastatin, fluvastatin, and lovastatin. These statins all target HMG-CoA reductase, but differ in terms of their potencies and PK properties. Although statins are generally well tolerated, the response to treatment varies greatly among individuals. Up to 7 percent of patients receiving statins develop muscle-related complications (ranging from aches, cramps, mild myalgia, to severe rhabdomyolysis) and require either dosage reduction or discontinuation of therapy. Cerivastatin (Baycol) was withdrawn from the U.S. market in 2001 because of nearly 100 cases of rhabdomyolysis-related death. Various mechanistic hypotheses have been proposed to explain statin-induced muscle injury. These hypotheses range from pharmacodynamics (e.g., interference with interactions with target or mitochondrial function) to pharmacokinetics (e.g., variability in drug metabolism and transport) (33–36). However, the exact role of genetic polymorphisms in predicting adverse events associated with statin use is still unknown. A knowledge-guided candidate gene study may help us gain significant insight into the mechanism of statin-induced ADRs. PharmGKB can be used in the following fashion to help select candidate genes and variants in the study design for statin-induced ADRs (Figure 5.12).

Step 1. Candidate Gene Selection

Genes involved in pharmacokinetics or pharmacodynamics of the drugs are prime candidates for selection in a pharmacogenomic study. PharmGKB has multiple statin PK/PD pathways that summarize the complex multigenic influences on statin drug response. The PK pathways describe genes involved in the absorption, distribution, metabolism, and excretion of statins, whereas the PD pathways illustrate the physiological effects of the drug, its mechanism of action, and possible side effects. Each gene in the PK or PD pathway can serve as a candidate gene for a pharmacogenetic study of that drug. The detailed information on a specific gene can be retrieved by clicking on the gene symbol. The pathway for each drug can be found on the drug page under the **Pathways** tab. If no pathway is currently available for the drug of interest, another place to retrieve the gene-drug relationship is under the **Related Genes** tab. This section includes genes that have an established relationship to the drug of interest from published articles. Similarly, we can start from the **Related Diseases** page to look for candidate genes to follow up in the ADR study. For a statin-induced ADR, we can search for diseases such as "Myalgia" or "Muscular diseases" and then go to the specific disease page to find pathways and related genes under the **Pathways** and **Related Genes** tabs. This will potentially offer additional candidate genes from a different perspective that may not have been previously known to be related to statin response.

In addition to the method using only curated information available at PharmGKB, we also provide a link to an expanded "candidate gene finder tool" under the "PGx tool box" (http://www.cbs.dtu.dk/services/PGx_pipeline-1.0) on the home page. This tool was developed by using PharmGKB gene-drug relationships as well as external protein-protein interaction information and drug-drug similarity measurements to help prioritize pharmacogene candidates. By comparing an input drug (or a set of input drugs) and its (their) putative indications with a repository of known gene-drug interactions, the PGx pipeline ranks the entire genome for their likelihood of being involved in the pharmacogenetic response to the inputted drug(s) (37).

Step 2. Significant Variants Selection

Once a candidate gene list is assembled, the next step is to generate a list of functionally significant SNPs for the genes of interests. If a PharmGKB **VIP** page is available for the gene, the VIP will describe the most important variants and haplotypes for the gene of interest. Alternatively, the scientist can go to the PharmGKB gene page, under the **Variants** tab, to browse through the variant table to look for variants with "star" annotation under the **Annotated Variant** curation level column. In addition, SNPs that lead to changes in amino acid composition, activity of the protein, or expression of the gene

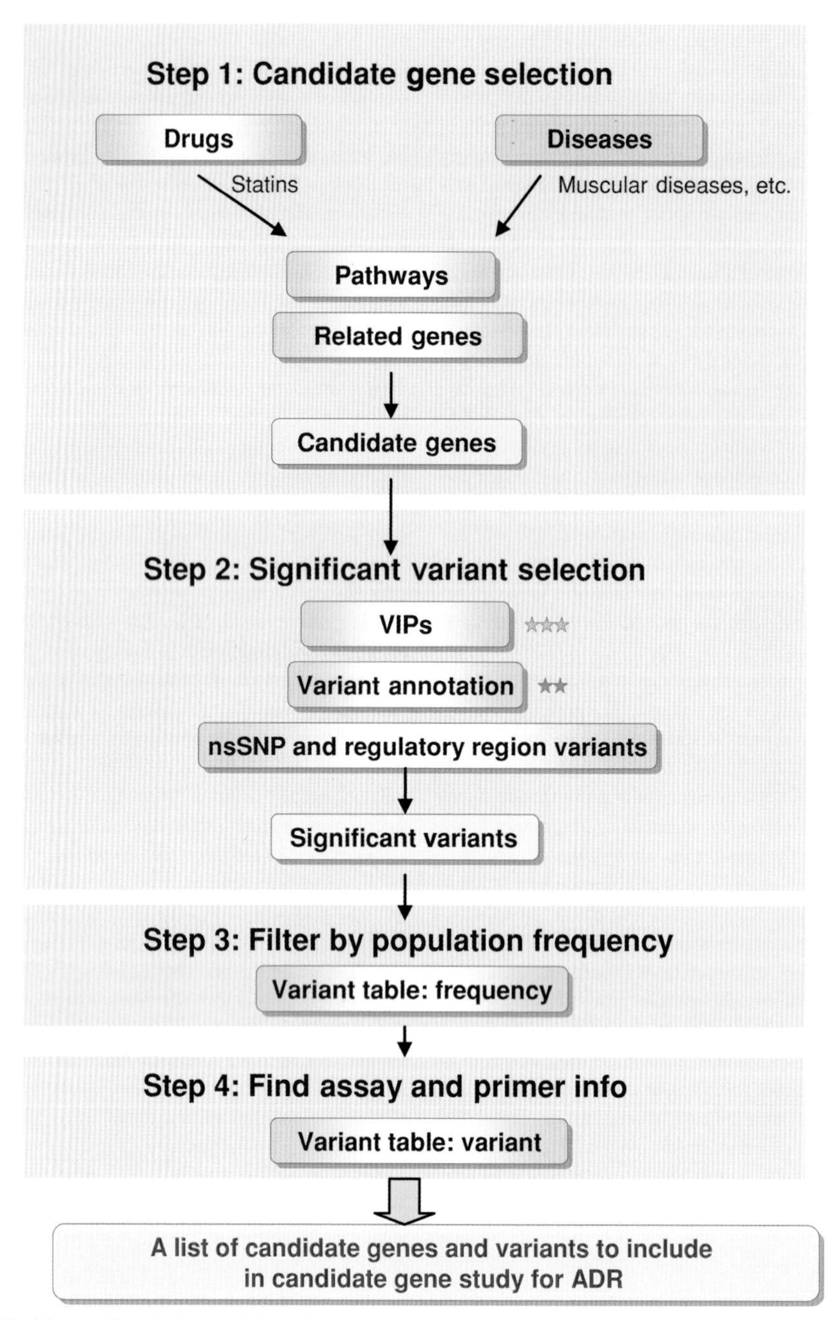

Figure 5.12. Workflow for knowledge-guided study design using PharmGKB candidate gene approach to study statin-induced ADRs (muscle-related complications).

and SNPs that reside in the regulatory regions of the gene are good candidates to be included for the study.

Step 3. Selection of Variants with Desirable Population Frequency

If the association study will only include a certain ethnic population (e.g., Asian or white), variants with extremely low frequency in the desired population can be further screened out by using frequency information on PharmGKB. The allele **frequency** information can be found under the frequency column of the variant table via the **Expanded Variants View** button. Clicking on the frequency value will display the breakdown of frequencies by racial categories. Additional frequency information can also be found in allele frequency databases such as ALFRED (http://alfred.med.yale.edu/).

Step 4. Find Assay and Primer Information for the Chosen Variants

If PharmGKB has genotype data submitted for the chosen variant, our users can find information such as **assay** methods and **primers** by clicking on the *variants* in the PharmGKB variant table. For instance, in many cases, if the TaqMan assay was used to genotype a specific drug-metabolizing enzyme variant, PharmGKB has a direct link to ordering information at Applied Biosystems to help the user locate the material required for the study.

Study design is an important aspect of any scientific investigation. By iterating through these steps, a scientist can rapidly compile a list of candidate genes and important variants to follow up in the candidate gene study to explain and predict the adverse effect profiles of the drug. This can potentially save scientists tremendous time and effort in their literature-mining process during the study design stage.

CONCLUSION

Pharmacogenomics is a rapidly evolving field with great potential to help clinicians select the most appropriate drug and doses in treatment of their patients. Advances in high-throughput genotyping and sequencing technologies and the availability of international publicly available databases are enabling new advances in pharmacogenomics and bring us closer to the long-awaited era of personalized medicine. PharmGKB is designed to provide centralized access to comprehensive knowledge and significant datasets that are central to pharmacogenomic research and to provide link outs to other relevant resources. We will continue to curate knowledge about how human genetics has an impact on drug response phenotypes. We will also create and provide access to more informatics tools that will be useful for understanding the mechanism of drug response. By aggregating, integrating, and annotating the latest findings in pharmacogenomic research, PharmGKB serves as a valuable resource, not only for basic science researchers, but also for the broader community, including clinicians, pharmacists, students, educators, and the general public, as well.

ACKNOWLEDGMENTS

This work is supported by the National Institutes of Health, National Institute of General Medical Sciences (U01GM61374).

REFERENCES

1. Marsh S & McLeod HL. Pharmacogenomics: from bedside to clinical practice. *Hum Mol Genet.* 2006;**15** (spec no 1):R89–93.
2. Wilke RA, Lin DW, Roden DM, Watkins PB, Flockhart D, Zineh I, Giacomini KM, & Krauss RM. Identifying genetic risk factors for serious adverse drug reactions: current progress and challenges. *Nat Rev Drug Discov.* 2007;**6**:904–16.
3. Cheok MH & Evans WE. Acute lymphoblastic leukaemia: a model for the pharmacogenomics of cancer therapy. *Nat Rev Cancer.* 2006;**6**:117–29.
4. Innocenti F & Ratain MJ. Pharmacogenetics of irinotecan: clinical perspectives on the utility of genotyping. *Pharmacogenomics.* 2006;**7**:1211–21.
5. Ndegwa S. Pharmacogenomics and warfarin therapy. *Issues Emerg Health Technol.* 2007; Oct:1–8.
6. Galperin MY. The Molecular Biology Database Collection: 2008 update. *Nucleic Acids Res.* 2008;**36**:D2–4.
7. Klein TE, Chang JT, Cho MK, Easton KL, Fergerson R, Hewett M, Lin Z, Liu Y, Liu S, Oliver DE, Rubin DL, Shafa F, Stuart JM, & Altman RB. Integrating genotype and phenotype information: an overview of the PharmGKB project. Pharmacogenetics Research Network and Knowledge Base. *Pharmacogenomics J.* 2001;**1**:167–70.
8. Gong L, Owen RP, Gor W, Altman RB, & Klein TE. PharmGKB: an integrated resource of pharmacogenomic data and knowledge. *Curr Protoc Bioinformatics.* 2008;Chapter 14:Unit 14.17.
9. Sangkuhl K, Berlin DS, Altman RB, & Klein TE. PharmGKB: understanding the effects of individual genetic variants. *Drug Metab Rev.* 2008;**40**:539–51.
10. McLeod HL. Drug pathways: moving beyond single gene pharmacogenetics. *Pharmacogenomics.* 2004;**5**:139–41.
11. Weinshilboum RM & Wang L. Pharmacogenetics and pharmacogenomics: development, science, and translation. *Annu Rev Genomics Hum Genet.* 2006;**7**:223–45.
12. Altman RB. PharmGKB: a logical home for knowledge relating genotype to drug response phenotype. *Nat Genet.* 2007;**39**:426.
13. Joshi-Tope G, Gillespie M, Vastrik I, D'Eustachio P, Schmidt E, de Bono B, Jassal B, Gopinath GR, Wu GR, Matthews L, Lewis S, Birney E, & Stein L. Reactome: a knowledge base of biological pathways. *Nucleic Acids Res.* 2005;**33**:D428–32.
14. Kanehisa M, Goto S, Kawashima S, Okuno Y, & Hattori M. The KEGG resource for deciphering the genome. *Nucleic Acids Res.* 2004;**32**:D277–80.
15. Salomonis N, Hanspers K, Zambon AC, Vranizan K, Lawlor SC, Dahlquist KD, Doniger SW, Stuart J, Conklin BR, & Pico AR. GenMAPP 2: new features and resources for pathway analysis. *BMC Bioinformatics.* 2007;**8**:217.
16. Eichelbaum M, Altman RB, Ratain M, & Klein TE. New feature: pathways and important genes from PharmGKB. *Pharmacogenet Genomics.* 2009;**19**:403.
17. Yang J, Bogni A, Schuetz EG, Ratain M, Dolan ME, McLeod H, Gong L, Thorn C, Relling MV, Klein TE, & Altman RB. Etoposide pathway. *Pharmacogenet Genomics.* 2009;**19**:552–3.

18. Poulter GL, Rubin DL, Altman RB, & Seoighe C. MScanner: a classifier for retrieving Medline citations. *BMC Bioinformatics.* 2008;**9**:108.
19. Garten Y & Altman RB. Pharmspresso: a text mining tool for extraction of pharmacogenomic concepts and relationships from full text. *BMC Bioinformatics.* 2009;**10**(suppl 2):S6.
20. Wishart DS, Knox C, Guo AC, Shrivastava S, Hassanali M, Stothard P, Chang Z, & Woolsey J. DrugBank: a comprehensive resource for in silico drug discovery and exploration. *Nucleic Acids Res.* 2006;**34**:D668–72.
21. Owen RP, Altman RB, & Klein TE. PharmGKB and the International Warfarin Pharmacogenetics Consortium: the changing role for pharmacogenomic databases and single-drug pharmacogenetics. *Hum Mutat.* 2008;**29**:456–60.
22. The International Warfarin Pharmacogenetics Consortium, Klein TE, Altman RB, Eriksson N, Gage BF, Kimmel SE, Lee MT, Limdi NA, Page D, Roden DM, Wagner MJ, Caldwell MD, & Johnson JA. Estimation of the warfarin dose with clinical and pharmacogenetic data. *N Engl J Med.* 2009;**360**:753–64.
23. Cheok MH, Lugthart S, & Evans WE. Pharmacogenomics of acute leukemia. *Annu Rev Pharmacol Toxicol.* 2006;**46**:317–53.
24. Marsh S. Cancer pharmacogenetics. *Methods Mol Biol.* 2008;**448**:437–46.
25. Scripture CD, Figg WD, & Sparreboom A. The role of drug-metabolising enzymes in clinical responses to chemotherapy. *Expert Opin Drug Metab Toxicol.* 2006;**2**:17–25.
26. den Boer ML & Pieters R. Microarray-based identification of new targets for specific therapies in pediatric leukemia. *Curr Drug Targets.* 2007;**8**:761–4.
27. Stegmaier K. Genomic approaches in acute leukemia. *Best Pract Res Clin Haematol.* 2006;**19**:263–8.
28. Pui CH, Robison LL, & Look AT. Acute lymphoblastic leukaemia. *Lancet.* 2008;**371**:1030–43.
29. Gut J & Bagatto D. Theragenomic knowledge management for individualised safety of drugs, chemicals, pollutants and dietary ingredients. *Expert Opin Drug Metab Toxicol.* 2005;**1**:537–54.
30. Need AC, Motulsky AG, & Goldstein DB. Priorities and standards in pharmacogenetic research. *Nat Genet.* 2005;**37**:671–81.
31. Sherry ST, Ward MH, Kholodov M, Baker J, Phan L, Smigielski EM, & Sirotkin K. dbSNP: the NCBI database of genetic variation. *Nucleic Acids Res.* 2001;**29**:308–11.
32. Wheeler DL, Barrett T, Benson DA, Bryant SH, Canese K, Chetvernin V, Church DM, Dicuccio M, Edgar R, Federhen S, Feolo M, Geer LY, Helmberg W, Kapustin Y, Khovayko O, Landsman D, Lipman DJ, Madden TL, Maglott DR, Miller V, Ostell J, Pruitt KD, Schuler GD, Shumway M, Sequeira E, Sherry ST, Sirotkin K, Souvorov A, Starchenko G, Tatusov RL, Tatusova TA, Wagner L, & Yaschenko E. Database resources of the National Center for Biotechnology Information. *Nucleic Acids Res.* 2008;**36**:D13–21.
33. Link E, Parish S, Armitage J, Bowman L, Heath S, Matsuda F, Gut I, Lathrop M, & Collins R. SLCO1B1 variants and statin-induced myopathy – a genomewide study. *N Engl J Med.* 2008;**359**:789–99.
34. Mangravite LM & Krauss RM. Pharmacogenomics of statin response. *Curr Opin Lipidol.* 2007;**18**:409–14.
35. Pasanen MK, Neuvonen M, Neuvonen PJ, & Niemi M. SLCO1B1 polymorphism markedly affects the pharmacokinetics of simvastatin acid. *Pharmacogenet Genomics.* 2006;**16**:873–9.
36. Tiwari A, Bansal V, Chugh A, & Mookhtiar K. Statins and myotoxicity: a therapeutic limitation. *Expert Opin Drug Saf.* 2006;**5**:651–66.
37. Hansen N, Brunak S, & Altman R. Generating genome-scale candidate gene lists for pharmacogenomics. *Clin Pharmacol Ther.* 2009;**86**:183–9.

DrugBank

6

David S. Wishart

ELECTRONIC DATABASES IN THE LIFE SCIENCES

During the past ten years, the life sciences (i.e., biology, medicine, and pharmaceutical research) have evolved from being largely low-throughput, observational disciplines to primarily high-throughput, data-driven disciplines. In other words, life sciences are becoming a "data science." Thanks to advances in DNA sequencing, medical imaging, robotic sample handling, and high-throughput screening, it is possible to generate as much data in a day-long experiment as it might have taken for an entire scientific career. For instance, a single eight-hour sequencing run on a DNA pyrosequencer can generate enough sequence data to fill a 1,000-page book (1, 2). The resulting genome sequence could be automatically annotated in a few hours yielding an enormous volume of information that could easily occupy ten large telephone books (3, 4).

Our capacity to generate gigabytes of information on a daily basis is having a profound impact on the way that scientific information is being disseminated or delivered. Although most scientific data are still presented in scientific journals and most high-level scientific knowledge is still published in textbooks, it is becoming increasingly obvious that the paper-publishing industry cannot keep up with the pace of scientific advancement and the quantity of data that the scientific community would like to publish. Fortunately the World Wide Web (i.e., the Web) has come to the rescue. The Web makes it possible to publish and disseminate huge quantities of information quickly and inexpensively. Not only has the Web helped to save scientific publishing, it has also led to the development of a new and very important kind of scientific archive: the electronic database. Electronic databases are Web-accessible archives that contain scientifically important data that are either too voluminous to publish in a book or journal or in a format that is incompatible with paper publication. Electronic databases such as GenBank (5), the Protein Data Bank (6), or PubMed allow information to be continuously updated through the contributions of thousands of scientists or dozens of curators who continuously upload and deposit data into these resources. Electronic databases also allow their data to be searched, accessed, or displayed in ways that were simply not possible through a paper journal or a leather-bound book. Indeed, the emergence of electronic, Web-accessible databases has to be considered one of the more significant developments in the field of life sciences.

Online or electronic databases not only facilitate rapid data retrieval and exchange, they also encourage knowledge exchange. Knowledge is distinguished from data in that knowledge is the information or wisdom gained from a critical assessment of raw data. In fact, there is a growing trend in biology and medicine to enrich the textual or numerical content of many "first-generation" databases (such as GenBank or the Protein Data Bank) with detailed annotations and expert commentary to create "second-generation" knowledge bases or encyclopedias. These resources attempt to consolidate multiple data sources into a single repository, while at the same time providing current, informative, and authoritative summaries on large numbers of related topics. Two examples of knowledge bases that are particularly relevant to the fields of pharmacogenetics and pharmacogenomics are PharmGKB (7) and DrugBank (8). A detailed description of PharmGKB has already been provided in Chapter 5. The present chapter focuses on DrugBank and describes how it can facilitate pharmacogenomic and pharmaceutical research.

DrugBank SYNOPSIS

First released in 2006 (8) and then updated again in 2008 (9), DrugBank (www.drugbank.ca) is a comprehensive, fully searchable electronic database that links sequence, structure, and mechanistic data about drug molecules with sequence, structure, and mechanistic data about their drug targets. DrugBank was one of the first electronic databases to combine cheminformatic (i.e., chemical) tools and data with bioinformatic (i.e., molecular biological) tools and data. Indeed, the explicit linkage between drugs and drug targets is one of the main features that has made DrugBank particularly popular. Likewise, the presentation of drug and drug target data in synoptic DrugCards (in analogy to library cards or study flash cards) has helped make the database particularly easy to view and navigate.

To facilitate higher-level navigation and querying, the DrugBank Web site contains many built-in tools and a variety of customized features for viewing, sorting, querying, and extracting drug or drug target data. These include a number of higher-level database-searching functions such as a local BLAST (10) sequence search (SeqSearch) that supports both single and multiple protein sequence queries (for drug target searching), a Boolean text search (TextSearch) for sophisticated text searching and querying, a chemical structure search utility (ChemQuery) for structure matching and structure-based querying, and a relational data extraction tool (Data Extractor) for performing complex queries.

The BLAST search (SeqSearch) is particularly useful for drug discovery applications because it can potentially allow users to quickly and simply identify drug leads from newly sequenced pathogens. Specifically, a new sequence, a group of sequences, or even an entire proteome can be searched against DrugBank's database of known drug target sequences by pasting the FASTA-formatted sequence (or sequences) into the SeqSearch query box and pressing the "submit" button. A significant hit can reveal the name(s) or chemical structure(s) of potential drug leads that may act on that query protein (or proteome). The structure similarity search tool (ChemQuery) can be used in a similar manner to SeqSearch. For instance, users may sketch a chemical structure or paste a SMILES string (SMILES strings are simple text-string representations or "shorthand" for describing chemical structures [11]) of a possible drug lead or a drug that appears to be causing an adverse reaction into the ChemQuery window. After submitting the query, the database launches a structure similarity search that looks for common substructures from the query compound that match DrugBank's database of known drug or druglike compounds. High-scoring hits are presented in a tabular format with hyperlinks to the corresponding DrugCards. The ChemQuery tool allows users to quickly determine whether their compound of interest acts on the desired protein target or whether the compound of interest may unexpectedly interact with unintended protein targets.

In addition to these search features, DrugBank also provides a number of general browsing tools for exploring the database and several specialized browsing tools such as Pharma Browse and Geno Browse for more specific tasks. For instance, Pharma Browse is designed to address the needs of pharmacists, physicians, and medicinal chemists who tend to think of drugs in clusters of indications or drug classes. This particular browsing tool provides navigation hyperlinks to more than seventy drug classes, which in turn list the U.S. Food and Drug Administration (FDA)-approved drugs associated with the drugs. Each drug name is then linked to its respective DrugCard. Geno Browse, in contrast, is specifically designed to address the needs of geneticists or those specialists interested in specific drug-single-nucleotide polymorphism (SNP) relationships. This browsing tool provides navigation hyperlinks to more than sixty different drugs, which in turn list the target genes, SNPs, and the physiological effects associated with these drugs.

As of July 2008, DrugBank contained detailed information on 1,480 FDA-approved drugs corresponding to 28,447 brand names and synonyms. This collection includes 1,281 synthetic small-molecule drugs, 128 biotech drugs, and 71 nutraceutical drugs or supplements. DrugBank also contains information on the 1,669 different targets (protein, lipid, or DNA molecules) and metabolizing enzymes with which these drugs interact. In addition, the database maintains data on 187 illicit drugs (i.e., those legally banned or selectively banned in most developed nations) and 64 withdrawn drugs (those removed from the market because of safety concerns). Chemical, pharmaceutical, and biological information about these classes of drugs is extremely important, not only in understanding their adverse reactions, but also in being able to predict whether a new drug entity may have unexpected chemical or functional similarities to a dangerous or highly addictive drug.

DrugBank updates are typically released every six months. The updates include information on newly approved drugs, corrections or updates on old drugs/drug targets, the addition of new data fields, and enhancements or improvements to DrugBank's interface or search utilities. These updates, as well as nearly all of the database content, are freely available in downloadable flat files. This accessibility has greatly facilitated a number of large-scale in silico studies on the prediction of drug targets (12), on the molecular characterization of pathological pathways (13), and on global characterization of gene expression and drug response data (14).

AN ILLUSTRATED TOUR OF DrugBank

DrugBank has many viewing features and software tools that can be used in a wide variety of ways. Although it is impossible to review all of these in this chapter, a short illustrated tour of the database may help the reader understand a little bit more about the database and its

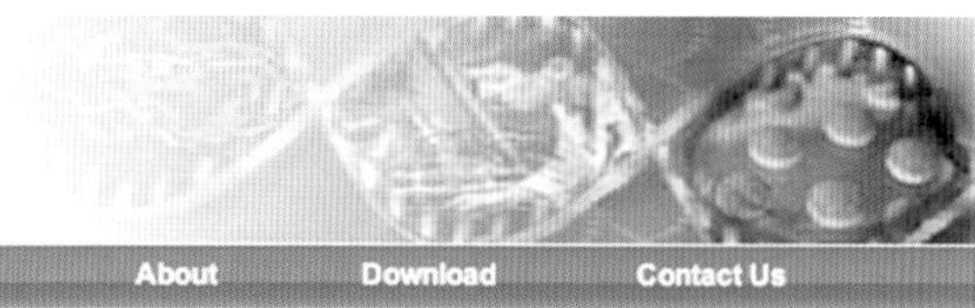

Figure 6.1. A screenshot showing the DrugBank home page with the main pull-down menus. Menu tabs with additional submenus are viewable or selectable by pointing the mouse over the down arrow to the right of the menu name.

capabilities. As with any online tool, DrugBank has a home page with a hyperlinked blue menu bar located near the top of the page with six clickable titles: Home, Browse, Search, About, Download, and Contact Us (Figure 6.1). The Browse, Search, and About menu tabs contain additional submenus that are accessible/viewable by dragging the mouse pointer over the main menu name. Under Browse, there are submenus for DrugBank Browse, Pharma Browse, and Geno Browse. Under Search, there are submenus for ChemQuery, TextQuery, SeqSearch, and Data Extractor. Under About, submenus for Details, Citing DrugBank, Release Notes, What's New, Statistics, Source Data, Data Field Explanations, and Other Databases are provided. Clicking any of the menu or submenu names will launch a new window with the appropriate function or selected view. DrugBank's blue menu bar, which appears near the top of every DrugBank Web page, allows users to easily navigate to the different browsing and search utilities in the database. Below this menu bar is a text box with the phrase "Search DrugBank for." This text search utility, which is the most commonly used search feature in DrugBank, is also displayed near the top of nearly every DrugBank Web page. Essentially all text (or sequence) queries in DrugBank are typically done by typing or pasting text into the text boxes and activated by pressing a "Search" or "Submit" button. For instance, if one typed "tricyclic" into the Search DrugBank for text box, then clicked on the All Text Fields check box, and then pressed the Search button, a four-column table should appear within a few seconds consisting of a list of almost all known tricyclic antidepressant drugs as well as other tricyclic molecules (Figure 6.2). The first column displays the DrugBank accession number (which is hyperlinked), the second column displays the drug's common name, the third column displays the chemical formula, and the fourth column displays the molecular weight. Directly below these data, the query word is also highlighted in the selected DrugCard field(s) from which it was retrieved.

Because the spelling of many drug names, chemical compound names, and protein names is often difficult or non-intuitive, DrugBank also supports an "intelligent" text search where alternative spellings to misspelled or incompletely entered names are automatically provided. Figure 6.3 shows an example of a user entering the word "Cinemet" (a common misspelling the drug "Sinemet") and DrugBank suggesting the correct spelling for this anti-Parkinson drug. Note that for DrugBank's spelling

Search DrugBank for: tricyclic (Search) (Reset)
☑ Common Name ☑ Synonym and Brand Name ☐ All Text Fields

DrugBank Search Results

DrugBank is searching for "*tricyclic* " ...

Summary for query "*tricyclic*":
Text Search found 56 matches.
(Some matches may be to HTML tags which may not be shown.)

Accession No	Common Name	Chemical Formula	Molecular Weight
DB00176	Fluvoxamine	C15H21F3N2O2	318.3347
	...norepinephrine receptors than tricyclic antidepressant...		
DB00196	Fluconazole	C13H12F2N6O	306.2708
	Amitriptyline The imidazole increases the effect and toxicity of the tricyclic		
DB00215	Citalopram	C20H21FN2O	324.3919
	...in preference to tricyclic antidepressants, which aggravate...		
DB00234	Reboxetine	C19H23NO3	313.3908
	...extent to the tricyclic...		
DB00247	Methysergide	C21H27N3O2	353.4580
	...Yamada J: The tricyclic antidepressant clomipramine increases...		
DB00285	Venlafaxine	C17H27NO2	277.4018
	...but, unlike the tricyclics and similar to...		
DB00321	Amitriptyline	C20H23N	277.4033
	tricyclic antidepressant with anticholinergic and sedative properties. It appears to prevent the re-uptake of norepinephrine and		

Figure 6.2. A screenshot showing the results of querying the DrugBank database using the "Search DrugBank for" text query tool. The input query was "tricyclic" with the All Text Fields check box being checked off.

Search DrugBank for: (Search) (Reset)
☑ Common Name ☑ Synonym and Brand Name ☐ All Text Fields

DrugBank Search Results

DrugBank is searching for "*Cinemet* " ...

Did you mean Sinemet?

Figure 6.3. A screenshot showing how DrugBank will suggest alternate spellings of drug names if an incorrect spelling is used. In this case, a user typed "Cinemet," and DrugBank suggested "Sinemet" (an anti-Parkinson drug). Note that for DrugBank's spelling correction utility to work, the "All Text Fields" check box must remain unchecked.

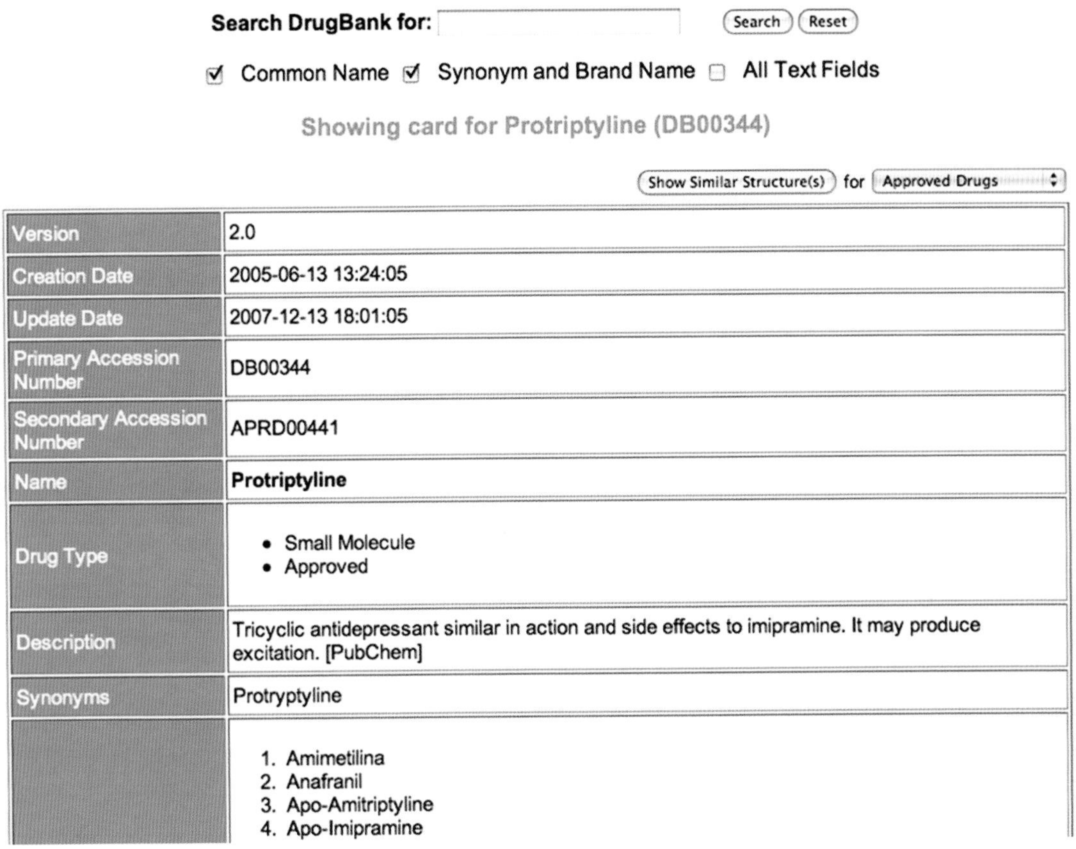

Search DrugBank for: [ ] (Search) (Reset)

☑ Common Name ☑ Synonym and Brand Name ☐ All Text Fields

Showing card for Protriptyline (DB00344)

(Show Similar Structure(s)) for [Approved Drugs]

Version	2.0
Creation Date	2005-06-13 13:24:05
Update Date	2007-12-13 18:01:05
Primary Accession Number	DB00344
Secondary Accession Number	APRD00441
Name	**Protriptyline**
Drug Type	• Small Molecule • Approved
Description	Tricyclic antidepressant similar in action and side effects to imipramine. It may produce excitation. [PubChem]
Synonyms	Protryptyline
	1. Amimetilina 2. Anafranil 3. Apo-Amitriptyline 4. Apo-Imipramine

Figure 6.4. A screenshot of a DrugBank DrugCard for the drug Protriptyline.

correction utility to work, the All Text Fields check box must remain unchecked.

Clicking on the hyperlinked drug name or on a DrugCard button will launch a new table containing a more detailed description of the drug of interest (Figure 6.4). Each DrugCard contains two columns, with one column corresponding to data field names or titles (given on the left side with a blue background) and the other corresponding to drug-specific descriptors (given on the right, with a white background). Each DrugCard entry contains more than 100 data fields with half of the information devoted to drug/chemical data and the other half devoted to drug target or protein data (Table 6.1). The data fields follow a very specific order, with drug nomenclature being given first, followed by physical property data, then chemical structural data, pharmacological data, and finally drug target or target protein data. If a drug has more than one biomolecular target, the protein and genetic data fields are repeated for each protein target. In addition to providing comprehensive numeric, sequence, and textual data, each DrugCard also contains hyperlinks to other databases, abstracts, digital images, and interactive applets for viewing molecular structures.

These hyperlinks can be activated by simply clicking the appropriate hyperlinked word, image, or button. For instance, clicking on the KEGG Compound ID hyperlink will launch a new window to the KEGG Web page for the drug of interest (Figure 6.5). Clicking on the PharmGKB hyperlink will launch the PharmGKB page for the drug of interest (Figure 6.6). To view a two-dimensional image of a drug of interest, one simply scrolls down to the data field called MOL File Image and clicks on the hyperlinked button called View 2D Structure. This launches ChemAxon's MarvinView Structure Viewing Java applet. After a few seconds an image of the drug should appear in the applet window (Figure 6.7). This structure display applet allows the user to interactively view, rotate, edit, export, or zoom in on the structure. The editing/exporting functions can be activated by double clicking on the structure image itself (this launches a separate, editable view) or by clicking on the right mouse button (which activates control menus superimposed over the image). A three-dimensional structure of a drug of interest can also be viewed by clicking on the hyperlinked button View 3D Structure contained in the PDB File Calculated Image field. This launches the ChemAxon MarvinView interactive three-dimensional

Table 6.1. A Summary of the Main DrugBank Data Fields.

Database Version	*Water Solubility (Pred)*	*Organisms Affected*
DrugCard Creation Date	State	Phase 1 Metabolizing Enz.
DrugCard Update Date	LogP (Exp)	Phase 1 Enzyme Sequences
DrugBank Accession #	LogP (Pred)	Phase 1 Enzyme SProt IDs
Old DrugBank Accession #	LogS (Exp)	Phase 1 Enzyme GeneName
Common or Generic Name	Caco2 Permeability (Exp)	Drug Targets (Names)
Drug Type	pKa/Isoelectric Point	Drug Target (1) Name
Drug Description	Mass Spectrum	Drug Target Synonyms
Drug Synonyms	MOL File (Image)	Drug Target Gene Name
Drug Brand Names	MOL File (Text)	Drug Target Prot. Seq.
Drug Brand Mixtures	SDF File (Text)	Drug Target # Residues
Chemical or IUPAC Name	PDB File (Text)	Drug Target MW
Chemical Structure	PDB File (Image)	Drug Target pl
CAS Registry #	PDB File (Exp)	Drug Target GO Class
InChi Identifier	PDB File (Exp-Image)	Drug Target Functions
KEGG Drug ID	Isomeric SMILES	Drug Target Reaction
KEGG Compound ID	Canonical SMILES	Drug Target Pfam Function
PubChem Compound ID	Drug Category	Drug Target Signal Seqs
PubChem Substance ID	ATC Codes	Drug Target Transmemb.
ChEBI ID	AHFS Codes	Drug Target Essentiality
PharmGKB ID	Indication	Drug Target GenBank ID
HET (PDB) ID	Pharmacology	Drug Target SwissProt ID
GenBank ID (if Biotech)	Mechanims of Action	Drug Target SProt Name
Drug ID Number (DIN)	Absoprtion	Drug Target PDB ID
RxList Link	Toxicity	Drug Target PDB (Text)
PDRhealth Link	Biotransformation	Drug Target PDB (Image)
Wikipedia Link	Half Life	Drug Target Cell Location
FDA Label	Dosage Forms	Drug Target Gene Sequence
MSDS Link	Patient Information	Drug Target GenBank ID
Synthesis Reference	Contraindications	Drug Target GeneCard ID
Average Molecular Weight	Interactions	Drug Target HGNC ID
Monoisotopic Weight	Drug-Drug Interactions	Drug Target Locus
Melting Point	Food-Drug Interactions	Drug Target SNPs
Water Solubility (Exp)	Drug References	Drug Target References

structure viewing applet (Figure 6.8). Right-clicking the mouse or double clicking the image will create a similar list of editing and viewing functions as seen for the View 2D Structure. Scrolling further down any given DrugCard will reveal many other data fields containing detailed pharmaceutical, pharmacological, and clinical data (Drug Category, Indication, Pharmacology, Mechanism of Action, Contraindications, Absorption, Toxicity, Half Life, Interactions, etc.) as well as hyperlinks to different online Drug References (RxList, PDRhealth, Wikipedia). Scrolling even further down, one will eventually see a field labeled Phase 1 Metabolizing Enzyme followed shortly thereafter by a field separator titled Drug Target 1. These enzymatic and drug target data mark the beginning of the molecular biological data for this drug.

As can be seen from Table 6.1, the biological data about most drug targets and drug-metabolizing enzymes are quite extensive. For drug targets, these data include information on the names, sequences (DNA and protein), physical properties, function(s) of the protein, and any reactions or pathways that it is known to participate in. It also includes multiple database links, chromosomal/locus information, and SNP data. For drug-metabolizing enzymes DrugBank provides information on their names, sequences, and associated SNPs. This molecular biological information is presented both in terms of protein information and genetic information. The protein-based information is particularly useful to protein chemists, structural biologists, and medicinal chemists, whereas the genetic information is most

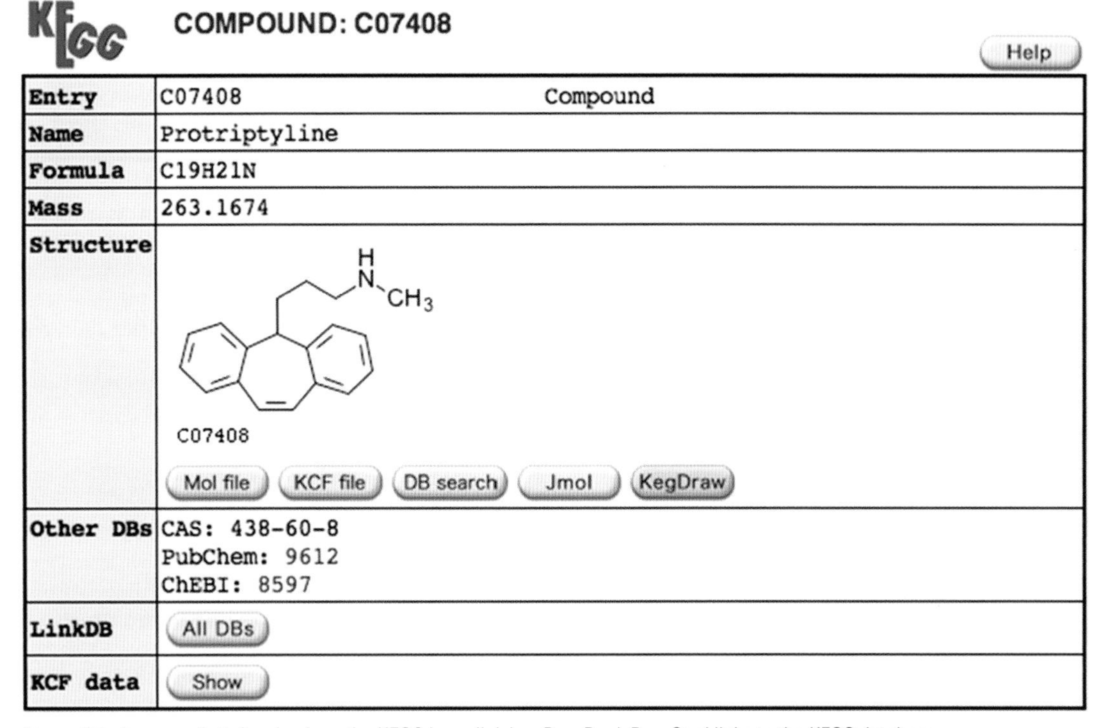

Figure 6.5. A screenshot showing how the KEGG hyperlink in a DrugBank DrugCard links to the KEGG database.

useful to pharmacogeneticists and pharmacogenomics specialists.

DrugBank AND PHARMACOGENOMICS

In addition to its general utility as a general drug encyclopedia, DrugBank also contains several tables, data fields, or data types that are particularly useful for pharmacogenomic or pharmacogenetic studies. These include synoptic descriptions of a given drug's pharmacology as well as its mechanism of action, contraindications, toxicity, phase I metabolizing enzymes (name, protein sequence, and SNPs), and associated drug targets (names, protein sequence, DNA sequence, chromosome location, locus number, and SNPs). The information contained in DrugBank's pharmacology, mechanism of action, contraindications, and toxicity fields often includes details about any known adverse reactions. This may include descriptions of known phase I or phase II enzyme interactions, alternate metabolic routes, or the existence of secondary drug targets. Secondary drug targets represent

PharmGKB
The Pharmacogenetics and Pharmacogenomics Knowledge Base
Search PharmGKB: ? Go
Home Search Submit Help PGRN Contributors Sign In Feedback

protriptyline

Overview | Properties | Non-curated Publications | Downloads/Cross-references

Overview

PharmGKB Accession Id:	PA451168
Generic Names:	None
Trade Names:	Triptil; Vivactil

feedback sitemap privacy policy usage conditions citing pharmgkb

Figure 6.6. A screenshot showing how the PharmGKB hyperlink in a DrugBank DrugCard links to the PharmGKB database.

DrugBank: Displaying MOL File (Protriptyline)

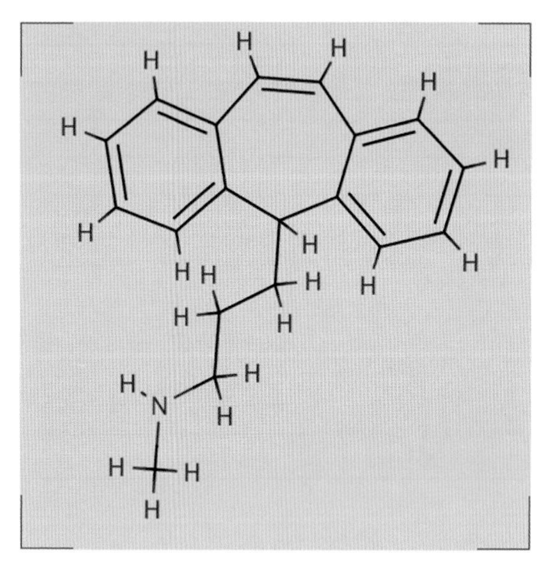

Figure 6.7. A screenshot of a drug structure (Protriptyline) as displayed by pressing the View 2D Structure button on the Protriptyline DrugCard.

proteins (or other macromolecules) that are different than the primary target for which the drug was initially designed or targeted. Some drugs may have five or more targets, of which only one might be relevant to treating the disease. DrugBank uses a relatively liberal interpretation of drug targets to help identify these secondary drug targets. That is, the database curators define a drug target as any macromolecule identified in the literature that binds, transports, or transforms a drug. The binding or transformation of a drug by a secondary drug target or an "off-target" protein is one of the most common causes for unwanted side effects or adverse drug reactions (ADRs) (15). In contrast, there are a few cases where a secondary drug target can lead to synergistic effects that may enhance the potency of a drug (16). By providing a fairly comprehensive listing of secondary drug targets (along with their SNP information and other genetic data), DrugBank is potentially able to provide additional insight into the underlying causes of a patient's response to a given drug.

DrugBank: Displaying PDB File (Protriptyline)

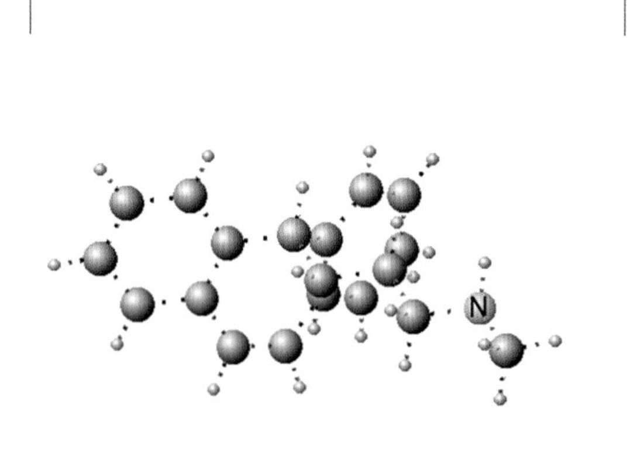

Figure 6.8. A screenshot of a three-dimensional rendering of the same drug structure as displayed by pressing the View 3D Structure button on the Protriptyline DrugCard.

Perhaps the information in DrugBank that is most directly relevant to pharmacogenomics research is the detailed sequence and SNP data that it contains on known drug-metabolizing enzymes and known drug targets. In particular, DrugBank provides detailed summary tables about each of the SNPs for each of the drug targets or drug-metabolizing enzymes that have been characterized by various SNP-typing efforts, such as the SNP Consortium (17) and HapMap (18). At present, DrugBank contains information on 26,292 coding (exon) SNPs and 73,328 noncoding (intron) SNPs derived from known drug targets. It also has data on 1,188 coding SNPs and 8,931 noncoding SNPs from known drug-metabolizing enzymes. An in-house program, called SNP-Updater, is run before each semiannual DrugBank release to assemble and format DrugBank's SNP data. SNP-Updater compiles mapping, validation, and population frequency data from dbSNP (19) and HapMap corresponding to every intron and exon for each of the 1,600+ human drug targets and metabolizing enzymes in DrugBank. The program then reformats the data into more easily viewed synoptic tables. By clicking on the "Show SNPs" hyperlink listed beside either the metabolizing enzymes or the drug target SNP field, the SNP summary table can be viewed (Figure 6.9). These tables include (1) the reference SNP ID (with a hyperlink to dbSNP); (2) the allele variants; (3) the validation status; (4) the chromosome location and reference base position; (5) the functional class (synonymous, nonsynonymous, untranslated, intron, and exon); (6) mRNA and protein accession links (if applicable); (7) the reading frame (if applicable); (8) the amino acid change (if existent); (9) the allele frequency as measured in African, European, and Asian populations (if available); and (10) the sequence of the gene fragment with the SNP highlighted in a red box.

The purpose of these SNP tables is to allow one to go directly from a drug of interest to a list of potential SNPs that may contribute to the reaction or response seen in a given patient or in a given population. In particular, these SNP lists may serve as hypothesis generators that allow SNP or gene characterization studies to be somewhat more focused or targeted. At the same time, DrugBank's

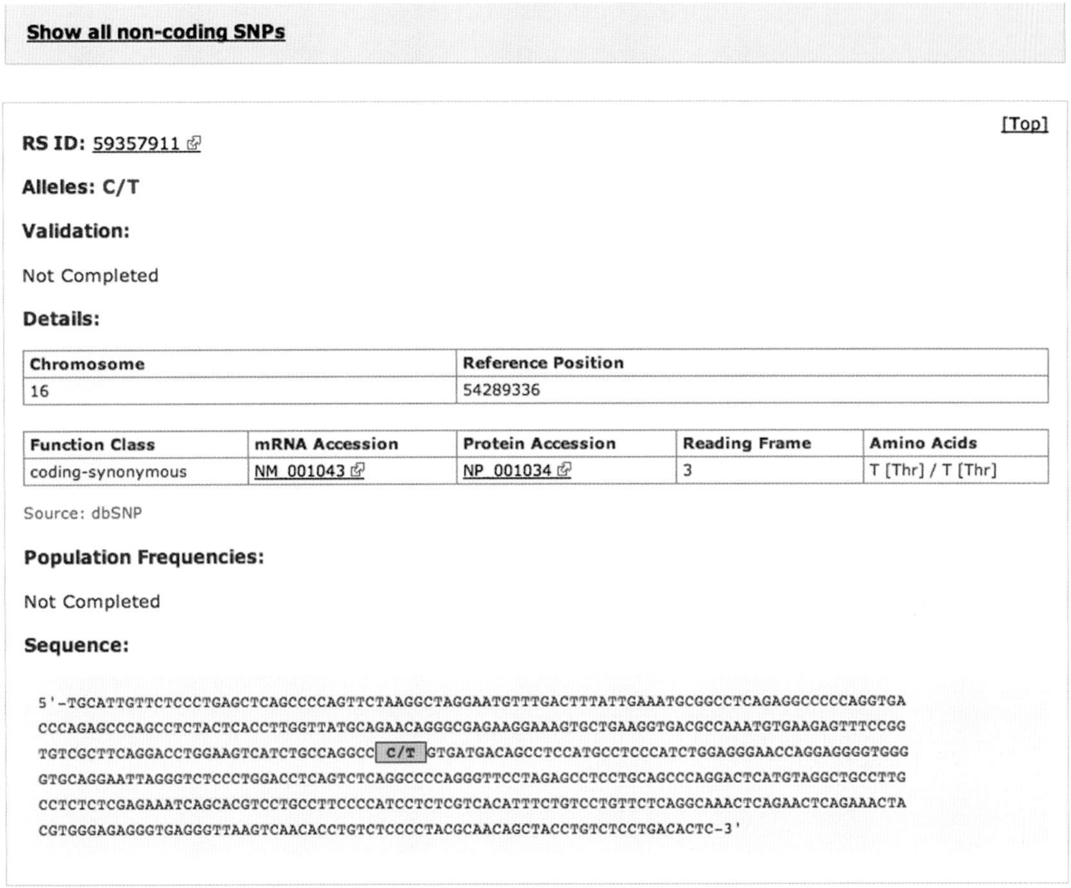

Show all non-coding SNPs

[Top]

RS ID: 59357911

Alleles: C/T

Validation:

Not Completed

Details:

Chromosome	Reference Position
16	54289336

Function Class	mRNA Accession	Protein Accession	Reading Frame	Amino Acids
coding-synonymous	NM_001043	NP_001034	3	T [Thr] / T [Thr]

Source: dbSNP

Population Frequencies:

Not Completed

Sequence:

```
5'-TGCATTGTTCTCCCTGAGCTCAGCCCCAGTTCTAAGGCTAGGAATGTTTGACTTTATTGAAATGCGGCCTCAGAGGCCCCCAGGTGA
CCCAGAGCCCAGCCTCTACTCACCTTGGTTATGCAGAACAGGGCGAGAAGGAAAGTGCTGAAGGTGACGCCAAATGTGAAGAGTTTCCGG
TGTCGCTTCAGGACCTGGAAGTCATCTGCCAGGCC C/T GTGATGACAGCCTCCATGCCTCCCATCTGGAGGGAACCAGGAGGGGTGGG
GTGCAGGAATTAGGGTCTCCCTGGACCTCAGTCTCAGGCCCCAGGGTTCCTAGAGCCTCCTGCAGCCCAGGACTCATGTAGGCTGCCTTG
CCTCTCTCGAGAAATCAGCACGTCCTGCCTTCCCCATCCTCTCGTCACATTTCTGTCCTGTTCTCAGGCAAACTCAGAACTCAGAAACTA
CGTGGGAGAGGGTGAGGGTTAAGTCAACACCTGTCTCCCCTACGCAACAGCTACCTGTCTCCTGACACTC-3'
```

Figure 6.9. A screenshot of DrugBank's SNP summary table.

SNP tables may also be used to interpret the results of SNPChip or AmpliChip CYP450 microarray tests (20) for those patients showing an unusual response to a given drug. By comparing the experimentally obtained SNP results with those listed in DrugBank for that drug (and its drug targets), it may be possible to ascertain which polymorphism for which drug target or drug-metabolizing enzyme may be contributing to an unusual drug response. These database-derived SNP suggestions obviously may require additional experimental validation to prove their causal association.

As part of its July 1, 2008 release, DrugBank now includes two tables that provide much more explicit information on the relationship between drug responses/reactions and gene variant or SNP data. The two tables, which are accessible from the Geno Browse submenu located on DrugBank's Browse menu bar, are called SNP-FX (short for SNP-associated effects) and SNP-ADR (short for SNP-associated adverse drug reactions). SNP-FX contains data on the drug, the interacting protein(s), the "causal" SNPs or genetic variants for that gene/protein, the therapeutic response or effects caused by the SNP-drug interaction (improved or diminished response, changed dosing requirements, etc.), and the associated references describing these effects in more detail. SNP-ADR follows a format similar to SNP-FX, but the clinical responses are restricted only to ADRs. SNP-FX contains literature-derived data on the therapeutic effects or therapeutic responses for more than seventy drug-polymorphism combinations, whereas SNP-ADR contains data on adverse reactions compiled from more than fifty drug-polymorphism pairings. All of the data in these tables is hyperlinked to drug entries from DrugBank, protein data from SwissProt, SNP data from dbSNP, and bibliographic data from PubMed. A screen shot of the SNP-ADR table is shown in Figure 6.10.

As seen in Figure 6.10, these tables provide consolidated, detailed, and easily accessed information that clearly identifies those SNPs that are known to affect a given drug's efficacy, toxicity, or metabolism. For instance, in the case of 5-fluorouracil (a common anticancer drug), it can be seen that polymorphisms in no fewer than five different proteins are responsible for a

***DrugBank* Geno Browse**

-SNP-ADR- SNP-FX

Drug Name	Interacting Gene/Enzyme	SNP(s)	Adverse Reaction	Reference(s)
5-Fluororuacil (DB00544)	Orotate phophoribosyltransferase, Uridine 5'-monophosphate synthase (Gene symbol = OPRT or UMPS) Swissprot P11172	rs1801019 (C Allele)	Neutropenia, diarrhea	Ichikawa W, Takahashi T, Suto K, Sasaki Y, Hirayama R: Orotate phosphoribosyltransferase gene polymorphism predicts toxicity in patients treated with bolus 5-fluorouracil regimen. Clin Cancer Res. 2006 Jul 1;12(13):3928-34. [PubMed]
5-Fluororuacil (DB00544)	Thymidylate synthase (Gene symbol = TYMS) Swissprot P04818	TSER*2 rs34743033	Neutropenia, diarrhea	Ichikawa W, Takahashi T, Suto K, Sasaki Y, Hirayama R: Orotate phosphoribosyltransferase gene polymorphism predicts toxicity in patients treated with bolus 5-fluorouracil regimen. Clin Cancer Res. 2006 Jul 1;12(13):3928-34. [PubMed]
5-Fluoruracil (DB00544)	Dihydropyrimidine dehydrogenase (Gene symbol = DPYD) Swissprot Q12882	rs1801265 (C allele) rs1801159 (G allele)	Nausea, vomiting, reduced white cell count	Zhang H, Li YM, Zhang H, Jin X: DPYD*5 gene mutation contributes to the reduced DPYD enzyme activity and chemotherapeutic toxicity of 5-FU: results from genotyping study on 75 gastric carcinoma and colon carcinoma patients. Med Oncol.

Figure 6.10. A screenshot of DrugBank's SNP-ADR table showing the information contained on drugs, genes, SNPs, and ADRs.

range of hematological adverse reactions. Because of the relatively small number of SNP-drug associations that have been compiled so far, these tables do not have the extensive searching and sorting tools found in many of DrugBank's other tables or views. However, with the number of reported SNP-drug interactions rapidly growing, and with the interest of pharmacogenomics and SNP-typing increasing, it can be expected that upcoming releases of DrugBank will contain additional database linkages, improved search tools, and much more data on those SNPs, copy number variants, and mutations that have been convincingly proven to affect a given drug's efficacy, toxicity, or metabolism.

CONCLUSIONS

DrugBank is a dual-purpose, bioinformatics-cheminformatics knowledge base with a strong focus on quantitative, analytic, or molecular-scale information about both drugs and drug targets. In particular, DrugBank combines the data-rich molecular biology content normally found in manually curated sequence databases such as SwissProt and UniProtKB (21) with the equally rich data found in medicinal chemistry textbooks and chemical reference handbooks. DrugBank's comprehensive collection of pharmacological, metabolic, and genetic variation data allows it to serve as a "one-stop shop" for a wide range of pharmacogenomic queries. This chapter is intended to show readers how DrugBank can be used as an enabling tool to identify and explain ADRs, to identify secondary drug targets (for understanding off-target effects), to identify SNPs for drug-metabolizing enzymes, to assess ethnic variations in certain pharmacologically important SNPs, to compare or track genetic variations in different drug targets, to generate testable hypotheses for SNP or mutation hunting, and to interpret SNPChip or AmpliChip array data.

Although these applications certainly are useful for pharmacological and pharmacogenomic research, it is also clear that DrugBank still could be significantly improved with regard to the information that it provides. In particular, the inclusion of more drug metabolism information and drug metabolic pathways would certainly improve the linkage between SNP data and ADRs. Likewise, closer integration of DrugBank with SNPChip and AmpliChip array data (including support for the direct reading and processing of these array data) would certainly help with the interpretation and processing of these data. DrugBank currently does not include data on copy number variants and splice variants of drug-metabolizing enzymes or drug targets that may also contribute to genetically determined drug responses. These and other changes or additions will appear with upcoming releases of DrugBank.

PROBLEMS

1. An individual has recently been identified that has a nonfunctional form of the CYP2D6 enzyme. Use DrugBank to find out which drugs he or she will likely have trouble metabolizing.
2. Some ADRs arise from food-drug interactions. Some foods contain compounds that inhibit the action of cytochrome P450 enzymes. For instance, grapefruit contains naringin, bergamottin, and dihydroxybergamottin that are known CYP450 inhibitors. Use DrugBank to find out which drugs are affected by the consumption of grapefruit juice.
3. Using the SNP-ADR and SNP-FX Tables in DrugBank, try to identify which class (or classes) of drugs produce the most significant ADRs. Why do you think this class of drugs leads to so many ADRs?
4. Certain SNPs appear with different frequencies in different populations or ethnic groups. Using SNP-ADR and SNP-FX, identify at least three drugs that would likely have ethnicity-dependent dosing requirements.
5. A polymorphism has been identified in the multidrug resistance protein 1 that affects morphine pain relief (see the entry for morphine in SNP-FX). What other pain-relief drugs might this SNP affect? Try to use DrugBank's structure search utilities to get the answer.

REFERENCES

1. Moore MJ, Dhingra A, Soltis PS, Shaw R, Farmerie WG, Folta KM, & Soltis DE. Rapid and accurate pyrosequencing of angiosperm plastid genomes. *BMC Plant Biol.* 2006;**6**:17.
2. Mao C, Evans C, Jensen RV, & Sobral BW. Identification of new genes in Sinorhizobium meliloti using the Genome Sequencer FLX system. *BMC Microbiol.* 2008;**8**:72.
3. Stothard P & Wishart DS. Automated bacterial genome analysis and annotation. *Curr Opin Microbiol.* 2006;**9**:505–10.
4. Van Domselaar GH, Stothard P, Shrivastava S, Cruz JA, Guo A, Dong X, Lu P, Szafron D, Greiner R, & Wishart DS. BASys: a web server for automated bacterial genome annotation. *Nucleic Acids Res.* 2005;**33**(Web Server issue): W455–9.
5. Benson DA, Karsch-Mizrachi I, Lipman DJ, Ostell J, & Wheeler DL. GenBank. *Nucleic Acids Res.* 2008;**36**(database issue):D25–30.
6. Kouranov A, Xie L, de la Cruz J, Chen L, Westbrook J, Bourne PE, & Berman HM. The RCSB PDB information portal for structural genomics. *Nucleic Acids Res.* 2006;**34**(database issue):D302–5.
7. Hernandez-Boussard T, Whirl-Carrillo M, Hebert JM, Gong L, Owen R, Gong M, Gor W, Liu F, Truong C, Whaley R, Woon M, Zhou T, Altman RB, & Klein TE. The pharmacogenetics and pharmacogenomics knowledge base: accentuating the knowledge. *Nucleic Acids Res.* 2008;**36**(database issue):D913–18.
8. Wishart DS, Knox C, Guo AC, Shrivastava S, Hassanali M, Stothard P, Chang Z, & Woolsey J. DrugBank: a comprehensive resource for in silico drug discovery and exploration. *Nucleic Acids Res.* 2006;**34**(database issue):D668–72.
9. Wishart DS, Knox C, Guo AC, Cheng D, Shrivastava S, Tzur D, Gautam B, & Hassanali M. DrugBank: a knowledge base for drugs, drug actions and drug targets. *Nucleic Acids Res.* 2008;**36**(database issue):D901–6.
10. Altschul SF, Madden TL, Schäffer AA, Zhang J, Zhang Z, Miller W, & Lipman DJ. Gapped BLAST and PSI-BLAST: a new generation of protein database search programs. *Nucleic Acids Res.* 1997;**25**:3389–402.
11. Weininger D. SMILES, a chemical language and information system. 1. Introduction to methodology and encoding rules. *J Chem Inf Comput Sci.* 1988;**28**:31–6.
12. Kuhn M, Campillos M, Gonzalez P, Jensen LJ, & Bork P. Large-scale prediction of drug-target relationships. *FEBS Lett.* 2008;**582**:1283–90.
13. Pache RA, Zanzoni A, Naval J, Mas JM, & Aloy P. Towards a molecular characterization of pathological pathways. *FEBS Lett.* 2008;**582**:1259–65.
14. Kutalik Z, Beckmann JS, & Bergmann S. A modular approach for integrative analysis of large-scale gene-expression and drug-response data. *Nat Biotechnol.* 2008; **26**:531–9.
15. Shoshan MC & Linder S. Target specificity and off-target effects as determinants of cancer drug efficacy. *Expert Opin Drug Metab Toxicol.* 2008;**4**:273–80.
16. Hooper DC. Quinolone mode of action. *Drugs.* 1995; **49**(suppl 2):10–15.
17. Thorisson GA & Stein LD. The SNP Consortium website: past, present and future. *Nucleic Acids Res.* 2003;**31**:124–7.
18. International HapMap Consortium. A second generation human haplotype map of over 3.1 million SNPs. *Nature.* 2007;**449**:851–61.
19. Sherry ST, Ward MH, Kholodov M, Baker J, Phan L, Smigielski EM, & Sirotkin K. dbSNP: the NCBI database of genetic variation. *Nucleic Acids Res.* 2001;**29**:308–11.
20. Heller T, Kirchheiner J, Armstrong VW, Luthe H, Tzvetkov M, Brockmoller J, & Oellerich M. AmpliChip CYP450 GeneChip: a new gene chip that allows rapid and accurate CYP2D6 genotyping. *Ther Drug Monit.* 2006;**28**:673–7.
21. Boutet E, Lieberherr D, Tognolli M, Schneider M, & Bairoch A. UniProtKB/Swiss-Prot: the manually annotated section of the UniProt KnowledgeBase. *Methods Mol Biol.* 2007;**406**:89–112.

7 Ethical Considerations for Pharmacogenomics: Privacy and Confidentiality

Sandra Soo-Jin Lee

Discovering how individual genetic variation influences why some people respond to therapeutics but others show little improvement or have serious side effects lies at the heart of the promise of pharmacogenomics to improve health care. Realizing this goal, however, will depend on building a robust infrastructure that supports genomic medicine. These changes include developing affordable and ubiquitous genomic sequencing capability, creating expansive DNA sample sets annotated with detailed phenotypic information, educating health care providers on developments in genomic medicine, integrating diagnostic tools into clinical decision making, and attending to a complex array of social, economic, and ethical concerns that accompany each of these shifts in biomedicine. Achieving these major endeavors will depend on ensuring that the public not only understands the goals of pharmacogenomics, but also that the public actively supports and participates in basic and clinical research.

Public approval of pharmacogenomics research will depend on assurances that the risks will be minimal and that the benefits of its integration into clinical care will result in improved health care. A major issue of concern over the collection, storage, and distribution of human DNA is the risk associated with potential breaches of confidentiality and privacy and the threat of stigmatization and/or discrimination against individuals by insurance companies, employers, and others. The principle of protecting an individual's genetic information in the course of research and/or clinical care is a deeply held principle in the United States. Personal medical history has remained a sensitive and legally protected category of information. The special status of privacy of medical information has led to confidentiality protections as in physician-patient privilege. To attend to these requirements, researchers and health providers must develop strategies that secure individual genetic information and personal medical history that also enable basic and translation research.

This chapter will address the rationale for the special treatment of genetic information and the reasons why future development of pharmacogenomics requires careful consideration of issues of privacy and confidentiality. This discussion will focus on the trajectory of pharmacogenomics research and its clinical translation and the approaches used by researchers and health care personnel in protecting genetic and phenotypic data. In addition, this chapter will review the related issues of data sharing and returning secondary results, and federal and state statutes and their implications for the governance of databanks.

GENETIC EXCEPTIONALISM

In a context where the public is bombarded with new information on genetic variants associated with a broad range of conditions from premature baldness to childhood obesity, there is growing potential for the misconception that genetics determines most human conditions and traits. Reports of results from genome-wide association studies (GWAS) has further reified the importance of genes and oversimplified understanding of complex human diseases that overemphasizes the role of any one genetic variant. In reality, genes often play a limited role in the pathophysiology of these complex human conditions and traits. Genes are mediated by a whole host of environmental factors that ultimately determine an individual's likelihood of developing any specific phenotype. In fact, those who challenge the principle of genetic exceptionalism argue that, in reifying genetic contributions to disease and other conditions, other types of information, such as age, income, or immigration status, which may have a larger impact on health status, are occluded from view. A 2006 report by the Nuffield Council on Bioethics suggests that, given the similarities between genetic and other forms of personal information, it is a mistake to assume that genetic information is qualitatively different.[1] A major challenge in mitigating the tendency to overstate the role of genetics stems from the difficult, yet fundamental question of what, exactly, a gene is. Historically, the meaning of the gene has shifted from its early concept as a discrete unit of heredity in the nineteenth century to the mid-twentieth century framework of a physical molecule to the more recent concept of "subroutines" in an integrated network.[2] The evolving definition of what constitutes genes can have significant implications for why genetic information deserves a heightened level of scrutiny and how it is to be protected.

Genetic exceptionalism builds on the assumption that one's genetics have higher predictive value than other nongenetic medical information. Genes are understood as not only determinative, but also as immutable, not the result of individual choice or behavior. In this framework, the demand for greater protection of individual genetic information emerges from a moral perspective that those who have lost out in nature's genetic lottery do not deserve their bad fortune any more (or perhaps deserve it even less) than those who have been favored by the genetic dice deserve their good fortune. An individual should not be penalized for having a mutation that confers a greater risk for cystic fibrosis or rewarded for genes associated with increased height. A parallel can be drawn between genetic information and racial identity. In both cases, special protection is afforded by our legal statutes that make it illegal to discriminate on qualities that are believed to be given.

Those that defend genetic exceptionalism also cite the relevance of an individual's genes for biological relatives, where individual information would be highly informative for family members.[3] Although individuals can exercise their autonomy in choosing whether or not to have genetic testing, the results of their tests may have serious implications for their biological relatives, who would have never had an opportunity to weigh in on the initial decision. Do these family members have the right to receive these test results? Is there an obligation on the part of the participant, researcher, or health care provider to provide this information? And, if yes, under what circumstances? One could argue that genetics is only one type of information that may confer risk to family members. Other nongenetic medical information may also fall into this category. For example, a family history of lung disease may be as informative of traits that run in families, such as smoking or residence near environmental hazards. In addition, it is unclear what value genetic information may provide for the majority of diseases and conditions for which genetic associations may be drawn. Aside from cases of simple Mendelian genetic diseases such as Huntington's disease, it is unclear how individuals should interpret tests for genetic markers and whether such information will provide predictive or useful health information.

The significance placed on genetics is easily understood within the context of our technoscientific culture. The gene has become a powerful symbolic icon in our society in which genetic information is believed to be uniquely identifying, holding great significance for individuals in defining who and what they are.[4] The clichéd explanation of "it must be in their genes" for individuals displaying unusual musicality or athleticism reveals the entrenched view that certain human traits are destined in our genetic code.[5] Although there are other sources of uniquely identifying information such as fingerprints, voice recordings, and photographs, the cultural significance of genetic information has resulted in a growing architecture of statutes and policies aimed at its protection.

In addressing data protection, the international organization, the United National Educational, Scientific and Cultural Organization (UNESCO) states that human genetic data have a special status and that due consideration should be given and, where appropriate, special protection should be afforded to human genetic data and to biological samples.[6] Building on the notion of genetic exceptionalism, genetic protocols are beginning to underscore the importance of privacy and confidentiality by requiring researchers to provide the potential impact of research on others such as family members, risks for stigma or discrimination, predictive power for future illness, and the psychological impact of predictive information.

REGULATORY PROTECTION OF GENETIC INFORMATION PRIVACY AND CONFIDENTIALITY

Privacy is a broad concept that encompasses a person's right to protection from others accessing their own person and their personal spaces; third-party interference with personal choices, especially in intimate spheres such as procreation; and ownership of materials and information derived from persons.[7] In the context of pharmacogenomics, a central question is whether genetic information can be considered personal property. So far, only one state – Alaska – has defined genetic information as personal property in their statutes. However, many of the policies aimed at protecting individuals against genetic discrimination proceed with this tacit assumption.

Before the passage of the Genetic Information Nondiscrimination Act (GINA) [H.R. 493, S. 358] in 2008, there were no federal statutes that specifically prohibited using genetic information to discriminate against individuals. In 1996, Congress passed the Health Insurance Portability and Accountability Act (HIPAA), which applies to employer-based and commercially issued group health insurance. HIPAA is the only federal law that directly addresses genetic discrimination. The act prohibits group health plans from using genetic information as a basis for denying or limiting eligibility for coverage or for increasing premiums. It also limits exclusions for preexisting conditions to twelve months and prohibits exclusions if the individual had previously been covered for a particular condition. It stipulates that unexpressed genetic conditions will not be considered as a preexisting condition. Although the statute regulates insurance companies in this way, the stipulations of this statute do not extend to employers and to decisions on whether to offer health coverage to their employees.

In 2002, Congress passed the HIPAA National Standards to Protect Patients' Personal Medical Records, which is aimed at protecting personal health information maintained by health care providers and institutions. The standards, though not specific to genetic information, regulate the use and release of private health information and provide patients with broader rights to access their medical records and to be informed of who has had access to them. In addition, the standards establish specific protocol and restrictions to records by researchers.

Before HIPAA, other statutes such as the Americans with Disabilities Act (ADA) and the Rehabilitation Act of 1973 provided for some employment protections. The ADA prohibits discrimination against persons with symptomatic genetic disabilities; however, this does not extend to individuals with unexpressed genetic conditions. The statutes do not protect individuals from being asked by potential employers for genetic information as a condition of an offer of employment nor does it protect workers from requirements to provide medical information that is considered related to a particular job.

A legal test of the ADA and its coverage of genetic testing occurred in 2002 when the Equal Employment Opportunity Commission (EEOC) filed a suit against the Burlington Northern Santa Fe (BNSF) Railroad for surreptitiously testing its employees for the rare genetic condition of hereditary neuropathy with liability to pressure palsies (HNPP). HNPP is believed to cause carpal tunnel syndrome, and BNSF asserted by testing its employees that it could determine whether the high incidence of repetitive-stress injuries was work related. In addition to genetic testing for HNPP, company-sponsored physicians were instructed to screen for several other medical conditions, including diabetes and alcoholism. BNSF employees examined by company physicians were not told that they were being genetically tested, and one employee who refused testing was threatened with possible termination. On behalf of BNSF employees, the EEOC argued that the tests were unlawful under the ADA, contending that the tests did not generate information that was critical to the question of whether employees could perform their work, and that any condition of employment based on such tests would be illegal discrimination based on a medical disability. The lawsuit was settled quickly, with BNSF agreeing to pay the 36 employees involved in the lawsuit a total of $2.2 million.

The passage of GINA addresses many concerns over privacy and potential discrimination by both employers and health insurers. It is the first federal law to protect Americans from being treated unfairly because of differences in their DNA that may affect their health. The statute prohibits group and individual health insurers from using a person's genetic information in determining eligibility or premiums and an insurer from requesting or requiring that a person undergo a genetic test. The provision also prohibits employers from using a person's genetic information in making employment decisions such as hiring, firing, job assignments, or any other terms of employment, or requesting, requiring, or purchasing genetic information about persons or their family members.[8] The legislation, however, does have several restrictions. It does not cover members of the military nor does it cover life, disability, or long-term care insurance.

In addition to the federal statute, states have provided their own varying levels of protection of genetic information. Five states – Alaska, Colorado, Florida, Georgia, and Louisiana – define genetic information as personal property; however, only Alaska extends this definition to the physical DNA samples themselves. A far greater number of states provide limited protection that requires individual consent for personal access to genetic information or individual consent for performing genetic tests or obtaining, accessing, retaining, or disclosing genetic information. Eighteen states do not offer any regulations. These are Alabama, Connecticut, Indiana, Iowa, Kansas, Kentucky, Maine, Mississippi, Montana, North Carolina, North Dakota, Ohio, Oklahoma, Pennsylvania, Tennessee, West Virginia, Wisconsin, and Wyoming (Table 7.1).[9]

BIOBANKING AND SCALING UP

The full integration of personalized medicine will depend on the development of a network of databases that make available a large number of genetic profiles to researchers. Large numbers are important in order to provide sufficient power in studies attempting to detect the often small effects of individual genes involved in the onset of complex disease and traits like drug response. Several countries have begun national databases to achieve this goal.[10,11] Francis Collins, the previous director of the National Human Genome Research Institute (NHGRI), has suggested that the United States follow suit.[12] Even with the creation of these large datasets, a significant challenge for pharmacogenomics research is the annotation of these large datasets with extensive phenotypic information. A central goal of pharmacogenomics research is to "correlate genotype with

Table 7.1. State Requirement for Individual Consent

State	*Perform Genetic Test*	*Obtain Genetic Information*	*Retain Genetic Information*	*Disclose Genetic Information*
Alaska	X	X	X	X
Arizona	X			X
Arkansas				X
California				X
Colorado				X
Delaware		X	X	X
Florida	X			X
Georgia	X		X	X
Hawaii				X
Illinois				X
Louisiana				X
Maryland				X
Massachusetts	X			X
Michigan	X			
Minnesota		X	X	X
Missouri				X
Nebraska	X			
Nevada		X	X	X
New Hampshire				
New Jersey				X
New Mexico		X	X	X
New York	X		X	X
Oregon		X	X	X
Rhode Island				X
South Carolina	X			X
South Dakota	X			
Texas				X
Vermont	X			X
Virginia				X
Washington				X

phenotype" in the context of pharmacotherapy; however, few public databases provide the broad array of phenotypic information that would be helpful in discovering genetic associations related to drug development.

Phenotype is not a single entity, but a set of parameters. These parameters are highly variable, and their characterization is subject to the individual practices of researchers. Inconsistency in the inclusion and exclusion criteria for key phenotypes, for example, diabetes, high blood pressure, or race, may contribute to low replication rates of most genetic association studies of initial findings. Guidelines on standardizing this critical information are needed to address this potential problem, and scientists have called for agreement on an optimal set of phenotypic variables and their definitions to accompany each DNA sample.[13,14] In an effort to meet this challenge, the NHGRI has sponsored the PhenX project, which seeks to achieve consensus on twenty phenotypes and environmental exposures to be used in GWAS studies and to provide standard measures for the scientific community.[15]

In addition to standardization, pharmacogenomic development and integration depend on comprehensive access to genetic profiles by physicians to determine the likelihood of adverse drug reactions, side effects, and efficacy. One of the difficulties is that this information is likely to change over time. In continually updating phenotypic parameters, data more than a few years old will rapidly diminish in value in generating new hypotheses for research. Newly discovered parameters that come to be regarded as mandatory for the evaluation of a disease of interest may be absent from the original dataset. The only ways to obtain values for these parameters would be to approach the investigator who originated the data and to seek access to those patients in the dataset if still available. In many cases this will not be possible, either because the data needed are time-dependent (e.g., drug levels) or because institutional review boards may have required personal, identifying information to be delinked from research samples, which would prohibit follow-up.

The need for new and updated phenotypic information often requires ongoing contact with research participants and patients; however, this may increase risks to their privacy. This represents a fundamental conflict where research interests may present greater exposure to individuals and conflict with participants' interests for maintaining personal privacy. Policies are needed to address this tension and are critical to the future trajectory of pharmacogenomics research and development.

DATA SHARING AND THE RISK TO PRIVACY

One of the major challenges to coordinated research efforts in pharmacogenomics is the sharing of data across institutions and national borders. Great difficulty remains in obtaining genetic information with highly detailed phenotypic information, and the desire for greater efficiency and utility of genetic data may compete with the ethical concerns of maintaining individual privacy and public trust. Individuals are often sampled in a variety of contexts – in their physician's office, at the hospital, at a research laboratory – and, although great efforts may be taken to deidentify, encrypt, or delink personal information, there are substantial fears that the multiple locations at which samples are obtained, stored, and then redistributed for basic research present serious challenges to maintaining anonymity and confidentiality.

In their review, Lunshof et al. list the broad range of possible threats to privacy and confidentiality in genomic research that may occur even with tiered policies on access.[16] These threats include reidentification of individual information after deidentification, which may occur by using a combination of surnames, genotype, and geographical information, as well.[17] Other possible approaches include inferring phenotype from genotype through information such as stature, hair, eye or skin color,[18] or identification of a first-degree relative,[19] in DNA and RNA. There is also the real possibility of theft or loss of data storage devices, such as laptops, that are often used to hold research data. These threats are exacerbated by the increasing availability of aggregate data in public, private, and state-controlled databases including clinical biobanks and databases, population biobanks and databases, research biobanks and databases with academia and industry, and forensic biobanks and databases in data-sharing practices.

These potential threats are a source of concern for the public. Research indicates that participants are often conflicted; although they are committed to biomedical research and are comfortable with their information being shared within the medical community, they fear that their information might be leaked beyond researchers.[20] When presented with the choice of restricting access to their information, most participants' concerns over privacy outweigh their sense of altruism in participating in research to improve health, which is more likely to lead them to refuse public data release in data sharing.[21] Several guidelines on data sharing have emerged to address these concerns. For example, the NHGRI GWAS Data Sharing Policy requires that, when phenotypic information accompanies genetic data, this information should be released to databases with restricted access, which applies to the release of all data from National Institutes of Health-funded or supported GWAS into the restricted database, dbGap.[22] Current policies make a distinction between information that is solely genetic as opposed to genetic information that is accompanied by phenotypic data. In the case of whole-genome information, public access to genetic information is allowable with specific consent for data sharing. However, if phenotypic information is included, policies recommend only controlled access with specific consent for data sharing. In the case of information on individual gene variants that are included in sets of aggregated data, public access with general consent for future use of genotypic data is possible; however, scholars suggest that when phenotypic data are added to these datasets, they should be subject to controlled access that requires obtaining general consent for future use from participants.[23]

STRATEGIES FOR PROTECTING PRIVACY

In addressing concerns over privacy and confidentiality, researchers have used several different strategies that offer varying levels of protection (Table 7.2). A common practice is to assign a unique symbol – such as a number or a combination of letters and numbers – to each participant's set of information. Personally identifying information such as an individual's name or address are stripped from the participant's dataset, essentially "blinding" researchers from linking the identities of their participants with specific genotypic and phenotypic data. The coded number becomes the only way of referring to specific datasets. The lists of codes that link to personally identifying information are held by a designated member of the research staff and are kept in a locked and secure location. Datasets may be coded in this way once, twice, or three times depending on the level of security desired. For example, triple-coded data involve coding the datasets and then coding these codes separately, where each code key is held by different personnel in different locations. These extra efforts are intended to decrease the likelihood that personal identities can be relinked with specific datasets.

Researchers may wish to permanently delete personally identifying information from samples and *anonymize* samples by stripping samples of the identifying information. By doing so, however, researchers will not be able to recontact participants or to reconsent

Table 7.2. Types of Privacy Measures

Privacy Measures	*Procedures*
Blinded or single-coded	A code, often a uniquely assigned number, is attached to a sample and personal information is delinked. The list of codes is held by a research staff member and stored in a secure location.
Double-blinded or double-coded	Samples are coded and then given codes that are kept by a third party not directly involved with the research. Identification of the samples can only be retrieved by bringing together the two codes.
Triple coding	Three steps of coding involving using three different codes for an individual set of information that are held by three different research personnel. These steps are taken to ensure that no single person is able to reconstruct the complete key code list or any single participant's identity.
Anonymized	All personally identifying information is permanently delinked from the samples, destroyed, and rendered inaccessible to the researchers.
Anonymous	Samples are received by researchers already stripped of personally identifying information. Researchers are unable to recontact participants.
Identified	Personal information including names, address, social security numbers, or other individual data remain linked with samples.
Open consent	Participants consent to unrestricted redisclosure of data originating from a confidential relationship, namely their health records, and to unrestricted disclosure of information that emerges from any future research on their genotype-phenotype dataset, the information content of which cannot be predicted. No promises of anonymity, privacy, or confidentiality are made.

individuals for future research. In some cases, researchers receive *anonymous* samples that have already been stripped of identifying information.[24] One question that has emerged is that, if data have been deidentified but include large amounts of genetic information, are the individuals still considered "human subjects?" The answer has important implications for how researchers go about obtaining informed consent and meeting ethical requirements for safeguards. Some have urged that genomic sequencing studies should be recognized as human subjects research regardless of whether the data have been delinked from personally identifying information and brought unambiguously under the protection of existing federal legislation.[25] However, this interpretation is not unanimous. The U.S. Office of Human Research Protections does not consider data or biospecimens collected for one purpose and delinked from personally identifying information and key-coded as human subjects research if the data are not "individually identifiable" when used secondarily for research on questions different from the original research.

Several scholars have argued that, given the trajectory of genomic research and development, promises to participants for complete privacy and confidentiality may be impossible to keep. Technological developments allow the identification of individuals based on ever refined genomic markers. The American Society of Human Genetics has issued statements that indicate that genetic information alone may be better indicators of individual identity than conventional demographic information.[26] This has led some scholars and scientists to suggest a new approach of asking participants for *open consent.* Instead of attempting to maintain privacy and confidentiality, in open consent participants would agree to make all of their genotypic and phenotypic information, including their medical history, available to researchers for current and future research. Open consent is used in the Personal Genome Project, in which an individual can participate only if they agree to share their personal data for research purposes.[27] By advocating full disclosure of personal information, advocates of open consent sidestep the challenges of mitigating the risk to privacy in genomic research and the ongoing difficulty of creating lasting safeguards.

RESPONSIBILITIES FOR SECONDARY INFORMATION

Several scholars have argued that, unlike other forms of genetic testing, pharmacogenomic testing presents fewer risks for stigmatization and potentially more usable information in predicting individual drug response. Whereas other genetic tests may require complicated reproductive decisions or provide sensitive information on relative risk for disease, pharmacogenomic testing is perceived as less threatening by the public. However, recent studies indicate that pharmacogenomics research may involve genes that are not only significant for drug response, but may be associated with other health-related human traits and disease. In a meta-review of 555 pharmacogenomic studies, 53 percent of the forty-two variants studied by Henrikson et al. reported significant association with disease risk in at least two published studies.[28] For example, a study of Alzheimer's disease identifies genes that are predictive of the disease and therapeutic efficacy.[29] It is known that therapeutic response in Alzheimer's disease is genotype specific, where apolipoprotein E 4/4 carriers are the worst responders to conventional treatments.[30]

An issue of concern is that the scope of predictive value of pharmacogenetic tests will not have been discovered at the time that individuals are recruited for specific clinical trials and may only emerge after a trial has been completed. Such secondary risk information represents the unexpected consequence of tests that are used to predict drug response, adverse events, and dosage requirements. The current informed consent process does not address the return of such information to research participants. Without sufficient safeguards that protect individual privacy and confidentiality, the participants may be vulnerable to potential discrimination and/or stigmatization for being at increased risk for a host of unforeseen conditions. The question remains as to what responsibilities researchers have to study participants in informing them of these possibilities in the absence of specific data. This creates a new challenge that stems from the unknown potentiality of genomic research and reveals the shortcomings of the current informed consent process in accounting for these scenarios.

Although some ethicists have suggested that genetic results specific to individuals should unconditionally be reported back to participants,[31] several issues complicate this policy. For example, not all results will be readily comprehensible to participants, they may be only exploratory, without validation or replication in the literature, or they may not accompany clinical interventions. Is there an ethical obligation to return only genetic risk information in a context where few therapeutics are available for the particular disease or condition? The World Health Organization suggests that certain conditions must be met before results are returned to participants. These include that the data have been instrumental in identifying a clear clinical benefit to identifiable individuals; the disclosure of the data to the relevant individuals will avert or minimize significant harm to those individuals; and there is no indication that the individuals in question would prefer not to know.[32] In a context where several policies have emerged, it is important to recognize that, at the international level, there is growing consensus that there is an ethical duty to return individual genetic research results if there is evidence that the results are valid, significant, and beneficial. Yet, at the same time, the right of the research participant not to know must also be taken into consideration.[33]

GOOD GOVERNANCE AND PRIVACY PROTECTION

With the proliferation of genetic datasets and the creation of DNA biobanks, it is increasingly important to address the broad range of issues that concern the multiple interests of those involved in pharmacogenomic research and development. One challenge is the multiple approaches that are taken in different contexts and the difficulty in reconciling variations in policies across national borders.[34] Those responsible for governing the sampling, storage, and distribution of DNA samples will require systematic, yet flexible policies that take into account the complex issues related to protecting the privacy of participants and supporting current and future research projects.

To be most effective, data-sharing policies should be developed in a manner that reflects the full range of stakeholders and the multiple ways in which their varying interests intersect. Foster and Sharp provide a systematic analysis of the different stakeholder interests to identify a range of implications that should be considered when deciding on genomic data-sharing policies. A comprehensive, multiplexed analysis of the interplay of stakeholders and their interests can yield a more balanced, contextualized evaluation of genomic data sharing that single-issue and single-stakeholder perspectives cannot. Although there are many different stakeholder interests in genomic data sharing, these interests can be divided into three broad categories: first-party producers of genomic data, second-order users of genomic data, and third parties who are affected by future uses of genomic data.[35]

Pharmacogenomics research will often focus on group differences, in particular, social groups defined by racial and ethnic identities.[36] In using racial and ethnic categories, researchers have been advised to follow guidelines that will minimize the risk of misinterpretation,[37] but strategies for incorporating diverse populations into the research process are needed. One approach taken by the International HapMap Project is community engagement where populations from which individuals were recruited for sampling were invited to a series of dialogues with researchers. The goals of this process were created to foster an open exchange of information, concerns, and recommendations around the use of DNA samples. The aims were to share community views about the ethical, social, and cultural issues the project raised; to provide input into such matters as how the samples from their locality would be collected and described; and to generate ideas of how to ensure that the communities remained informed about how the HapMap and the samples are being used and about findings from future studies.[38] This approach may be useful in addressing the full range of privacy and confidentiality issues that face researchers and clinicians in pharmacogenomics. To engage the specific concerns that any one stake-holding group may have regarding the sampling, storage, and distribution of their DNA, new models will be needed to ensure ethical governance of genetic information.

DISCUSSION QUESTIONS

1. What is genetic exceptionalism?
2. What is GINA?
3. What is the difference between anonymized and open consent?
4. What are the ethical concerns around secondary information?
5. What are the goals of community engagement?

ENDNOTES

1 Reischl J, Schroder M, Luttenberger N, Petrov D, Chumann B, Ternes R, & Sturzebecher S. Pharmacogenetic research and data protection – challenges and solutions. *Pharmacogenomics J.* 2006;**6**:225–33.

2 Gerstein MB, Bruce C, Rozowsky JS, Zheng D, Du J, Korbel JO, Emanuelsson O, Zhang ZD, Weissman S, & Synder M. What is a gene, post-ENCODE? History and updated definition. *Genome Res.* 2007;**17**:669–81.

3 McGuire AL, Fisher R, Cusenza P, Hudson K, Rothstein MA, McGraw D, Matteson S, Glaser J, & Henley DE. Confidentiality, privacy, and security of genetic and genomic test information in electronic health records: points to consider. *Genet Med.* 2008;**10**:495–9.

4 Keller EF. *The Century of the Gene.* Cambridge, MA: Harvard University Press; 2000.

5 Nelkin D & Lindee S. *DNA Mystique. The Gene as a Cultural Icon.* New York: Freedman; 1995.

6 Reischl J, Schroder M, Luttenberger N, Petrov D, Chumann B, Ternes R, & Sturzebecher S. Pharmacogenetic research and data protection – challenges and solutions. *Pharmacogenomics J.* 2006;**6**:225–33.

7 Anderlik MR & Rothstein MA. Privacy and confidentiality of genetic information: what rules for the new science. *Ann Rev Genomics Hum Gen.* 2001;**2**:401–33.

8 Hudson KL, Holohan MK, & Collins FS. Keeping pace with the times: The Genetic Information Nondiscrimination Act of 2008. *N Engl J Med.* 2008;**358**:2661–3.

9 National Conference of State Legislatures. State Genetic Privacy Laws. http://www.ncsl.org/programs/health/genetics/prt.htm, accessed July 8, 2011.

10 Fan CT, Lin JC, & Lee CH. Taiwan Biobank: a project aiming to aid Taiwan's transition into a biomedical island. *Pharmacogenomics.* 2008;**9**:235–46.

11 Tutton R, Kaye J, & Hoeyer K. Governing UK Biobank: the importance of ensuring public trust. *Trends Biotechnol.* 2004;**22**:284–5.

12 Collins FS. Plenary talk. Presented at: Annual Meeting of the American Society of Human Genetics, Salt Lake City, Utah, October 25–29, 2005.

13 Saavedra JM. The challenge of genetic studies in hypertension. *Circ Res.* 2007;**100**:1389–93.

14 Ginsburg GS, Shah SH, & McCarthy JJ. Taking cardiovascular genetic association studies to the next level. *J Am Coll Cardiol.* 2007;**50**: 930–2.

15 National Human Genome Research Institute. PhenX Project. https://www.phenx.org/Default.aspx?tabid/36, accessed July 8, 2011.

16 Lunshof JE, Chadwick R, Vorhaus DB, & Church GM. From genetic privacy to open consent. *Nat Rev Genet.* 2008;**9**:406–11.

17 Motluk A. Anonymous sperm donor traced on internet. *New Scientist News Service (Magazine Issue 2524).* 3 November, 2005. http://www.newscientist.com/article/mg18825244.200-anonymous-sperm-donor-traced-on-internet.html, accessed July 8, 2011.

18 Markoff J. Researchers find way to steal encrypted data. *The New York Times.* 2008. http://www.nytimes.com/2008/02/22/technology/22chip.html, accessed July 8, 2011.

19 Kohane IS & Altman RB. Health-information altruists – a potentially critical resource. *N Engl J Med.* 2005;**353**:2074–7.

20 Sterling R, Henderson GE, & Corbie-Smith G. Public willingness to participate in and public opinions about genetic variation research: a review of the literature. *Am J Public Health.* 2006;**96**:1971–8.

21 McGuire AL, Hamilton JA, Lunstroth R, McCullough LB, & Goldman A. DNA data sharing: research participants' perspectives. *Genet Med.* 2008;**10**:46–53.

22 Policy for sharing of data obtained in NIH-supported or conducted genome-wide association studies (GWAS). http://grants.nih.gov/grants/guide/notice-files/NOT-OD-07-088.html, accessed July 8, 2011.

23 McGuire AL. 1000 genomes: on the road to personalized medicine. *Per Med.* 2008;**5**:195–7.

24 Joly Y, Knoppers BM, & Nguyen MT. Stored tissue samples: through the confidentiality maze. *Pharmacogenomics J.* 2005;**5**:2–5.

25 McGuire AL & Gibbs RS. Genetics: no longer identified. *Science.* 2006;**312**:370.

26 The American Society of Human Genetics. *ASHG Response to NIH on Genome-Wide Association Studies.* ASHG Policy Statement Archives 2006. http://www.ashg.org/pages/statement_nov3006.shtml, accessed July 8, 2011.

27 The Personal Genome Project. Are guarantees of genomic anonymity realistic? http://arep.med.harvard.edu/PGP/Anon.htm, accessed July 8, 2011.

28 Henrikson NB, Burke W, Veenstra DL. Ancillary risk information and pharmacogenetic tests: social and policy implications. *Pharmacogenomics J.* 2008;**8**:85–9.

29 Pritchard A, Harris J, Pritchard CW, St Clair D, Lemmon H, Lambert JC, Chartier-Harlin MC, Hayes A, Thaker U, Iwatsubo T, Mann DM, & Lendon C. Association study and meta-analysis of low-density lipoprotein receptor related protein in Alzheimer's disease. *Neurosci Lett.* 2005; **382**(3):221–6.

30 Cacabelos R. Pharmacogenomics in Alzheimer's disease. *Methods Mol Biol.* 2008;**448**:213–357.

31 Fernandez CV, Kodish E, & Weijer C. Informing study participants of research results: an ethical imperative. *IRB: Ethics Hum Res.* 2003;**25**: 12–19.

32 World Health Organisation. Genetic databases: assessing the benefits and the impact on human and patient rights. Geneva: World Health Organisation; 2003. http://www.law.ed.ac.uk/ahrb/publications/online/whofinalreport.doc, accessed July 8, 2011.

33 Knoppers BM, Joly Y, Simard J, & Durocher F. The emergence of an ethical duty to disclose genetic research results: international perspectives. *Eur J Hum Genet.* 2006;**14**:1170–8.

34 Zika E, Schulte in den Baumen T, Kaye J, Brand A, & Ibarreta D. Sample, data use and protection in biobanking in Europe: legal issues. *Pharmacogenomics.* 2008;**9**:773–81.

35 Foster MW & Sharp RR. Share and share alike: deciding how to distribute the scientific and social benefits of genomic data. *Nat Rev Genet.* 2007;**8**:633–9.

36 Lee SS-J. "Racializing drug design": pharmacogenomics and implications for health disparities. *Am J Public Health.* 2005;**95**:2133–8.

37 Lee SS-J, Mountain J, Koenig B, Altman R, Brown M, Camarillo A, Cavalli-Sforza L, Cho M, Eberhardt J, Feldman M, Ford R, Greely H, King R, Markus H, Satz D, Snipp M, Steele C, & Underhill P. The ethics of characterizing difference: guiding principles on using racial categories in human genetics. *Genome Biol.* 2008;**9**:404.1–4.

38 Rotimi C, Leppert M, Matsuda I, Zeng C, Ahang H, Adebamowo C, Ajayi I, Aniagwu T, Dixon M, Fukushima Y, Macer D, Marshall P, Nkwodimmah C, Peiffer A, Royal C, Suda E, Zhao H, Wang VO, & McEwen J. The International HapMap Consortium. Community engagement and informed consent in the International HapMap Project. *Community Genet.* 2007;**10**:186–98.

8 Informed Consent in Pharmacogenomic Research and Treatment

Mark A. Rothstein

Informed consent is an ethical and legal requirement for research on human subjects and the treatment of patients, as well. The process of informed consent provides an opportunity for the disclosure of material information and a structure for shared decision making that respects the autonomy of subjects and patients to decide whether to participate in research and to plan the course of their treatment. This chapter will describe the key elements of informed consent in research and clinical care in the context of pharmacogenetics and pharmacogenomics.

It is important to distinguish research from clinical care. Research is defined as "a systematic investigation, including research development, testing and evaluation, designed to develop or contribute to generalizable knowledge" (45 CFR § 46.102(d)). By contrast, clinical care or "practice" has been defined as "interventions that are designed solely to enhance the well-being of an individual patient or client and that have a reasonable expectation of success" (Belmont Report [1], 3). Although the definitions have been in place for several decades, new technologies and methodologies increasingly are blurring the distinctions. For example, translational research, case studies, outcomes research, quality improvement, and public health applications often present definitional challenges.

The primary goals of informed consent are the protection of the welfare of subjects and patients and the promotion of their autonomy (Berg et al. [2], 18). The key elements of informed consent are the provision to individuals with relevant information and the ability to act on the information free from external constraints. The rise of autonomy and patients' rights in health care has not been without critics who contend that, even though autonomy is a value to be given consideration, it does not automatically override other values, such as beneficence (3, 4). For example, in the research setting, an asserted overemphasis on the rights of subjects could undermine research and prevent the discovery of treatments and cures that would benefit society.

INFORMED CONSENT IN RESEARCH

General Principles

The principle of informed consent for research can be traced directly to the Nuremberg Trials after World War II, part of which dealt with the fiendish experiments conducted by Nazi doctors in concentration camps. The Nuremberg Code, published in 1947, expressly recognized informed consent as its first principle of ethical research.

> The voluntary consent of the human subject is absolutely essential. This means that the person involved should have the legal capacity to give consent; should be so situated as to be able to exercise free power of choice, without the intervention of any element of force, fraud, deceit, duress, overreaching, or other ulterior form of constraint or coercion; and should have sufficient knowledge and comprehension of the elements of the subject matter involved as to enable him to make an understanding and enlightened decision.

The next important international document on research ethics was the World Medical Association's Declaration of Helsinki (1964, latest revision in 2000). Although originally drafted as an ethical guide to physician-investigators, it has been adopted as law by several countries. Other international documents on research ethics include those developed by the United Nations Educational Scientific and Cultural Organization, the World Health Organization, and the European Commission. In addition, numerous countries, including the United States, have adopted detailed regulations setting forth substantive rules and procedural requirements that must be followed in certain research with human subjects.

The ethical foundations of regulating biomedical research in the United States were developed by the National Commission for the Protection of Human Subjects of Biomedical and Behavioral Research (National Commission). The National Commission, in its often-cited *Belmont Report*, stated that there are three basic ethical principles that should govern research with human subjects: respect for persons, beneficence, and justice (Belmont Report [1], 4). A key element of respect for persons is autonomy, and informed consent is a mechanism to promote autonomy. "Respect for persons requires that subjects, to the degree that they are capable, be given the opportunity to choose what shall or shall not happen to them. This opportunity is provided when adequate standards for informed consent are satisfied" (Belmont Report [1], 5).

The reports of the National Commission directly led to the promulgation of the federal regulations for the protection of human subjects (45 CFR 46, Subpart A). An important procedural element that must be satisfied as a precondition of research is favorable review of the protocol by a neutral (i.e., not including the investigators) board. In the United States, these entities are known as institutional review boards (IRBs). According to the federal research regulations, an IRB must have at least five members of varied backgrounds, including at least one member who is not affiliated with the institution where the research will be conducted (45 CFR § 46.107). For a protocol to be approved, the IRB must determine that:

1. The risks to subjects are minimized;
2. The risks to subjects are reasonable in relation to anticipated benefits;
3. Selection of subjects is equitable;
4. Informed consent will be sought from each prospective subject;
5. Informed consent will be appropriately documented;
6. When appropriate, the data will be monitored to ensure the safety of subjects; and
7. When appropriate, there are adequate provisions to protect the privacy of subjects and the confidentiality of data (45 CFR § 46.111(a)).

Although informed consent to participate in research is indicated by the signature of the research subject on an informed consent document, informed consent should not be regarded merely as the act of obtaining the individual's signature. Informed consent is a *process* during which information is disseminated to the prospective subject, the individual's questions and concerns are addressed, and the individual decides, free from duress or coercion, whether to participate in the research.

The criteria for informed consent for research differ to some degree in different countries, but the informed consent requirements of the United States are typical. The Federal Regulation for the Protection of Human Subjects is often referred to as the "Common Rule" because it has been adopted by virtually all federal government agencies that sponsor research (45 CFR Part 46, Subpart A). According to this regulation, the informed consent document provided to each potential subject must include the following:

1. A statement that the study involves research, an explanation of the purposes of the research and the expected duration of the subject's participation, a description of the procedures, and identification of any procedures that are experimental;
2. A description of any reasonably foreseeable risks or discomforts to the subject;
3. A description of any benefits to the subject or to others that may reasonably be expected from the research;
4. A disclosure of appropriate alternative procedures or courses of treatment, if any, that might be advantageous to the subject;
5. A statement describing the extent, if any, to which confidentiality of records identifying the subject will be maintained;
6. For research involving more than minimal risk, an explanation of whether compensation and/or medical treatment are available;
7. An explanation of whom to contact for answers to pertinent questions or to report a research-related injury; and
8. A statement that participation is voluntary and the subject may discontinue participation at any time (45 CFR § 46.116).

A separate but similar set of regulations applies to research involving drug development for approval by the Food and Drug Administration (21 CFR Parts 50, 56). Special rules also apply to research involving the following classes of vulnerable subjects of research: fetuses, pregnant women, and in vitro fertilization (45 CFR Subpart B); prisoners (45 CFR Subpart C); and children (45 CFR Subpart D).

Besides complying with regulations protecting human subjects of research, investigators conducting research with individually identifiable health information also must comply with the Health Insurance Portability and Accountability Act (HIPAA) Privacy Rule (45 CFR Parts 160, 164). The Privacy Rule applies to health care providers, health plans, and health clearinghouses (entities that convert billing records into standard formats). Covered entities that engage in research using individually identifiable health information ("protected health information") must obtain an authorization from the individual for the use and disclosure of his or her information. An authorization for research must be written in plain language and contain the following elements: (1) a description of the information to be used or disclosed; (2) an indication of how long the authorization will be in effect; (3) a statement that the authorization may be revoked and an indication of how to do so; (4) a statement that the information, if released to an entity not covered by the Privacy Rule, may be redisclosed and that it will no longer be protected; (5) a signature of the individual; and (6) an indication that the individual has been provided with a copy of the authorization. It is permissible to have the informed consent and authorization combined into a single document, but many institutions prefer to have separate documents.

The HIPAA Privacy Rule only applies to individually identifiable health information, and the Privacy Rule contains detailed provisions for the deidentification of health information (45 CFR 164.514(a)). Because much of the research value of health information could be lost in the deidentification process, the Privacy Rule provides for the use of a "limited dataset" for research, public health, or health care operations. Protected health information may be used without obtaining an authorization if the recipient of the information (e.g., the researcher) signs a data use agreement with the custodian of the health information setting forth the permitted uses and disclosures of the information and all direct identifiers (e.g., name, Social Security number) are removed (45 CFR 164.514(e)). An IRB or privacy board also may waive the authorization requirement in appropriate circumstances, as detailed in the Privacy Rule (45 CFR 164.512(i)).

Pharmacogenomic Research

Biobanks

Pharmacogenomics uses genome-wide analytical tools to identify the role of genetic factors in individual variability to pharmaceuticals. Large repositories of biological samples (or "biobanks") are needed to identify the genotype-phenotype associations underlying individual variability. Biobank-enabled research represents a new research paradigm in several respects. For example, it involves shifting the focus of the research from a single, current protocol to many, future protocols; from concerns about physical harms to nonphysical harms (e.g., stigma, discrimination) to the research subjects; and from focusing on individual risks and benefits to individual and group risks and benefits.

Informed consent for research involving a biobank depends to a great extent on whether the biobank already exists or is being created to be used prospectively for research (5). Tens of millions of samples originally obtained as pathology specimens already are stored in biobanks; some are fifty or more years old. In general, these tissue and other specimens were collected or retained without informed consent – and certainly not informed consent for pharmacogenomic research. If the extant samples are anonymous, then informed consent is unnecessary. If they are identified or identifiable, the issue is whether obtaining informed consent is feasible on the basis of the age, number of the samples, and the intended use. In cases where obtaining additional consent is impossible or infeasible, IRBs will often approve the research pursuant to a waiver, especially if individual identifiers are removed from the specimens.

Informed consent for individuals donating samples to establish a biobank is more complicated (6). The main problem is that the samples are being collected for unknown, future uses. Collecting samples for future research is deemed to be research. IRBs traditionally have been reluctant to approve a research proposal unless the protocol describes with specificity the use of the samples. A biobank, however, by definition, collects and retains samples to be used for multiple future research projects. IRBs have increasingly approved "tiered consent" for biobanks. This means that, when the sample is donated, the individual selects from a menu of possible uses (e.g., psychiatric research, cancer research) for which consent is given. Tiered consent or consent to research by category is considered sufficiently specific to satisfy the requirements of the Common Rule. Under the HIPAA Privacy Rule, however, authorization or waiver is needed for each new use of individually identifiable health information. The lack of harmonization between the Common Rule and the Privacy Rule continues to be an area of great concern (7).

Clinical Research

Pharmacogenomic research also represents a new model of clinical trials in which more data are generated from fewer research subjects who are genotype-matched and homogeneous (8). In terms of informed consent, the main issue is whether research subjects are willing to participate in such trials. The two key elements of informed consent in these circumstances are that genetic testing will be required before enrollment and that the pharmacogenomic-based drug may be used to replace standard therapy.

Perhaps the most important disclosure of the informed consent process concerns the confidentiality of genetic information. It is common for consent forms to contain language that "confidentiality will be protected to the extent permitted by law" or to list ways in which the confidentiality of health information will be protected. These "boilerplate" statements are inadequate when describing the confidentiality of genetic information. Genetic information can increasingly be reidentified through linking with available databases and through other means (9, 10). The reidentifiability issue needs to be part of the informed consent process, so that potential research subjects can decide whether the risks of participation outweigh the potential benefits.

Group Harms

Individual genetic variations of pharmacogenomic significance often are not distributed equally throughout the population because of endogamy, ancestral migration patterns, geographic isolation, founder effect, and other principles of population genetics. Many of these variations have differential frequency among subpopulations socially defined by race or ethnicity. Consequently, there is great potential for stigmatization and discrimination when an increased risk of an undesirable health or behavioral condition is associated with a specific population group, especially when the group is a racial or ethnic minority in a particular society. These groups are said to be "vulnerable." A socially vulnerable population also can be based on a shared somatic mutation or epigenetic mark (e.g., based on occupational or environmental exposure). Any social identity, such as language, religion, or geographic origin, also could be accorded undue biological significance simply on the basis of genotype (11).

Because of the social risks of genomic research, investigators should undertake special efforts to reduce or eliminate the risk of harm to research subjects, and nonsubjects, as well, who are members of the same racial or ethnic group. In particular, researchers should do the following:

1. Before undertaking any research, the investigators should consult with leaders and members of affected subpopulations about the research, including the goals, methods, potential findings, and their implications.
2. Investigators should be careful about inclusion and exclusion criteria for studies, and avoid the use of convenience samples that have been obtained from a subpopulation group.
3. Investigators should make special efforts to ensure that informed consent documents and other aspects of the study (e.g., recruitment, medical examinations, and return of specimens) are developed with due regard for social sensitivities.
4. In publications and public pronouncements, investigators should be careful not to overgeneralize about the findings or place needless emphasis on the particular study group.
5. Investigators should provide for health screening or interventions, where appropriate, for vulnerable individuals identified in the study as being at risk.
6. For research findings with commercial value, investigators should consider benefit sharing or similar measures.

INFORMED CONSENT TO TREATMENT

General Principles

Informed consent to treatment developed as a legal doctrine before becoming a central part of medical ethics and medical practice. One of the classic formulations of a patient's right to self-determination with regard to bodily integrity was articulated by Judge (later Justice) Cardozo in 1914: "Every human being of adult years and sound mind has a right to determine what shall be done with his own body; and a surgeon who performs an operation without his patient's consent commits an assault, for which he is liable in damages" (12). Although setting forth a requirement of "consent," Cardozo did not articulate the principles of *informed* consent that would later characterize the doctrine.

The principle of informed consent as an issue in American medicine emerged in the late 1950s and early 1960s (Faden and Beauchamp [13], 86). At the same time, "truth telling" (accurately and completely discussing with patients their diagnosis and prognosis) became recognized as an important value in medicine. Both truth telling and informed consent are based on respect for patient autonomy and a model of shared decision making. Together, they replaced the paternalistic notion that physicians know best what information should be shared with patients and what treatments are medically indicated (14). The American Medical Association Code of Ethics declares the following:

> Social policy does not accept the paternalistic view that the physician may remain silent because divulgence might prompt the patient to forego needed therapy. Rational, informed patients should not be expected to act uniformly, even under similar circumstances, in agreeing to or refusing treatment (American Medical Association [15], sec. 8.08).

In general, there are five components to informed consent: (1) competence (adult patients are presumed to be competent); (2) disclosure (what is disclosed is based on a standard of materiality); (3) understanding (the information must be provided in a culturally appropriate manner and in ways best calculated to make it comprehensible); (4) voluntariness (any "slanting" of the alternatives or overt coercion is unethical); and (5) consent (the affirmative indication from the patient) (Beauchamp and Childress [16], 79). With regard to the specific elements to be disclosed, the following five types of information must be conveyed to the patient: (1) diagnosis, (2) recommended intervention, (3) benefits, (4) risks, and (5) alternatives. Informed consent is not required in an emergency if the individual is unable to give or refuse consent (when consent is legally presumed) and in cases where a patient expressly declines the right, such as by telling the physician in advance to use his or her best judgment without consulting with the patient.

Pharmacogenomics-Based Therapies

From the standpoint of informed consent, the main problem with the use of pharmacogenomics-based medications is that a genetic test must be used before prescribing the appropriate drug. Genetic testing for the purpose of prescribing a pharmacogenomic medication differs from clinical genetic testing, primarily because it is of a narrower scope and, by itself, is likely to lack any independent, conclusive diagnostic or predictive value. The test simply reveals the polymorphism on which the differential drug response is expected. Nevertheless, without careful explanation during the informed consent process, these substantial differences may not be conveyed to the patient or the patient may not understand them.

Part of the informed consent process for genetic testing includes informing the individual about the risks associated with other genetic information that might be disclosed by the testing. For example, genetic testing for the ApoE4 allele might be used to determine the individual's likely response to statin drugs used to reduce the risk of heart disease (17). The presence of the same allele, however, also has been associated with an increased risk of Alzheimer's disease (18). The principle of pleiotropy means that a single allele or a single-nucleotide polymorphism can affect the likelihood of the individual expressing two or more different conditions. Informed consent might even require disclosing the risk of pleiotropy even before another condition is linked to the genotype for which genetic testing is sought.

There is some concern about possible coercion where, as a condition of treatment (e.g., in clinical practice guidelines or payment policies), a genetic test is required. The consent to the test is not voluntarily given, the argument goes, if genetic testing is a precondition to treatment (Secretary's Advisory Committee [19], 44). Although this may be true in a narrow sense, it is customary and essential in many spheres of medical practice to require one or more tests before a certain procedure, including blood transfusion, organ transplantation, and chemotherapy. The alternative, prescribing in the absence of genetic testing, would needlessly expose the patient to harm. The fact that the test is "genetic," without more, should not be dispositive. Informed consent (including genetic testing) should be seen as part of the overall process of prescribing medication, just as informed consent to surgery includes testing routinely required to ensure that the surgery will be safe and efficacious.

Even with adequate informed consent, a genetic test for pharmacogenomic purposes could still result in psychological or social harms to the patient. At the outset, a genetic test might indicate that the patient is more likely to be unresponsive to treatment. For patients with cancer or other serious conditions, such information is likely to have a devastating effect. In addition, a pharmacogenomic test could lead to the individual being classified as "more difficult to treat" or "more expensive to treat." Such a categorization could adversely affect the patient's eligibility or premium levels for individual health insurance, disability insurance, or other insurance products (20). These possibilities should be disclosed in the informed consent process. Nevertheless, the likelihood of such discrimination occurring, based on the research to date, is quite low. Thus, even though the risk of social harm from genetic testing should be disclosed, the pretest counseling should avoid describing dire social consequences as inevitable. The result of exaggerating the social risks could be that many individuals decide not to avail themselves of a tangible medical benefit to avoid an often hypothetical social risk.

The second ethical issue raised by informed consent for pharmacogenomic products in the clinical setting is the correlation between genotype and race and ethnicity. Because of ancestral patterns of migration and reproduction, genetic variations of pharmacogenomic significance often correlate (to varying degrees) with race and ethnicity. Thus, pharmacogenomic medicine raises the specter of race-based medicine. Although research on this issue remains ongoing, anecdotal evidence suggests that some minority patients may decline to take the drug genetically indicated for them if it is considered the "minority" drug. To these patients, because minority medical care has historically often translated into inferior care, they would simply prefer to take the drug the "white patients" are given.

There is an even larger societal issue raised by "race-based" medicine, and that is the risk of reifying race as a biological concept (21). The likely prospect of direct-to-consumer advertising of race-specific pharmaceuticals increases this risk. For many members of the public, if race is associated with genetic variation and genetic variation is associated with specially tailored medications, then "race" must have compelling biological significance. Public education and caution in marketing stratified medications are essential to avoid misunderstandings and unintended negative social consequences, such as stigmatization and discrimination.

Social and Economic Harms

The most immediate concern about generating potentially sensitive health information is that the information be protected from intentional or inadvertent disclosure by health care facilities to individuals and entities beyond the health care setting. Although it is important to ensure the security of health information from wrongful disclosure, most of the social harms about which individuals are concerned stem from the *lawful* disclosure of health information (22). Such disclosures, made in response to a patient's voluntary-but-compelled authorization, are made at least 25 million times annually in the United States (23). As a lawful condition of employment, life insurance, and other financial and governmental transactions, the individual can be required to release relevant (or all) health records, including the results of genetic tests. Electronic health records have the ability to link numerous disparate records of a single individual, and there is currently no technology available that can aggregate or filter the information so that only data relevant to the reason for the request (e.g., the individual's ability to perform a particular job) are disclosed. Thus, there is an increasing likelihood that disclosures based on compelled authorizations will result in more sensitive (including genetic) health information being disclosed.

For some individuals, the social harm is the mere fact of disclosure – having social contacts, business associates, and even strangers learning about sensitive health information. For other individuals, the harm occurs if and when the information is used to their detriment in tangible ways. In the latter situation, there are two main concerns. First is the concern that the non-health care entities receiving the information will draw *incorrect* conclusions about the significance of the health information and unjustifiably disqualify the individual from access to employment, life insurance, or some other benefit or transaction. Second is the concern that the entities will draw *correct* conclusions about the health implications of the information, but the result will be to disqualify the individual from access to a societal relationship or transaction of importance and to which the individual believes he or she has some entitlement. Consequently, genetic "nondiscrimination" is less about genetics than the underlying "entitlement."

CONCLUSION

Informed consent is a foundational principle of ethical research involving human subjects and the ethical treatment of patients. Informed consent promotes autonomy and facilitates informed choosing about participating in research and shared decision making with regard to treatment. As applied to pharmacogenomics research, informed consent issues are raised by the donation of biological specimens for biobanks, genetic testing, and participating in genotype-matched trials. As applied to pharmacogenomic treatment, informed consent issues are raised by the possibility of genetic testing revealing genotypes with possible pleiotropic effects, discovery of information about reduced efficacy of drugs, the risk status of family members, and potential individual and group-based harms.

Although complete disclosure of information regarding pharmacogenomics is essential to informed decision making, health care providers supplying the information should present the material in a neutral manner – neither unduly alarming the individual about possible risks nor glossing over possible adverse social consequences. As the field of pharmacogenomics becomes an increasingly important part of research and treatment, the issues surrounding informed consent are likely to change to reflect scientific developments, while still maintaining their ethical significance.

DISCUSSION QUESTIONS

1. Do you think informed consent is more important in the research or clinical setting?
2. How do you think new scientific developments in pharmacogenomics will affect informed consent?
3. There are only about 2,000 genetic counselors in the United States, and most work at academic medical centers. Who will provide the counseling and information-sharing part of informed consent in the clinical setting?
4. Do you think that the health care finance system will limit individual choice with regard to pharmacogenomics, thereby undermining informed consent?
5. How, if at all, can the group-based harms associated with pharmacogenomics be eliminated or reduced?

REFERENCES

1. National Commission for the Protection of Human Subjects of Biomedical and Behavioral Research. *The Belmont Report: Ethical Principles for the Protection of Human Subjects of Research*, http://ohsr.od.nih.gov/guidelines/belmont.html, June 29, 2011.
2. Berg JW, Appelbaum PS, Lidz CW, & Parker LS. *Informed Consent: Legal Theory and Clinical Practice*. 2nd ed. New York: Oxford University Press; 2001.
3. Pellegrino ED & Thomasma DC. *For the Patient's Good: The Restoration of Beneficence in Health Care*. New York: Oxford University Press; 1988.
4. Wolf, S. Shifting paradigms in bioethics and health law: the rise of a new pragmatism. *Am J Law Med*. 1994;**20**:395–415.
5. Weir RF & Olick RS. *The Stored Tissue Issue: Biomedical Research, Ethics, and Law in the Era of Genomic Medicine*. New York: Oxford University Press; 2004.
6. Rothstein MA. Expanding the ethical analysis of biobanks. *J Law Med Ethics*. 2005;**33**:89–101.
7. Rothstein MA. Research privacy under HIPAA and the Common Rule. *J Law Med Ethics*. 2005;**33**:154–9.
8. Manasco PK & Arledge TE. Drug development strategies. In: Rothstein MA, ed. *Pharmacogenomics: Social, Ethical, and Clinical Dimensions*. Hoboken, NJ: John Wiley & Sons; 2003.
9. Lowrance WW & Collins FS. Identifiability in genomic research. *Science*. 2007;**317**:600–2.
10. McGuire AL & Gibbs RA. No longer de-identified. *Science*. 2006;**312**:370–1.

11. Foster MW & Sharp RR. Race, ethnicity, and genomics: social classifications as proxies of biological heterogeneity. *Genome Res.* 2002;**12**:844–50.
12. Schloendorff v. Society of New York Hospitals. 105 N.E. 92 (N.Y. 1914).
13. Faden RR & Beauchamp TL. *A History and Theory of Informed Consent.* New York: Oxford University Press; 1986.
14. Katz J. *The Silent World of Doctor and Patient.* Baltimore: Johns Hopkins University Press; 2002.
15. American Medical Association. *Code of Medical Ethics: Current Opinions with Annotations.* 2006–2007 ed. Chicago: American Medical Association; 2006.
16. Beauchamp TL & Childress JF. *Principles of Biomedical Ethics.* 5th ed. New York: Oxford University Press; 2001.
17. Carmena R, Roederer G, Mailloux H, Lussier-Cacan S, & Davignon J. The response to lovastatin in patients with heterozygous familial hypercholesterolemia is modulated by apolipoprotein E polymorphism. *Metabolism.* 1993;**42**:895–901.
18. Corder E, Saunders A, & Strittmatter W. Gene dose of apolipoprotein E type 4 allele and the risk of Alzheimer's disease in late onset families. *Science.* 1993;**261**:921–3.
19. Secretary's Advisory Committee on Genetics, Health, and Society. *Realizing the Potential of Pharmacogenomics: Opportunities and Challenges.* http://oba.od.nih.gov/oba/SACGHS/reports/SACGHS_PGx_report.pdf 2007. (June 26, 2011).
20. Rothstein MA. Epilogue: policy prescriptions. In: Rothstein MA, ed. *Pharmacogenomics: Social, Ethical, and Clinical Dimensions.* Hoboken, NJ: John Wiley & Sons; 2003.
21. Kahn JD. From disparity to difference: how race-specific medicines may undermine policies to address inequalities in health care. *South Calif Interdisciplinary Law J.* 2005; **15**:105–30.
22. Rothstein MA & Talbott MK. Compelled disclosure of health information: protecting against the greatest potential threat to privacy. *JAMA.* 2006;**295**:2882–5.
23. Rothstein MA & Talbott MK. Compelled authorizations for disclosure of health records: magnitude and implications. *Am J Bioethics.* 2007;**7**:38–45.

9

Legal Trends Driving the Clinical Translation of Pharmacogenomics

Barbara J. Evans

The Human Genome Project has fueled rapid progress in pharmacogenomics over the past decade. Genetic variability of drug response was suspected in the mid-twentieth century, but scientists only now are gaining the ability to explain this variability and harness it to improve health outcomes. These advances are forcing health care professionals and regulators to reevaluate the duties they owe to the people they serve and protect. Patients with identical symptoms may have genetic differences that carry important implications for the best choice of therapy[1] and for the physician's standard of care. Clinical trial subjects who seem comparable under traditional trial selection criteria like age, gender, or severity of underlying disease[2] may not be comparable at all when genetic differences are taken into account.[3] Studies that gloss over these differences may mislead the public about the safety and effectiveness of the drugs they consume, yet regulators routinely rely on such studies when approving new drugs.[4] Additional steps are needed to protect the public. The very notion of "the public" is evolving as pharmacogenomics divides the population into multiple subgroups. Regulators who once needed only to protect a single, homogeneous "public" now face a situation where their decisions visit disparate impacts on various subgroups within the broader public they are seeking to protect. This shift in perspective already is reshaping the legal and commercial environment for medical products and health care.

This chapter explores trends that are emerging as law adapts to the era of personalized genomic medicine. The focus is on how the broad contours of regulations affecting drugs and medical devices are changing and how these changes may affect clinicians, patients, medical researchers, and people who participate in biomedical research. Important reforms already are in progress, and more can be expected in coming years. One concrete example is the Food and Drug Administration Amendments Act of 2007 (FDAAA),[5] which instituted a major program to modernize medical product regulation by the U.S. Food and Drug Administration (FDA). Related reforms are on the drawing board or already are being implemented in various nations around the world.

This chapter does not dwell on legal minutiae of how specific nations are choosing to update their laws. Instead, it identifies a common set of legal challenges that all nations will confront as pharmacogenomics makes its way from the laboratory to the clinic to the community. Different nations ultimately may adopt different legal solutions to these challenges, but the challenges themselves are global in character because they reflect a science-driven shift in perspective. A goal of this chapter is to acquaint readers with options that are available for addressing these challenges so that they can become active and knowledgeable participants in the search for fair and workable solutions.

REGULATION IN THE ERA BEFORE PHARMACOGENOMICS

It is helpful to start by reviewing several core principles of drug regulation as it existed late in the twentieth century, before pharmacogenomics entered its period of rapid progress. Pharmacogenomics, by challenging these principles, sparked the process of modernization now underway. After 1960, all industrial nations that regulate drugs adopted approaches that share a common feature: before a new drug can be sold to the public, regulators conduct a data-driven review of the drug's risks and benefits. The United States embraced this approach in 1962 when Congress amended the Food, Drug, and Cosmetic Act[6] to require substantial evidence of efficacy as well as safety.[7,8] In response, the FDA implemented the three-phase premarket clinical study process that continues in use today.[9,10] (Figure 9.1[11]) The FDA

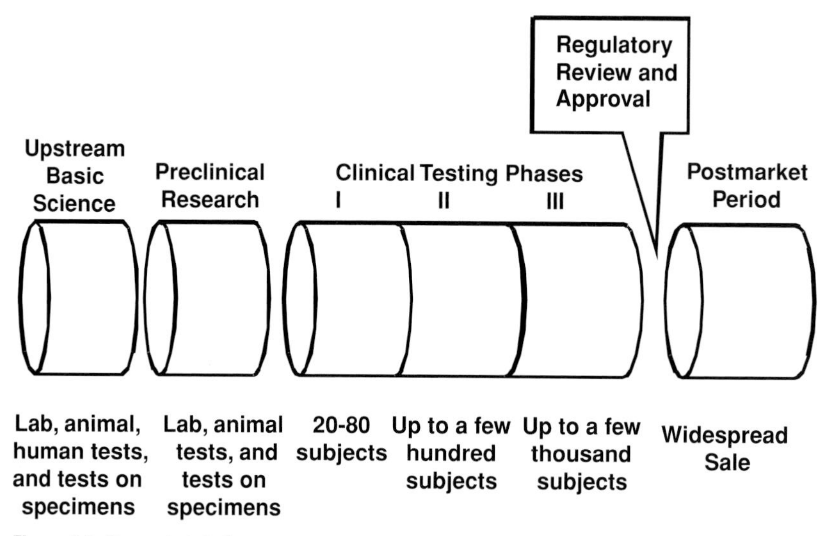

Figure 9.1. Premarket study process.

and its counterpart regulators in nations that participate in the International Conference on Harmonisation devised standard methodologies for conducting premarket risk-benefit analysis of new drugs.[12]

Pharmacoeconomic analysis was not part of this mid-twentieth-century approach to drug regulation. Internationally, the modern trend is for regulators to require cost-effectiveness data during their premarket review processes, although wide variation still exists in how regulators apply these data in their actual decision making.[13] In the United Kingdom, where 95 percent of drugs are purchased with public funds, the National Center for Health and Clinical Excellence (NICE) conducts rigorous analysis of cost-effectiveness to help decide which drugs should be provided through the National Health Service.[14,15] Some nations' regulators collect economic information without systematically incorporating it into their approval processes. The United States was slow to join this modern trend, although comparative effectiveness research now is receiving increased attention and funding. Drug approval decisions are separate from reimbursement decisions in the United States, with the latter decisions spread among many different private insurers and state and federal agencies. As a result, economic data are not required during the drug approval process. Risk-benefit analysis along the dimensions of safety and effectiveness became a common feature of twentieth-century drug regulation throughout the world, but analysis of cost-effectiveness and comparative effectiveness has not yet been universally embraced.

The twentieth-century approach to drug regulation concentrated research and regulatory oversight in the premarket period before drugs are approved.[16] This pattern is apparent in 2005 statistics for the FDA, which had 1,000 employees working to review the few dozen new drug applications that come before the agency each year, whereas only 100 professional employees were involved in postmarket monitoring for the approximately 3,000 prescription drugs and 8,000 over-the-counter medications already on the market.[17] In many countries, including the United States, regulators had little power to require ongoing studies after drug approval.[18,19] A core assumption of twentieth-century regulation was that premarket studies are capable of detecting a drug's material risks and benefits before it goes on the market. This assumption grew questionable because of multiple incidents where new safety risks came to light after drug approval.[20]

As shown in Figure 9.1, premarket drug studies typically include preclinical investigations[21] (laboratory and animal studies) and three phases of clinical studies in humans.[22] The required phase I and phase II studies often are called "trials" but they usually are not randomized, controlled clinical trials (RCTs) in a true sense of the word.[23,24] Phase III drug trials are RCTs, typically lasting 1 to 4 years and including up to 10,000 subjects,[25,26] although 600 to 3,000 would be more typical.[27,28] In most instances, no more than a few hundred subjects actually consume the test drug for longer than 3 to 6 months.[29,30] Obviously, such trials are not likely to detect risks and benefits that are rare or that occur only after prolonged consumption of a drug.[31–33] Boxed warnings later have to be added to approximately 10 to 20 percent of approved drugs.[34,35] FDA has had to withdraw approvals for about 75 drugs since the late 1960s[36] or about 2 percent of the approvals it issued during that same period.[37] This experience has been similar in other nations.[38] Some authors note an upward trend in recent years[39] and estimate the current withdrawal rate at approximately 4 percent.[40] These statistics are not surprising,

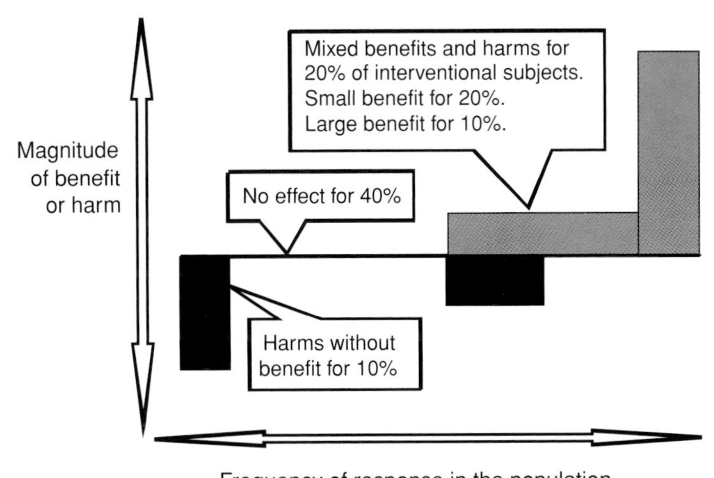

Figure 9.2. Heterogeneous response within the interventional group in a clinical trial.

given the relatively small size and brief duration of premarket studies on which regulators base drug approvals. Unfortunately, the public and many physicians do not understand the methodological limitations of premarket drug studies and often overestimate the reliability of risk-benefit data available at the time drugs are approved.[41,42]

Another core assumption of twentieth-century drug regulation was that people who are alike in terms of traditional trial selection criteria (i.e., broad demographic and clinical criteria such as age, gender, disease state, and comorbidities) are comparable to each other, so that variations in their drug response can be ascribed to random chance.[43] During the twentieth century, this assumption justified regulators' reliance on *average* statistics to characterize the safety and effectiveness of drugs. Clinical drug trials always have observed wide variations in individual response to drugs.[44] Figure 9.2[45] schematically depicts the range of responses that might be observed in the interventional group of a phase III drug trial – that is, among the people who took the test drug.

The horizontal axis indicates how often various responses occurred among people taking the drug. The vertical axis records the magnitude of benefits and harms that various members of that group experienced. In this graph, 10 percent of the interventional subjects (at left) experienced harm (in black) without receiving any therapeutic benefit. A sizeable group of people in the middle of the chart were nonresponders who were not hurt, but also were not helped by the drug. The remaining subjects derived some benefit from the drug (shown in gray). From left to right, some subjects had mixed benefits and harms; others escaped adverse effects and benefited slightly; still others (at far right) had a strong favorable reaction.

Twentieth-century drug approval criteria, which continue in use today, require proof of safety and effectiveness only at the level of the average person who took the drug during premarket studies.[46] A drug can be deemed "safe" if it exhibits a favorable ratio of benefits and risks. This condition is met if the regulator thinks the aggregate benefits in the interventional group (the total gray area in Figure 9.2) outweigh the collective harms (the total black area) in that same group. A drug can be deemed effective if it shows *any* therapeutic benefit. This condition is met if the gray area in Figure 9.2 is larger than the corresponding gray area observed in the control group who took the different treatment (or a placebo) with which the drug was being compared. A few test subjects who experience large benefits (such as the people on the far right side of Figure 9.2) may cause the drug to appear effective on average.[47] Both safety and efficacy are judged in terms of average effects.

The drug in Figure 9.2 was not safe and effective for many of the trial subjects who took it. The drug's labeling would need to list the types of harm observed. However, labeling does not have to disclose the skewed distribution of risks and benefits: for example, how few of the research subjects actually benefited; how many subjects received harms without benefit; how many were nonresponders; and how the magnitude of harms and benefits varied within the group.[48] Labeling suppresses virtually all of the granular information in Figure 9.2 in favor of average findings. A core assumption of twentieth-century regulation was that average responses are meaningful indicators of a drug's safety and effectiveness for individuals.[49] Another core assumption was that members of the public and physicians are not capable of applying granular risk-benefit information and need a paternalistic regulator to distill information into average statistics that are easy to grasp even if misleading.

Advances in pharmacogenomics have undercut the assumption that heterogeneity of drug response reflects random chance.[50] This, in turn, has pulled the legs out from under twentieth-century drug regulation. "As scientific knowledge advances, ability to classify is improved. Today's homogeneous group may be considered heterogeneous tomorrow."[51] Random chance explains some of the variation in Figure 9.2, but much of it reflects individual genetic and metabolic differences that, increasingly, can be explained and predicted. Average statistics are misleading if there are a priori reasons to expect individual variation.[52,53]

Why did the world's regulators embrace average statistics in the first place? Two factors explain this decision.[54] The first was scientific in nature. Fifty years ago, when today's drug approval processes were designed, factors explaining variable drug response were poorly understood. Treating the variations as random may have been an appropriate response to this ignorance, although, arguably, much more could have been done to study the problem and to educate the public about it.[55] The U.S. Congress heard testimony in 1979 that most FDA-approved drugs were effective in only 35 percent to 70 percent of patients,[56] yet a large segment of the public has continued to believe that approved drugs carry no risk[57] and are usually effective. The second factor was political compromise. Few people today remember that Congress, in 1962, instructed FDA *not* to concern itself with individual variations in drug response and to focus instead on average safety and effectiveness.[58] In the Senate Report to the 1962 Drug Amendments, Congress expressed its intent that "substantial evidence" of safety and efficacy should not imply "identical results for different patients"[59] and that physicians, rather than drug regulators, should be responsible for deciding what is effective at the level of individual patients.[60] This reflected a pragmatic compromise with key interest groups that had vigorously opposed letting the FDA regulate drug efficacy at all. Those stakeholders included drug manufacturers concerned that the proposed efficacy requirements would be unachievable; the medical profession, which traditionally had been responsible for assessing drug efficacy and, in 1962, lobbied heavily against governmental regulation of efficacy;[61] and state medical practice regulators wary of federal incursions on their terrain. This compromise set the course of drug regulation for the next fifty years in the United States and in many other nations where regulators were influenced by the FDA's approaches.

THE CHALLENGE OF EVIDENCE DEVELOPMENT

Clinical translation of pharmacogenomics requires thousands of small decisions: a decision by an insurer to pay for this or that pharmacogenetic test; decisions by regulators to require pharmacogenetic information in drug labeling or to allow the sale of a pharmacogenetic test; decisions by professional organizations to include pharmacogenetic testing in their clinical guidelines; even decisions by courts to hold clinicians liable when they fail to test patients whose injuries might thus have been avoided. Decisions favoring pharmacogenomics will not happen without high-quality evidence.[62–64] Many beneficial tests are underutilized because they lack evidentiary support to overcome these hurdles. Even when pharmacogenomic tests do enter the clinic, there often are unresolved questions about their clinical validity and utility. This situation adversely affects clinical decision making and health outcomes and creates a risk that poorly performing tests may be overutilized.

A central problem for biomedical scientists and regulators is how best to generate the necessary evidence. The perceived quality of medical evidence varies depending on the study design.[65–67] In the latter half of the twentieth century, RCTs (and meta-analyses of RCTs) came to be viewed as the highest-quality medical evidence, and many regulators adopted evidentiary standards that require RCTs as the basis for their decision making.[68] An advantage of RCTs is that they "can distinguish treatment effects from the noise of human variability,"[69] yet, where pharmacogenomics is concerned, the variability itself is the matter of interest. When the "noise" is the focus of scientific and regulatory attention, are RCTs still the best source of evidence? *The answer depends not just on the level of scientific certainty each methodology can provide, but also on its ethical acceptability and its logistical and economic feasibility.*

In an age of personalized medicine, there may not be a single source of evidence or methodological "gold standard" that is ideal for answering all questions about all medical products for all genomes. Pharmacogenomics has been described as "the right drug, at the right dose, for the right patient." Generating evidence of individual and subgroup benefits and risks requires "the right methodology, at the right time, for the right product."[70] "What is the best evidence?" is a question that will have to be answered on a case-by-case basis. The question has two subparts: *when* to generate evidence (premarket vs. postmarket)[71] and which methodology to use.[72]

Timing of Research

One of the significant trends now emerging is an increase in research and regulatory effort during the postmarket phase of the pharmaceutical life cycle. This trend will challenge clinicians to assimilate continuously evolving evidence from postmarket studies into their clinical practices. They will face day-to-day ethical quandaries, such as whether to inform patients about an emerging risk that

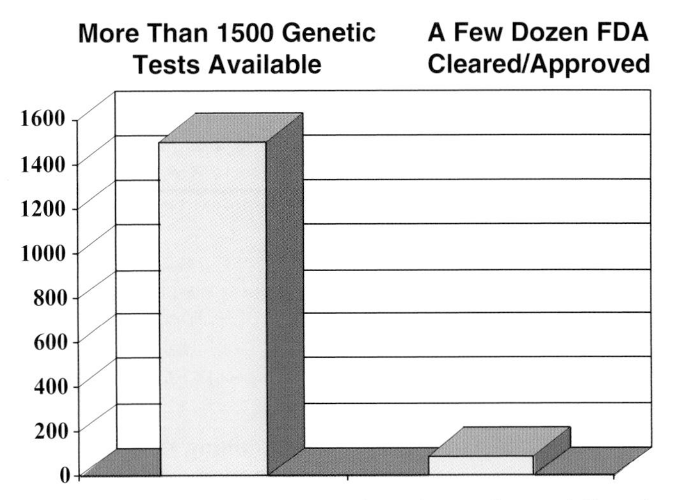

Figure 9.3. FDA oversight of genetic tests (including genetic tests of all types).

is not yet well substantiated, or how to portray the degree of uncertainty about a drug's benefits and risks. In particular, how should the uncertainties of pharmacogenetic testing be communicated to patients, when insurance coverage for such tests is spotty and many patients may be paying for tests from their household budgets? The line between "treatment" and "research" will grow blurrier in future years as studies to assess – and to improve – safety and efficacy continue throughout the commercial life of a drug. This trend also has implications for academic medicine. Institutions that develop infrastructure for postmarket studies (e.g., by navigating the privacy, ethical, and technical barriers to assemble large interoperable health data networks or research-ready tissue repositories) will enjoy competitive advantages in the future environment for funding of postmarket studies. Just as contracts to conduct premarket clinical drug studies became a major source of funding for academic medicine during the 1980s and 1990s, postmarket studies are poised to play a similar role in coming years. Institutions that have positioned themselves to garner funding for traditional premarket clinical trials will still have work, but institutions that tailor their infrastructure for the emerging wave of postmarket studies may overtake them or even pass them by.

Several factors support this trend. While it may seem desirable, from a public-safety standpoint, to require exhaustive pharmacogenomic study of drugs and rigorous validation of pharmacogenetic tests *before* they go on the market, genuine obstacles to this approach exist.[73] Pharmacogenomic discovery does not fit well into the traditional twentieth-century pathways of product development[74,75] and technology assessment.[76] There are logistical and economic barriers to imposing drug-like premarket clinical trial requirements on the in vitro diagnostic (IVD) tests used to screen patients before administration of a drug.[77] Many tests have short product life cycles,[78] with new-generation tests replacing older ones in as few as 12 to 18 months.[79] When technology is rapidly evolving, RCTs may fail to deliver timely results.[80] Tests may be obsolete by the time they are fully validated. Test innovators often are thinly capitalized entrepreneurs[81] lacking resources to bear the costs and delays of rigorous premarket assessment. There have been calls for public funding of clinical trials to validate all tests destined for clinical use,[82] but, to date, adequate funds have not been made available. A major concern, then, is how to protect the public from poorly validated tests and inaccurate claims about the benefits.

Clinical trials have never been the norm for evaluating IVD tests.[83] In the United States, which reviews tests as rigorously as any country does, the FDA regulates genetic and molecular diagnostic tests as medical devices under the 1976 Medical Device Amendments.[84] If a device is novel or poses significant risks, it falls under FDA's premarket approval requirements[85] that frequently require at least one premarket RCT.[86] A less-rigorous clearance process[87] applies to devices that present less novelty and risk.[88]

Many genetic tests, including pharmacogenomic tests, come to market through this clearance process. In general, it does not require RCTs, although some data are required to support the risk classification and to validate any analytical or clinical claims the sponsor intends to make about the test. As shown in Figure 9.3, more than 90 percent of genetic tests available in the United States are not regulated by the FDA.[89] They are laboratory-developed tests regulated under a different law, the Clinical Laboratory Improvement Amendments of 1988,[90] which does not require a data-driven regulatory review to confirm clinical validity and utility before a test can be used in clinical care. Even when the FDA does regulate a test, off-label use is allowed,[91] and well-validated tests can be diverted into new uses for which their performance characteristics are unknown.

In 1997 and in 2000, U.S. government advisory bodies recommended that all genetic tests destined for clinical use should receive evidence-driven, external regulatory review to confirm analytical and clinical validity and clinical utility.[92,93] International bodies made similar recommendations. The implementation of this recommendation proved unworkable and, in 2008, a successor body[94] modified it somewhat. It proposed that tests be stratified according to the risk they pose, so that regulatory oversight would be concentrated on the higher-risk tests.[95] Details of this approach (such as the criteria for characterizing risks and appropriate review levels for each risk category) have not been worked out as of this writing.[96] Now and in the foreseeable future, clinicians will be faced with applying tests amid gaps in evidence as to whether the tests actually improve health outcomes.

There has been similar difficulty developing evidence of the safety and efficacy of drugs at the individual and subgroup levels. In theory, clinical trials could be expanded to include multiple subgroups stratified according to every factor believed to influence drug response.[97] However, the required trials would be very large and could pose major costs and delays in getting new drugs to patients.[98] Under the current norms of voluntary informed consent, it might well be impossible to enroll the required number of trial subjects. Even if that were not a problem, it is not possible to identify ahead of time all the factors that may influence response to a particular drug, so there still would be unexpected variability of response after drug approval. Enrolling some genetic subgroups might pose ethical problems; for example, if there is a basis to suspect that they are at heightened risk of adverse effects. In short, there are practical limits to how much pharmacogenetic information can be generated in the premarket period before drugs are approved.[99]

Because of these problems, it is likely that drugs will continue to be approved based on average indicators of safety and effectiveness, with subgroup effects deferred for study during the postmarket period. This reality is reflected in FDAAA, the recent statute modernizing U.S. drug regulation.[100] Congress rejected calls to focus additional research and regulatory effort into the premarket period. The FDAAA leaves the FDA's existing premarket study requirements unchanged; in other words, they were not expanded. Instead, the FDAAA calls for a significant expansion of postmarket evidence development.

This approach is consistent with historical patterns of pharmacogenomic discovery. To date, there have been a few notable examples of prospective codevelopment of drugs and screening tests. Such an example was trastuzumab (Herceptin™),[101] for which the targeting mechanism was understood early in drug development before clinical drug studies began. However, this situation is not the norm. Factors explaining individual response to drugs often are not apparent until late in clinical trials or even after the drug is approved and consumed by large numbers of people.[102] This makes the postmarket phase of the drug life cycle the more promising period for pharmacogenomic discovery and translation of the discoveries. Accordingly, the future trend will be for regulators to increase postmarket research requirements and oversight efforts.

Determining the "Best" Methodology for Postmarket Studies

Once a drug is approved and moves into wide clinical use, various options exist for developing further evidence of its risks and benefits and for validating pharmacogenetic tests for use together with the drug. In addition to postmarket RCTs, there is a possibility of conducting various types of observational studies that follow health outcomes in patients who actually have taken a drug or have been screened using a particular test. RCTs and observational studies offer distinct advantages and drawbacks from a purely methodological standpoint. A major question for regulators, insurers, and other decision makers is when to accept alternative postmarket study methodologies and when to insist on postmarket RCTs.[103] *These choices will affect the cost and speed of evidence generation and, hence, the pace of clinical translation of pharmacogenomics.*

Postmarket RCTs produce high-quality evidence that is essential in certain contexts.[104] Unfortunately, RCT evidence often is unavailable because of logistical, ethical, and economic problems.[105,106] For example, many clinicians regard randomization as ethically problematic even during premarket trials.[107] These ethical concerns with RCTs intensify when studying an approved drug that already has substantial data to support its safety and effectiveness.[108] Once a drug moves into commercial use, it can be logistically difficult to enroll trial subjects, because patients who want the drug need not enroll in a trial to get it.[109–111] Off-label prescribing is lawful for most drugs in most countries and provides a pathway of access even for unapproved uses. It is not always practicable or economically feasible to conduct RCTs of the scale or duration that would be required to supply timely information about rare subgroup effects or about long-term risks and benefits. Postmarket observational studies that harness data from large numbers of patients already taking a drug may detect a risk more quickly than a postmarket clinical trial could have done,[112] may provide better insights into how a drug or test performs in the real-world clinical setting,[113] or may be ethically superior when randomization is problematic.[114]

Clinical trials and observational studies both have value during the postmarket period.[115,116] A recent Institute of Medicine workshop on genomic innovation discussed the need for flexible evidentiary standards that draw on multiple study methodologies.[117] The FDAAA adopted this view, expanding the FDA's authority to require postmarket RCTs,[118] while, at the same time, calling for alternative study methodologies to be considered first, before a postmarket RCT can be ordered.[119] Observational studies appear set to become a major source of postmarket evidence in the future. The regulator was given significant flexibility to decide the best way to clarify the risks and benefits of an approved drug, after receiving advice from an expert advisory committee through a public process[120] that has yet to be clarified.

Some nations are technically well positioned to conduct large-scale observational studies as a result of having centrally coordinated national health care delivery and payment systems that contain rich repositories of data.[121] In recent years, the first signals of new drug safety problems often have been detected by investigators in such nations.[122] Nations that lack large, interoperable health databases risk falling behind in a world where observational studies will be important sources of evidence to support both innovation and translation of discoveries. The United States, with its fragmented system of health care delivery and payments, is particularly at risk in this regard. There is growing acceptance that this fragmentation has the potential to harm patients and to produce economic waste. What has been less understood is that this fragmentation also is inimical to innovation, discovery, and translation to the extent that these will, in the future, depend on a nation's capacity to conduct large-scale observational studies. To help remedy this problem, the FDAAA authorized creation of a 100-million-person data network,[123] the Sentinel System,[124] to support postmarket drug safety surveillance[125] and advanced drug safety studies,[126] which are broadly defined and potentially could include pharmacoepidemiological studies of effectiveness as well as safety.[127]

All nations – even those that already have data readily available in platforms that support postmarket analysis of drug-related risks and benefits – face a common set of nontechnical challenges in this environment. The most obvious one is to develop suitable ethical and privacy frameworks that facilitate research and public health uses of patients' health data while maintaining public trust. The human subject protection frameworks now applied in all drug-innovating nations trace their origins to the 1960s and 1970s, when regulators around the world were first integrating clinical trial requirements into their regulations. Not surprisingly, these frameworks were designed primarily to address ethical issues in interventional research (clinical trials).[128] They were not specifically designed for two types of research that have gained prominence more recently: research with tissue specimens and with networked health databases. Tissue-based research is crucial in pharmacogenomics, because tissues are used both in basic scientific studies of factors that affect drug response and in the generation of evidence to support regulatory approvals for new pharmacogenetic tests.[129] Observational research advanced quickly after 1980 in response to various stimuli,[130–132] most notably, the improvements in information technology.[133] It relies on large administrative databases, including claims databases maintained by payers and insurers, and clinical data from actual health care encounters in which the patients typically have not consented to research or regulatory uses of their data.[134–136]

In an attempt to adapt their human-subject protections for these later developments, most nations have simply made minor adjustments to their existing consent-based frameworks. When applied to observational studies, these frameworks may fail to afford optimal protection. The subjects of observational studies often are "data-only subjects" – that is, they are patients whose previously collected health and insurance claims data are used in studies in which the patients will have no other involvement. Typically, the patients have consented to the medical treatments they received but probably have not consented to later studies that use their data. In many cases, it is not practicable to recontact former patients to obtain their consent. Even when it is practicable, existing informed consent frameworks may underprotect these data-only subjects in some respects, overprotect them in other respects, or provide misguided protections that fail to address the true concerns people feel about having their data used in observational studies. This problem will require further attention in coming years.

The underlying purpose of informed consent is to let research subjects control the type and amount of risk to which they are exposed. In interventional research, the nature of the study directly affects the risk. For example, it is material for a research subject to know whether the proposed study is of a drug that affects the endocrine system as opposed to a drug that is believed not to do so. Existing consent frameworks tend to dwell on providing detailed, protocol-specific information of this sort, because it bears directly on the subjects' risk exposure. However, this is not true in the case of data-only research subjects. The risks they face are not physical but privacy-related. Privacy-related risks depend on data-security arrangements, system architecture, and procedures for handling and disclosing data. The precise nature of the study being performed may have no bearing on the privacy-related risks it poses. Traditional informed consent regulations

Table 9.1. Types of Decisions That Are Critical to Clinical Uptake of Pharmacogenomics

Reimbursement approvals: decisions by governmental health care programs and private insurers to pay for pharmacogenetic testing or to condition payment for certain drugs on pharmacogenetic test results.
Labeling decisions: decisions by medical product regulators to include pharmacogenetic information in drug labeling or to allow a pharmacogenetic test to claim that it can be used to improve the prescribing of a certain drug.
Medical device approvals and clearances for pharmacogenetic tests: decisions to allow particular tests to be marketed for use in patient care.
Clinical practice guidelines: decisions made within the medical profession or by professional groups or medical-practice regulators to recommend the use of pharmacogenetic testing in particular contexts of patient care.
The tort standard of care: decisions by courts whether physicians and manufacturers of drugs and tests can be held liable for particular patterns of use (or nonuse) of pharmacogenetic tests or for failure to provide adequate information about the genetic basis of variability in patients' responses to drugs.

overprotect data-only subjects by requiring disclosure of information about the purposes of a study – information that often is irrelevant as the subjects attempt to understand their privacy risks. At the same time, these regulations fail to require uniform, standardized disclosures about the data-security arrangements that will apply, nor do the regulations establish a clear, understandable set of privacy conditions that a data-only subject always can count on. A more fundamental redesign of yesterday's ethical and regulatory frameworks is needed in the context of modern observational studies. This work already is underway and will be continuing in coming years.

Observational research traditionally required bringing data together into a large, central database.[137] This raised privacy risks because people's sensitive health data actually had to be transmitted from one place to another. The modern trend is toward using distributed data networks.[138] In other words, people's health data are left wherever they currently are (e.g., with the peoples' health insurers or health care providers) rather than moved to a central location. Questions are sent out to these various locations, and only deidentified summary answers are sent back for compilation at a central location.[139] This approach offers advantages from the standpoint of privacy.[140,141] Distributed data architectures of this sort have been adopted for FDA's Sentinel System and for pharmacoepidemiological systems now being developed in Canada and Europe.[142] However, even with this approach, some identifying information ultimately may need to be shared among the participating data sources to correlate records and achieve longitudinal linkage of data, and this raises privacy concerns.[143–145] Thus, although observational research does not subject patients to physical risk, it is not without risks to persons whose data are used.

When data are used for public health purposes, as would be the case for drug safety surveillance and other public health uses of large data networks, individual informed consents and privacy authorizations are not required under the Health Insurance Portability and Accountability Act of 1996.[146] Privacy Rule[147] and the Federal Policy for the Protection of Human Subjects (Common Rule).[148] This situation implies a need for strong data network governance structures to protect – and maintain the trust of – people whose data will be held in the networks.[149,150] The Institute of Medicine recently explored a governance-oriented approach to privacy protection, emphasizing system security and trusted intermediary arrangements.[151] Under this type of approach, for example, certain institutions would be certified to receive existing identifiable data from different sources, merge them into a common database or network structure, and then prepare deidentified datasets for researchers to use. Such arrangements have the potential to promote large-scale information-based research while still protecting the privacy of people whose data are used.

Another common legal challenge that nations will face in coming years will be to define evidentiary standards – that is, to decide which study methodologies should be required for which purposes. Table 9.1 lists various types of decisions that will have to be made as pharmacogenomics moves into widespread clinical use. These various purposes may call for different levels of evidence. The process of clinical translation risks being delayed by a "chicken-and-egg" problem: regulators and other decision makers, familiar with twentieth-century evidentiary standards, tend to demand RCTs before making decisions that promote wider clinical use of pharmacogenomics. For reasons already discussed, data from RCTs often are not available. Yet, until clinical translation occurs, there is no good body of outcomes data to support alternative evidentiary approaches such as large-scale observational studies. There is a risk that clinical translation may stagnate, with therapies lacking RCT evidence to move into the clinic and thus lacking clinical data to support the use of alternative study methods.

This evidentiary problem demands systematic study in its own right. A new evidentiary science is needed to support decision makers as they evaluate the question, "What is the best evidence for various purposes?" Their decisions will call for a framework to support rigorous comparisons among the available sources of evidence (e.g., RCT vs. observational studies in large, interoperable data networks), taking account of the incremental quality of scientific evidence each provides, its timeliness, and its costs in economic, logistical, ethical, and privacy terms. Without such a framework, clinical translation may remain mired in delays as regulators insist on forms of evidence that cannot feasibly be supplied.

COMMUNICATING AND APPLYING THE EVIDENCE

Once evidence is generated, two additional challenges remain: (1) to communicate continuously evolving risk-benefit information to clinicians and patients, and (2) to promote prescribing practices that conform to the best evidence available at any point in time.[152] Various reforms will be needed to meet these challenges.

Communicating the Evidence

Continuously evolving risk-benefit information implies that drug regulators need to be able to require labeling changes in the period after drugs are approved. This issue was addressed, for example, in recent amendments that strengthened the FDA's authority to require labeling changes during the postmarket period.[153] Similar changes can be expected in any nation that does not already grant its regulators the power to require postmarket labeling changes. However, this alone will not fully solve the problem. As they translate pharmacogenomics into the clinic, clinicians will need information that is more timely and nuanced than traditional printed labeling can provide.[154] Doctors, many of whom have had little training in genetics, will be expected to use, interpret, and advise patients about an expanding array of genetic and biomarker screening tests.[155] They will need current, application-specific information about which test to use with which drug and how to vary prescribing in response to test results.[156]

Drug and test labeling rarely meet this need. Very few cross-labeled drugs and tests currently exist.[157] Cross-labeling means that the drug's labeling identifies a specific screening test to use with the drug and provides instructions on how to apply test results when prescribing the drug.[158] When pharmacogenetic information is included in drug labeling at all, it often takes the form of a nonspecific warning that patient response may vary on the basis of certain genetic or other factors. Specific advice about how to test for those factors may be lacking. There are practical barriers to improving the amount of detail in drug labeling when multiple tests exist, when test technology is rapidly evolving, when there is incomplete evidence about the clinical impact of using a particular test, and when manufacturers are concerned about the commercial and liability impacts of linking their products to those of unrelated manufacturers.[159,160]

In the future, traditional labeling will continue to exist but will be supplemented by electronic communication media capable of supplying nuanced, detailed information on a more timely basis than traditional labeling can do.[161] Such systems already are being developed[162,163] and have the potential to provide timelier feedback of emerging evidence, including signals of potential risks that are not yet well enough verified to warrant labeling changes and, when available, information relevant to various population subgroups. Optimal risk-communication strategy is set to become a thorny policy issue in coming years. Premature communication of unsubstantiated risk signals could be commercially damaging to product manufacturers, who cite a danger that patients might be discouraged from taking beneficial drugs. On the other hand, it is paternalistic to keep physicians and patients ignorant of risk signals "for their own good." In a world where more and timelier risk information is available, clinicians will undoubtedly bear the brunt of helping patients make sense of large volumes of data. Clinicians also will have to consider the liability impact of failing to stay abreast of the evolving information or failing to apply it appropriately.

Promoting Compliance with Evidence

Simply communicating information about subgroup risks and benefits will not improve clinical outcomes unless the information actually is used to guide day-to-day prescribing decisions. Labeling changes traditionally have had minimal impact on clinicians' actual prescribing practices.[164,165] Similar problems of adherence are likely to continue even as electronic communication media are more widely adopted. One of the most difficult policy issues nations are facing is what to do when clinicians ignore well-substantiated evidence that certain subgroups of patients should not take a drug.

During the twentieth century, regulators adopted permissive policies on off-label use that have allowed physicians to disregard instructions and warnings included in product labeling.[166] Such policies made sense in the past, when regulators were confining their focus to the average safety and effectiveness of drugs.[167] Regulators were unable to offer insights about individual and subgroup safety and efficacy, so these matters were simply left to the clinician's judgment. However, as regulators start to provide more and better pharmacogenetic information, should clinicians still be allowed this same degree of latitude to disregard warnings and instructions in labeling? If not, what mechanisms should be used to promote physicians' compliance? As the science of pharmacogenomics improves, regulators will be forced to confront these questions.

When there are known factors that cause individual drug response to vary, a permissive policy on off-label use may cease to be appropriate.[168] When pharmacogenetic information is available, some off-label uses may be beneficial, whereas others may be quite risky or wasteful (e.g., giving a drug to patients known to be at high risk of adverse events or nonresponse). The off-label use policy itself will need to become more nuanced for drugs that have well-understood variability of individual response.[169] Certain "bad" off-label uses – such as giving a drug to a patient who clearly is not a good candidate to take it – may need to be discouraged or banned even as other off-label uses are tolerated.[170,171] Various mechanisms are available for discouraging these bad off-label uses: for example, clinical guidelines and practice standards maintained by the medical profession; denial of reimbursement by insurers and payers; increased tort liability for physicians who fail to adhere to sound prescribing practices; and even outright restrictions on off-label prescribing. In practice, nations are adopting a mix of these approaches. Different nations undoubtedly will arrive at different approaches best adapted to their particular legal systems. Gatekeeping by payers and insurers appears likely to play a prominent role in many nations, including the United States. This approach involves refusing to reimburse certain drugs unless screening tests indicate that the patient is a suitable candidate to take the drug.

The larger issue here is that pharmacogenomics is progressively blurring the line between medical product and medical practice regulation.[172] Drug manufacturers and regulators, increasingly, will be supplying information – such as whether a screening test should be used when prescribing a drug – that affects medical practice, setting up conflict at the interface between the medical profession and regulators. These concerns are particularly intense as drug regulators contemplate actual use restrictions that ban or restrict certain off-label uses of drugs.[173] In the United States – a nation that traditionally has gone to great lengths to keep drug regulators from intruding on physicians' discretion – Congress recently moved in the direction of giving the FDA greater control over how drugs are used in the clinic. The FDAAA lets the FDA condition the sale of a drug on specific risk-management measures by requiring a Risk Evaluation and Mitigation Strategy (REMS)[174] either at the time a drug is approved or at a later point in the drug's commercial life.[175] In situations where a drug offers benefits but poses serious risks that otherwise would cause the drug to be withdrawn from the market, a REMS can include use restrictions.[176] These can include, for example, a requirement for patients to be screened with laboratory tests before receiving the drug.[177] Thus, the FDA now has legal authority to prevent a drug from being prescribed without first conducting pharmacogenetic testing. As of this writing, the FDA has not yet approved any REMS that requires pharmacogenetic testing, but the legal authority is there to do so in the future.[178] It is also noteworthy that the FDAAA defines drug-related risks in a way that includes not just injuries that are directly caused by a drug's toxicity, but also could include injuries caused by failure of a drug's efficacy.[179] This potentially opens the door for nonresponse, and traditional drug-related injuries, as well, to be addressed through REMS use restrictions.

The concern already exists that REMS use restrictions constitute regulation of medical practice.[180] This issue arose before the FDAAA was passed,[181] at which time the REMS provisions were defended as striking a realistic balance between managing risks and intruding on medical practice.[182] From a purely legal standpoint, REMS provisions do not regulate physicians. Only the drug manufacturer (rather than the physician) is subject to legal penalties when a REMS is violated.[183] For example, the drug can be deemed misbranded, which allows the FDA to pursue sanctions such as seizure of the drug.[184] Even if a physician would not face fines or other legal penalties for violating the use restrictions in a REMS, it is still true that the "REMS could have the effect of limiting the ability of physicians who might use the drug for off-label uses to get their hands on the drug for their patients."[185] Thus, REMS restrictions have the potential to amount to de facto regulation of medical practice by limiting the availability of certain drugs for certain uses. Also, from a practical standpoint, much of the burden of complying with REMS falls on physicians. For example, physicians will be responsible for doing paperwork required by REMS, and it is they who must order pregnancy tests before prescribing certain teratogenic medical products.[186] Thus, it is something of a legal fine point to say that physicians are not "regulated" by REMS.

This raises a more general question: What is the best way to enforce the appropriate clinical application of pharmacogenetic testing? In the United States, medical malpractice lawsuits continue to be the primary mechanism for enforcing safe prescribing practices by physicians.[187] Other nations rely less heavily on this approach. Many physicians resist the notion of expanding tort liability as a way to create incentives for pharmacogenetic testing. If that option seems unattractive, there are two major alternatives to promote safe prescribing of drugs, in situations where there are known strategies that could reduce risks. First, drug regulators can increase their reliance on use restrictions that govern how drugs are prescribed and implement direct regulatory mechanisms to enforce these restrictions against physicians. Alternatively, the medical profession can embrace the task of policing itself through clinical guidelines and the development of effective oversight mechanisms to promote compliance with those guidelines. At this time, the medical profession still has an opportunity to take charge of this issue. Should it fail to do so, it can be assumed that regulators and tort lawyers will fill the breach.

ENDNOTES

1 Greenfield S & Kravitz RL. Heterogeneity of treatment effects: subgroup analysis. In: Olsen LA, et al., eds. Roundtable on Evidence-Based Medicine, Institute of Medicine, The Learning Healthcare System: Workshop Summary. 2007 (hereinafter IOM, Learning Healthcare). http://books.nap.edu/openbook.php?record_id=11903, 113–23, at 113.

2 See 21 C.F.R. § 314.126(b)(4) (2009) (discussing selection of human participants and variables pertinent in ensuring comparability of test subjects and control groups).

3 Evans BJ. Seven pillars of a new evidentiary paradigm: The Food, Drug, and Cosmetic Act enters the genomic era. *Notre Dame Law Rev.* 2010;**85**:419–24. http://www.nd.edu/~ndlrev/archive_public/85ndlr2/Evans.pdf; see pages 469–72.

4 Id. at pages 474–5.

5 Food and Drug Administration Amendments Act, Pub L No. 110–85, 121 Stat. 823 (2007) (codified as amended at scattered sections of 21 U.S.C.) (hereinafter FDAAA).

6 Federal Food, Drug, and Cosmetic Act, Pub L No. 75–717, 52 Stat. 1040 (1938) (codified as amended at 21 U.S.C. §§ 301–399) (hereinafter FDCA).

7 Drug Amendments of 1962 (Harris-Kefauver Act), Pub L No. 87–781, 76 Stat. 780 (codified as amended at scattered sections of 21 U.S.C.). See 21 U.S.C. § 355(d).

8 Roberts BS & Bodenheimer DZ. The Drug Amendments of 1962: the anatomy of a regulatory failure. *Arizona State Law J.* 1982;581–614, at 585.

9 See Investigational New Drug Application, 21 C.F.R. § 312.21.

10 Levitt GM, Czaban JN, & Paterson AS. Human drug regulation. In: Adams DG, et al., eds. *Fundamentals of Law and Regulation: An In-Depth Look at Therapeutic Products.* Washington, D.C.: Food & Drug Law Institute, Vol. **2**, 159–204, 1999 (hereinafter Fundamentals of Law and Regulation); see pages 165–6.

11 Evans BJ. What will it take to reap the clinical benefits of pharmacogenomics? *Food Drug Law J.* 2006;**61**:753–94, at 756. Figure 9.1 is reprinted with permission of *The Food & Drug Law Journal.*

12 See, e.g., Int'l Conference on Harmonisation of Technical Requirements for Registration of Pharm. for Human Use. Doc. No. ICH E3, Guideline for Industry: Structure and Content of Clinical Study Reports. 1996. (July 5, 2011): http://www.fda.gov/downloads/Drugs/GuidanceComplianceRegulatoryInformation/Guidances/UCM073113.pdf

13 Dickson M, Hurst J, & Jacobzone S. *Survey of Pharmacoeconomic Assessment Activity in Eleven Countries.* Paris: Organization for Economic Cooperation and Development; 2003.

14 Dodds-Smith I & Bagley G. *Global Counsel Life Sciences Handbook 2004/05.* London: Practical Law Company; 2004:77–85.

15 Rawlins MD. NICE work – providing guidance to the British National Health Services. *N Engl J Med.* 2004;**351**:1383–5.

16 Evans BJ & Flockhart DA. The unfinished business of U.S. drug safety regulation. *Food & Drug Law J.* 2006;61:45–63, at 45, 53–54.

17 Kessler DA & Vladeck DC. A critical examination of the FDA's efforts to preempt failure-to-warn claims. *Georgetown Law J.* 2008; **96**:461–95, at 485.

18 U.S. Government Accountability Office. Drug Safety: Improvement Needed in FDA's Postmarket Decision-making and Oversight Processes. Doc. No. GAO-06-402. 2006 (hereinafter GAO, Drug Safety), http://www.gao.gov/new.items/d06402.pdf.

19 Committee on the Assessment of the U.S. Drug Safety System, Institute of Medicine. In: Baciu A, et al., eds. *The Future of Drug Safety.* 2007 (hereinafter IOM, Future of Drug Safety), http://books.nap.edu/openbook.php?record_id=11750, at 153.

20 Evans, supra note 3, at 429–31.

21 21 C.F.R. § 312.23(a)(8).

22 Levitt et al., supra note 10, at 160, 165–66.

23 Committee on Identifying & Preventing Medication Errors, Institute of Medicine. In: Aspden P, et al., eds. *Preventing Medication Errors.* 2007 (hereinafter IOM, Preventing Medication Errors), http://books.nap.edu/openbook.php?record_id=11623, at 55–6.

24 Friedman LM, Furberg CD, & DeMets DL. *Fundamentals of Clinical Trials.* 3rd ed. New York: Springer Science + Business Media, L.L.C; 1998:3–6.

25 IOM, Preventing Medication Errors, supra note 23, at 56.

26 Friedman et al., supra note 24, at 181.

27 IOM, Future of Drug Safety, supra note 19, at 36.

28 Furberg BD & Furberg CD. *Evaluating Clinical Research.* 2nd ed. New York: Springer Science + Business Media, L.L.C.; 2007:19.

29 IOM, Preventing Medication Errors, supra note 23, at 56.

30 Friedman et al., supra note 24, at 181.

31 Furberg & Furberg, supra note 28, at 8, 19.

32 IOM, Preventing Medication Errors, supra note 23, at 56.

33 Evans, supra note 3, at 444–50.

34 Duh MS, Greenberg PE, & Weinder PE. The role of epidemiology in drug safety litigations. *Food & Drug Law Institute Update.* 2007;**Nov-Dec**:31–4.

35 Lasser KE, Allen PE, Woolhandler SJ, et al. Timing of new black box warnings and withdrawals for prescription medications. *JAMA.* 2002;**287**:2215–20.

36 Wysowski DK & Swartz L. Adverse drug event surveillance and drug withdrawals in the United States, 1969–2002. *Arch Intern Med.* 2005; **165**:1363–9.

37 Evans, supra note 3, at 428–29.

38 Lexchin J. Drug withdrawals from the Canadian market for safety reasons, 1963–2004. *CMAJ.* 2005;**172**:765–7.

39 Pillans PI. Clinical perspectives in drug safety and adverse drug reactions. *Expert Rev Clin Pharmacol.* 2008;**1**:695–705, at 700.

40 Duh et al., supra note 34, at 31.

41 Institute of Medicine. *Understanding the Benefits and Risks of Pharmaceuticals,* reporter Leslie Pray. 2007 (hereinafter IOM, Understanding Benefits), http://www.nap.edu/openbook.php?record_id=11910, at page 7.

42 Melmon KL. Attitudinal factors that influence the utilization of modern evaluative methods. In: Gelijns AC, ed. *Modern Methods of Clinical Investigation.* Institute of Medicine: 1990, http://www.nap.edu/openbook.php?record_id=1550, 135–45, at page 142.

43 Evans, supra note 3, at 474–5.

44 Greenfield, Kravitz, supra note 1, at 114.

45 Evans, supra note 11, at 763. Figure 9.2 is used with permission of *Food & Drug Law Journal.*

46 Id. at 762–4.

47 Kent DM & Hayward RA. Limitations of applying summary results of clinical trials to individual patients: the need for risk stratification. *JAMA.* 2007;**298**:1209–12.

48 Evans, supra note 3, at 473.

49 Friedman et al., supra note 24, at 34.

50 Evans, supra note 3, at 472.

51 Friedman et al., supra note 24, at 34.

[52] IOM, Understanding Benefits, supra note 41, at 16, 50.

[53] Epstein RA. Why the FDA must preempt tort litigation: a critique of Chevron deference and a response to Richard Nagareda. *J Tort Law.* 2006;**1**:32.

[54] Evans, supra note 3, at 501–3.

[55] Id. at 474, 502.

[56] The Food and Drug Administration's Process for Approving New Drugs: Hearings Before the Subcommittee on Science, Research, & Technology of the House Committee on Science and Technology, 96th Congress. 1979, at page 211 (statement of Dr. Gilbert McMahon).

[57] IOM, Understanding Benefits, supra note 41, at 1, 7.

[58] Evans, supra note 3, at 501–2.

[59] Senate Report No. 87-1744, reprinted in 1962 U.S.C.C.A.N. 2884, at 2921.

[60] Id. at 2920–1.

[61] Grow JA. *The Legislative History of the 1962 Drug Amendments* (May 1, 1997), http://leda.law.harvard.edu/leda/data/189/jgrow.pdf, at page 32 (citing hearings before the Subcommittee on Antitrust and Monopoly of the Senate Committee on the Judiciary, pursuant to S. Res. 52 and 5. 1552, Pt. 4, 87th Cong. 1997 (1961) (statement of Eugene N. Bessley, President of Eli Lilly & Co. and Chairman of the Pharmaceutical Manufacturers Association, reprinted in 19 FDA, A Legislative History of the Federal Food, Drug, and Cosmetic Act and its Amendments, at 810–11 (1979)).

[62] The Lewin Group, Inc. *The Value of Diagnostics: Innovation, Adoption, and Diffusion into Healthcare,* July 2005: at ch. 5, 92–125.

[63] Evans, supra note 11, at 777, 780–1.

[64] Evans BJ. Distinguishing product and practice regulation in personalized medicine. *Clin Pharmacol Ther.* 2007;**81**:288–93, at 291–2.

[65] Green SB & Byar DP. Using observational data from registries to compare treatments: the fallacy of omnimetrics. *Stat Med.* 1984;**3**: 361–70.

[66] Evidence-Based Medicine Working Group. Evidence-based medicine: a new approach to teaching the practice of medicine. *JAMA.* 1992;**268**:2420–5.

[67] U.S. Department of Health & Human Services, Office of Public Health & Science, Office of Disease Prevention and Health Promotion, Preventive Services Task Force. Guide to Clinical Preventive Services: Report of the U.S. Preventive Services Task Force. 2d ed. Washington, D.C.: U.S. Government Printing Office; 1996, http://odphp.osophs.dhhs.gov/pubs/guidecps/pcpstoc.htm, accessed July 2, 2011.

[68] Evans, supra note 3, at 435–9.

[69] Woodcock J & Lesko LJ. Pharmacogenetics – tailoring treatment for the outliers. *N Engl J Med.* 2009;**360**:811–13, at 811.

[70] Evans, supra note 3, at 488.

[71] Roundtable on Translating Genomic-Based Research for Health, Institute of Medicine. *Diffusion and Use of Genomic Innovations in Health and Medicine: Workshop Summary,* reporter Lyla M. Hernandez. 2008 (hereinafter IOM, Diffusion), http://books.nap.edu/openbook.php?record_id=12148, at page 62.

[72] Evans, supra note 3, at 486–91.

[73] National Cancer Policy Forum, Institute of Medicine. Developing Biomarker-Based Tools for Cancer Screening, Diagnosis, and Treatment: Workshop Summary, reporters Margie Patlak & Sharyl Nass. 2006 (hereinafter IOM, Biomarkers), http://books.nap.edu/openbook.php?record_id=11768, at 70–1.

[74] Evans, supra note 11, at 755–61.

[75] U.S. Department of Health and Human Services. FDA, Drug-Diagnostic Co-development Concept Paper. April 2005, http://www.fda.gov/downloads/Drugs/ScienceResearch/ResearchAreas/Pharmacogenetics/UCM116689.pdf.

[76] IOM, Biomarkers, supra note 73, at 60.

[77] Evans, supra note 3, at 466–7.

[78] IOM, Biomarkers, supra note 73, at 77.

[79] Evans, supra note 11, at 781.

[80] Friedman et al., supra note 24, at 8.

[81] Evans, supra note 11, at 789–90.

[82] IOM, Biomarkers, supra note 73, at 73.

[83] Aronson N. Assessing technology for use in health and medicine. In: IOM Diffusion, supra note 71, 29–33, at 30–1.

[84] Medical Device Amendments, Pub L No. 94-295, 90 Stat. 539 (1976) (codified as amended at 15 U.S.C. § 55 and scattered sections of 21 U.S.C.).

[85] See 21 U.S.C. § 360e; 21 C.F.R. §§ 814, 860.7.

[86] Holstein HM & Wilson EC. Developments in medical device regulation. In: Fundamentals of Law and Regulation, supra note 10, at 257–97; see page 282 (noting that evidentiary standards for devices subject to premarket approval are potentially less rigorous than for drugs, but that FDA often requires RCTs).

[87] See FDCA § 510(k); 21 U.S.C. § 360(k); 21 C.F.R. §§ 807.81–807.100.

[88] Nagareda RA. FDA preemption: when tort law meets the administrative state. *J Tort Law.* 2006;**1**:art. 1, at 8–12; accessible at http://www.bepress.com/jtl/vol1/iss1.

[89] Figure 9.3 reflects data from U.S. Department of Health and Human Services, Secretary's Advisory Comm. on Genetics, Health, & Soc'y, U.S. Dep't of Health & Human Servs., U.S. System of Oversight of Genetic Testing: A Response to the Charge of the Secretary of Health and Human Services. April 2008 (hereinafter SACGHS Oversight Report), at 39, *available at* http://oba.od.nih.gov/oba/SACGHS/reports/SACGHS_oversight_report.pdf.

[90] Clinical Laboratory Improvement Amendments of 1988, Pub L No. 100-578, 102 Stat. 2903 (codified as amended at 42 U.S.C. § 263a (2006)); see also 42 C.F.R. pt. 493.

[91] IOM Biomarkers, supra note 73, at 70.

[92] Task Force on Genetic Testing, Joint Nat'l Insts. of Health-Dep't of Energy Working Group on Ethical, Legal, & Social Implications of Human Genome Research, Promoting Safe and Effective Genetic Testing in the United States, at ch. 2 (Neil A. Holtzman & Michael S. Watson eds., 1997), *available at* http://www.genome.gov/10001733.

[93] Secretary's Advisory Comm. on Genetic Testing, Nat'l Insts. of Health, Enhancing the Oversight of Genetic Tests (2000), *available at* http://www4.od.nih.gov/oba/sacgt/reports/oversight_report.pdf. see Executive Summary at x.

[94] SACGHS Oversight Report, supra note 89, at 39, 65–7.

[95] Id. at 10–11.

[96] Id. at 136.

[97] Evans, supra note 3, at 475.

[98] Califf RM. Clinical trials bureaucracy: unintended consequences of well-intentioned policy. *Clin Trials.* 2006;**3**:496–502, at 498.

[99] Evans, supra note 3, at 453–7, 475.

[100] Id. at 477–9.

[101] Genentech. Herceptin (Trastuzumab) development timeline. http://www.gene.com/gene/products/information/oncology/herceptin/timeline.html.

[102] Evans, supra note 11, at 756–8.

[103] IOM Diffusion, supra note 71, at 62.

[104] IOM, Future of Drug Safety, supra note 19, at 105–6.

[105] IOM Diffusion, supra note 71, at 64.

[106] Furberg & Furberg, supra note 28, at 29.

[107] Friedman et al., supra note 24, at 48.

[108] Furberg & Furberg, supra note 28, at 29.

[109] IOM, Biomarkers, supra note 68, at 71.

[110] Weisman H, et al. Broader post-marketing surveillance for insights on risk and effectiveness, in IOM Learning Healthcare, supra note 1, at 128–34; see page 132.

[111] Friedman et al., supra note 24, at 7–8.

[112] U.S. Food and Drug Administration, U.S. Department of Health and Human Services. Proceedings, Sentinel Network Public Meeting, March 7, 2007 (hereinafter FDA, March 7 Proceedings), http://www.fda.gov/downloads/Safety/FDAsSentinelInitiative/ucm116513.pdf; see page 70 (statement of Richard Platt).

[113] Weisman et al., supra note 104, at 129.

[114] Evans, supra note 3, at 456–8.

[115] IOM, Future of Drug Safety, supra note 19, at 105–6.

[116] Furberg & Furberg, supra note 29, at 17.

[117] IOM Diffusion, supra note 71, at 82.

[118] See 21 U.S.C. § 355(o)(3) (empowering the FDA to require post-market studies or clinical trials of drugs with known or suspected safety problems); 21 U.S.C. § 352(z) (letting drugs be treated as misbranded if the manufacturer fails to carry out the required post-market studies).

[119] See 21 U.S.C. § 355(o)(3)(D) (requiring the FDA to consider alternatives, such as observational studies and studies using FDA's post-market risk identification and analysis system (known as the Sentinel System) before a postmarket clinical trial can be ordered).

[120] 21 U.S.C. § 355(k)(4)(C).

[121] Evans BJ. Authority of the Food and Drug Administration to require data access and control use rights in the Sentinel Data Network. *Food & Drug Law J.* 2010;**65**:67–112, at 74–75.

[122] U.S. Food and Drug Administration, U.S. Department of Health and Human Services. Proceedings, Sentinel Network Public Meeting, March 8, 2007, http://www.fda.gov/ohrms/dockets/dockets/07n0016/07n-0016-tr00002.pdf; see page 42 (statement of Dr. Robert M. Califf); id. at 23–4 (statement of Dr. Miles Braun).

[123] 21 U.S.C. §§ 355(k)(3)-(4); see, 21 U.S.C. § 355(k) (3)(B)(ii).

[124] U.S. Department of Health and Human Services. FDA, The Sentinel Initiative: National Strategy for Monitoring Medical Product Safety. May 2008, http://www.fda.gov/downloads/Safety/FDAsSentinelInitiative/UCM124701.pdf, accessed July 4, 2011.

[125] 21 U.S.C. § 355(k)(3)(C)(i)(I)-(VI).

[126] 21 U.S.C. § 355(k)(4)(A)(ii).

[127] Evans BJ. Congress' new infrastructural model of medical privacy. *Notre Dame Law Rev.* 2009;**84**:585–654, http://www3.nd.edu/~ndlrev/archive_public/84ndlr2/Evans.pdf; see pages 588, 601–4.

[128] Id. at 635–6.

[129] Evans BJ & Meslin EM. Encouraging translational research through harmonization of FDA and common rule informed consent requirements for research with banked specimens. *J Legal Med.* 2006;**27**:119–66, at 122.

[130] Agency for Healthcare Research and Quality (AHRQ). Outcomes Research Fact Sheet. http://www.ahrq.gov/clinic/outfact.htm, accessed July 5, 2011.

[131] Brenneman FD, et al. Outcomes research in surgery. *World J Surg.* 1999;**23**:1220–3.

[132] Evans, supra note 3, at 438–9.

[133] Irony T. Evolving methods: evaluating medical device interventions in a rapid state of flux. In: IOM, Learning Healthcare, supra note 1, at 93–9; see page 95.

[134] Institute of Medicine, Panel on Performance Measures and Data for Public Health Performance Partnership Grants. Data and Information Systems: Issues for Performance Measurement. In: Perrin EB, Durch JS, & Skillman SM, eds. *Health Performance Measurement in the Public Sector: Principles and Policies for Implementing an Information Network* 1999:84–93.

[135] Weissberg J. Use of large system databases, in IOM, Learning Healthcare, supra note 1, at 46–50.

[136] Brenneman et al., supra note 131, at 1220.

[137] Diamond CC, Mostashari F, & Shirky C. Collecting and sharing data for population health: a new paradigm. *Health Affairs.* 2009; **28**:454–66, at 456.

[138] Id. at 460.

[139] Platt R, et al. The New Sentinel Network – improving the evidence of medical-product safety. *N Engl J Med.* 2009;**361**:645–47.

[140] Evans, supra note 127, at 606.

[141] Evans, supra note 121, at 76–8.

[142] Id. at 74, 78.

[143] Evans, supra note 127, at 594–5.

[144] Rothstein MA. Health privacy in the electronic age. *J Legal Med.* 2007;**28**:487–501, at 489.

[145] Terry NP & Francis LP. Ensuring the Privacy and Confidentiality of Electronic Health Records, U. Ill. L. Rev. 2007;2007:681–735, at 700.

[146] Pub L No. 104-191, 110 Stat. 1936 (1996) (codified as amended at scattered sections of 18, 26, 29, 42 U.S.C.).

[147] 45 C.F.R. pts. 160, 164.

[148] 45 C.F.R. pt. 46, subpt. A., §§46.101–124.

[149] Evans, supra note 127, at 597–8.

[150] Evans, supra note 121, at 102–9.

[151] Committee on Health Research and the Privacy of Health Information: The HIPAA Privacy Rule; Institute of Medicine. Beyond the HIPAA Privacy Rule: Enhancing Privacy, Improving Health Through Research, eds. Nass SJ, Levit LA, & Gostin LO. 2009.

[152] Evans, supra note 3, at 503–15.

[153] 21 U.S.C. § 355(o)(4) (letting FDA order labeling changes after notice and period for response and discussions with the manufacturer).

[154] Evans, supra note 11, at 780–2.

[155] Evans, supra note 3, at 504–5.

[156] FDA, March 7 Proceedings, supra note 112, at 7–8 (statement of Dr. Janet Woodcock).

[157] Evans, supra note 11, at 780.

[158] Id. at 785–7.

[159] FDA/DIA Workshop: Combination Products and Mutually Conforming Labeling. May 10, 2005, http://www.fda.gov/CombinationProducts/MeetingsConferences-Workshops/ucm116623.htm.

[160] Evans, supra note 3, at 505–7.

[161] Id. at 507–8.

[162] See FDA, Postmarket Drug Safety Information for Patients and Providers. http://www.fda.gov/Drugs/DrugSafety/Postmarket-DrugSafetyInformationforPatientsandProviders/default.htm.

[163] 21 U.S.C. § 355(r).

[164] Smalley W, Shatin D, Wysowski, DK, et al. Contraindicated use of cisapride: impact of Food and Drug Administration regulatory action. *JAMA.* 2000;**284**:3036–9.

[165] Woosley RL, & Rice G. A new system for moving drugs to the market. *Issues Sci Technol Online.* Winter 2005, http://www.issues.org/21.2/woosley.html.

[166] See, e.g., 37 Fed. Reg. at 16,504 (July 30, 1972) (stating that FDA-approved labeling is not intended to impede the physician's exercise of judgment concerning what is best for the patient or to impose liability for prescribing decisions that are at odds with drug labeling).

[167] Evans, supra note 3, at 509.

[168] Id.

[169] Evans, supra note 11, at 784.

[170] Id. at 785.

[171] IOM, Understanding Benefits, supra note 41, at 16–17.

[172] Evans, supra note 64, at 288–93.

[173] Evans, supra note 3, at 510–15.

[174] 21 U.S.C. §§ 355(p), 355–1.

[175] Id. at. § 355–1(a)(1) – (2).

[176] Id. at § 355–1(f).

[177] Id. at § 355–1(f)(3)(D).

[178] See U.S. Department of Health and Human Services, U.S. Food and Drug Administration. Approved Risk Evaluation and Mitigation Strategies (REMS). http://www.fda.gov/Drugs/DrugSafety/PostmarketDrugSafetyInformationforPatientsandProviders/ucm111350.htm. Last visited November 22, 2009.

[179] Evans, supra note 127, at 601–2.

[180] Evans, supra note 3, at 521–3.

[181] Gottlieb S. Drug safety proposals and the intrusion of federal regulation into patient freedom and medical practice. *Health Affairs.* 2007;**26**:664–77, at 665–8.

[182] Enzi MB & Kennedy EM. Risk management and intrusions on medical practice: striking a balance. *Health Affairs.* 2007;**26**:678–80. http://content.healthaffairs.org/cgi/content/abstract/26/3/678.

[183] 21 U.S.C. §§ 355(p), 333(f).

[184] 21 U.S.C. § 352(y).

[185] Wilenzick M, Lietzan E, & Doebler S. Interview with Gerald Masoudi, Outgoing Chief Counsel of the U.S. Food and Drug Administration. American Health Lawyers Association Membership Briefing. February, 2009, at 2 (quoting Mr. Masoudi), http://www.healthlawyers.org/Members/PracticeGroups/LS/memberbriefings/Documents/Masoudi%20interview_Final.pdf.

[186] Gever J. FDA to step up regulation of extended-release opioids. *MedPage Today.* February 9, 2009.

[187] Evans & Flockhart, supra note 16, at 51.

PART II

Therapeutic Areas

Oncologic Drugs

10

Uchenna O. Njiaju and M. Eileen Dolan

Pharmacogenomics is of particular importance in oncology, a medical subspecialty characterized by rapidly lethal diseases and drugs with narrow therapeutic indices and significant toxicities. Identification of individuals likely to respond to or experience toxicity from a given chemotherapeutic agent, will have significant impact on outcomes, particularly in the field of oncology. Several models currently exist for discovery of pharmacogenomic markers in oncology. Phenotypic variations may range from variability in response as measured by survival or time to progression, to variability in toxicity in individuals treated with a particular agent. Measurements of toxicity can be a challenge to quantify in individuals because of interobserver variability. Lymphoblastoid cell lines (LCLs) and the NCI60 bank of tumor cell lines have been used as models for clinical phenotypes. To date, there are several examples of germline polymorphisms and somatic mutations that predict likelihood of response and/or toxicity from chemotherapeutic agents. A pattern of interethnic variability in response and toxicity has been observed for some chemotherapeutic agents, and the associated field of pharmacoethnicity is likely to contribute to our understanding of pharmacogenomics.

Pharmacogenomics has found extensive application in the field of oncology and is likely to remain an important tool in the race toward personalized medicine. Individualization of therapies is of particular importance in oncology because of several unique features of cancer treatment. First, most oncologic therapies have potential for organ toxicity and typically give rise to an array of potential life-threatening side effects. For example, taxanes are highly efficacious against malignancies of the lung, breast, ovary, and head and neck, but are also associated with significant toxicities such as myelosuppression and peripheral neuropathy. Identification of individuals unlikely to respond to taxane therapy a priori will be tremendously important in therapeutic decision making, because alternative therapies can be considered, thereby reducing the likelihood of unnecessary toxicity. Second, many oncologic diseases progress rapidly and are generally lethal in the absence of effective therapy. Consequently, prompt diagnosis and early institution of efficacious therapies is of paramount importance. In the absence of knowledge about predictors of response, individuals could be subjected to therapies to which their tumors might not respond, resulting in further disease progression. With more advanced disease and organ dysfunction, some therapies may no longer be given safely and may only serve a palliative rather than a curative role. For example, a five-year period of adjuvant tamoxifen therapy following successful treatment of early-stage estrogen receptor (ER)-positive breast cancer in a premenopausal woman is associated with a reduction in the rate of disease recurrence and mortality. A poor metabolizer phenotype results in insufficient conversion of tamoxifen to endoxifen and an increased risk of disease relapse and progression (1). Affected individuals may be better served by alternative antiestrogen maneuvers such as the combination of ovarian ablation and aromatase inhibitor therapy. Third, most chemotherapeutic agents have a fairly narrow therapeutic index (see Figure 10.1). The therapeutic index of a drug compares the dose that produces toxicity with the dose that produces the desired effect and, as such, provides a measure of the drug's safety. Given the significant likelihood of an adverse effect even within the therapeutic window, treatment with a particular agent is best reserved for individuals likely to respond, with the careful weighing of risks and benefits and informed decision making on the part of the patient. Finally, the expenses associated with oncologic therapies necessitate avoidance of therapy-related morbidity that further increases the likelihood of hospitalization and overall cost of care. For example, trastuzumab is an agent used in the treatment of HER2neu-positive breast cancer, is typically infused on a three-weekly schedule for at least one year, and may cost as much as $70,000 for a full course of therapy (2). Given all the aforementioned, it is not surprising that current pharmacogenomic research

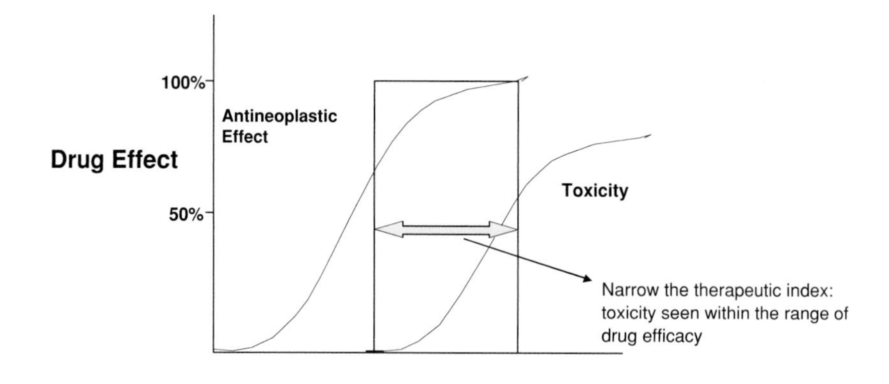

Figure 10.1. Illustration of therapeutic index.

is dominated by investigation of variability in response to, and toxicity from, oncologic therapies (3).

PHARMACOGENOMIC DISCOVERY IN ONCOLOGY

Germline Polymorphisms and Somatic Mutations

Unlike pharmacogenomic research in other medical subspecialties, variability in response to chemotherapeutic agents may be studied for association with variants in germline DNA, somatic DNA, or both. Neoplastic cells are characterized by malignant transformation, a process that involves significant alteration in the cellular genetic material, including changes in oncogene and tumor suppressor gene expression (4). Downstream effects include alteration in cellular signaling, proliferation, and apoptosis. These somatic mutations cause a change in the genotype of the resultant cancer cell, and are not present in nonmalignant cells. In contrast, germline variants are found in all cells, may be passed on to offspring, and can be expected to be identical in all cells of the same individual (see Figure 10.2).

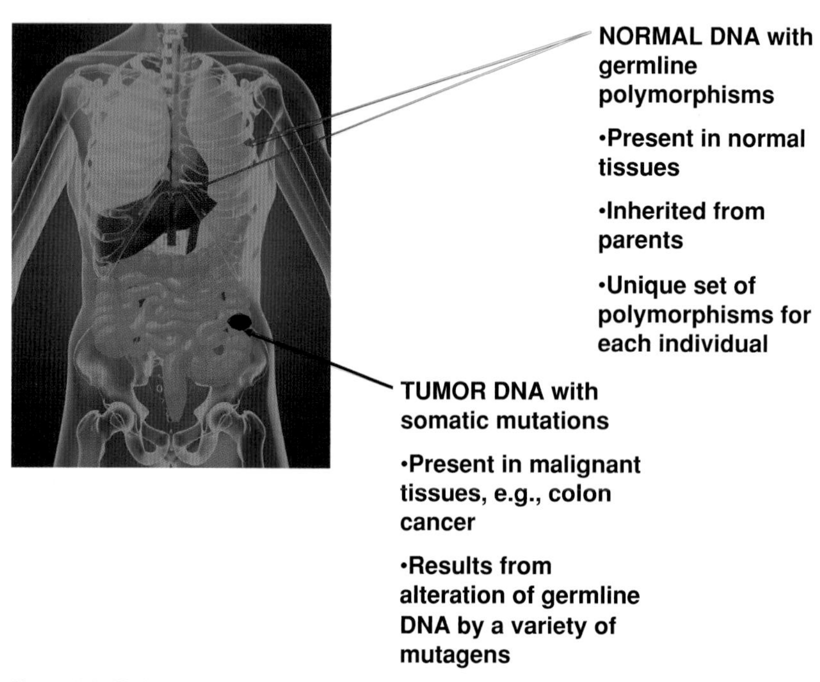

Figure 10.2. Distinguishing features of somatic mutations and germline polymorphisms.

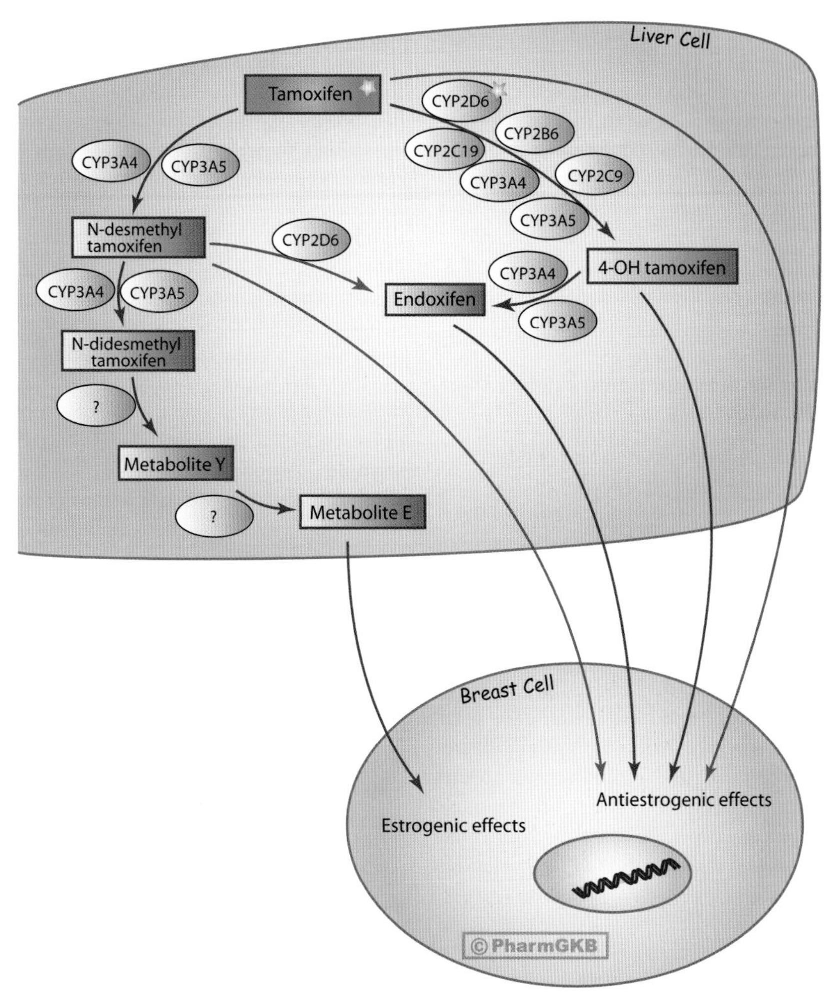

Figure 10.3. EGFR signaling pathway. Reproduced with permission from PharmGKB and Stanford University (68).

Somatic and germline mutations occurring in the same gene may give rise to different phenotypes. For example, the *TP53* gene is a tumor suppressor gene involved in several human malignancies. Germline mutations give rise to the Li-Fraumeni syndrome, which is characterized by predisposition to early-onset cancers such as breast carcinoma, sarcomas, brain tumors, and adrenocortical carcinomas. On the other hand, somatic mutations in *TP53* occur in almost every type of human cancer at rates as high as 38 percent to 50 percent, and tend to be more frequent when the cancer is of advanced stage or aggressive behavior (5). In addition, it has been demonstrated that somatic mutations can be highly variable even within the same tumor (6) and change as the tumor evolves (7). Because the efficacy of an antineoplastic agent is related to its ability to exert an effect on malignant tissues, studies of the variability in intended drug effect have historically relied on the analysis of somatic DNA and gene expression within the tumor (8). For example, mutations in the epidermal growth factor receptor (EGFR) gene have been found to correlate with response to EGFR tyrosine kinase inhibitors (Figure 10.3). In contrast, toxicity results from the effects on normal tissues and therefore is most likely predicted by germline polymorphisms and their effects on drug pharmacokinetics or pharmacodynamics.

Although germline polymorphisms in general are thought to affect drug toxicity, there is growing evidence that germline polymorphisms may also predict chemotherapeutic response (9). For example, host genetic variations are associated with treatment response for childhood acute lymphoblastic leukemia (ALL), with polymorphisms related to leukemia cell biology and host drug disposition associated with the lower risk of residual disease (10). In addition, genotypes in ethnic Asian patients with non-small cell lung cancer (NSCLC) are predictive of response to gemcitabine (11). Although variation exists, germline DNA and matched somatic DNA may also have significant concordance in variants of pharmacogenetic genes (12), a feature that supports a possibility of correlation between germline polymorphisms and intended drug effects. Given the relative ease with which germline DNA may be collected from healthy individuals and that it remains the same throughout an individual's life, there is great value in determining germline variants contributing to either response or toxicity.

Phenotypes in Pharmacogenomic Studies of Chemotherapeutic Agents

A wide range of phenotypes may be studied for chemotherapeutic agents both in the clinical and preclinical setting. Clearly, cancer patients are the optimal system for identifying genetic variants contributing to chemotherapeutic drug response and toxicity. The problem with human studies in oncology is the rarity in which a homogeneous patient population receives the same dose of a single chemotherapeutic agent. For measurements of chemotherapeutic toxicity, interobserver variability and difficulty in obtaining quantitative measurements is another factor. To get around these issues and avoid confounders such as comorbidities, concomitant medications, and diet, preclinical cell-based models have been developed that provide a useful discovery tool. Critical to the process is the validation of these markers in a clinical setting.

Clinical Phenotypes

Clinical phenotypes may be binary, such as death and survival, or continuous, such as bone marrow suppression. Because side effects may be subjective, a toxicity-grading system is often used. The Common Terminology Criteria for Adverse Events was developed by the National Cancer Institute and is the predominant system used to describe the severity of adverse events in oncology clinical trials (13). A myriad of toxicities may be associated with oncologic therapies and range from relatively benign to severe and life threatening. Examples of benign side effects include nausea, vomiting, alopecia, fatigue, and anorexia, all of which may be seen with most chemotherapeutic agents. Such side effects are often easy to control with symptomatic therapies like antiemetics, and typically do not have a major influence on the choice of chemotherapeutic agents. On the other hand, examples of severe toxicities include peripheral neuropathy from taxanes and epothilones, cardiomyopathy from anthracyclines and HER2-targeted agents, profound myelosuppression from several agents, central nervous system toxicity from ifosfamide and 5-fluorouracil (5-FU), and hemorrhagic cystitis from cyclophosphamide. The suppression of bone marrow cells can be considered a quantitative phenotype; however, the degree of suppression often depends on the frequency at which the measurement is taken. Peripheral neuropathy may be debilitating by affecting an individual's ability to perform activities of daily living, whereas cardiomyopathy and central nervous system toxicity may be life threatening. The latter category of side effects is of significant clinical importance, because their occurrence typically necessitates a change of therapy.

The measurement of the response to chemotherapy may take several forms, giving rise to a range of clinical response phenotypes for use in pharmacogenomic studies. For solid tumors, the response to therapy may take the form of a decrease in the size of a tumor mass (partial response) or the complete disappearance of a lesion (complete response). In addition, the lack of change in a lesion may signify stable disease. For standardization, the Response Evaluation Criteria in Solid Tumors guidelines are most commonly used, and use unidimensional measurements of target lesions before and after therapy to evaluate for a complete response, partial response, or progressive disease (14).

For hematologic malignancies, the response to therapy is assessed differently because such neoplasms are associated with the presence of aberrant cells in the blood and/or bone marrow rather than with distinct masses evident on radiologic imaging. For example, chronic myelogenous leukemia (CML) is characterized by the presence of large numbers of neoplastic myeloid cells bearing the Philadelphia chromosome or t(9;22) translocation that gives rise to the *BCR-ABL1* chimeric gene (15). Imatinib is standard therapy for CML and induces a high rate of hematologic and cytogenetic response. A complete hematologic response is defined as the attainment of a white blood cell count of $<10{,}000/\mu L$, no immature granulocytes and <5 percent basophils, and a platelet count of $<450{,}000/\mu L$ with a nonpalpable spleen. In contrast, a complete cytogenetic response is defined by the complete disappearance of Philadelphia chromosome positive cells. A major molecular response is present when the ratio of BCR-ABL transcript to housekeeping genes is ≤ 0.1 percent on an international scale (16). Additional parameters such as rate of generation of metabolites and overall drug exposure, as measured by the area under the drug pharmacokinetic curve, are related to drug pharmacokinetics and may also be used as phenotypes in clinical pharmacogenomic studies (17).

Preclinical Cellular Phenotypes

The selection of molecular phenotypes in cell lines that accurately reflect clinical drug response is a major challenge. The appropriate phenotype usually depends on the mechanism of action of the drug, and on the clinical phenotype of interest (18). For example, anticancer drugs are intended to cause growth inhibition, cell death, or apoptosis; therefore, measuring cellular apoptosis or cell growth inhibition across a range of drug dosages is generally performed (19). Another phenotype to consider is measurement of the conversion of parent drug to active metabolite. This has been effectively analyzed in the case of methotrexate glutamation (20) and the chemotherapeutic drug cytarabine, in which the amount of active metabolite (AraCTP) was associated with a specific genotype within an important drug-metabolizing gene (21).

There are a number of advantages to using cell lines derived from individuals for pharmacogenomic discovery. Cells can be grown under identical conditions, allowing the genetic contributions toward a specific phenotype to be tested in a well-controlled, isolated system without the confounders present in vivo. Cell lines offer ease of experimental manipulation and unlimited resources to study pharmacodynamic effects that would be considered unsafe in healthy volunteers. Despite the advantages of this ex vivo system, there are limitations to using cell lines to identify pharmacogenetic effects. These include the following: (1) Few cell lines are available from nonmalignant tissue, making it difficult to find an appropriate ex vivo system to study toxicities such as neurotoxicity, cardiovascular toxicity, and nephrotoxicity, to name a few. (2) Most cell lines do not have a cytochrome P450 system, making it difficult to study the pharmacokinetics of drugs that require metabolic conversion. (3) Transformation of lymphoblasts into LCLs or tumors into tumor cell lines could introduce phenotypic changes, which may result in expression differences with regard to the phenotype of study.

Approaches in Chemotherapy Pharmacogenomic Research

Whether using cell lines, animals, or humans for pharmacogenomic studies, there are primarily two approaches, the candidate gene approach, in which a gene or pathway is identified as potentially important based on literature evidence and then subjected to further study, and a hypothesis generating approach, in which the whole genome is considered and no assumptions are made about what genes are important. The sequencing of the human genome and the genetic resource provided by The International HapMap Project have allowed researchers to greatly expand the focus of pharmacogenomic studies to more routinely perform genome-wide studies. A major advantage to the genome-wide approach is the enormous amount of information gained; however, along with that information comes false-positive findings as a result of multiple testing on a large scale. A major advantage of the genome-wide approach is that it opens up the possibility of identifying previously unknown genetic variants that contribute to chemotherapy-induced cytotoxicity.

Candidate Gene Approach in Oncology

In this method, a single gene or genes within a pathway known to be important in the pharmacokinetics or pharmacodynamics of a particular drug are examined for genetic variability and compared with phenotypic variation. Such an approach has met with success in elucidating the pharmacogenetics of chemotherapeutic agents (Table 10.1). Some successful examples of candidate gene studies include (1) genetic variations in thiopurine methyltransferase (*TPMT*) associated with increased risk for severe myelosuppression after 6-mercaptopurine (6-MP) treatment (22); (2) *UGT1A1*28* associated with a decrease in UGT1A1 expression and increased risk of severe neutropenia when irinotecan is administered (23); and (3) lack of response to tamoxifen in *CYP2D6* poor metabolizers (24). Importantly, the candidate gene approach depends on the presence of a small number of alleles seen in a significant fraction of the general population that gives rise to a major alteration in drug effectiveness. In addition, the approach is most successful if the gene involves a key step in the drug metabolic pathway. Because such a scenario is not applicable to all chemotherapeutic agents, and because chemotherapy-induced response and toxicity are most likely multigenic traits, a broader approach is important in accurately elucidating genetic predictors of response and toxicity (25).

Genome-Wide Association Studies in Oncology

A genome-wide approach (genome-wide association study, or GWAS) takes the whole genome into consideration and uses an unbiased method to generate candidate genes that may be further subjected to functional evaluation and validation (26, 27). GWAS has been facilitated by completion of the sequencing of the human genome and the International HapMap projects. The Human Genome Project was launched in 1990 and completed in 2004, and served to provide an accurate sequence of the human genome as a foundation for genetic studies of disease and response to drugs (28). The International HapMap Project was initiated in 2002 to characterize common variations in DNA sequence among four different populations and to construct haplotype maps (29) but has been extended to study eleven additional populations.

Genome-wide studies of germline DNA have been used in generating genomic predictors of response to a variety of chemotherapeutic agents. As a result of the large number of polymorphisms studied for association, GWAS may result in a high false-discovery rate. This important limitation may be curtailed by validation of findings in an independent set of similarly treated cells or patients. Nevertheless, a few GWAS findings have been correlated with results obtained by using a candidate gene approach. For example, cytarabine is a chemotherapeutic

Table 10.1. Pharmacogenomics, of Chemotherapeutic Agents Derived by a Candidate Gene Approach

Tumor Type	*Chemotherapeutic Agent(s)*	*Molecular Target*	*Pharmacogenomic Issues of Importance*	*Clinical Significance*
Breast cancer	Tamoxifen	ER	At least seventy CYP2D6 allelic variants exist and give rise to poor, intermediate, extensive, and ultrarapid metabolizers with progressively higher concentrations of the active metabolite, endoxifen.	Breast cancer patients homozygous for the null allele, CYP2D6*4 have shorter time to relapse and worse disease-free survival than those with CYP2D6*4/*1 or CYP2D6*1/*1 genotypes. No CYP2D6*4/*4 patients experienced moderate or severe hot flashes, whereas approximately 20% of women with CYP2D6*4/*1 or CYP2D6*1/*1 do experience hot flashes.
Colorectal cancer	Irinotecan Monoclonal antibodies against EGFR (EGFR-I), e.g., cetuximab, panitumumab	Topoisomerase I EGFR	The UGT1A1*28 allele, characterized by seven TA repeats in the promoter region, results in reduced activity of the UGT1A1 enzyme, with consequent accumulation of the toxic metabolite, SN-38. Gain-of-function mutations in the KRAS gene involved in downstream signaling result in bypassing of the EGFR signaling pathway.	Several studies show that individuals homozygous for the UGT1A1*28 allele are more predisposed to late irinotecan toxicity manifesting as diarrhea, neutropenia, or both. Patients with KRAS-mutated tumors are resistant to therapy with EGFR-I agents.
NSCLC	EGFR-tyrosine kinase inhibitors (EGFR-TKI), e.g., gefitinib and erlotinib	EGFR-tyrosine kinase	Somatic mutations in EGFR gene result in altered function of the associated tyrosine kinase.	Somatic mutations in EGFR in NSCLC are highly correlated with response to gefitinib and erlotinib, particularly among nonsmoking Asian females.
Multiple malignancies	5-FU	Dihydropyrimidine dehydrogenase Thymidylate synthetase	Several sequence variations in the dihydropyrimidine dehydrogenase gene (DPYD) have been described; however, only the relatively infrequent DPYD*2A and DPYD*13 alleles have been consistently associated with DPD deficiency. Variants such as a 6 base pair insertion and deletion polymorphism in the 3′-untranslated region, and a variable number of tandem repeats in the promoter-enhancer region, lead to increased expression of the thymidylate synthetase gene (*TS*).	Mutations resulting in DPD deficiency result in increased likelihood of potentially life-threatening toxicity from 5-FU. High *TS* gene expression variants are associated with decreased survival in colorectal cancer patients treated with 5-FU.
A variety of hematologic malignancies, e.g., childhood and adult ALL, childhood AML, childhood non-Hodgkin's lymphoma	6-MP and 6-thioguanine (6-TG)	Purine analogs, antagonists to endogenous purines required for DNA synthesis in the S-phase of the cell cycle	Azathioprine is converted to 6-MP. Both 6-MP and 6-TG are catabolized by thiopurine methyltransferase (TPMT). Seventeen mutant TPMT alleles have been described, some of which give rise to intermediate or low enzyme activity.	Low TPMT activity results in failure to catabolize purine analogs, with consequent life-threatening toxicity, including myelosuppression.

agent used in the treatment of hematologic malignancies such as adult acute myeloid leukemia (AML). The rate-limiting step in cytarabine catabolism is catalyzed by the enzyme deoxycytidine kinase (DCK). Earlier research demonstrated that low levels of DCK mRNA in blast cells correlated with poorer outcome as shown by a shorter disease-free and overall survival (30). In addition, clinical studies showed that low intracellular levels of cytarabine in leukemia cells resulted in similarly poor outcomes (31, 32). Consistent with clinical findings, a GWAS study using LCLs showed that single-nucleotide polymorphisms (SNPs) within the gene *DCK* resulted in increased enzyme levels and heightened sensitivity to cytarabine (21).

Advances in molecular biology and bioinformatics coupled with the development of methods for high-throughput analysis have made it feasible to study large numbers of individuals in GWAS. Such studies may be conducted clinically by genotyping individuals with a variable drug response; however, there are important considerations in clinical GWAS. First, because variation in response to most clinically administered drugs depends on the combined contribution of multiple genes, one must consider sample size for a study to have adequate statistical power. Clinical GWAS studies are expensive and time consuming, and require large numbers of patients and infrastructure to obtain reliable clinical phenotype data. Establishing a prospective cohort can take years because of the time required to meet regulatory requirements, to accrue a population of sufficient size, and for follow-up analysis. Although samples from retrospective clinical trials require fewer resources, in general, they are not powered to answer specific pharmacogenetic questions. This problem is further compounded by the need for multiple large patient cohorts to enable both discovery and replication studies. In addition, confounding factors such as comorbidities, dosage, timing of drug administration, and diet are difficult to standardize and cannot be easily separated from genomic contributions to variation in drug response. Uncontrolled confounders, including population stratification or admixture, can bias measured effect estimates of genotype-phenotype relationships. Finally, pharmacogenetic discovery for highly toxic drugs, such as chemotherapeutics and certain antiviral agents, poses additional challenges because these drugs cannot be administered to healthy individuals for classical genetic studies. For the reasons just mentioned, some researchers have turned to the use of human cell-based models for pharmacogenetic discovery and validation studies (18).

Cell-Based Models in Chemotherapy Pharmacogenomic Discovery

For oncologic research, the examples of cell lines used include those from healthy individuals, and those derived from tumors from humans. As described earlier, the DNA and certainly expression of genes can vary considerably between normal tissue and tumor. Lymphoblastoid cell lines and the NCI60 bank of tumor cell lines are the most frequently used cell-based models for pharmacogenomic discovery. The NCI60 bank of cancer cell lines is derived from multiple human tumors and contains somatic DNA. As part of the International HapMap Project, LCLs were collected in phases from distinct world populations, including whites from Utah (CEU), Yorubas from Nigeria (YRI), Chinese from Beijing (CHB), and Japanese from Tokyo (JPT). LCLs are commercially available, genotyped (with many being sequenced through the 1000 Genome Project), and, for the CEU samples, also are part of large pedigrees. The NCI60 bank of tumor cell lines are derived from diverse human malignancies including those of the brain, blood and bone marrow, breast, colon, kidney, lung, ovary, prostate, and skin. For cell-based models, phenotypes such as growth inhibition, cell death through apoptosis, generation of an active metabolite, and biochemical activities have been used (18).

LCL Model

Use of the LCL model allows pharmacogenomic research to be conducted with cells from healthy, related individuals for whom inclusion in chemotherapeutic drug studies would not be feasible because of ethical considerations. Cell lines in culture may be treated with a range of drug concentrations for a set period of time to obtain cellular growth rate inhibition or apoptosis. In addition, parameters such as concentration at which 50 percent growth inhibition occurs (IC_{50}) or area under the percentage survival-concentration curve (AUC) can be used as a single value representing the degree of cellular sensitivity to the drug. Lymphoblastoid cell lines are prepared by Epstein-Barr virus (EBV) transformation of peripheral blood mononuclear cells, which results in the immortalization of B-lymphocytes and the ability to proliferate indefinitely (33). In particular, during the past few years, GWAS using the EBV-transformed LCLs (e.g., the HapMap samples) have demonstrated the feasibility of integrating genotypic data (e.g., >3.1 million SNPs) with cytotoxicities of anticancer agents; for example, 5-FU, docetaxel, etoposide, cisplatin, daunorubicin, carboplatin, cytarabine, and gemcitabine (21, 34–37). Table 10.2 lists some of the discoveries made by using this cell-based model. One of the main criticisms of the use of LCLs for pharmacogenomic discovery of variants contributing to chemotherapeutic toxicity is the effect of confounders (variation in cellular growth rate, baseline EBV copy number, and ATP levels) on phenotypes measured or cellular growth rate (38). Although studies have shown that baseline ATP or EBV copy number was not significantly correlated with

Table 10.2. Examples of Pharmacogenomic Discoveries Using LCLs

Chemotherapeutic Agent	*Examples of Related Gene(s)*	*Significant SNPs and Loci*	*Pharmacogenomic Discovery*	*Reference*
5-FU	TYMS	rs2847153 and rs2853533 9q13-q22	2 SNPs in TYMS, the gene encoding thymidylate synthetase, are significantly associated with 5-FU cytotoxicity in HapMap CEPH LCLs. Genome-wide linkage analyses demonstrate a quantitative trait locus (QTL) at 9q13-q22 which influences 5-FU cytotoxicity in HapMap LCLs.	(59), (35)
Docetaxel	N/A	5q11-21, 9q13-22	Genome-wide linkage analyses demonstrates 2 QTLs on chromosomes 5q11-21 and 9q13-q22 which influence docetaxel cytotoxicity in HapMap LCLs.	(35)
Etoposide	UVRAG, SEMA5A, SCL7A6, PRMT7	22 SNPs with 15 located within introns of the aforementioned genes (rs10079862, rs571826, rs16882871, rs10213926, rs2135071, rs3777359, rs369459, rs446732, rs421548, rs442173, rs486947, and rs268478 in SEMA5A; rs7116263 in UVRAG; rs11644360 in SLC7A6; and rs3785125 in PRMT7) and 1 (rs1127773) located in a 3′-untranslated region of SLC7A6	Linkage-directed association demonstrates 22 SNPs that are significantly associated with etoposide cytotoxicity at one or more treatment concentrations in HapMap CEPH LCLs.	(60)
Cisplatin	CDH13, ZNF659, LRRC3B, PITX2, LARP2	20 SNPs including 10 located within the 5 aforementioned genes (rs17758876 in CDH13; rs17041972, rs17624452, rs17041968, and rs2278782 in PITX2; rs4834232 in LARP2; rs1026686 and rs3860575 in SNF659; and rs17018468 and rs7652737 in LRRC3B), and 10 nongenic SNPs (rs7131224, rs7113868, rs7119153, rs7949504, rs11944754, rs1028074, rs12795809, rs10510534, rs7683488, and rs6848982)	Linkage-directed association analysis demonstrates that 20 SNPs are associated with cisplatin cytotoxicity in HapMap CEPH LCLs, with 4 of those explaining 10% of variation in apoptosis.	(61)
Daunorubicin	HNRPD, CYP1B1	rs1551315, rs12052523, rs2195830, rs623360, rs10083335, rs3750518.	2 SNPs (rs120525235 and rs3750518) were significant predictors of transformed daunorubicin IC_{50} in a validation set of HapMap CEPH LCLs, whereas 6 SNPs predicted 29% of variation of transformed daunorubicin IC_{50}. rs3750518 acts by altering HNRPD gene expression. Additionally, rs10932125 genotype was associated with CYP1B1 expression and transformed daunorubicin IC_{50}.	(62)

Chemotherapeutic Agent	*Examples of Related Gene(s)*	*Significant SNPs and Loci*	*Pharmacogenomic Discovery*	*Reference*
Carboplatin	GPC5	rs1031324 and rs1993034	2 SNPs are significantly associated with carboplatin cytotoxic phenotypes at all concentrations and IC_{50} through an effect on GPC5 gene expression in HapMap LCLs from Yoruba individuals probably by a distant-acting effect because the SNPs are not located within the GPC5 gene.	(63)
Cytarabine	G1T1, SCL25A37, P2RX1, CCDC24, RPS6KA2, SSH2, LOC399491, ANPEP, SOD3	rs17808412, rs2775139, rs17795186, rs368182 in Caucasian LCLs; rs938562, rs10906723, rs2430853, rs10181725, rs10193059 in Yoruba LCLs	4 SNPs explain 51% of variability in sensitivity to cytarabine in HapMap cell lines from white individuals, while 5 SNPs explain 58% of variation in HapMap cell lines from Yoruba individuals, by affecting expression of the aforementioned target genes.	(21)
Gemcitabine and cytarabine	IQGAP2 and TGM3	rs3797418 and rs6082527	A SNP in IQGAP2 (rs3797418) is significantly associated with variation in multiple gene expression as well as both gemcitabine and cytarabine IC_{50} in ethnically defined "Human Variation Panel" LCLs. A second SNP in TGM3 (rs6082527) is associated with gene expression and gemcitabine IC_{50}.	Li L, et al. 2009 (43)

cellular growth rate or drug-induced cytotoxicity, cellular growth rate and drug-induced cytotoxicity were significantly, directly related for a number of chemotherapeutic agents. Importantly, cellular growth rate is under appreciable genetic influence ($h^2 = 0.30$–0.39). Not surprisingly, a percentage of SNPs that significantly associate with drug-induced cytotoxicity also associate with cellular growth rate ($P \leq 0.0001$). Studies using LCLs for pharmacologic outcomes should therefore consider that a portion of the genetic variation explaining drug-induced cytotoxicity is mediated via heritable effects on growth rate (39).

NCI60 Cell Lines

The NCI60 is a bank of sixty human cancer cell lines derived from several malignant tissues including those in the brain, blood and bone marrow, breast, colon, kidney, lung, ovary, prostate, and skin. NCI60 cell lines have been screened for the cytotoxic effect of more than 40,000 compounds, and results are publicly available (40). In addition, they have been used to study genetic predictors of response to several chemotherapeutic agents. For example, the pharmacogenomics of gemcitabine has been studied in such a model (41). A number of studies have correlated baseline gene expression with sensitivity to specific compounds in the NCI60 panel (42). Table 10.3 lists some of the discoveries made using the NCI60 cell lines. An important limitation in the use of cancer cell lines for pharmacogenomic research lies in the small sample size that is available and that the sixty cell lines comprises nine different tissue types. Nevertheless, tumor cell lines may be used for functional studies of predictive genes/loci derived from clinical studies or studied in LCLs. For example, RNA interference technology may be used to upregulate or downregulate gene function in relevant tumor cell lines in order to study variability in drug response resulting from altered gene expression (43).

PHARMACOGENOMICS OF IMPORTANCE IN ONCOLOGY

Germline Polymorphisms

Historically, such variants have been found to predominantly affect pharmacokinetic pathways resulting in altered levels of active drugs and/or metabolites.

Table 10.3. Examples of Pharmacogenomic Discoveries Using NCI60 Cell Lines

Class(es) of Chemotherapeutic Agents Studied	*Related Gene(s)*	*Significant SNPs*	*Pharmacogenomic Discovery*	*Reference*
Taxanes, topoisomerase inhibitors, antimetabolites, N7 alkylating agents.	ERCC2, ERCC5, and GSTP1	rs13181, rs17655, rs1695	Cytotoxicity of taxanes is markedly dependent on ERCC2 genotype; ERCC5 genotype is important only for taxanes; and GSTP1 polymorphisms are relevant for other drug classes.	(64)
Gemcitabine	CDC5L, EPC2, POLS, and PARP1	rs525043, rs2279653, rs6739555, rs2293464	SNPs in 4 genes are significantly associated with gemcitabine sensitivity.	(41)
Erlotinib, geldanamycin, topoisomerase I and II inhibitors, alkylating agents	EGFR	rs2227983 EGFR-216G>T	rs2227983 is associated with lower sensitivity to alkylating agents, whereas -216G>T variants are associated with increased sensitivity to erlotinib and reduced sensitivity to geldanamycin, topoisomerase I inhibitors, and alkylating agents.	(65)
Antimetabolites	MTHFR	MTHFR-A1298C	Cells homozygous for the mutant allele (CC at MTHFR-A1298C) are more sensitive to cyclocytidine, cytarabine, and floxuridine than those with AA or AC.	(66)
Alkylating agents and topoisomerase I inhibitors	MDM2	SNP309	SNP309 is significantly associated with increased sensitivity to alkylating agents and topoisomerase I inhibitors in cells with wild-type TP53 gene.	(67)

Germline polymorphisms have typically predicted toxicity from chemotherapeutic agents, although there is emerging evidence that such variants may also predict response. A few clinically relevant examples are detailed here.

CYP2D6 Genotype and Tamoxifen

Tamoxifen is a selective estrogen receptor modulator (SERM) that exerts agonist effects on uterine endometrium and antagonist effects on breast tissues (44). It is the most widely used antiestrogen therapy in adjuvant treatment of ER-positive breast cancer in premenopausal women, and also has efficacy in preventing invasive and noninvasive breast cancer in women at highest risk. The National Surgical Adjuvant Breast and Bowel Project P1 trial (45) demonstrated a 49 percent reduction in the incidence of invasive breast cancer among high-risk women treated with a five-year course of tamoxifen. As a prodrug, the activity of tamoxifen depends on conversion to a number of active metabolites. In hepatocytes, it is metabolized by cytochrome P450 enzymes to *N*-desmethyltamoxifen and 4-hydroxytamoxifen (see Figure 10.4). Oxidation of both metabolites results in the synthesis of 4-hydroxy-*N*-desmethyltamoxifen (endoxifen). The antiestrogenic effects of tamoxifen and its metabolites depend on interaction with the ERs. Although both 4-hydroxytamoxifen and endoxifen have significant affinity for ER, endoxifen plasma concentrations are five- to tenfold higher, and so it is believed to be the more important metabolite (1).

CYP2D6 is one of the cytochrome enzymes involved in tamoxifen metabolism, and it is encoded by the *CYP2D6* gene on chromosome 22q13.1. Several polymorphisms are known to be present in the gene, and more than seventy-five variant alleles have been reported (46). Null alleles such as CYP2D6*4 and *6 are particularly important, because homozygosity for them results in a poor metabolizer phenotype, a condition in which CYP2D6 enzyme activity is negligible. Clinical trials have demonstrated that affected patients have lower endoxifen levels and poorer outcomes as shown by worse relapse-free time and disease-free survival in comparison with patients lacking null alleles (24). Concurrently,

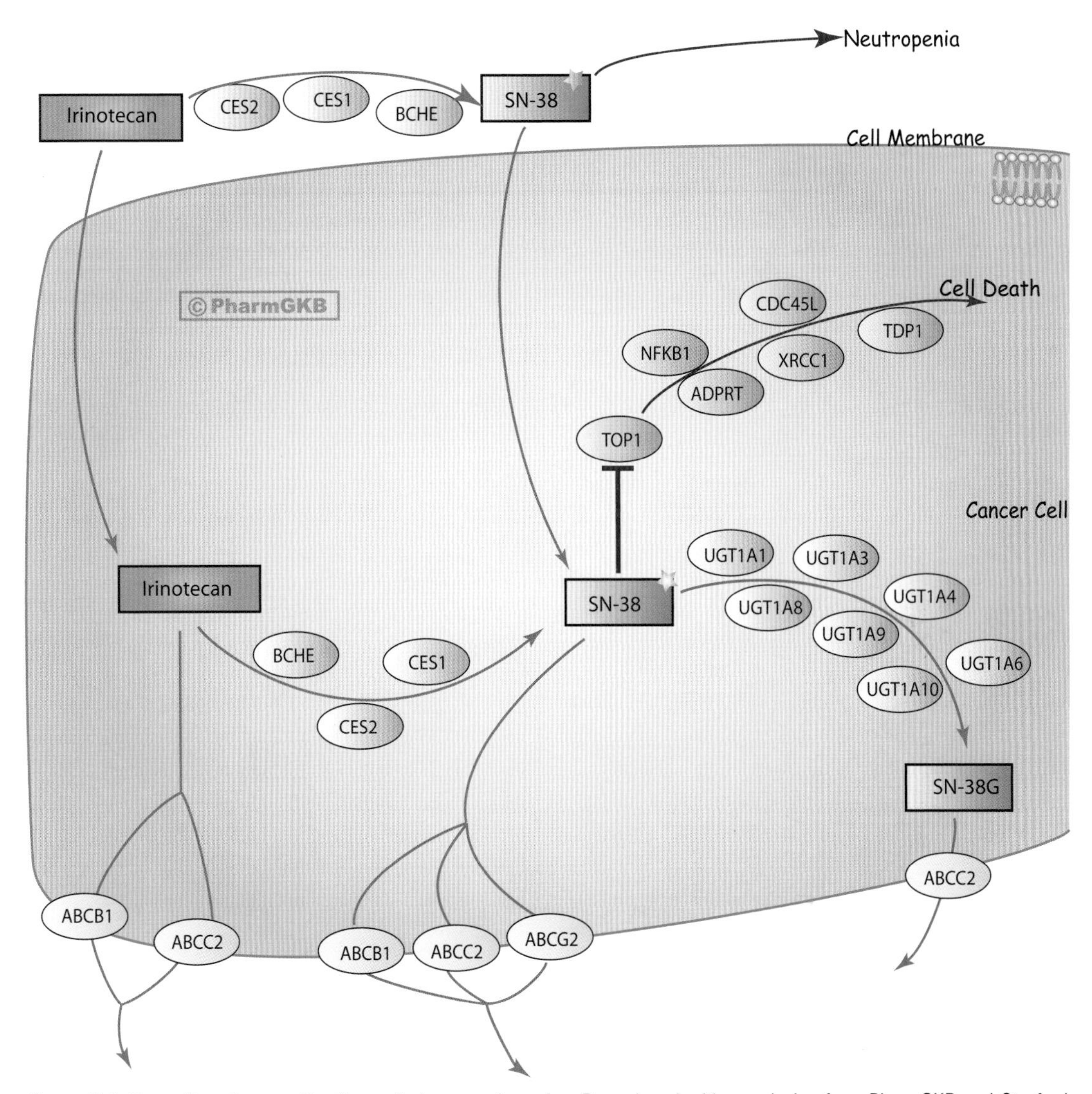

Figure 10.4. Tamoxifen pharmacokinetics and pharmacodynamics. Reproduced with permission from PharmGKB and Stanford University (68).

patients with homozygosity for the null alleles are less likely to experience hot flashes, a common side effect of tamoxifen. CYP2D6 genotyping is clinically available and is listed in the U.S. Food and Drug Administration (FDA) table of valid genomic biomarkers (47).

UGT1A1 and Irinotecan Therapy

Irinotecan is a topoisomerase I inhibitor and acts by binding reversibly to a topoisomerase I-DNA complex to induce double-strand DNA breaks that lead to cell death in the S-phase of the cell cycle (48). Irinotecan is indicated for treatment of metastatic colorectal carcinoma in combination with 5-FU and leucovorin, and may also be used in combination with cetuximab (49). Conversion of irinotecan to a more potent metabolite, SN-38, results from the activity of serum carboxylesterases. Degradation of SN-38 is mediated by uridine diphosphate-glucuronyltransferase 1A1 (UGT1A1), resulting in the formation of the glucuronide conjugate, SN-38-glucuronide (Figure 10.5). SN-38 is largely responsible for toxicities of irinotecan such as neutropenia and diarrhea (50).

Inadequate UGT1A1 activity results in accumulation of SN-38 and increased likelihood of irinotecan toxicity (51). Several polymorphisms in the *UGT1A1* gene have been reported to date (52). The homozygous genotype of *UGT1A1*28* has a frequency as high as 10 percent to 20 percent in some ethnic groups and has correlated with a high frequency of delayed irinotecan toxicity in clinical trials (53). Based on those findings, the irinotecan label was modified to incorporate the role of UGT1A1*28

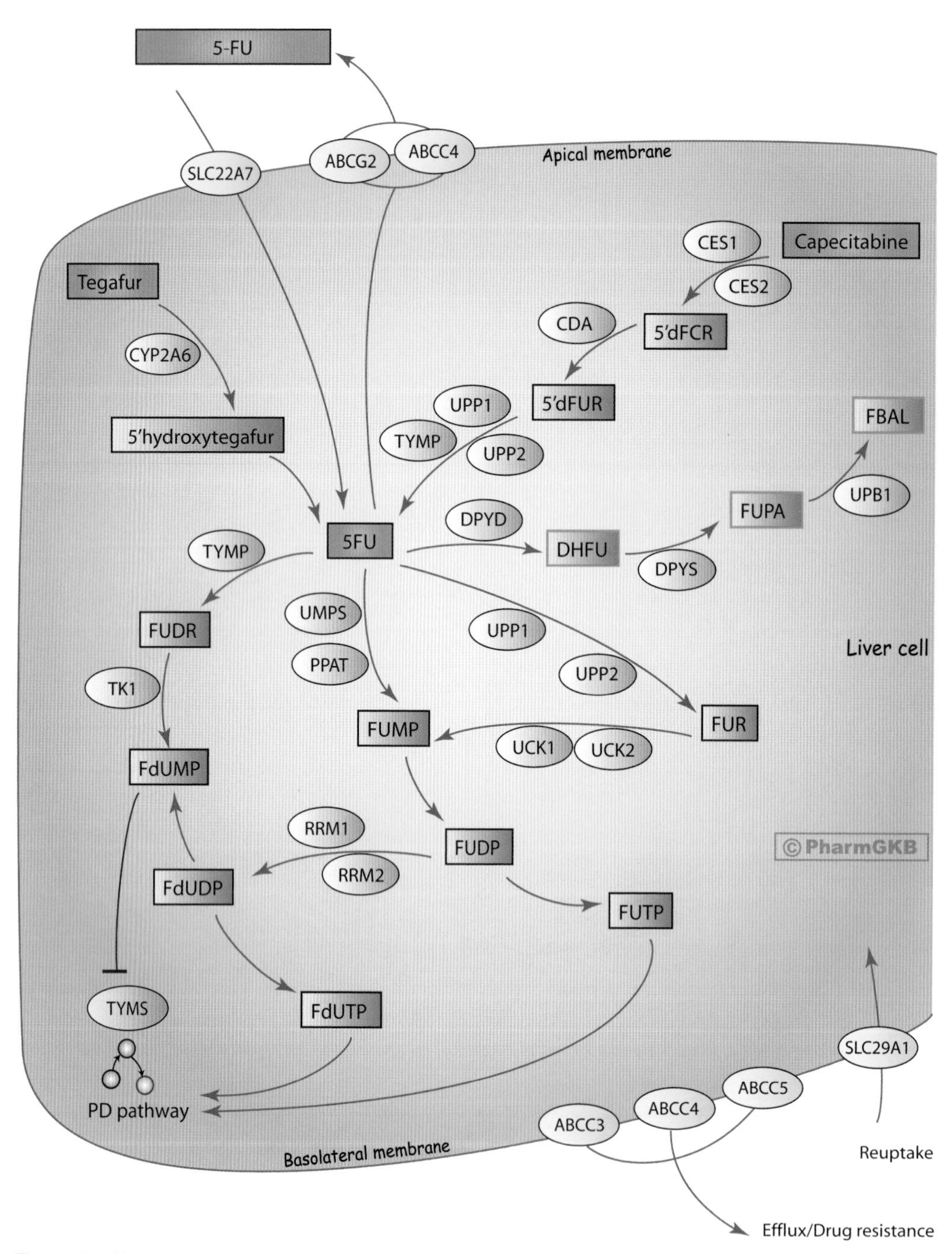

Figure 10.5. Irinotecan pharmacokinetics. Reproduced with permission from PharmGKB and Stanford University (68).

polymorphism in predicting severe neutropenia with irinotecan therapy.

Dihydropyrimidine Dehydrogenase (DPYD) and Thymidylate Synthetase (TS) Polymorphisms and 5-FU Therapy

An important drug in the treatment of colorectal cancer, 5-FU, is a pyrimidine antimetabolite that acts by inhibition of thymidylate synthetase, an enzyme involved in the synthesis of dTMP in the DNA synthetic pathway. 5-FU has efficacy in the treatment of other solid malignancies, most notably those of the breast and head and neck. An oral prodrug formulation of 5-FU, capecitabine, is also in clinical use. Common side effects of 5-FU include myelosuppression, diarrhea, stomatitis, and hand-and-foot syndrome.

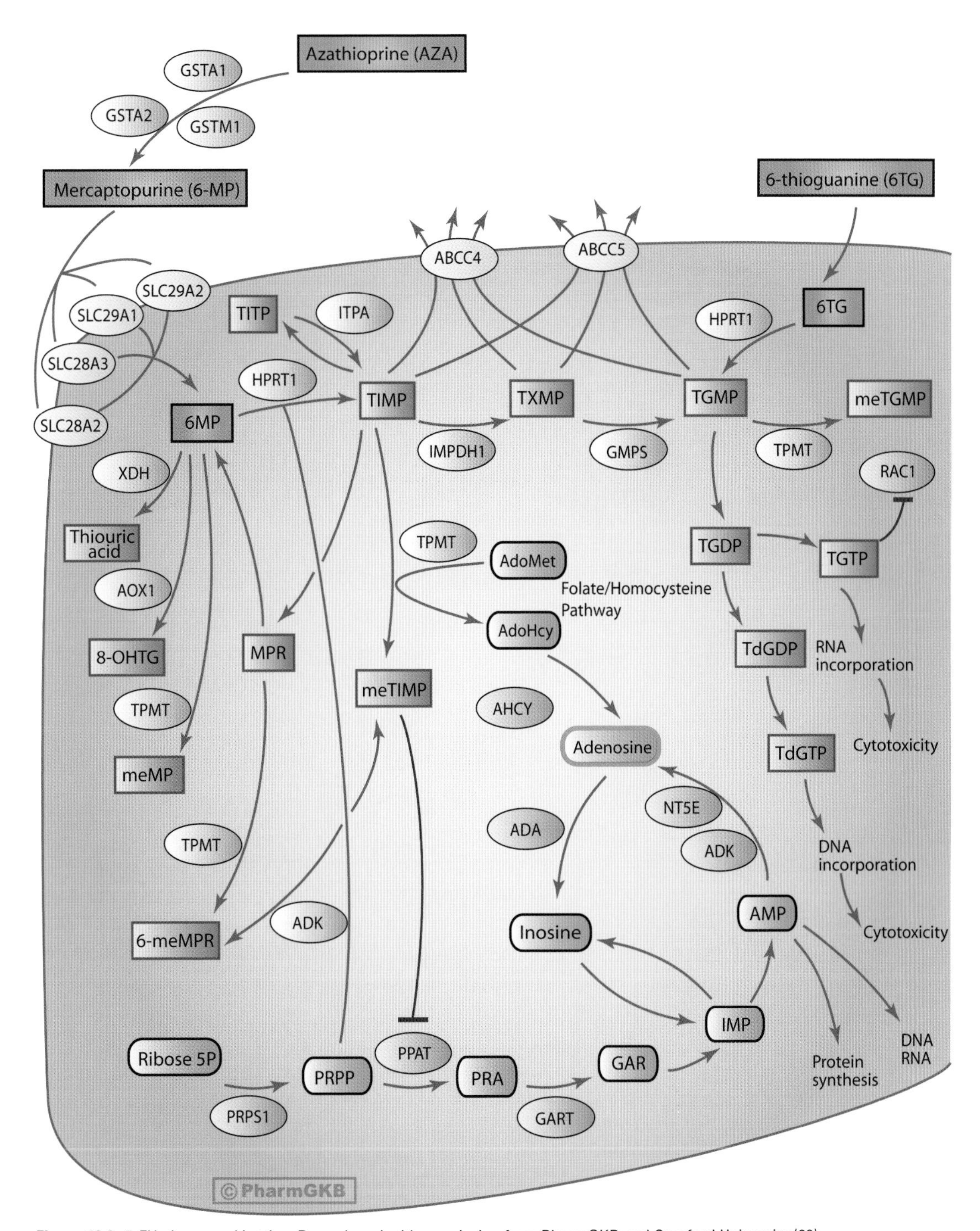

Figure 10.6. 5-FU pharmacokinetics. Reproduced with permission from PharmGKB and Stanford University (68).

The catabolic pathway of 5-FU involves the activity of DPD, an enzyme encoded by the *DPYD* gene on chromosome 1p22 (Figure 10.6). More than 80 percent of 5-FU is metabolized by DPD, and levels of the enzyme show significant interindividual variability. It has been estimated that 3 percent to 5 percent of the population is partially DPD deficient, whereas 0.2 percent is completely deficient (54). Several polymorphisms of uncertain significance have been reported in the *DPYD* gene; however, the *DPYD*2A* variant has been seen in 40 percent to 50 percent of individuals with partial or complete DPD deficiency (55). In pediatric oncology,

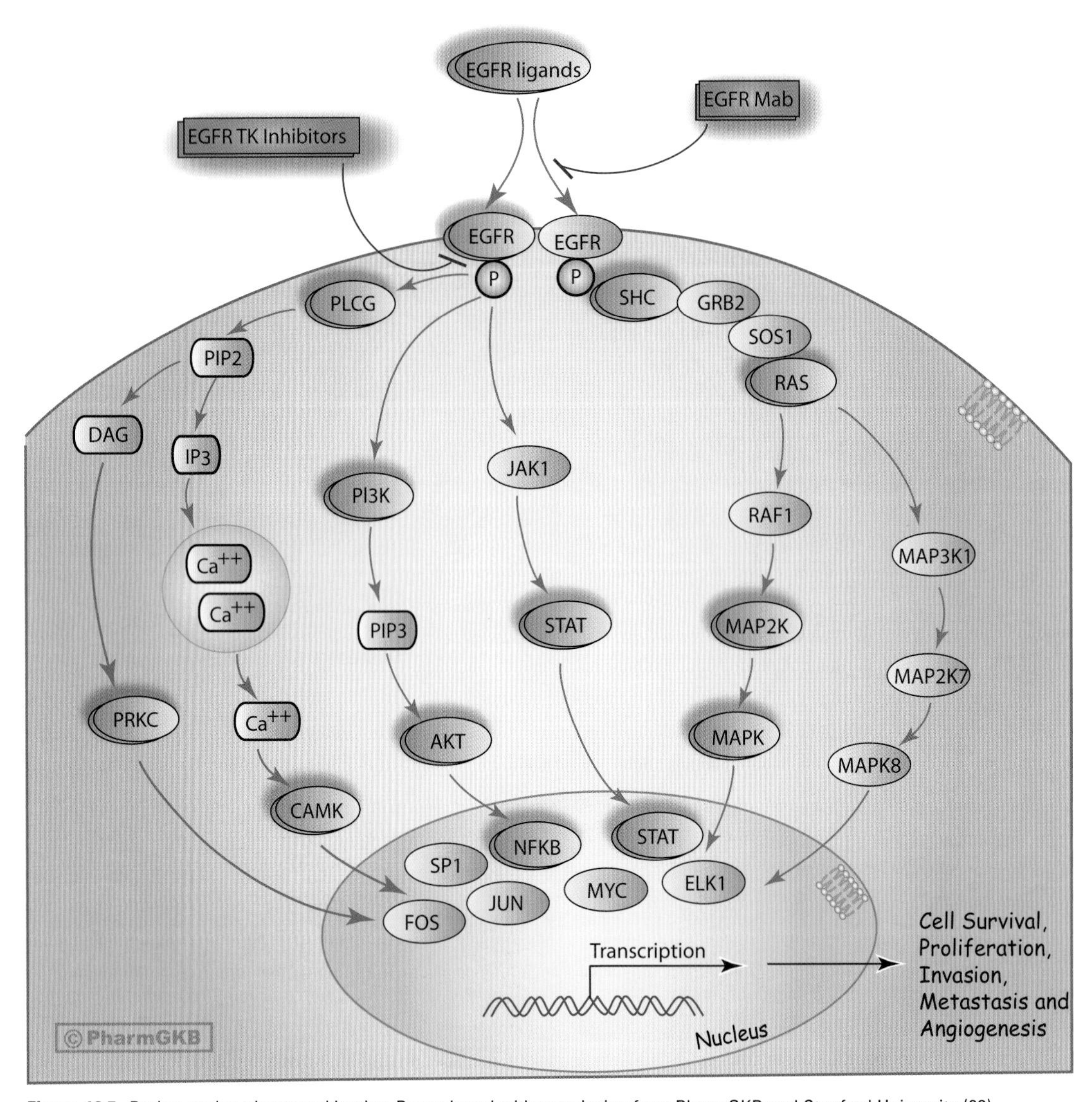

Figure 10.7. Purine analog pharmacokinetics. Reproduced with permission from PharmGKB and Stanford University (68).

severe neurologic toxicity has been seen with complete DPD deficiency (56). Although such a state is almost invariably associated with a heightened risk of 5-FU toxicity, studies have shown that *DPYD* mutations do not always have an effect on DPD enzyme activity (55). As a result, decreased levels of DPD, rather than *DPYD* mutations, have been included in the FDA table of valid genomic biomarkers.

In addition to DPD, genetic variability in *TS*, the gene encoding thymidylate synthetase, is associated with outcomes in patients with colorectal cancer treated with 5-FU. Variants such as a six base pair insertion and deletion polymorphism in the 3′-untranslated region, and variable number tandem repeat polymorphisms in the promoter-enhancer region, lead to increased *TS* gene expression. In clinical studies, these high-expression variants have correlated with decreased survival in patients treated with 5-FU (55).

TPMT and Purine Analogs

Thiopurines such as 6-MP and 6-thioguanine (6-TG) are used in treatment of hematologic malignancies such as AML, ALL, and non-Hodgkin's lymphoma. Both drugs require activation by hypoxanthine-guanine phosphoribosyl transferase to exert cytotoxic effects by inhibiting DNA synthesis in the S-phase of the cell cycle. As analogs of the naturally occurring purines, they act as antagonists, thereby blocking DNA and RNA synthesis. A significant catabolic pathway for both 6-MP and 6-TG involves conversion by TPMT to inactive metabolites, 6-methyl-MP and 6-methyl-TG (Figure 10.7). In the absence of TPMT activity, the active agents accumulate, increasing the likelihood of toxicity. Like many other chemotherapeutic agents, thiopurines have a narrow therapeutic index. The most concerning side effect seen is myelosuppression with increased risks of infection and bleeding. Several low-activity alleles are known to exist at the *TPMT* gene locus on chromosome 6p22.3, and may give rise to a heterozygous state with intermediate activity (10 percent of white individuals) or a homozygous state with negligible (0.3 percent of whites) activity (57). Examples of such low-activity alleles include *TPMT*2*, *TPMT*3A*, and *TPMT*3C*, which together account for 80 percent to 95 percent of cases of intermediate or low enzyme activity. Affected individuals are at heightened risk for potentially life-threatening myelosuppression as a consequence of thiopurine therapy. Furthermore, thiopurines and TPMT activity provide a successful example of genotype-driven chemotherapy dosing with recommendations for lower doses (30 percent to 50 percent) in patients with intermediate TPMT levels (57).

PHARMACOGENOMICS AND PHARMACOETHNICITY

Pharmacogenomics challenges the "one size fits all" approach that has dominated medical oncology care for decades. Of particular importance and relevance to global oncology practice is a pattern of interethnic differences in response to chemotherapeutic agents, as exemplified by the heightened sensitivity to EGFR-TKI seen among young, Asian, nonsmoking females with NSCLC. Although such differences are multifactorial, the evidence for a significant genetic component is growing. This area of study has been referred to as pharmacoethnicity and may be defined as ethnic diversity in drug response or toxicity (58). The objective of pharmacogenomics is to identify individuals most susceptible to a particular drug; however, the populations most sensitive to specific drugs may be enriched in the genetic variants associated with drug sensitivity. In low-resource settings, pharmacoethnicity will be of particular importance because sparse resources may be preferentially channeled toward purchase and administration of specific chemotherapeutic agents.

CASE PRESENTATIONS

Case 1: CYP2D6 Genotype and Response to Tamoxifen Therapy

A thirty-year-old postmenopausal woman sought medical care for a lump in the left breast. A mammogram revealed a suspicious left breast mass, and a core biopsy revealed malignant cells. Breast magnetic resonance imaging showed a single 2.2 × 2.6 cm mass in the upper outer quadrant of the left breast. She underwent lumpectomy and sentinel lymph node biopsy. Pathologic analysis of the resected specimen revealed a grade III infiltrating ductal carcinoma measuring 2.6 cm in greatest diameter. The tumor had estrogen (ER) and progesterone receptor (PR) expression, as well as gene amplification. None of four sentinel lymph nodes had metastatic disease. An Oncotype DX assay revealed a high recurrence score indicating a high likelihood of disease recurrence. As a result, a decision was made to treat with adjuvant chemotherapy in addition to antiestrogen therapy. She received four cycles of docetaxel and cyclophosphamide followed by local breast irradiation. Afterward, she was started on letrozole, an aromatase inhibitor. Unfortunately, she experienced severe joint aches that persisted even after letrozole was replaced by exemestane. After learning that tamoxifen was another option, the patient requested CYP2D6 genotyping to exclude the presence of a null allele.

Case 2: UGT1A1 Genotype and Irinotecan Toxicity

A seventy-six-year-old man was evaluated for abdominal cramping and a lower gastrointestinal bleed. Colonoscopy revealed a cecal mass that was biopsied with a finding of moderately differentiated adenocarcinoma of the colon. He underwent laparotomy and hemicolectomy. Pathologic analysis revealed a moderately differentiated colonic adenocarcinoma measuring 3 cm. The tumor penetrated the colon wall and extended into pericolic fat. Fourteen of fifty-seven pericolic lymph nodes had metastatic disease. Subsequently, staging computed tomography (CT) scans demonstrated multiple liver metastases. He was started on a combination chemotherapy with 5-FU, leucovorin, oxiliplatin, and bevacizumab. He received a total of six cycles and had a complete response with no further metastatic lesions evident on a repeat CT scan. Six months later, CT scans revealed new hepatic lesions and pathologic retroperitoneal adenopathy. Because his initial therapy had been administered less than twelve months earlier, he was started on a different regimen consisting of 5-FU, leucovorin, and irinotecan. After the first cycle, he developed profuse diarrhea, dehydration, and neutropenia that required inpatient management. Studies ruled out an infectious etiology. The same happened with the second and third cycles, even though the dose of irinotecan had been reduced by 25 percent for the latter. Genetic testing revealed homozygosity for the UGT1A1*28 polymorphism.

SUMMARY POINTS

- Chemotherapeutic agents are associated with significant toxicities and narrow therapeutic indices, and, for some drugs, high costs, making the field of pharmacogenomics of anticancer agents extremely important.
- A distinguishing feature of pharmacogenomic research in oncology is the consideration of both somatic DNA and germline DNA.
- Both candidate gene and genome-wide approaches have been used successfully in the study of pharmacogenetics of oncologic therapies.
- Although germline polymorphisms and somatic mutations have historically correlated with toxicity and response, respectively, there is emerging evidence to support a role for germline variants in predicting response.
- Pharmacoethnicity is an area that focuses on interethnic variability in drug responses and is likely to be an important focus in future chemotherapy pharmacogenomic research.

REFERENCES

1. Hoskins JM, Carey LA, & McLeod HL. CYP2D6 and tamoxifen: DNA matters in breast cancer. *Nat Rev Cancer.* 2009;**9**:576–86.
2. Fleck LM. The costs of caring: who pays? who profits? who panders? *Hastings Cent Rep.* 2006;**36**:13–17.
3. Holmes MV, Shah T, Vickery C, Smeeth L, Hingorani AD, & Casas JP. Fulfilling the promise of personalized medicine? Systematic review and field synopsis of pharmacogenetic studies. *PLoS.* 2009;**4**:e7960.
4. Devereux TR, Risinger JI, & Barrett JC. Mutations and altered expression of the human cancer genes: what they tell us about causes. *IARC Sci Publ.* 1999;19–42.
5. Olivier M, Hollstein M, & Hainaut P. TP53 mutations in human cancers: origins, consequences, and clinical use. *Cold Spring Harb Perspect Biol.* 2010;**2**:a001008.
6. Chung JH, Choe G, Jheon S, Sung S-W, Kim TJ, Lee KW, Lee JH, & Lee C-T. Epidermal growth factor receptor mutation and pathologic-radiologic correlation between multiple lung nodules with ground-glass opacity differentiates multicentric origin from intrapulmonary spread. *J Thorac Oncol.* 2009;**4**:1490–5.
7. Bell DW. Our changing view of the genomic landscape of cancer. *J Pathol.* 2010;**220**:231–43.
8. Ikediobi ON. Somatic pharmacogenomics in cancer. *Pharmacogenomics J.* 2008;**8**:305–14.
9. Bernig T & Chanock S. Challenges of SNP genotyping and genetic variation: its future role in diagnosis and treatment of cancer. *Expert Rev Mol Diagn.* 2006;**6**:319–31.
10. Yang JJ, Cheng C, Yang W, Pei D, Cao X, Fan Y, Pounds SB, Neale G, Trevino LR, French D, Campana D, Downing JR, Evans WE, et al. Genome-wide interrogation of germline genetic variation associated with treatment response in childhood acute lymphoblastic leukemia. *JAMA.* 2009;**301**:393–403.
11. Soo RA, Wang LZ, Ng SS, Chong PY, Yong WP, Lee SC, Liu JJ, Choo TB, Tham LS, Lee HS, Goh BC, & Soong R. Distribution of gemcitabine pathway genotypes in ethnic Asians and their association with outcome in non-small cell lung cancer patients. *Lung Cancer.* 2009;**63**:121–7.
12. McWhinney SR & McLeod HL. Using germline genotype in cancer pharmacogenetic studies. *Pharmacogenomics.* 2009;**10**:489–93.
13. Trotti A, Colevas AD, Setser A, Rusch V, Jacques D, Budach V, Langer C, Murphy B, Cumberlin R, Coleman CN, & Rubin P. CTCAE v3.0: development of a comprehensive grading system for the adverse effects of cancer treatment. *Semin Radiat Oncol.* 2003;**13**:176–81.
14. Therasse P, Arbuck SG, Eisenhauer EA, Wanders J, Kaplan RS, Rubinstein L, Verweij J, Van Glabbeke M, van Oosterom AT, Christian MC, & Gwyther SG. New guidelines to evaluate the response to treatment in solid tumors. European Organization for Research and Treatment of Cancer, National Cancer Institute of the United States, National Cancer Institute of Canada. *J Natl Cancer Inst.* 2000;**92**:205–16.
15. Santos FP & Ravandi F. Advances in treatment of chronic myelogenous leukemia – new treatment options with tyrosine kinase inhibitors. *Leuk Lymphoma.* 2009;**50**(suppl 2):16–26.
16. Kantarjian H, Schiffer C, Jones D, & Cortes J. Monitoring the response and course of chronic myeloid leukemia in the modern era of BCR-ABL tyrosine kinase inhibitors: practical advice on the use and interpretation of monitoring methods. *Blood.* 2008;**111**:1774–80.
17. Thorn CF, Klein TE, & Altman RB. Pharmacogenomics and bioinformatics: PharmGKB. *Pharmacogenomics.* 2010;**11**:501–5.
18. Welsh M, Mangravite L, Medina MW, Tantisira K, Zhang W, Huang RS, McLeod H, & Dolan ME. Pharmacogenomic discovery using cell-based models. *Pharmacol Rev.* 2009;**61**:413–49.
19. Shukla SJ & Dolan ME. Use of CEPH and non-CEPH lymphoblast cell lines in pharmacogenetic studies. *Pharmacogenomics.* 2005;**6**:303–10.
20. Masson E, Relling MV, Synold TW, Liu Q, Schuetz JD, Sandlund JT, Pui CH, & Evans WE. Accumulation of methotrexate polyglutamates in lymphoblasts is a determinant of antileukemic effects in vivo. A rationale for high-dose methotrexate. *J Clin Invest.* 1996;**97**:73–80.
21. Hartford CM, Duan S, Delaney SM, Mi S, Kistner EO, Lamba JK, Huang RS, Dolan ME. Population-specific genetic variants important in susceptibility to cytarabine arabinoside cytotoxicity. *Blood.* 2009;**113**:2145–53.
22. Lennard L, Lilleyman JS, Van Loon J, & Weinshilboum RM. Genetic variation in response to 6-mercaptopurine for childhood acute lymphoblastic leukaemia. *Lancet.* 1990;**336**:225–9.
23. Hoskins JM, Goldberg RM, Qu P, Ibrahim JG, & McLeod HL. UGT1A1*28 genotype and irinotecan-induced neutropenia: dose matters. *J Natl Cancer Inst.* 2007;**99**:1290–5.
24. Goetz MP, Rae JM, Suman VJ, Safgren SL, Ames MM, Visscer DW, Reynolds C, Couch FJ, Lingle WL, et al. Pharmacogenetics of tamoxifen biotransformation is associated

with clinical outcomes of efficacy and hot flashes. *J Clin Oncol.* 2005;**23**:9312–18.
25. Hartford CM & Dolan ME. Identifying genetic variants that contribute to chemotherapy-induced cytotoxicity. *Pharmacogenomics.* 2007;**8**:1159–68.
26. Huang RS & Ratain MJ. Pharmacogenetics and pharmacogenomics of anticancer agents. *CA Cancer J Clin.* 2009;**59**:42–55.
27. Zhang W & Dolan ME. Impact of the 1000 genomes project on the next wave of pharmacogenomic discovery. *Pharmacogenomics.* 2010;**11**:249–56.
28. International Human Genome Sequencing Consortium. Finishing the euchromatic sequence of the human genome. *Nature.* 2004;**431**:931–45.
29. International HapMap Consortium. The International HapMap Project. *Nature.* 2003;**426**:789–96.
30. Galmarini CM, Thomas X, Graham K, El Jafaari A, Cros E, Jordheim L, Mackey JR, & Dumontet C. Deoxycytidine kinase and cN-II nucleotidase expression in blast cells predict survival in acute myeloid leukaemia patients treated with cytarabine. *Br J Haematol.* 2003;**122**:53–60.
31. Estey E, Plunkett W, Dixon D, Keating M, McCredie K, & Freireich EJ. Variables predicting response to high dose cytosine arabinoside therapy in patients with refractory acute leukemia. *Leukemia.* 1987;**1**:580–3.
32. Raza A, Gezer S, Anderson J, Lykins J, Bennett J, Browman G, Goldberg J, Larson R, Vogler R, & Preisler HD. Relationship of [3H]Ara-C incorporation and response to therapy with high-dose Ara-C in AML patients: a Leukemia Intergroup study. *Exp Hematol.* 1992;**20**:1194–1200.
33. Ling PD & Huls HM. Isolation and immortalization of lymphocytes. *Curr Protoc Mol Biol.* 2005;Chapter 28:Unit 28.22.
34. Dolan ME, Newbold KG, Nagasubramanian R, Wu X, Ratain MJ, Cook EH Jr, & Badner JA. Heritability and linkage analysis of sensitivity to cisplatin-induced cytotoxicity. *Cancer Res.* 2004;**64**:4353–6.
35. Watters JW, Kraja A, Meucci MA, Province MA, & McLeod HL. Genome-wide discovery of loci influencing chemotherapy cytotoxicity. *Proc Natl Acad Sci USA.* 2004;**101**:11809–14.
36. Huang RS, Duan S, Shukla SJ, Kistner EO, Clark TA, Chen TX, Schweitzer AC, Blume JE, & Dolan ME. Identification of genetic variants contributing to cisplatin-induced cytotoxicity by use of a genomewide approach. *Am J Hum Genet.* 2007;**81**:427–37.
37. Duan S, Bleibel WK, Huang RS, Shukla SJ, Wu X, Badner JA, & Dolan ME. Mapping genes that contribute to daunorubicin-induced cytotoxicity. *Cancer Res.* 2007;**67**:5425–33.
38. Choy E, Yelensky R, Bonakdar S, Plenge RM, Saxena R, De Jager PL, Shaw SY, Wolfish CS, Slavik JM, Cotsapas C, Rivas M, Dermitzakis ET, Cahir-McFarland E, Kieff E, Hafler D, Daly MJ, & Altshuler D. Genetic analysis of human traits in vitro: drug response and gene expression in lymphoblastoid cell lines. *PLoS Genet.* 2008;**4**:e1000287.
39. Stark AL, Zhang W, Mi S, Duan S, O'Donnell PH, Huang RS, & Dolan ME. Heritable and non-genetic factors as variables of pharmacologic phenotypes in lymphoblastoid cell lines. *Pharmacogenomics.* 2010;**10**:505–12.
40. Shoemaker RH. The NCI60 human tumour cell line anticancer drug screen. *Nat Rev Cancer.* 2006;**6**:813–23.
41. Jarjanazi H, Keifer J, Savas S, Briollais L, Tuzmen S, Pabalan N, Ibrahim-Zada I, Mousses S, & Ozcelik H. Discovery of genetic profiles impacting response to chemotherapy: application to gemcitabine. *Hum Mutat.* 2008;**29**:461–7.
42. Mori S, Chang JT, Andrechek ER, Potti A, & Nevins JR. Utilization of genomic signatures to identify phenotype-specific drugs. *PLoS.* 2009;**4**:e6772.
43. Li L, Fridley BL, Kalari K, Jenkins G, Batzler A, Weinshilboum RM, & Wang L. Gemcitabine and arabinosylcytosin pharmacogenomics: genome-wide association and drug response biomarkers. *PLoS.* 2009;**4**:e7765.
44. Jensen EV & Jordan V. The estrogen receptor: a model for molecular medicine. *Clin Cancer Res.* 2003;**9**:1980–9.
45. King MC, Wieand S, Hale K, Lee M, Walsh T, Owens K, Tait J, Ford L, Dunn BK, Constantino J, Wickerham L, Wolmark N, & Fisher B. Tamoxifen and breast cancer incidence among women with inherited mutations in BRCA1 and BRCA2. *JAMA.* 2001;**286**:2251–6.
46. Ingelman-Sundberg M, Sim SC, Gomez A, & Rodriguez-Antona C. Influence of cytochrome P450 polymorphisms on drug therapies: pharmacogenetic, pharmacoepigenetic and clinical aspects. *Pharmacol Ther.* 2007;**116**:496–526.
47. Table of valid genomic biomarkers in the context of approved drug labels. 2009. http://www.fda.gov/Drugs/ScienceResearch/ResearchAreas/Pharmacogenetics/ucm083378.htm. Accessed March 3, 2010.
48. Furuta T. Pharmacogenomics in chemotherapy for GI tract cancer. *J Gastroenterol.* 2009;**44**:1016–25.
49. Wong SF. Cetuximab: an epidermal growth factor receptor monoclonal antibody for the treatment of colorectal cancer. *Clin Ther.* 2005;**27**:684–94.
50. Ando Y, Saka H, Asai G, Sugiura S, Shimokata K, & Kamataki T. UGT1A1 genotypes and glucuronidation of SN-38, the active metabolite of irinotecan. *Ann Oncol.* 1998;**9**:845–7.
51. Innocenti F, Undevia SD, Iyer L, Chen PX, Das S, Kocherginsky M, Karrison T, Janisch L, Ramirez J, Rudin CM, Vokes EE, & Ratain MJ. Genetic variants in the UDP-glucuronosyltransferase 1A1 gene predict the risk of severe neutropenia of irinotecan. *J Clin Oncol.* 2004;**22**:1382–8.
52. Nagar S & Remmel RP. Uridine disphosphoglucuronosyltransferase pharmacogenetics and cancer. *Oncogene.* 2006;**25**:1659–72.
53. Ramchandani RP, Wang Y, Booth BP, Ibrahim A, Johnson JR, Rahman A, Mehta M, Innocenti F, Ratain MJ, & Gobburu JV. The role of SN-38 exposure, UGT1A1*28 polymorphism, and baseline bilirubin level in predicting severe irinotecan toxicity. *J Clin Pharmacol.* 2007;**47**:78–86.
54. Walter A, Johnstone E, Swanton C, Midgley R, Tomlinson I, & Kerr D. Genetic prognostic and predictive markers in colorectal cancer. *Nat Rev Cancer.* 2009;**9**:489–99.
55. Yen JL & McLeod HL. Should DPD analysis be required prior to prescribing fluoropyrimidines? *Eur J Cancer.* 2007;**43**:1011–16.
56. Van Kuilenburg AB, Vreken P, Abeling NG, Bakker HD, Meinsma R, Van Lenthe H, De Abreu RA, Smeitink JA, Kayserili H, Apak MY, et al. Genotype and phenotype in patients with dihydropyrimidine dehydrogenase deficiency. *Hum Genet.* 1999;**104**:1–9.

57. Sahasranaman S, Howard D, & Roy S. Clinical pharmacology and pharmacogenetics of thiopurines. *Eur J Clin Pharmacol.* 2008;**64**:753–67.
58. O'Donnell PH & Dolan ME. Cancer pharmacoethnicity: ethnic differences in susceptibility to the effects of chemotherapy. *Clin Cancer Res.* 2009;**15**:4806–14.
59. Peters EJ, Kraja AT, Lin SJ, Yen-Revollo JL, Marsh S, Province MA, & McLeod HL. Association of thymidylate synthase variants with 5-fluorouracil cytotoxicity. *Pharmacogenet Genomics.* 2009;**19**:399–401.
60. Bleibel WK, Duan S, Huang RS, Kistner EO, Shukla SJ, Wu Z, Badner JA, & Dolan ME. Identification of genomic regions contributing to etoposide-induced cytotoxicity. *Hum Genet.* 2009;**125**:173–80.
61. Shukla SJ, Duan S, Badner JA, Wu X, & Dolan ME. Susceptibility loci involved in cisplatin-induced cytotoxicity and apoptosis. *Pharmacogenet Genomics.* 2008;**18**:253–62.
62. Huang RS, Duan S, Kistner EO, Bleibel WK, Delaney SM, Fackenthal DL, Das S, & Dolan ME. Genetic variants contributing to daunorubicin-induced cytotoxicity. *Cancer Res.* 2008;**68**:3161–8.
63. Huang RS, Duan S, Kistner EO, Hartford CM, & Dolan ME. Genetic variants associated with carboplatin-induced cytotoxicity in cell lines derived from Africans. *Mol Cancer Ther.* 2008;**7**:3038–46.
64. Le Morvan V, Bellott R, Moisan F, Mathoulin-Pelissier S, Bonnet J, & Robert J. Relationships between genetic polymorphisms and anticancer drug cytotoxicity vis-a-vis the NCI-60 panel. *Pharmacogenomics.* 2006;**7**:843–52.
65. Puyo S, Le Morvan V, & Robert J. Impact of EGFR gene polymorphisms on anticancer drug cytotoxicity in vitro. *Mol Diagn Ther.* 2008;**12**:225–34.
66. Sasaki S, Kobunai T, Kitayama J, & Nagawa H. DNA methylation and sensitivity to antimetabolites in cancer cell lines. *Oncol Rep.* 2008;**19**:407–12.
67. Liu W, He L, Ramirez J, & Ratain MJ. Interactions between MDM2 and TP53 genetic alterations, and their impact on response to MDM2 inhibitors and other chemotherapeutic drugs in cancer cells. *Clin Cancer Res.* 2009;**15**: 7602–7.
68. Klein TE, Chang JT, Cho MK, Easton KL, Fergerson R, Hewett M, Lin Z, Liu Y, Liu S, Oliver DE, Rubin DL, Shafa F, Stuart JM, & Altman RB. Integrating genotype and phenotype information: an overview of the PharmGKB Project. *Pharmacogenomics J.* 2001;**1**:167–70.

Pharmacogenetics and Pharmacogenomics of Cardiovascular Disease

11

Daniel Kurnik and C. Michael Stein

SCOPE OF CARDIOVASCULAR THERAPEUTICS

An estimated 81 million American adults (37.1 percent) have cardiovascular disease (CVD), including coronary artery disease, hypertension, congestive heart failure, and stroke (1). CVD accounts for 36.3 percent of all deaths in the United States (2,400 deaths daily) and is the most common cause of death in developed countries. These chronic diseases usually require lifelong drug treatment, and medications for their treatment or prevention are among the most commonly prescribed drugs worldwide.

Hypertension, congestive heart failure, and other CVDs are often considered as discrete entities, but are, in fact, complex and heterogeneous syndromes mediated by many different pathophysiological mechanisms that eventually result in a similar clinical picture. As a result, clear and reproducible identification of environmental and genetic factors that contribute to these multifactorial diseases is challenging. Adding to the challenges, there is large variability, only a portion of which is genetically determined, among patients in their responses to a drug for a particular disease.

Cardiovascular Disease Therapeutics and Genetic Variation

Many variables such as age, sex, renal or hepatic function, comorbidity, and genetic factors can affect therapeutic responses and adverse effects of a drug, When the drug effect can be measured by a readily quantifiable surrogate marker (e.g., reduction in blood pressure), the distribution of the magnitude of drug response is typically unimodal, suggesting that many factors contribute to the variability in response, each of which may by itself have a small effect.

Conversely, a bi- or trimodal distribution of drug response in a population suggests a strong effect of a few genetic or environmental factors and typifies an early example of pharmacogenetics – the discovery of a hereditary trait determining the rate of metabolism of the blood pressure-lowering drug debrisoquine (Figure 11.1). This observation led to the discovery of many genetic variants in cytochrome P450 2D6 (CYP2D6), the phase I enzyme mediating debrisoquine hydroxylation. These genetic variants result in differences in the metabolic capacity of CYP2D6 and thus determine drug clearance, plasma concentrations, and, consequently, the effects of CYP2D6 substrate drugs (Figure 11.1) (2).

Subsequently, many studies have further defined the genetic contribution to heterogeneity in cardiovascular therapeutics. For a few drugs, the genetic contribution is clear, reproducible, and potentially clinically important (e.g., warfarin), but for most the data are less clear. In this chapter, we summarize current knowledge in a rapidly changing field.

Approach to Pharmacogenetics and Cardiovascular Drugs

Pharmacogenetic research regarding cardiovascular drugs has been based mainly on a hypothesis-driven, candidate gene approach: genetic variants of proteins known to interact with a drug's pharmacokinetic or pharmacodynamic pathways are examined for association with altered drug response. This approach requires detailed knowledge of such pathways to identify particular candidate genes. Specifically, variants in genes affecting absorption, distribution, metabolism, and excretion of a given drug can contribute to between-subject pharmacokinetic variability, and variants in genes encoding drug targets and associated signaling and regulatory pathways can affect pharmacodynamic variability (gene by drug interaction). In addition, genes associated with the pathophysiology and baseline risk of a disease have been investigated as pharmacogenetic predictors for drug response, even if not directly linked to a particular drug's

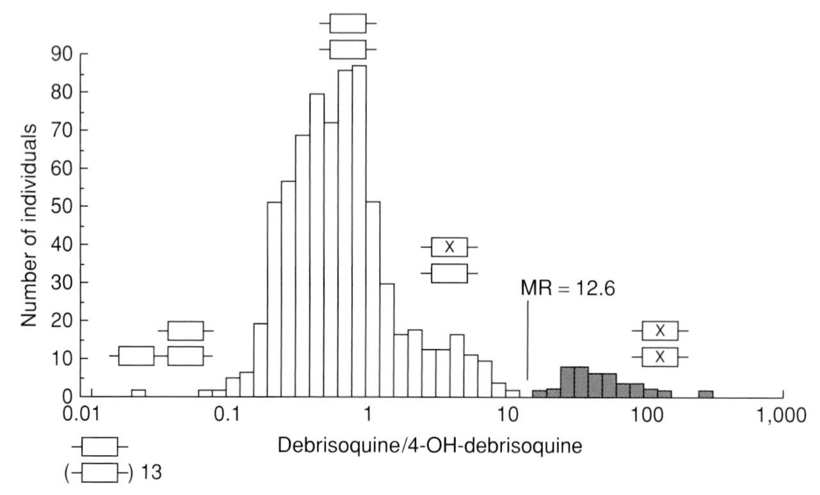

Figure 11.1. Distribution of the urinary debrisoquine metabolic ratio in 757 healthy Swedish subjects with schematic presentation of CYP2D6 genotypes. An X in an allele indicates a detrimental mutation. CYP2D6 mediates the hydroxylation of debrisoquine, and the ratio of debrisoquine and its hydroxyl metabolite in the urine reflects CYP2D6 activity. The distribution of the debrisoquine metabolic ratio is trimodal: small ratios represent ultrafast metabolizers (reflecting *CYP2D6* gene duplications), intermediate ratios represent normal metabolizers (reflecting one or two normally functioning alleles), and high ratios represent slow metabolizers (reflecting two reduced-function alleles). Adapted from (2).

PK-PD pathways. The rationale behind this approach is that treatment efficacy of a drug may depend on the pathophysiology and baseline risk of the disease (gene by disease by drug interaction). Furthermore, if a candidate gene variant results in functional changes in in vitro and in vivo models (e.g., decreased or increased function or expression in transfected cell cultures or genetically manipulated animals), there is greater biological plausibility for that variant to affect drug responses in humans.

However, for many drugs that have been used for decades to treat CVD, the precise mechanisms of action remain controversial (e.g., β-adrenergic receptor [AR] antagonists [β-blockers] and thiazide diuretics in the treatment of hypertension). Therefore, a systematic, unbiased search for pharmacogenetic markers may identify new genes not previously implicated in a drug's PK-PD pathway that affect its treatment effect. Accordingly, genome-wide association studies (GWAS) will play an increasingly important role in identifying such new gene-drug interactions.

HYPERTENSION

Disease Burden and Pharmacotherapy

Of every ten U.S. adults, approximately three (33 percent) have hypertension, two of whom (61 percent) receive treatment (1). Hypertension is a multifactorial syndrome in which pathophysiological disturbances in different regulatory mechanisms (e.g., renin-angiotensin-aldosterone system [RAAS], sympathetic nervous system, and renal sodium handling) result in a common phenotype – high blood pressure. Drug treatment for hypertension is effective in lowering blood pressure, reducing target-organ damage, and prolonging survival. However, only a third of patients diagnosed with hypertension in the United States (35 percent) achieve adequate blood pressure control (1). First-line antihypertensive drugs include diuretics, angiotensin-converting enzyme inhibitors (ACE-Is), angiotensin receptor blockers (ARBs), calcium channel blockers (CCBs), and β-blockers. Response to antihypertensive therapy is variable, and more than two-thirds of patients will require two or more drugs for adequate blood pressure control. Because treatment response and adverse effects can be predicted only poorly in an individual patient, therapy typically involves a trial-and-error approach with different doses, drug classes, and drug combinations, a lengthy process that contributes to lower adherence and thus lower rates of blood pressure control. Genetic factors predicting drug response could potentially help guide pharmacotherapy.

Pharmacogenetics of Antihypertensive Therapy

Blood pressure is 40 percent to 55 percent heritable (1); however, only a few uncommon genetic mutations have been identified that by themselves cause rare forms of hypertension, and thus their contribution to blood pressure in the general population is very small. Instead, hypertension is a polygenetic disease, involving many variants that individually contribute only a little. Pharmacogenetic studies examining responses to a particular antihypertensive drug have investigated candidate variants in pathways of blood pressure regulation and in the PK-PD pathways of particular drugs. Many reports of associations between genetic variants and response to a drug are unconfirmed or have not been reproduced reliably.

β-Blockers

β-Blockers are widely used in patients with CVD, in particular, for congestive heart failure, coronary artery disease, and hypertension. This class comprises more than twenty different drugs that vary in their selectivity for β_1- versus β_2-ARs, distribution (lipophilicity), and elimination (metabolism or renal excretion). Although the exact mechanism of blood pressure reduction during chronic treatment is not fully understood, inhibition of adrenergic stimulation of the heart and of renin secretion in the kidneys is thought to contribute. The primary target of β-blockers in CVD is the β_1-AR, whose secondary signaling pathways and regulatory proteins include G-proteins and G-protein-related receptor kinases (GRKs).

There are two common nonsynonymous polymorphisms in the coding area of *ADRB1*, the gene encoding the β_1-AR. The Arg389Gly variant (minor allele frequency [MAF] for whites, 0.24–0.24) results from a substitution of glycine for arginine in the intracellular carboxy-terminal receptor loop, a region important for G-protein coupling (3). The Ser49Gly variant (MAF in whites, 0.12–0.16) results from the substitution of glycine for serine in the extracellular, amino-terminal receptor loop. Studies in transfected cells show that the Gly389 variant has impaired binding (coupling) to Gs, resulting in reduced response to receptor activation by an agonist and to receptor blockade by an antagonist (3). The Gly49 variant results in increased receptor downregulation (desensitization) after prolonged activation by agonist. Thus, the Ser49 and Arg389 alleles characterize a hyperfunctional β_1-receptor associated with increased signal transduction. Concordant with the in vitro findings, studies in healthy subjects under strictly controlled conditions have shown that carriers of the Arg389 and the Ser49 alleles are more responsive to β-blockade than are the respective minor alleles (4).

Several studies have examined the impact of these common functional genetic variants on blood pressure response to β-blocker therapy in hypertension (4). In some, carriers of the major alleles (Arg 389 and Ser49) were more responsive than carriers of the minor alleles (Gly 389 and Gly 49), and a rank order of blood pressure responsiveness could be established for genotype combinations (Ser49-Arg389 > Gly49-Arg389 > Ser49-Gly389 haplotype) (5). In fact, the carriers of hyporesponsive haplotypes did not significantly reduce blood pressure after β-blockade. However, not all studies have found an effect of the *ADRB1* genotype and response to β-blockers, suggesting that the genetic component of response may be overshadowed by other clinical variables (4). The potential pharmacogenetic significance of *ADRB1* haplotypes in hypertensive patients was suggested by the INVEST study (6). Among 5,895 patients, the Ser49-Arg389 haplotype was associated with higher all-cause mortality (hazard ratio [HR], 3.66; 95% confidence interval [CI], 1.68–7.99; $P = 0.001$). However, this higher mortality was offset by treatment with the β-blocker atenolol (HR, 2.31; 95% CI, 0.82–6.55; $P = 0.11$), but not with the CCB verapamil (HR, 8.58; 95% CI, 2.06–35.8; $P = 0.003$). These findings suggest that, in hypertensive patients carrying the Ser49-Arg389 haplotype, β-blockers may be the treatment of choice. However, confirmatory studies will be required.

Other candidate genes in the pharmacodynamic or blood pressure regulatory pathways that were associated with the therapeutic effect of β-blockers include variants in the G-protein subunits (*GNAS1* and *GNB3*), β_2- and α_2-ARs (*ADRB2* and *ADRA2A*), endothelial nitric oxide synthase, the low-density lipoprotein (LDL) receptor gene (*LDLR*), and adducin (*ADD1*); however, these associations have not been adequately validated in confirmatory studies and remain inconclusive.

CYP2D6 genotype (grouped into poor, intermediate, and extensive metabolizer class) affects the clearance of β-blockers that are CYP2D6 substrates (7). For example, plasma concentrations of metoprolol can be four to six times higher in CYP2D6 poor metabolizers in comparison with extensive metabolizers (7). However, a number of small studies found no clear association between genotype and therapeutic or adverse effects of such β-blockers. More recently, in a population-based study (Rotterdam Study) (8), metoprolol had a greater effect on heart rate in patients with poor-metabolizer *CYP2D6* genotypes, and some studies have suggested a relationship between CYP2D6 poor-metabolizer status and more frequent adverse effects (7). The clinical significance of these findings, however, is unclear, because, in clinical practice, β-blockers are usually titrated gradually to achieve the target heart rate.

Diuretics

Thiazide diuretics are among the oldest and most effective drugs recommended as first-line treatment for hypertension. Thiazide diuretics block the Na^+-Cl symporter in the distal convoluted tubule, resulting initially in plasma volume contraction and reduced cardiac output. However, as is the case with β-blockers, the exact mechanism of blood pressure reduction during chronic thiazide therapy remains unclear.

Adducin is a cell membrane skeleton protein composed of α- and β-subunits. A common genetic variation in the α-adducin subunit (*ADD1* Gly460Trp) affects ion transport in the renal tubular cell, presumably by changes in the actin cytoskeleton, resulting in increased renal tubular reabsorption of sodium. The Trp460 variant has been associated with renal sodium retention, a greater prevalence of hypertension, in particular, the salt-sensitive form, and a greater risk of stroke. In some studies, the 460Trp allele was associated with greater blood pressure reduction by thiazide diuretics, and with a greater protective effect of thiazides, but not other antihypertensive medications, against a combined end point of myocardial infarction (MI) and stroke. However, a number of recent large prospective or population-based pharmacogenetic studies (e.g., INVEST-GENES, PHARMO registry) (9, 10) did not reproduce these findings.

Atrial natriuretic peptide (ANP) is a hormone that controls electrolyte and fluid volume homeostasis through its diuretic action. In the large GenHAT study ($n = 39{,}000$), the genetic substudy of ALLHAT, a variant in the gene encoding the ANP precursor, atrial natriuretic precursor A (NPPA), affected response to antihypertensive treatment (11). In hetero- or homozygous carriers of the minor *NPPA* T2238C allele, approximately 40 percent of the population, treatment with the diuretic chlorthalidone resulted in greater reduction in blood pressure after six months and in lower risk of outcomes (coronary heart disease, CVD, and all-cause mortality) than treatment with amlodipine. In contrast, among patients homozygous for the major allele (TT), chlorthalidone was associated with a higher risk for disease outcomes and mortality. However, the drug × genotype interactions were not statistically significant after correction for multiple comparisons (11).

Other gene variants that have been associated with response to diuretics include the ACE I/D variant, *GNB3* C825T, *AGTR1* A1166C, but these findings have not been consistently confirmed.

The Renin-Angiotensin-Aldosterone System

The angiotensin-converting enzyme (ACE) is involved in the regulatory pathways of three hormones that play important roles in blood pressure regulation: ACE cleaves angiotensin I to angiotensin II, a potent vasoconstrictor, which itself stimulates the adrenal release of the mineralocorticoid, aldosterone. In addition, ACE degrades bradykinin, a potent vasodilator. Thus, inhibition of ACE by ACE-I results in reduced concentrations of angiotensin II and aldosterone and increased concentrations of bradykinin, accounting for their blood pressure-lowering effect.

An important candidate in RAAS pharmacogenetics is a common 287-base pair insertion/deletion variant in intron 16 of the ACE gene, with the deletion allele having a frequency of 0.35 to 0.61 in different ethnic groups. In early studies, the deletion variant was associated with higher plasma and tissue ACE concentrations and an increased risk for MI. Later studies showed inconsistent associations with clinical CVD, and a recent meta-analysis concluded that there is evidence for a modest positive association between the deletion allele and coronary artery disease (12). Similarly, many pharmacogenetic studies have examined the association of blood pressure response to ACE-I therapy by the *ACE* I/D genotype, but the results are inconsistent. In by far the largest study (GenHAT), among 38,000 patients, the *ACE* Ins/Del genotype did not affect response to the ACE-I, lisinopril, or any of the other drugs studied (chlorthalidone, amlodipine, and doxazosin), and it was not associated with any CVD outcome (13). Possible reasons for these disparate results include the hypothesis that the intronic Ins/Del variant itself does not confer any functional differences, but is a marker for another functional variant with which it is in variable linkage disequilibrium.

Other genetic variants that have been linked to therapeutic response to ACE-I include variants in the angiotensin receptor (AGTR1), but these results remain largely unconfirmed.

ARBs do not block the synthesis of angiotensin II, but block its biological action by inhibiting its physiological receptor (angiotensin II receptor type 1, AGTR1). In largely unconfirmed findings, the therapeutic effect of ARBs has been associated with the *ACE* I/D variant and variants in the genes coding for angiotensinogen, the angiotensin II type 1 receptor (*AGTR1*), and for CYP11B2 (aldosterone synthase). The effect size of the pharmacogenetic component of response in general was small.

Calcium Channel Blockers

Voltage-sensitive calcium channels mediate the entry of extracellular calcium into vascular smooth muscle cells, resulting in vasoconstriction and thus blood pressure elevation. Blocking these channels with CCBs reduces intracellular calcium and mediates vasodilation, a reduction in total peripheral resistance, and thus blood pressure. The large conductance, calcium- and voltage-sensitive potassium channel beta1 subunit (KCNMB1) is a protein associated with calcium sensitivity and blood pressure regulation, and was therefore a candidate gene for CCB pharmacogenetics. In one large study (INVEST, $n = 5{,}979$), genetic variants in *KCNMB1* affected time to blood pressure control and intensity of concomitant antihypertensive therapy, and the incidence of the combined CVD outcome (mortality, nonfatal MI, and stroke) in patients treated with verapamil, but not with atenolol. A separate analysis of the same cohort suggested that variation in *CYP3A5*, coding for an enzyme that contributes to the metabolism of verapamil, might influence response to verapamil. In blacks and Hispanics (ethnic groups in whom functional alleles are found in 81.4 percent and

43.1 percent of people in comparison with 16.8 percent in whites), patients with two functional alleles, and presumably greater CYP3A5-mediated verapamil clearance, had a smaller drug response than those carrying one or no functional allele.

HEART FAILURE

Heart failure (HF) is characterized by the activation of multiple counterregulatory mechanisms aimed at preserving cardiac function; for example, neurohormonal activation of sympathetic nervous system, RAAS, and natriuretic peptides, resulting in retention of salt and water and increased cardiac output. Eventually, however, these mechanisms contribute to disease progression. Pharmacological treatment of HF targets these diverse deleterious pathways and includes β-blockers, ACE-I and ARBs, and diuretics. There is a great variability in outcomes of patients with HF. Thus, pharmacogenetic markers associated with the therapeutic efficacy of these drugs in other diseases such as hypertension, have also been studied in HF.

ACE-I and ARBs

A number of genetic variants in proteins of the RAAS have been identified as prognostic markers in HF. The common ACE deletion variant was associated with higher serum concentrations of ACE, angiotensin II, and aldosterone, and shorter survival, reflecting chronically increased RAAS activity. In the GRACE study, the deletion allele was also associated with greater responsiveness to treatment with ACE-I which markedly attenuated the negative association of the deletion allele with survival (14). In fact, among patients receiving high-dose ACE-I treatment, there was no longer a difference in survival by *ACE* I/D genotype. This example illustrates a genotype-disease-drug interaction: a genetic marker associated with disease progression can also be a marker for greater drug response. Therefore, in patients on drug treatment, the gene effect on disease progression could be masked.

The information about the pharmacogenetics of ARBs and aldosterone-antagonists in HF is limited. Candidates include variants in the genes coding for ACE, angiotensinogen, the angiotensin II receptor types 1 and 2 (AGTR1/AGTR2), and a promoter variant in the aldosterone synthase gene (CYP11B2), associated with higher aldosterone production, hypertension, coronary artery disease, and progression and reduced survival in HF in some studies.

β-Blockers

HF is associated with increased sympathetic tone, which ultimately contributes to disease progression and death. Because β-blockers reduce heart rate and contractility in normal hearts, β-blocker therapy was long considered contraindicated in HF; however, some β-blockers improve symptoms and prolong survival in chronic HF. The candidates for pharmacogenetic HF studies include the genes encoding for ARs (β_1-, β_2-, and α_{2C}-ARs), signaling pathways (e.g., G-protein β_3 subunit), and mediators of receptor desensitization (e.g., GRK5) (7).

The common Arg389Gly variant in the β_1-AR encodes a hypofunctional receptor with reduced response to β-blockers in vitro and in hypertensive patients (4). Concordant with those findings, some studies showed that HF patients with reduced ejection fraction who carried the Gly allele had a smaller improvement in cardiac function (assessed by the surrogate marker of the ejection fraction) compared with Arg homozygotes (7). The Beta Blocker Evaluation of Survival Trial evaluated the use of bucindolol in 2,708 patients with advanced HF (New York Heart Association class III to IV). Although bucindolol-treated patients overall did not have a survival benefit in comparison with patients receiving placebo, there was a strong interaction between Arg389Gly genotype and bucindolol treatment (15). Patients who carried the hyporesponsive Gly389 allele did not benefit from bucindolol therapy, whereas Arg389 homozygotes receiving bucindolol had a marked improvement in survival and hospitalization rates in comparison with patients with the same genotype receiving placebo (HR, 0.66; 95% CI, 0.50–0.88). A pharmacogenetic subgroup analysis of another controlled trial (MERIT-HF) did not find an association of *ADRB1* genotypes with the primary outcome (all-cause mortality or hospitalization); however, this analysis has been criticized for analyzing placebo and treatment groups together, not allowing the assessment of treatment by genotype interactions. However, other studies with metoprolol and carvedilol also found no effect of *ADRB1* genotypes on survival in HF.

Acting synergistically with the *ADRB1* Arg389 allele, a deletion variant in the adrenergic α_{2C}-AR was associated with a higher risk of HF in a case-control study (16) and a higher risk for death or cardiac transplant in a cohort of HF patients (17), but other studies did not confirm a detrimental role of the deletion allele in HF. The effect of the combination of variants in the α_{2C}- and β_1-ARs on responses to β-blockers will be of interest (7).

The GRK5 phosphorylates cardiac β-ARs, resulting in receptor uncoupling from G-proteins and thus receptor desensitization. An SNP (rs17098707) resulting in the *GRK5* Gln41Leu variant is common in African Americans (MAF approximately 25 percent) but not in whites (MAF 1 percent to 2 percent) and is associated with increased β_1-AR desensitization in vitro and in transgenic mouse models (18). Black patients with HF carrying the GRK5 Leu41 allele had a longer transplantation-free survival than Gln41 homozygotes, compatible with chronic endogenous desensitization of β_1-adrenergic signaling pathways. Treatment with β-blockers increased survival in Gln41 homozygotes but did not affect Leu41 carriers, so that β-blocker therapy neutralized the genotypic differences in prognosis (18). Additional confirmatory studies will be required.

Only a few studies have examined the effect of other adrenergic pathway genetic variants on drug response in HF, and positive (e.g., with variants in the norepinephrine reuptake transporter) and negative reports await confirmation in larger studies (7).

LIPID-LOWERING DRUGS

Drug treatment with lipid-lowering drugs, especially statins, reduces morbidity and mortality in patients with hyperlipidemia and in patients with normal plasma lipids and increased CVD risk. Every 10 percent reduction in serum cholesterol is estimated to reduce total and coronary artery disease mortality by approximately 11 percent and 15 percent, respectively. A number of drug classes improve the plasma lipid profile (decrease LDL cholesterol and triglycerides and increase high-density lipoprotein cholesterol), and statins (β-hydroxy-β-methylglutaryl-coenzyme A [HMG-CoA] inhibitors) have unequivocally been shown to reduce cardiovascular and total mortality. Thus, we will focus on the pharmacogenetics/pharmacogenomics of statins.

Statins

The lipid-lowering response to statin therapy is highly variable, requiring different doses and drug combinations in different patients. Pharmacogenetic markers have been sought for the prediction of both therapeutic efficacy and the occurrence of adverse events (statin-induced myopathy). Because the efficacy of lipid-lowering drugs is highest in subjects with higher pretreatment cholesterol levels, genetic variants associated with higher plasma lipids have been candidate genes for pharmacogenetic studies. Other candidate genes include variants in genes encoding relevant PK enzymes (metabolizing enzymes and drug transporters) and drug targets.

Pharmacokinetic Candidate Genes

Statins differ in their physicochemical properties (e.g., lipophilicity) that substantially affect their PK profile. Effects of a certain genetic variant will depend on the contribution of the respective gene to a particular statin's PK pathway, and therefore vary across the drug class.

For most statins, drug transporters (e.g., SLCO1B1, SLC15A1, ABCB1, and ABCC2) facilitate absorption, hepatic uptake, and biliary and renal excretion. Metabolism, often to active metabolites, is mediated by different CYPs in the intestine and the liver (e.g., CYP3A4, CYP2C9) and phase II conjugation, further complicating the PK-PD relation. Furthermore, systemic exposure (often measured as the area under the curve [AUC] of plasma concentration) may not reflect exposure of the target organ, the liver; indeed, a decrease in hepatic uptake (reduced activity of drug transporters) may be associated with increased plasma AUC, resulting in an inverse relationship between therapeutic effect and plasma AUC.

The drug transporter solute carrier organic transporter (SLCO) 1B1 mediates the active uptake of many statins into hepatocytes. In a number of small studies, genetic variants in *SLCO1B1* were associated with changes in statin plasma AUC, but the association of these pharmacokinetic changes with clinical outcomes was unclear. A recent GWAS in patients who developed myopathy during high-dose simvastatin treatment (80 mg/day) (19) revealed that a T>C variant in exon 6 of *SLCO1B1* (rs4149056), resulting in the Val174Ala polymorphism (MAF of approximately 15 percent in Europeans), represented a haplotype associated with an increased risk for myopathy: patients with the CT and CC genotypes had odds ratios of 4.5 (95% CI, 2.6–7.7) and 16.9 (95% CI, 4.7–61.1), respectively, and approximately 60 percent of all cases of myopathy could be attributed to this *SCLO1B1* variant. In keeping with the proposed mechanism, in previous studies the C allele was also associated with higher statin plasma concentrations and slightly less therapeutic efficacy in LDL reduction.

Similarly, in the ACCESS study, genotypes and haplotypes of MDR1 (ABCB1), a drug transporter mediating intestinal absorption and thus bioavailability of statins, had a small effect on LDL response to statins (20). Although functional genetic variants in CYP3A, 2D6, and 2C9 have been associated with changes in the clearance of their substrate statins or with their adverse effects, the association with therapeutic efficacy is less clear, possibly because many metabolites still have lipid-lowering activity.

Pharmacodynamic Candidate Genes

Genetic variants in numerous proteins involved in lipid homeostasis have been associated with lipid disorders or have been identified as cardiovascular risk factors. Many of these variants have also been studied in pharmacogenetic studies of responses to statin therapy, including variants in genes coding for HMG-CoA reductase (HMGCR), the drug target of statins; LDLR; apolipoprotein E (ApoE); lipoprotein lipase; squalene synthase;

cholesterol transfer protein; and transcription factors such as sterol-regulatory element binding proteins and peroxisome proliferator-activated receptors.

One example of the candidate gene approach in statin pharmacogenetics is ApoE. ApoE is a constituent of lipoproteins that mediates their interaction with the LDL receptor. The three major genetic variants (ε_2 to ε_4) in *apoE* have different affinities for the LDL receptor, resulting in differences in lipoprotein clearance and upregulation (*apoE2*) or downregulation (*apoE4*) of HMGCR. Accordingly, in some studies *apoE2* and *apoE4* were associated with greater and lesser statin response, respectively.

However, many associations found in small studies were not reproduced in other studies or meta-analyses. In two larger studies, the Pravastatin Inflammation/CRP Evaluation (PRINCE) and ACCESS study, numerous genetic variants in multiple candidate genes were studied, but only two highly linked intronic single-nucleotide polymorphisms (SNPs) in *HMGCR* were associated with reduced therapeutic efficacy of pravastatin (22 percent less reduction in total cholesterol), whereas variants in *apoE* and *ABCB1* had only small, clinically insignificant effects (20). In a recent analysis of the TNT study in 5,700 patients randomly assigned to low- or high-dose atorvastatin, a combined GWAS/candidate gene approach confirmed that, among common SNPs (MAF $\geq$ 5.0 percent), only variants in *apoE* (most significantly, rs7412; MAF = 5.5 percent) were associated with the lipid-lowering effect of atorvastatin (21). A rare variant in *PCSK9* and SNPs in *HMGCR* were also associated with LDL response, but with a smaller effect size. Clinical variables (age, gender) affected LDL response to a similar extent as the strongest genetic predictors, and the authors concluded that genetic markers are currently not useful in predicting LDL response to atorvastatin.

ANTIPLATELET DRUGS AND ORAL ANTICOAGULANTS

Two antiplatelet drugs are in common use for the prevention and treatment of CVD: aspirin and clopidogrel. They inhibit platelet aggregation through different mechanisms. Aspirin irreversibly acetylates a serine at position 529 of the cyclo-oxygenase-1 (COX-1) enzyme that is responsible for the formation of thromboxane from arachidonic acid. Clopidogrel is a prodrug whose active metabolite irreversibly blocks the platelet adenosine diphosphate (ADP) $P2Y_{12}$ receptor. Substantial variability exists among individuals in their response to aspirin and clopidogrel, and the concept of "aspirin resistance" and "clopidogrel resistance" has emerged to describe individuals with suboptimal platelet inhibition after receiving these drugs. There is substantial interest in identifying a genetic component to resistance to antiplatelet drugs because therapy could be tailored appropriately.

Clopidogrel

The formation of the active metabolite from the inactive parent compound clopidogrel is catalyzed in vitro by several CYP450 enzymes including CYP3A4, CYP2C19, CYP2B6, and CYP2C9. The relative contribution of each in vivo is unclear. However, drug interactions with inhibitors of particular CYPs and studies in subjects with genetic variants of the CYPs of interest have provided clues.

Earlier studies showed that coadministration of omeprazole, a CYP2C19 inhibitor, attenuates the effect of clopidogrel on platelet aggregation, presumably by inhibiting the metabolism of the prodrug, clopidogrel, to its active metabolite. In a recent study, healthy volunteers carrying a loss-of-function variant of *CYP2C19* (most commonly *2), comprising 30 percent, 40 percent, and 55 percent of white, black, and Asian populations, respectively, had lower plasma concentrations of the active clopidogrel metabolite and reduced antiplatelet effects ex vivo (22). Moreover, clinical studies confirmed the impact of *CYP2C19* genotype on clinical outcomes (22, 23). In patients after acute coronary syndrome, each *CYP2C19* loss-of-function allele increased the risk for the composite outcome (cardiovascular death, nonfatal MI, or stroke) by about 50 percent, and in patients after percutaneous coronary intervention, in whom clopidogrel is considered especially important to prevent stent thrombosis, the hypofunctional alleles were associated with an approximately threefold increase in the incidence of in-stent thrombosis.

Genetic variations in other P450 enzymes implicated in clopidogrel activation (e.g., CYP3A4/5, CYP2C9, and CYP2B6) or in the drug target, the ADP $P2Y_{12}$ receptor, were not consistently associated with platelet aggregation or clinical outcomes. However, the C3435T variant in *ABCB1*, the P-glycoprotein drug transporter implicated in the absorption of clopidogrel at the intestinal brush border, has been associated with reduced clopidogrel absorption and a 1.7-fold greater cardiovascular event rate in patients after acute MI (23).

Coadministration of clopidogrel and ketoconazole, a prototypic CYP3A4/5 inhibitor, decreases the formation of the active metabolite of clopidogrel and attenuates platelet inhibition. In contrast, these effects are not seen with prasugrel, a newer ADP $P2Y_{12}$ receptor antagonist. However, genetic variations of *CYP3A4* and *CYP3A5* have not consistently been associated with altered clopidogrel response or with clinical outcomes.

In summary, the strongest evidence implicates genetic variation in *CYP2C19* as a contributor to variability in the response to clopidogrel. Alleles with decreased enzyme activity are associated with impaired formation of the active metabolite of clopidogrel, resulting in attenuated platelet inhibition and decreased clinical effectiveness.

Aspirin

Studies of genetic factors contributing to the variability in aspirin response have suffered from the lack of a uniform outcome measure. Aspirin response has variously been measured as the inhibition of the following measures: serum thromboxane formation, urinary thromboxane excretion, bleeding time, light transmission platelet aggregation, and clotting in a platelet function analyzer. Results using different outcomes have not always been concordant. A recent review (24) reported that the genetic contribution to aspirin response of fifty polymorphisms in eleven genes had been studied in 2,834 subjects. There were no significant associations with variants in COX-1, GP1a, P2Y1, and P2Y12 genes. There was a significant association between the P1A1/A2 polymorphism of GPIIIa and aspirin response in healthy subjects, but the effects were not significant when subjects with CVD were also considered. Thus, on balance, no clinically significant genetic contribution to aspirin response has been identified.

Warfarin

Oral anticoagulants, most commonly warfarin, are widely used for the prevention and treatment of thromboembolic diseases. Warfarin treatment is complicated by a high interindividual variability in dose requirements (up to twentyfold) and a narrow therapeutic window, so that comparatively small increases or decreases out of a therapeutic range markedly increase the risks of toxicity and inefficacy, respectively (25). Furthermore, there are many interactions with drugs and food resulting in considerable intraindividual variability in warfarin response, and a complicated PK-PD relationship between dose and the laboratory marker of therapeutic activity, the international normalized ratio (INR). Thus, warfarin is among the drugs most commonly associated with serious adverse events (i.e., serious bleeding) and tops the list of medications associated with drug-related deaths (25).

Since its introduction into clinical medicine more than sixty years ago, individualization of warfarin dosing has traditionally been performed by empiric dose titration. Warfarin is often started at an average dose (5 to 10 mg daily) for the first few days, and further doses are then adjusted according to repeated INR measurements (usually 1 to 2 per week) until the steady-state dose has been established empirically for a particular patient. However, with this trial-and-error approach a considerable portion of patients are under- or overanticoagulated during the first weeks of treatment, resulting in high rates of bleeding or thrombotic complications, especially during the early stages of warfarin treatment (25). Individualization of warfarin dosing with the use of clinical variables affecting warfarin sensitivity, for example, age, concomitant diseases and medications, and liver function, improves warfarin dose prediction somewhat. However, because these factors explain only approximately 20 percent of the interindividual variability in warfarin requirements, other factors that predict an individual's sensitivity to warfarin have been sought. Starting with the initial report by Aithal et al. in 1999, over the past decade, genetic variants in both the drug-target (pharmacodynamic effect) and drug-metabolizing enzymes (pharmacokinetic effect) have been shown to predict warfarin sensitivity (26). This has ushered in the prospect of a new era of pharmacogenetics-based warfarin dosing.

Warfarin blocks the enzyme vitamin K oxide reductase (VKORC1) and thus the generation of the active form of vitamin K, ultimately resulting in a reduction of activated coagulation factors. Genetic variants in VKORC1 affect warfarin sensitivity. In particular, the common *VKORC1* A haplotype, often represented by the tag SNP rs9934438 (-1639G>A), is associated with reduced VKORC1 expression and thus greater warfarin sensitivity (26). The haplotype A is associated with a more rapid rise of INR at the start of therapy, resulting in a shorter time to reach therapeutic and supratherapeutic INRs, and also in an estimated reduction in warfarin steady-state dose (25 percent for each copy of haplotype A) and a higher risk of bleeding, especially during the first months of therapy. The prevalence of haplotype A varies among different ethnicities (African Americans, 15 percent; whites, 42 percent; Asians, 90 percent), and the greater warfarin sensitivity in Asians can be attributed in part to these ethnic differences (26, 27).

The active *S*-enantiomer of warfarin is metabolized by the hepatic enzyme cytochrome P450 2C9 (CYP2C9). Two nonsynonymous SNPs result in reduced-function enzymes (*2, Arg11Cys, and *3, Ile359Leu) and are associated with reduced *S*-warfarin clearance. Thus, each *2 and *3 allele is associated with an estimated mean reduction of 17.5 percent to 19 percent and 28 percent to 33 percent in steady-state warfarin dose, respectively (25). Again, these alleles are also associated with a higher risk for supratherapeutic INRs, a longer time to steady state, and a three- to fourfold higher bleeding risk, especially during the induction period (26, 27).

Genetic variants in *VKORC1* and *CYP2C9* have been estimated to explain 30 percent to 35 percent of the interindividual variability in warfarin dose requirement (Figure 11.2) (28). Although other genetic variants (e.g., *VKORC1* Asp36Tyr, *CYP2C9*5*; variants in other proteins associated with the coagulation cascade, e.g., factor II, VII, γ-glutamyl carboxylase, calumenin) were

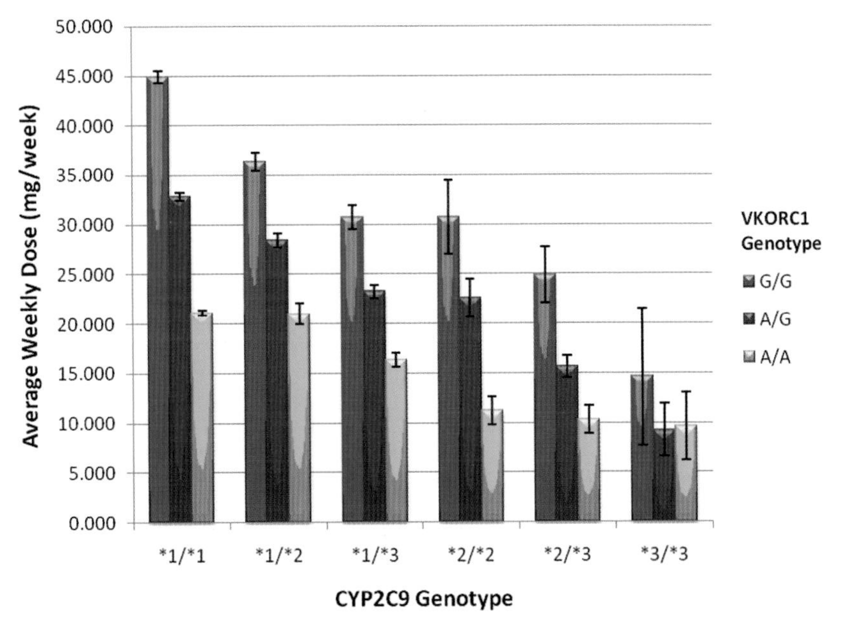

Figure 11.2. Genetic variants in CYP2C9 and VKORC1 affect warfarin dose requirement. Genetic variants in CYP2C9 (*2 and *3), the metabolizing enzyme of *S*-warfarin, and in VKORC1 (G-1639A), warfarin's drug target, affect warfarin weekly doses at steady state in an additive manner. For two otherwise similar patients, the predicted weekly warfarin dose varies from about 45 mg (in a patient homozygous for CYP2C9 *1 and VKORC1 G/G) to <10 (in a patient homozygous for CYP2C9*3 and VKORC1 A/A). Adapted from the International Warfarin Pharmacogenetic Consortium (IWPC), PharmGKB (27).

associated with warfarin sensitivity in some studies, the attributable effect was generally small, either because of a small biological effect or the rarity of the variant. Indeed, a recent genome-wide search did not identify other common genetic markers with large effects on warfarin sensitivity (29).

In comparison with models using clinical information only (age, body weight or height, concomitant disease, and medication), the inclusion of pharmacogenetic information (*VKORC1* and *CYP2C9* genotypes) has improved dose prediction for an individual patient considerably, raising the variability explained by the model (R^2) from approximately 20 percent to 45 percent to 55 percent (27, 28). Evidence-based algorithms predicting warfarin dose requirements based on clinical and pharmacogenetic factors are available (28). Thus, warfarin is an attractive candidate to become the first drug in which genetically based dosing could have a substantial impact on clinical outcomes: it is commonly used, has a narrow therapeutic window and substantial adverse events, and exhibits large interindividual variability in drug sensitivity, a large portion of which is explained by a few genetic markers. However, it is still unclear whether pharmacogenetics-based dosing translates into improved anticoagulant control. In particular, a considerable part of the warfarin sensitivity attributable to genetic variants is reflected in the early INR results, and the traditional INR-based titration may therefore capture much of the genetically mediated differences in warfarin sensitivity. Large randomized controlled trials are in progress to examine to what extent pharmacogenetics-based dosing algorithms improve clinical outcomes compared with traditional INR-based dose titration.

CONCLUSION

Cardiovascular pharmacogenetics is a rapidly expanding field. Early studies, often in small cohorts, focused on how few genetic variants in certain candidate genes affect response to cardiovascular medications. Some genetic variants appear to identify subgroups of patients at high risk for adverse outcomes and disease progression (gene by disease interaction), and these variants may therefore be markers for patients who would most benefit from drug therapy (gene by disease by drug interaction). Other variants directly affect a drug's PK-PD mechanisms (gene by drug interaction). Although promising, many of these earlier studies yielded inconsistent results, and many of the positive associations could not be reproduced in validation cohorts. Recently, advances in large-scale genotyping and the development of statistical methods for the analysis of such data have allowed a more systematic pharmacogenomic approach in large populations, examining simultaneously either numerous variants in many candidate genes or performing an unbiased genome-wide analysis. Although these approaches are not sensitive for the detection of rare variants or variants with small effect sizes, they allow the identification of new candidate genes and may thus shed new light on pharmacological pathways. So far, only a few examples have been identified in which a few genetic variants have a great effect on cardiovascular drug effect (e.g., warfarin), making genotype-based dosing feasible. Future studies will determine whether the combination of multiple genotypes, each with comparatively small effect size by itself, can collectively define genetic patterns that predict drug response in a clinically useful way.

REFERENCES

1. American Heart Association. Heart Disease and stroke statistics – 2008 update. americanheart.org/statistics. 2008. (June 29, 2011)
2. Bertilsson L. Metabolism of antidepressant and neuroleptic drugs by cytochrome p450s: clinical and interethnic aspects. *Clin Pharmacol Ther.* 2007; **82**:606–9.
3. Small KM, McGraw DW, & Liggett SB. Pharmacology and physiology of human adrenergic receptor polymorphisms. *Annu Rev Pharmacol Toxicol.* 2003;**43**:381–411.
4. Rosskopf D & Michel MC. Pharmacogenomics of G protein-coupled receptor ligands in cardiovascular medicine. *Pharmacol Rev.* 2008;**60**:513–35.
5. Liu J, Liu ZQ, Yu BN, et al. beta1-Adrenergic receptor polymorphisms influence the response to metoprolol monotherapy in patients with essential hypertension. *Clin Pharmacol Ther.* 2006;**80**:23–32.
6. Pacanowski MA, Gong Y, Cooper-Dehoff RM, et al. beta-adrenergic receptor gene polymorphisms and beta-blocker treatment outcomes in hypertension. *Clin Pharmacol Ther.* 2008;**84**:715–21.
7. Azuma J & Nonen S. Chronic heart failure: beta-blockers and pharmacogenetics. *Eur J Clin Pharmacol.* 2009;**65**:3–17.
8. Bijl MJ, Visser LE, van Schaik RH, et al. Genetic variation in the CYP2D6 gene is associated with a lower heart rate and blood pressure in beta-blocker users. *Clin Pharmacol Ther.* 2009;**85**:45–50.
9. Gerhard T, Gong Y, Beitelshees AL, et al. Alpha-adducin polymorphism associated with increased risk of adverse cardiovascular outcomes: results from GENEtic Substudy of the INternational VErapamil SR-trandolapril STudy (INVEST-GENES). *Am Heart J.* 2008;**156**:397–404.
10. van Wieren-de Wijer DB, Maitland-van der Zee AH, de Boer A, et al. Interaction between the Gly460Trp alpha-adducin gene variant and diuretics on the risk of myocardial infarction. *J Hypertens.* 2009; **27**:61–8.
11. Lynch AI, Boerwinkle E, Davis BR, et al. Pharmacogenetic association of the NPPA T2238C genetic variant with cardiovascular disease outcomes in patients with hypertension. *JAMA.* 2008;**299**:296–307.
12. Zintzaras E, Raman G, Kitsios G, Lau J. Angiotensin-converting enzyme insertion/deletion gene polymorphic variant as a marker of coronary artery disease: a meta-analysis. *Arch Intern Med.* 2008;**168**:1077–89.
13. Arnett DK, Davis BR, Ford CE, et al. Pharmacogenetic association of the angiotensin-converting enzyme insertion/deletion polymorphism on blood pressure and cardiovascular risk in relation to antihypertensive treatment: the Genetics of Hypertension-Associated Treatment (GenHAT) study. *Circulation.* 2005;**111**:3374–83.
14. McNamara DM, Holubkov R, Postava L, et al. Pharmacogenetic interactions between angiotensin-converting enzyme inhibitor therapy and the angiotensin-converting enzyme deletion polymorphism in patients with congestive heart failure. *J Am Coll Cardiol.* 2004;**44**:2019–26.
15. Liggett SB, Mialet-Perez J, Thaneemit-Chen S, et al. A polymorphism within a conserved beta(1)-adrenergic receptor motif alters cardiac function and beta-blocker response in human heart failure. *Proc Natl Acad Sci USA.* 2006;**103**:11288–93.
16. Small KM, Wagoner LE, Levin AM, Kardia SL, & Liggett SB. Synergistic polymorphisms of beta1- and alpha2C-adrenergic receptors and the risk of congestive heart failure. *N Engl J Med.* 2002;**347**:1135–42.
17. Kardia SL, Kelly RJ, Keddache MA, et al. Multiple interactions between the alpha 2C- and beta1-adrenergic receptors influence heart failure survival. *BMC Med Genet.* 2008;**9**:93.
18. Liggett SB, Cresci S, Kelly RJ, et al. A GRK5 polymorphism that inhibits beta-adrenergic receptor signaling is protective in heart failure. *Nat Med.* 2008;**14**:510–17.
19. **SEARCH Collaborative Group.** SLCO1B1 variants and statin-induced myopathy – a genomewide study. *N Engl J Med.* 2008;**359**:789–99.
20. Thompson JF, Man M, Johnson KJ, et al. An association study of 43 SNPs in 16 candidate genes with atorvastatin response. *Pharmacogenomics J.* 2005;**5**:352–8.
21. Thompson J, Hyde C, Wood S, et al. Comprehensive whole genome and candidate gene analysis for response to statin therapy in the TNT cohort. *Circ Cardiovasc Genet.* 2009; Apr;2(2):173-81.
22. Mega JL, Close SL, Wiviott SD, et al. Cytochrome p-450 polymorphisms and response to clopidogrel. *N Engl J Med.* 2009;**360**:354–62.
23. Simon T, Verstuyft C, Mary-Krause M, et al. Genetic determinants of response to clopidogrel and cardiovascular events. *N Engl J Med.* 2009;**360**:363–75.
24. Goodman T, Ferro A, & Sharma P. Pharmacogenetics of aspirin resistance: a comprehensive systematic review. *Br J Clin Pharmacol.* 2008;**66**:222–32.
25. Ansell J, Hirsh J, Hylek E, Jacobson A, Crowther M, & Palareti G. Pharmacology and management of the vitamin K antagonists: American College of Chest Physicians Evidence-Based Clinical Practice Guidelines (8th Edition). *Chest.* 2008;**133**:160S–98S.
26. Limdi NA & Veenstra DL. Warfarin pharmacogenetics. *Pharmacotherapy.* 2008;**28**:1084–97.
27. **International Warfarin Pharmacogenetics Consortium.** Estimation of the warfarin dose with clinical and pharmacogenetic data. *N Engl J Med.* 2009;**360**:753–64.
28. Washington University Medical Center. Warfarin dosing algorithm. www.warfarindosing.org. 2009. June 29, 2011.
29. Cooper GM, Johnson JA, Langaee TY, et al. A genome-wide scan for common genetic variants with a large influence on warfarin maintenance dose. *Blood.* 2008;**112**:1022–27.

12

Statin-Induced Muscle Toxicity

Russell A. Wilke, Melissa Antonik, Elenita I. Kanin, QiPing Feng, and Ronald M. Krauss

INTRODUCTION

β-Hydroxy-β-methylglutaryl-coenzyme A (HMG-CoA) reductase inhibitors (statins) are highly efficacious in the prevention of coronary artery disease. Although statins are generally considered safe, their use may be associated with musculoskeletal complaints that limit tolerance to treatment and, in the most extreme case, can lead to rhabdomyolysis. Both candidate gene and genome-wide association studies are being used to assess possible genetic susceptibility to statin-induced myopathy, with the recognition that this phenotype represents a broad spectrum of syndromes influenced by other drugs and disease states. In addition to potentially guiding statin therapy, the results of such studies may provide mechanistic insight into the critical cellular events linking statin use to muscle pathology in patients at risk.

Clinical Background

Multiple large clinical trials have demonstrated that statins (HMG-CoA reductase inhibitors) reduce the incidence of both primary and secondary coronary artery disease in patients at risk (1–4). Primary prevention trials have demonstrated that statin use can reduce the risk of a first major coronary event by more than 30 percent (3, 5). Secondary prevention trials reveal a risk reduction of similar magnitude (2). Aggressive intervention trials suggest that greater lipid lowering is associated with further reduction in risk (6).

Six drugs are currently available within this class (7). The first statin to be approved by the U.S. Food and Drug Administration was lovastatin in 1987, followed by simvastatin, 1988; pravastatin, 1991; fluvastatin, 1994; atorvastatin, 1997; and rosuvastatin, 2003. Cerivastatin was released in 1998 but was subsequently withdrawn from the market, in 2001, because of an increased frequency of muscle toxicity. During these few years, forty reported cases of muscle toxicity associated with cerivastatin use were fatal (8).

During the past two decades, the use of statin drugs has expanded dramatically, and statin-related muscle complications are becoming more fully characterized (7, 9–11). It remains difficult, however, to accurately determine the prevalence of statin-induced myotoxicity, in large measure, because of the wide variability in its clinical presentation (8, 12, 13). In this chapter, pharmacogenetic approaches for assessing susceptibility to statin-induced muscle toxicity will be discussed within the context of the full spectrum of this clinical phenotype. Statin pharmacokinetics and possible underlying pathophysiologic mechanisms are reviewed with an emphasis on the heterogeneity of the phenotype.

Rhabdomyolysis

Statin-related musculoskeletal complications include myalgia (focal or diffuse), myopathy (pain in the absence of inflammation), myositis (pain accompanied by an inflammatory response), and rhabdomyolysis (severe muscle damage accompanied by end-organ complications, especially kidney failure). Clinically, these diagnoses are separated by the use of two variables: muscle pain and circulating levels of a relatively nonspecific muscle enzyme, creatine kinase (CK). The CK level is often used as a marker for severity (14, 15).

Although some debate continues regarding the criteria necessary for establishing a diagnosis of statin-induced rhabdomyolysis, the most severe and potentially lethal form of this adverse drug reaction (14, 15), many clinicians consider an elevation in serum CK level greater than fiftyfold the upper limit of normal (ULN) to be necessary and sufficient to make the diagnosis. Graham and colleagues have published a now widely accepted stepwise approach for the identification of rhabdomyolysis cases in hospitalized patients (16). Their algorithm is based on procedural codes for hospital admission,

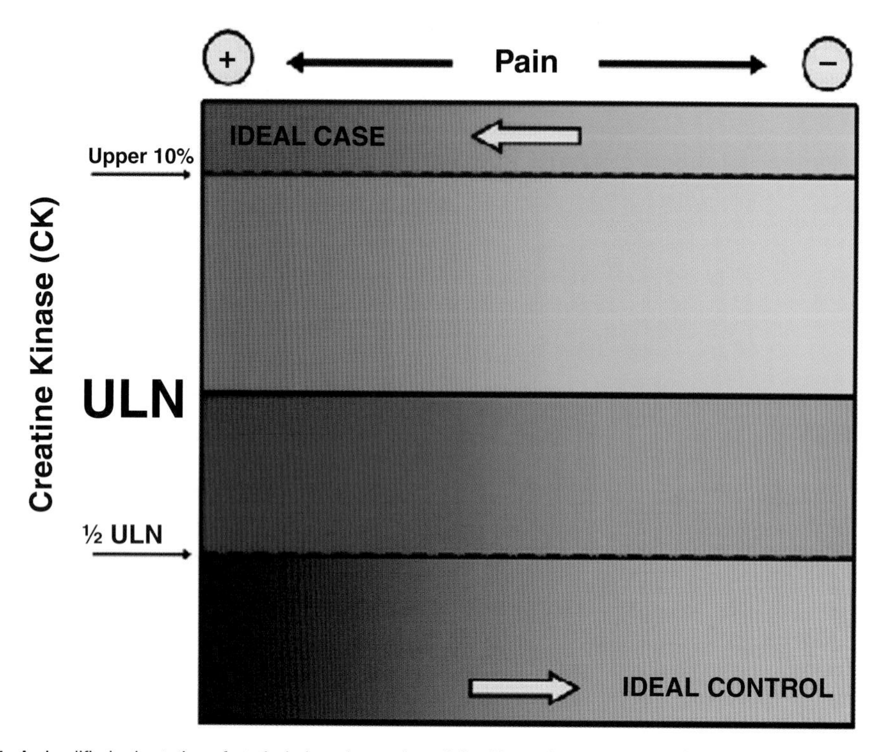

Figure 12.1. A simplified adaptation of statin-induced muscle toxicity. Two primary variables (pain and serum CK level) are often used clinically to define the various manifestations of this phenotype. Thresholds (horizontal) from (18).

discharge diagnoses (e.g., International Classification of Diseases, 9th Edition [ICD-9] code 791.3, for myoglobinuria), and clinical laboratory data (reflecting indices of kidney function, liver function, and skeletal muscle integrity). When these criteria were applied to the medical records of more than 250,000 patients treated with atorvastatin, pravastatin, or simvastatin, the observed rhabdomyolysis rate was 0.000044 events per person-year (16). Adverse event frequency was associated with age in this cohort. Statin users 65 years of age of older were found to have four times the risk of hospitalization due to rhabdomyolysis than were statin users younger than 65 years (odds ratio, 4.36; 95% confidence interval [CI], 1.5–14.1) (17).

Our recent published survey of statin-related muscle toxicity cases, obtained within the context of combined inpatient and outpatient medical records, yielded rhabdomyolysis rates that were equally low (18). From the medical records of 2,000,000 patients, only ten cases of rhabdomyolysis were identified (eight related to simvastatin, and two related to atorvastatin). With approximately 100,000 of these patients exposed to statins during the past decade, this represents an estimated frequency of 10^{-4}. Others have reported that rosuvastatin is associated with similarly low rates of rhabdomyolysis (at doses <40 mg) (5, 19). With such a low event rate, the study of underlying toxicodynamic mechanism(s), and the identification of genetic predictor(s), will be challenging (20).

Intermediate Myotoxicity

The clinical interpretation of CK level is complex (Figure 12.1). Statin-induced muscle damage is not always accompanied by leakage of muscle enzymes into the blood (14, 21–23). In a small series of patients, investigators at the Scripps Clinic reported that study subjects were able to discriminate between blinded statin therapy and placebo based on the presence or absence of a clinically recognizable pain syndrome (21). This pain occurred in the absence of elevated circulating CK levels. Furthermore, CK levels can be increased through physical activity, in the absence of any underlying pathology. In a recent study of seventy healthy young volunteers using no medication, strenuous exercise increased circulating CK levels from a baseline of 157 ± 11 (mean ± SE) units/L, to a peak elevation of 4,480 ± 705 (mean ± SE) units/L, 96 hours after exercise (24).

The literature currently supports at least four diagnostic strata, based solely on CK level, for use in the determination of myotoxicity grade. These are defined by (1) CK above ULN but less than threefold ULN (intermediate myotoxicity), (2) CK above threefold ULN but less than tenfold ULN (incipient myopathy), (3) CK above tenfold ULN but less than fiftyfold ULN (myopathy), and (4) CK above fiftyfold ULN (rhabdomyolysis). Although these thresholds can easily be superimposed on data from existing cohorts (as done within reference 18), it must be recognized that they are arbitrary and do not reflect the poor association of CK level with myopathic symptoms.

The prevalence of clinically significant statin-induced muscle symptoms short of rhabdomyolysis remains a matter of ongoing debate, and, of course, the prevalence varies according to diagnostic criteria (14, 15). In the context of statin monotherapy, myalgias have been reported to occur at a frequency of approximately 1 percent (10). These muscle symptoms are dose related, and their prevalence may be an order of magnitude greater at higher doses of the more potent statins. Myalgias accompanied by elevation in serum CK level occur at a much lower frequency (23, 25). If the definition of statin-induced myotoxicity is restricted to include a marked elevation in serum CK level, the frequency appears to be 0.1 percent or less for the currently available statins (11, 26). For lovastatin and simvastatin, the prevalence of mild to moderate CK elevation has been reported to be 6.4 cases per 1,000 patients (23). The frequency decreases to 1.6 cases per 1,000 patients if the definition is restricted to high CK elevation (23).

Dose Dependence

McLure and colleagues have quantified the frequency of statin-related muscle toxicity in one of the largest managed care populations in the United States (25). Using a relatively stringent definition of "myopathy" (CK level ≥10,000 units/L, plus a relevant ICD-9 diagnostic code) as the primary end point, those investigators observed an incidence rate that was 3.05-fold higher (95% CI, 1.13–8.21) in patients on statin monotherapy (i.e., compared with the general population). This relationship was dose dependent. The incidence rate for myotoxicity was 27.3-fold higher (95% CI, 14.6–50.8) in patients on high-dose statin therapy (i.e., equivalent to 40 mg of lovastatin daily or greater) than in the general population. All statins were considered, using literature-accepted equivalency ratios to adjust for differences in potency (25).

Until recently, the clinical impact of the dose dependence of statin myotoxicity may have been underappreciated. However, in 2005, the PRIMO study indicated that the frequency of mild to moderate muscle symptoms with high-dose statin therapy is higher than previously recognized (27). PRIMO was an observational study of adverse muscle symptoms in an unselected population of 7,924 hyperlipidemic patients receiving high-dose statin therapy in the context of usual care, in an outpatient setting in France. Of these 7,924 subjects, 10.5 percent reported muscle pain. For a drug class administered to millions of patients, the dose dependence of the relationship between exposure and myotoxicity will have striking public health implications. In 2011, the U.S. Food and Drug Administration therefore recommended *against* the use of high clinical doses (80 mg or greater) for simvastatin, the most commonly prescribed drug in this class.

To understand the frequency of this adverse drug reaction in a clinical practice-based setting, we recently surveyed the entire medical record of a large practice-based cohort residing in the Midwestern United States (18). Our approach, illustrated in Figure 12.2, has been published (20). From the records of nearly 2,000,000 unique patients served by a single comprehensive system of care, 213 cases of statin-induced myotoxicity were validated and enrolled in a population-based study of genetic risk determinants (http://www.pharmgkb.org/contributors/pgrn/parc_profile.jsp). Within this cohort, CK level was correlated with the clarity of the reported pain syndrome (18). In our analysis, which included atorvastatin, pravastatin, and simvastatin, the relationship between simvastatin dose and the severity of muscle damage was clearly dose dependent (18).

Factors Having an Impact on Drug Disposition

The pharmacokinetics of statins differ on a drug-by-drug basis (20, 28–31). Whereas some statins undergo extensive phase I oxidation (atorvastatin, fluvastatin, lovastatin, and simvastatin) (32, 33), others appear to undergo very little (pravastatin and rosuvastatin) (33, 34). Furthermore, the relevant oxidative enzymes can differ statin by statin (35). Atorvastatin is oxidized primarily by members of the cytochrome P450 (CYP) 3A family (36). Drugs known to inhibit CYP3A enzyme activity (e.g., itraconazole) have been associated with increased severity of atorvastatin-induced muscle damage (37). Although the CYP3A enzymes play a role in the metabolism of simvastatin, as well, data suggest a role for CYP2D6 in the oxidation of simvastatin metabolites (38–40). Fluvastatin, on the other hand, is metabolized to a great degree by CYP2C9 (28, 41).

Following oxidative phase I drug metabolism, many statins form hydroxyl intermediates (e.g., atorvastatin is converted to 2-OH atorvastatin and 4-OH atorvastatin [36]). These hydroxystatin derivatives are then subjected to additional modification, through phase II conjugation by UDP-glucuronosyltransferase-1 (UGT-1) (42). It is through a UGT-1-dependent interaction that the fibric

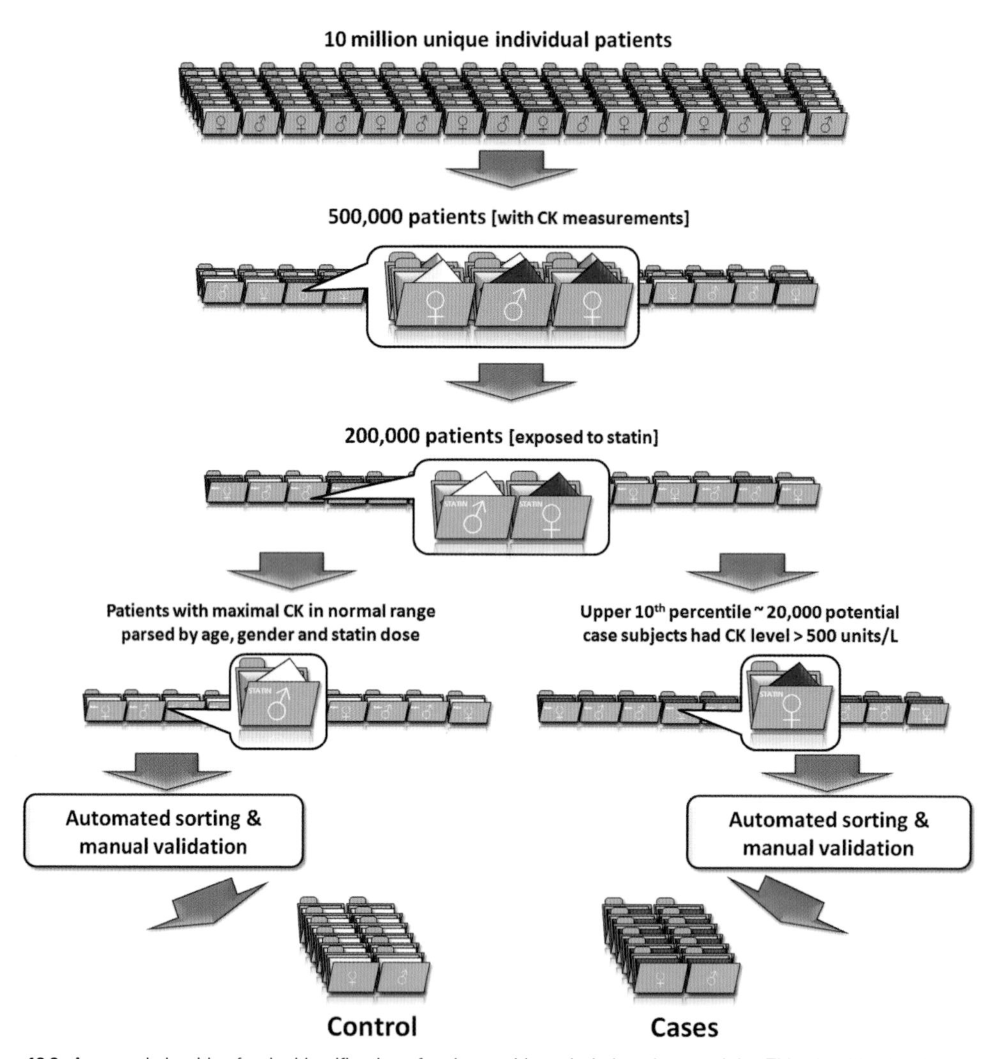

Figure 12.2. A general algorithm for the identification of patients with statin-induced myotoxicity. This algorithm has been used in the context of several large managed care organizations with comprehensive electronic medical records. Adapted and reprinted with permission from Macmillan Publishers Ltd: Wilke RA, Lin D, Roden DM, Watkins PB, Flockhart D, Zineh I, Giacomini KM, Krauss RM. Identifying genetic risk factors for serious adverse drug reactions – current progress and challenges. *Nature Reviews Drug Discovery*. 2007;6:904–16.

acid derivative, gemfibrozil, is thought to alter pharmacokinetics (and corresponding clinical outcomes) for a variety of statin drugs (16, 43). For example, by inhibiting the glucuronidation of simvastatin hydroxy acid (44), gemfibrozil is known to attenuate the biliary excretion of simvastatin and to place patients at increased risk for the development of statin-related muscle toxicity.

In general, statin disposition is altered in patients with comorbid liver or kidney disease (25, 45). In their rigorous assessment of dose-response, McClure and colleagues also observed that the relative risk for statin-related muscle toxicity was 4.3 (95% CI, 1.5–13) in the context of liver disease and 2.5 (95% CI, 1.3–5.0) in the context of kidney disease (25). Presumably, this increased risk is incurred because of perturbations in the clearance of parent drug and/or statin metabolite (33, 45–47).

Anticipated Challenges

As discussed elsewhere (20), the study of genetic predictors of statin-induced rhabdomyolysis has been limited by its extreme rarity and the interplay with nongenetic risk factors. However, because statin-induced muscle damage appears to occur along a clinical continuum, progress in studying genetic predictors of myopathy may depend on the extent to which intermediate phenotypes predict more severe myotoxicity. Several issues currently obscure the ability to answer this question. For example,

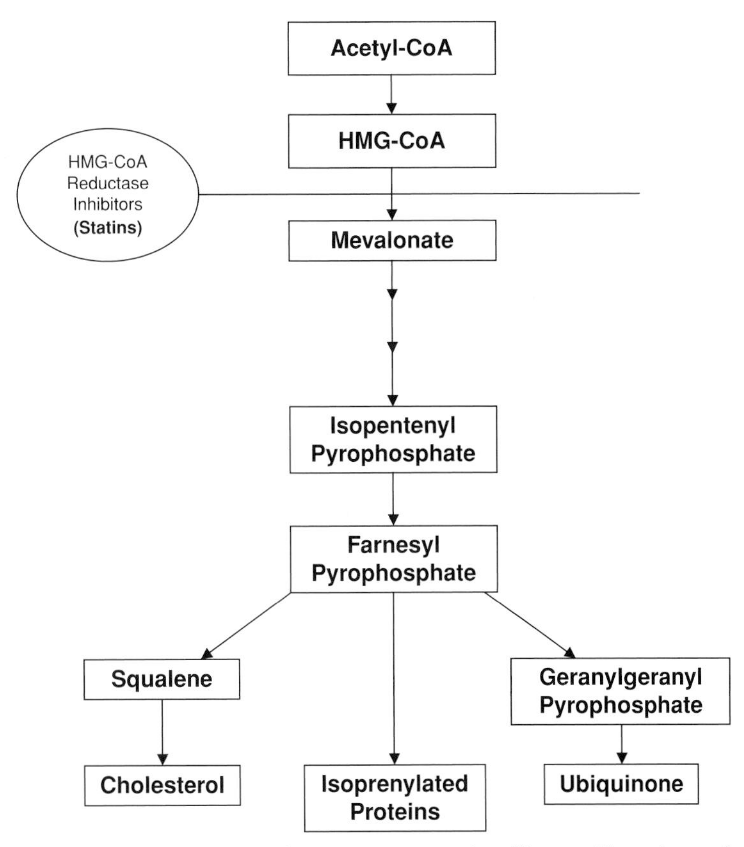

Figure 12.3. Cholesterol biosynthesis. This linear pathway generates a series of isoprenoids, and several of these metabolites prenylate key cellular components.

patients will often stop therapy because of pain before they manifest clinically recognizable muscle damage (i.e., before CK elevation). Furthermore, even in subjects with elevated serum CK levels, and despite the designation of myotoxicity strata based on CK, there is wide variability in the CK level used by practitioners to recommend cessation of therapy.

Another significant challenge has been the clinical observation that some patients may exhibit marked elevation in serum CK level in the absence of pain. Because most providers do not check CK levels as a matter of routine medication surveillance (i.e., in asymptomatic patients), CK elevation in the absence of pain may not come to clinical attention until other manifestations of myotoxicity (e.g., renal failure secondary to myoglobinuria) have become clinically apparent. Clearly, better biomarkers are needed, and the complexity of this phenotype will require a coordinated multidisciplinary effort to elucidate the underlying cellular mechanisms and the genetic determinants of risk. Efforts are underway to bank DNA on existing cases, standardize their phenotypic definition, rigorously characterize the genotype (e.g., through genome-wide resequencing), and improve analytical approaches (e.g., through the development of novel collapsing strategies for rare variants).

Mechanisms of Toxicity

The primary cellular mechanism linking statins to muscle damage has evaded investigators for more than a decade. At present, it remains unclear whether this adverse drug reaction represents an on-target effect (i.e., extension of the therapeutic mechanism) or an off-target effect (i.e., mediated by an as yet unrecognized intracellular signaling alteration). Several general mechanisms have been proposed (48). They include (1) depletion of intracellular cholesterol, (2) decreased intermediary metabolites, (3) reduced mitochondrial function, and (4) a shift in the balance between cell cycling and apoptosis.

Cholesterol Biosynthesis

Cholesterol is synthesized from acetylcoenzyme A, through a linear series of enzymatic steps (Figure 12.3) (49, 50). A rate-limiting enzyme is HMG-CoA reductase.

Mevalonic acid, the product of this reaction, then serves as the principal substrate for a series of several more enzymatic steps, generating a number of biologically active intermediates collectively referred to as isoprenoids (e.g., farnesyl pyrophosphate, geranylgeranyl pyrophosphate, and squalene) (51). Many isoprenoids are then attached covalently to key cellular components (52). As such, there is considerable interest in identifying network-based interactions that occur between isoprenoids and critical cellular signaling processes, such as the pathways involved in intermediary metabolism (e.g., glycolysis), cellular energy homeostasis (e.g., mitochondrial respiratory chain activity), and maintenance of cell viability (e.g., cell cycling vs. apoptosis) (50).

One successful approach to understanding these interactions has included transcriptional profiling. Using cultured myoblasts, and explants derived from skeletal muscle, Morikawa and colleagues showed that statins alter the expression of key gene products involved in cholesterol biosynthesis: HMG-CoA synthase-1, HMG-CoA reductase, farnesyl diphosphate synthase, and isopentenyl-diphosphate isomerase (53). Using a similar approach, Muhua and colleagues demonstrated that statins alter the expression of genes involved in both glycolysis and fatty acid oxidation: 6-phosphofructo-2-kinase, acetyl-CoA carboxylase β, and pyruvate dehydrogenase kinase, isozyme 4 (PDK4) (54). Because PDK4 inactivates pyruvate dehydrogenase, altered expression of PDK4 may serve as a cellular switch, diverting substrate toward glycolysis and away from fatty acid oxidation.

Mitochondrial Dysfunction

Other transcriptional profiling experiments have clearly implicated mitochondrial dysfunction in the pathogenesis of statin-induced myopathy. Recent gene expression studies conducted using cultured murine skeletal muscle have shown that lovastatin upregulates the expression of atrogin-1, a key gene involved in skeletal muscle atrophy (55). This finding has been reproduced in zebra fish embryos. In cultured mouse myotubes, atrogin-1 induction after lovastatin treatment was accompanied by distinct morphological changes known to accompany myotoxicity. Within these same experimental systems (zebra fish embryos and cultured mouse myotubes), overexpression of a transcriptional coactivator known to induce mitochondrial biogenesis ($PGC_{1\alpha}$) attenuated atrogin-1 induction, and prevented lovastatin-induced muscle damage. Statin-induced myotoxicity may therefore represent an alteration in the functional capacity of mitochondria within skeletal muscle.

Insight gained through transcriptional profiling has been consistent with prior structural and functional analyses of skeletal muscle conducted in humans exposed to statins (56). Phillips and colleagues have previously reported that myalgias, in the absence of elevated circulating CK levels, can be accompanied by measurable decrements in muscle strength, and by histological evidence of sarcomere disruption, as well (21). Similar findings have been reported independently by Troseid and colleagues (57), who studied four related patients with statin-induced myalgias in the context of normal serum CK levels. Two of the four subjects had electromyographic findings suggestive of muscle pathology, accompanied by histological evidence for mitochondrial changes. A third subject had similar findings, although milder and nondiagnostic. The fourth study subject had no mitochondrial changes. It is interesting that, in all four subjects, myalgias resolved clinically following cessation of the statin (57).

Additional functional studies, conducted on skeletal muscle biopsies, have confirmed this association between statin exposure and mitochondrial dysfunction. Laaksonen and colleagues have used skeletal muscle biopsies to assess mitochondrial respiratory chain enzyme activity in statin-treated subjects (58). Forty-eight patients with elevated low-density lipoprotein cholesterol were randomly assigned to atorvastatin, simvastatin, or placebo for eight weeks. The ratio of plasma lathosterol to cholesterol (a marker of endogenous cholesterol biosynthesis) significantly decreased in the statin-treated groups. Eighteen of these patients underwent skeletal muscle biopsy (six after atorvastatin, six after simvastatin, and six after placebo). Respiratory chain enzyme activity was reduced in the muscle of patients taking simvastatin (58), even though these patients had normal serum CK levels. Because these observations were recorded in subjects lacking myalgias, it is tempting to speculate that subtle changes in mitochondrial function occur very early in the myotoxic process, long before patients develop pain or CK elevation.

One proposed mechanism linking statins to mitochondrial function has been the altered prenylation of key components within the electron transport process. For example, ubiquinone contains an isoprenoid side chain synthesized via the cholesterol biosynthetic pathway (Figure 12.3), and ubiquinone is intimately involved in mitochondrial energy shuttling (59). Although circulating levels of ubiquinone have been reduced in some patients on statin therapy (60), it remains unclear whether tissue levels of ubiquinone are significantly reduced. The observation that ubiquinone levels within skeletal muscle can be altered by high-dose statin therapy (58) has led a number of investigators to advocate for the coadministration of ubiquinone (through the use of a food supplement commonly referred to as Coenzyme Q) in patients taking statins (60). However, evidence supporting the claim that coadministration of Coenzyme Q could mitigate against the development of statin-related muscle toxicity has been scant. A recent study by Young and colleagues randomized forty-four patients, who had previously failed statins due to myalgia, to receive simvastatin (40 mg) with Coenzyme Q supplementation versus placebo (61). No difference in myalgia score was observed between the two groups; Coenzyme Q supplementation did not improve statin tolerance (61). Therefore, at present, there is insufficient evidence to support a causative role for ubiquinone depletion in statin-induced myopathy, and routine Coenzyme Q supplementation for patients on statins is not recommended.

Cell Cycling

Check point genes are also modified through processes dependent on the presence of isoprenoid intermediates (62). Farnesyl transferase inhibitors are potent inhibitors of tumor cell growth, and their effect is augmented (in a multiplicative fashion), in vitro, by the addition of agents known to alter microtubule depolymerization (62). At least one pivotal modulator of cell survival, Ras, undergoes posttranslational farnesylation, and ras/raf/MEK/ERK pathways clearly modulate the activity of the cell cycle (63). In fact, nearly two decades ago, lovastatin was shown to arrest cells in the G1 phase of the cell cycle (64, 65). Since then, Sutter and colleagues have shown that both fluvastatin and pravastatin can arrest hepatocellular carcinoma cell lines in the G0/G1 phase of the cell cycle, an effect that was clearly dose dependent (66). Statins can also alter the activity of transcription factors in muscle tissue. Wada and colleagues have shown that statins can activate GATA-6, a member of the zinc finger transcription factor family characterized by their ability to bind the consensus sequence GATA (67). These factors play an important role in cellular events that regulate cell fate and differentiation. Although the recent work of Wada and colleagues was conducted in vascular smooth muscle (67), GATA-6 has also been implicated in the modulation of visceral smooth muscle differentiation (human bladder) (68) and cardiac myocyte differentiation (69). The observation that GATA-6 is downregulated during skeletal muscle regeneration raises the possibility that statin-induced changes in GATA-6 activity could influence the differentiation of skeletal muscle as well (70).

Recent transcriptional profiling experiments indicate that statins directly influence cell survival in skeletal muscle. The muscle biopsies characterized functionally and structurally by Laaksonen, as noted under "Mitochondrial Dysfunction," were also used for genome-wide expression studies (48, 71). More than 40,000 genes and splice variants were quantified in muscle obtained from six men receiving atorvastatin, five men receiving simvastatin (one of six cases yielded insufficient RNA), and six men receiving placebo (frequency-matched according to age). The same muscle biopsy specimens were then also characterized by lipidomics (liquid chromatography coupled to tandem mass spectrometry), quantifying 132 unique molecular lipid species (48, 71). Regression of lipidomic data on gene expression data for pathway-based signaling networks (multiple genes in combination, based on prior expert knowledge in annotated databases) confirmed the involvement of lipid-derived signaling pathways (e.g., prostanoid biosynthesis), and suggested a role for Ca^{2+}-dependent pathways known to modulate the balance between cell cycling and apoptosis (e.g., phospholipase C) (48, 71). These results have been validated indirectly by a second group studying statin-exposed subjects in the context of vigorous physical exercise (72). In this latter experimental paradigm, Urso and colleagues documented statin-induced alteration in the expression of fifty-six genes, and 20 percent of these gene products were involved in protein folding and apoptosis (72).

Apoptosis

To investigate the potential mechanism whereby statins could induce apoptosis in skeletal muscle, Johnson and colleagues have adapted a novel myotube culture model (73). In their experimental system, multiple drugs within the statin class induced apoptosis (as measured by TdT-mediated dUTP nick-end-labeling nuclei staining and caspase-3 activation), through a mechanism that was unrelated to cholesterol biosynthesis and independent of statin structure (i.e., not related to partition coefficient). The observed effect was reproduced by geranylgeranyl transferase inhibitors, and it was attenuated by replacement of mevalonic acid (73). In their system, statins clearly induced apoptosis at concentrations that suppressed the isoprenylation of Rap1a (a 21-kDa GTPase that serves as a substrate for geranylgeranylation). These observations suggest that statin-induced apoptosis in skeletal muscle may be transduced via the altered prenylation of critical signaling proteins (73).

Additional support for this claim comes from the work of Sutter and colleagues (66). Using hepatocellular carcinoma cell lines, they further demonstrated that both fluvastatin and pravastatin increase apoptosis through a reduction in mitochondrial membrane potential and activation of caspases 3 and 8 (66). Fluvastatin also dose-dependently decreased phosphorylation of the mitogenic and antiapoptotic ERK1/2 kinase, while dose-dependently increasing the phosphorylation of p38MAPK. Because both effects promote cell cycle arrest, it is tempting to speculate that statins attenuate the growth of malignancies (74). However, epidemiological data reported to date have not consistently revealed a reduction in cancer incidence (or death) in subjects exposed to statins in clinical trials (75).

Nuclear Abnormalities

As noted above, cholesterol biosynthesis generates a number of intermediate products with intracellular signaling capability. Several of these biologically active compounds interact with the nuclear lamina (52), a molecular scaffold that lies just inside the inner nuclear membrane and serves a critical functional role in the organization of heterochromatin. Composed of multiple proteins, this scaffold is built on a framework of lamin protein polymers. The constituent monomers are products of three lamin genes (LMNA, LMNB1, and LMNB2), and polymorphisms in these genes have been associated with a variety of clinically abnormal neuromuscular phenotypes, some containing myopathic components (76).

Lamin A is isoprenylated (by a farnesyl transferase), and this modification facilitates its nuclear transport and integration into the mature nuclear lamina matrix (76). Although phenotypes associated with lamin A gene defects can be extremely heterogeneous, one of the more common clinical presentations has been Emery-Dreifuss muscular dystrophy (EDMD), a syndrome characterized by myocardial dysfunction and skeletal muscle weakness. Although the EDMD phenotype exhibits locus heterogeneity (i.e., a small proportion of cases are X-linked), the majority of cases are autosomal dominant (EDMD-AD) and attributable to variation in the LMNA gene. EDMD-AD typically presents as gait change during childhood, but milder forms may not be clinically evident until adulthood, when affected individuals present with weakness and skeletal muscle atrophy (77). Circulating CK levels are typically elevated in these individuals, but they may remain within normal limits. This level of heterogeneity raises the question as to how often the clinical presentation of statin-induced myotoxicity represents a previously undiagnosed subclinical metabolic myopathy.

Recently, Vladutiu and colleagues reported allele frequencies for variants in candidate genes previously associated with inheritable metabolic myopathies, specifically within subjects experiencing myotoxicity while using a statin (78). Three candidate genes were characterized: carnitine palmitoyltransferase II (CPT2), myophosphorylase (MIM), and myoadenylate deaminase (AMPD). Compared with population-based controls, the number of minor allele carriers was increased thirteenfold for CPT2 and twentyfold for MIM. Although carrier status for AMPD was not significantly increased, homozygosity for AMPD variants was increased 3.25-fold with respect to nonmyotoxic statin-exposed controls (78). It is noteworthy that these associations were resolved in the context of marked phenotypic heterogeneity (i.e., CK levels ranged from normal to more than tenfold ULN). These results support our prior claim that characterization of intermediate myotoxicity will be an essential step in the identification of risk predictors for rhabdomyolysis (20).

Genetic Predictors

Statins undergo varying degrees of phase I oxidation, and the primary enzyme differs on a drug-by-drug basis (32, 34, 79). In 2005, we published the first genetic predictor of the severity of muscle toxicity related to atorvastatin use (29). In this retrospective observational cohort study ($n = 68$ cases), a splice variant in CYP3A5 was associated with the degree of CK elevation. However, the strength of this association depended on the presence or absence of concomitant medications known to interact with statins through mechanisms other than phase I oxidation (29). As noted earlier, many statins undergo additional modification through phase II conjugation by UGT-1, and this process can be disrupted by the concomitant administration of a fibric acid (42). Membrane transporters are also known to influence statin-related outcomes. Polymorphisms in candidate solute transporter genes have been associated with the altered hepatic uptake of pravastatin (80), and genes encoding membrane transporters have been implicated in statin-related drug interactions involving fibric acid derivatives as well (81–83). In the case of the organic anion transporter SLCO1B1, genetic variability has been associated with differential modulation of the cellular transport of statins (84).

Whole Genome Scanning

Recently, the SEARCH Collaborative Group in the United Kingdom applied a genome-scanning approach within the context of a nested case-control study to identify markers of statin-induced myotoxicity (85). The original SEARCH trial randomly selected more than 12,000 subjects, postmyocardial infarction, to receive either low-dose simvastatin (20 mg daily) or high-dose simvastatin (80 mg daily). During the course of the trial, forty-nine subjects within the high-dose (80 mg) arm developed myopathy (CK more than tenfold ULN with pain). An additional forty-nine subjects within the same treatment arm developed "incipient" myopathy (CK more than threefold ULN, with or without pain). Of these, eighty-five patients underwent whole genome scanning (with a platform containing 317,000 single-nucleotide polymorphisms), and the results were compared with genome-wide data from ninety frequency matched nonmyopathic controls (85). A single-nucleotide variant survived statistical correction for multiple testing: a base substitution in the SLCO1B1 gene (85). After genomic resequencing of SLCO1B1, the putative causative allele was retested for association in a subset of definite myopathy cases from the original study population, revealing an odds ratio for myopathy of 4.5 per copy of the variant allele (95% CI, 2.6–7.7) (64).

This association was subsequently replicated in individuals of similar ethnic composition in a second clinical trial based in the United Kingdom, the Heart Protection Study (HPS) (64). The HPS cohort consists of more than 20,000 subjects with known vascular disease, or vascular risk factors, randomly assigned to receive either 40 mg simvastatin daily or placebo. Twenty-four cases of myopathy (10 definite + 14 incipient) were identified in 10,269 participants receiving primary prevention with simvastatin. Of these, twenty-one were genotyped retrospectively for the variant identified in SEARCH (64). Within the HPS validation cohort, the relative risk was 2.6 per copy of the variant allele (95% CI, 1.3–5.0). Both cohorts (i.e., the discovery cohort and the validation cohort) represented controlled treatment trials (64). The observation that the strength of this association was markedly different between the discovery cohort (80 mg simvastatin) and the validation cohort (40 mg simvastatin) highlights the clinical importance of dose dependence (25). Hence, the odds ratio for this association may be further driven toward 1.0, when tested in the context of a practice-based cohort where patients are exposed to multiple doses. Efforts are underway to address this issue within the community (www.PharmGKB.org).

Outlook

The observation that mild myalgias occur in a significant minority of patients exposed to statins suggests that intermediate myotoxicity may be clinically relevant. If myalgias disrupt activities of daily living, patients may opt to switch to a different drug or discontinue statin therapy altogether (14, 15). Because statins are highly efficacious in reducing atherosclerotic coronary artery disease, the cessation of statin therapy may be accompanied by significant risk. Noncompliance with statin therapy has unequivocally been shown to increase morbidity and mortality in patients selected to receive these drugs for the primary prevention of cardiovascular disease (86). Thus, statin noncompliance has the potential to increase the overall cost burden on our health care infrastructure.

Statin-induced myotoxicity is an extremely heterogeneous clinical phenotype. Some gene variants will predict both mild and severe myotoxicity. However, only a subset of these variants will predict rhabdomyolysis. Although markers associated with rhabdomyosysis could be used, prospectively, to predict and prevent clinically severe myotoxicity, markers associated with the more intermediate phenotype will likely lead to a better understanding of the mechanisms underlying this adverse drug reaction.

REFERENCES

1. The Lipid Research Clinics Coronary Primary Prevention Trial results. I. Reduction in incidence of coronary heart disease. *JAMA.* 1984;**251**:351–64.
2. Randomised trial of cholesterol lowering in 4444 patients with coronary heart disease: the Scandinavian Simvastatin Survival Study (4S). *Lancet.* 1994;**344**:1383–9.
3. Shepherd J, Cobbe SM, Ford I, et al. Prevention of coronary heart disease with pravastatin in men with hypercholesterolemia. West of Scotland Coronary Prevention Study Group. *N Engl J Med.* 1995;**333**:1301–7.
4. Yee HS & Fong NT. Atorvastatin in the treatment of primary hypercholesterolemia and mixed dyslipidemias. *Ann Pharmacother.* 1998;**32**:1030–43.
5. Ridker PM, Danielson E, et al. JUPITER Study Group. Rosuvastatin to prevent vascular events in men and women with elevated C-reactive protein. *N Engl J Med.* 2008;**359**:2195–2207.
6. LaRosa JC, Grundy SM, et al. Treating to New Targets (TNT) Investigators. Intensive lipid lowering with atorvastatin in patients with stable coronary disease. *N Engl J Med.* 2005;**352**:1425–35.
7. Tobert JA. Lovastatin and beyond: the history of the HMG-CoA reductase inhibitors. *Nat Rev Drug Discov.* 2003;**2**:517–26.
8. Ballantyne CM, Corsini A, et al. Risk for myopathy with statin therapy in high-risk patients. *Arch Intern Med.* 2003;**163**:553–64.
9. Executive Summary of The Third Report of The National Cholesterol Education Program (NCEP) Expert Panel on Detection, Evaluation, And Treatment of High Blood Cholesterol in Adults (Adult Treatment Panel III). *JAMA.* 2001;**285**:2486–97.
10. Thompson PD, Clarkson P, & Karas RH. Statin-associated myopathy. *JAMA.* 2003;**289**:1681–90.
11. Waters DD. Safety of high-dose atorvastatin therapy. *Am J Cardiol.* 2005;**96**:69F–75F.
12. Black DM, Bakker-Arkema RG, et al. An overview of the clinical safety profile of atorvastatin (Lipitor), a new HMG-CoA reductase inhibitor. *Arch Intern Med.* 1998;**158**:577–84.
13. Bernini F, Poli A, et al. Safety of HMG-CoA reductase inhibitors: focus on atorvastatin. *Cardiovasc Drugs Ther.* 2001;**15**:211–18.
14. Thompson PD, Clarkson PM, & Rosenson RS. An assessment of statin safety by muscle experts. *Am J Cardiol.* 2006;**97**:69C–76C.
15. McKenney JM, Davidson MH, Jacobson TA, & Guyton JR. Final conclusions and recommendations of the National Lipid Association Statin Safety Assessment Task Force. *Am J Cardiol.* 2006;**97**:89C–94C.
16. Graham DJ, Staffa JA, Shatin D, et al. Incidence of hospitalized rhabdomyolysis in patients treated with lipid-lowering drugs. *JAMA.* 2004;**292**:2585–90.
17. Schech S, Graham D, Staffa J, Andrade SE, La Grenade L, Burgess M, Blough D, Stergachis A, Chan KA, Platt R, & Shatin D. Risk factors for statin-associated rhabdomyolysis. *Pharmacoepidemiol Drug Saf.* 2007;**16**:352–8.
18. Mareedu RK, Modhia FM, Kanin EI, Linneman JG, Kitchner T, McCarty CA, Krauss RA, & Wilke RA. Use of an electronic medical record to characterize cases of intermediate statin-induced muscle toxicity. *Prev Cardiol.* 2009;**12**(2):88–94.
19. Ferdinand KC. Rosuvastatin: a risk-benefit assessment for intensive lipid lowering. *Expert Opin Pharmacother.* 2005;**6**:1897–910.
20. Wilke RA, Lin DW, Roden DM, et al. Identifying genetic risk factors for serious adverse drug reactions: current progress and challenges. *Nat Rev Drug Discov.* 2007;**6**:904–16.

21. Phillips PS, Haas RH, Bannykh S, et al. Statin-associated myopathy with normal creatine kinase levels. *Ann Intern Med.* 2002;**137**:581–5.
22. Teichholz LE. Statin-associated myopathy with normal creatine kinase levels. *Ann Intern Med.* 2003;**138**:1008; author reply 1008–9.
23. Chan J, Hui RL, & Levin E. Differential association between statin exposure and elevated levels of creatine kinase. *Ann Pharmacother.* 2005;**39**:1611–16.
24. Yamin C, Amir O, Sagiv M, Attias E, Meckel Y, Eynon N, Sagiv M, & Amir RE. ACE ID genotype affects blood creatine kinase response to eccentric exercise. *J Appl Physiol.* 2007;**103**:2057–61.
25. McClure DL, Valuck RJ, Glanz M, Murphy JR, & Hokanson JE. Statin and statin-fibrate use was significantly associated with increased myositis risk in a managed care population. *J Clin Epidemiol.* 2007;**60**:812–18.
26. Pasternak RC, Smith SC, et al. ACC/AHA/NHLBI clinical advisory on the use and safety of statins. *Circulation.* 2002;**106**:1024–8.
27. Bruckert E, Hayem G, Dejager S, Yau C, & Begaud B. Mild to moderate muscular symptoms with high-dosage statin therapy in hyperlipidemic patients – The PRIMO study. *Cardiovasc Drugs Ther.* 2005;**19**:403–14.
28. Kirchheiner J & Brockmoller J. Clinical consequences of cytochrome P450 2C9 polymorphisms. *Clin Pharmacol Ther.* 2005;**77**:1–16.
29. Wilke RA, Moore JH, & Burmester JK. Relative impact of CYP3A genotype and concomitant medication on the severity of atorvastatin-induced muscle damage. *Pharmacogenet Genomics.* 2005;**15**:415–21.
30. Gibson DM, Bron NJ, et al. Effect of age and gender on pharmacokinetics of atorvastatin in humans. *J Clin Pharmacol.* 1996;**36**:242–6.
31. Wilke RA, Reif DM, et al. Combinatorial pharmacogenetics. *Nat Rev Drug Discov.* 2005;**4**:911–18.
32. Worz CR & Bottorff M. The role of cytochrome P450-mediated drug-drug interactions in determining the safety of statins. *Expert Opin Pharmacother.* 2001;**2**:1119–27.
33. Bottorff MB. Statin safety and drug interactions: clinical implications. *Am J Cardiol.* 2006;**97**: S27–S31.
34. Neuvonen PJ, Kantola T, & Kivisto KT. Simvastatin but not pravastatin is very susceptible to interaction with the CYP3A4 inhibitor itraconazole. *Clin Pharmacol Ther.* 1998;**63**:332–41.
35. Corsini A, Bellosta S, Baetta R, Fumagalli R, Paoletti R, & Bernini F. New insights into the pharmacodynamic and pharmacokinetic properties of statins. *Pharmacol Ther.* 1999;**84**:413–28.
36. Bullen WW, Miller RA, & Hayes RN. Development and validation of a high-performance liquid chromatography tandem mass spectrometry assay for atorvastatin, ortho-hydroxy atorvastatin, and para-hydroxy atorvastatin in human, dog, and rat plasma. *J Am Soc Mass Spectrom.* 1999;**10**:55–66.
37. Mazzu AL, Lasseter KC, Shamblen EC, Agarwal V, Lettieri J, & Sundaresen P. Itraconazole alters the pharmacokinetics of atorvastatin to a greater extent than either cerivastatin or pravastatin. *Clin Pharmacol Ther.* 2000;**68**:391–400.
38. Nordin C, Dahl ML, Eriksson M, & Sjoberg S. Is the cholesterol-lowering effect of simvastatin influenced by CYP2D6 polymorphism? *Lancet.* 1997;**350**:29–30.
39. Mulder AB, van Lijf HJ, Bon MA, et al. Association of polymorphism in the cytochrome CYP2D6 and the efficacy and tolerability of simvastatin. *Clin Pharmacol Ther.* 2001;**70**:546–51.
40. Geisel J, Kivisto KT, Griese EU, & Eichelbaum M. The efficacy of simvastatin is not influenced by CYP2D6 polymorphism. *Clin Pharmacol Ther.* 2002;**72**:595–6.
41. Kirchheiner J, Kudlicz D, et al. Influence of CYP2C9 polymorphisms on the pharmacokinetics and cholesterol-lowering activity of (-)-3S,5R-fluvastatin and (+)-3R,5S-fluvastatin in healthy volunteers. *Clin Pharmacol Ther.* 2003;**74**:186–94.
42. Jemal M, Ouyang Z, Chen BC, & Teitz D. Quantitation of the acid and lactone forms of atorvastatin and its biotransformation products in human serum by high-performance liquid chromatography with electrospray tandem mass spectrometry. *Rapid Commun Mass Spectrom.* 1999;**13**:1003–15.
43. Chang JT, Staffa JA, Parks M, & Green L. Rhabdomyolysis with HMG-CoA reductase inhibitors and gemfibrozil combination therapy. *Pharmacoepidemiol Drug Saf.* 2004;**13**:417–26.
44. Prueksaritanont T, Tang C, Qiu Y, Mu L, Subramanian R, & Lin JH. Effects of fibrates on metabolism of statins in human hepatocytes. *Drug Metab Dispos.* 2002;**30**:1280–7.
45. Kasiske L, Wanner C, & O'Neill WC. An assessment of statin safety by nephrologists. *Am J Cardiol.* 2006;**97**:S82–5.
46. Singhvi SM, Pan HY, Morrison RA, & Willard DA. Disposition of pravastatin sodium, a tissue-selective HMG-CoA reductase inhibitor, in healthy subjects. *Br J Clin Pharmacol.* 1990;**29**:239–43.
47. Sica DA & Gehr TW. 3-Hydroxy-3-methylglutaryl coenzyme A reductase inhibitors and rhabdomyolysis: considerations in the renal failure patient. *Curr Opin Nephrol Hypertens.* 2002;**11**:123–33.
48. Laaksonen R. On the mechanisms of statin-induced myopathy. *Clin Pharmacol Ther.* 2006;**79**:529–31.
49. Mangravite LM, Wilke RA, Zhang J, & Krauss RM. Pharmacogenomics of statin response. *Curr Opin Mol Ther.* 2008;**10**:555–61.
50. Wilke RA, Maredu RK, & Moore JH. The pathway less traveled – moving from candidate genes to candidate pathways in the analysis of genome-wide data from large scale pharmacogenetic association studies. *Curr Pharmacogenomics Pers Med.* 2008;**6**:150–9.
51. Liao JK. Isoprenoids as mediators of the biological effects of statins. *J Clin Invest.* 2002;**110**:285–8.
52. Baker SK. Molecular clues into the pathogenesis of statin-mediated muscle toxicity. *Muscle Nerve.* 2005;**31**:572–80.
53. Morikawa S, Murakami T, Yamazaki H, et al. Analysis of the global RNA expression profiles of skeletal muscle cells treated with statins. *J Atheroscler Thromb.* 2005;**12**:121–31.
54. Muhua LI. Biomarkers for statin-induced myopathy or rhabdomyolysis. novartis ag (lichtstrasse 35, basel, ch-4056, ch), novartis pharma gmbh (Brunner Strasse 59, Vienna, A-1230, AT). 2007.
55. Hanai JI, Cao P, Tanksale P, et al. The muscle-specific ubiquitin ligase atrogin-1/MAFbx mediates statin-induced muscle toxicity. *J Clin Invest.* 2007;**117**:3940–51.
56. Draeger A, Monastyrskaya K, et al. Statin therapy induces ultrastructural damage in skeletal muscle in patients without myalgia. *J Pathol.* 2006;**210**:94–102.

57. Troseid M, Henriksen OA, & Lindal S. Statin-associated myopathy with normal creatine kinase levels. Case report from a Norwegian family. *Apmis.* 2005;**113**: 635–7.
58. Paiva H, Thelen KM, Van Coster R, et al. High-dose statins and skeletal muscle metabolism in humans: a randomized, controlled trial. *Clin Pharmacol Ther.* 2005;**78**: 60–8.
59. Schaefer WH, Lawrence JW, Loughlin AF, et al. Evaluation of ubiquinone concentration and mitochondrial function relative to cerivastatin-induced skeletal myopathy in rats. *Toxicol Appl Pharmacol.* 2004;**194**:10–23.
60. Nawarskas JJ. HMG-CoA reductase inhibitors and coenzyme Q10. *Cardiol Rev.* 2005;**13**:76–9.
61. Young JA, Florkowski CM, et al. Effect of coenzyme Q(10) supplementation on Siwastatin-induced myalgia. *Am J Cardiol.* 2007;**100**:1400–3.
62. Moasser MM, Sepp-Lorenzino L, Kohl NE, et al. Farnesyl transferase inhibitors cause enhanced mitotic sensitivity to taxol and epothilones. *Proc Natl Acad Sci USA.* 1998;**95**:1369–74.
63. Berthier A, Lemaire-Ewing S, Prunet C, et al. 7-Ketocholesterol-induced apoptosis – Involvement of several pro-apoptotic but also anti-apoptotic calcium-dependent transduction pathways. *FEBS J.* 2005;**272**:3093–104.
64. Keyomarsi K, Sandoval L, Band V, & Pardee AB. Synchronization of tumor and normal cells from g-1 to multiple cell cycles by lovastatin. *Cancer Res.* 1991;**51**:3602–9.
65. Jakobisiak M, Bruno S, Skierski JS, & Darzynkiewicz Z. Cell cycle-specific effects of lovastatin. *Proc Natl Acad Sci USA.* 1991;**88**:3628–32.
66. Sutter AP, Maaser K, Hopfner M, Huether A, Schuppan D, & Scherubl H. Cell cycle arrest and apoptosis induction in hepatocellular carcinoma cells by HMG-CoA reductase inhibitors. Synergistic antiproliferative action with ligands of the peripheral benzodiazepine receptor. *J Hepatol.* 2005;**43**:808–16.
67. Wada H, Abe M, Ono K, et al. Statins activate GATA-6 and induce differentiated vascular smooth muscle cells. *Biochem Biophys Res Commun.* 2008;**374**:731–6.
68. Kanematsu A, Ramachandran A, et al. GATA-6 mediates human bladder smooth muscle differentiation: involvement of a novel enhancer element in regulating beta-smooth muscle actin gene expression. *Am J Physiol Cell Physiol.* 2007;**293**:C1093–102.
69. Zhao R, Watt AJ, Battle MA, Li JX, Bondow BJ, & Duncan SA. Loss of both GATA4 and GATA6 blocks cardiac myocyte differentiation and results in acardia in mice. *Dev Biol.* 2008;**317**:614–19.
70. Yan Z, Choi SD, Liu XB, et al. Highly coordinated gene regulation in mouse skeletal muscle regeneration. *J Biol Chem.* 2003;**278**:8826–36.
71. Laaksonen R, Katajamaa M, et al. A systems biology strategy reveals biological pathways and plasma biomarker candidates for potentially toxic statin-induced changes in muscle. *PLoS ONE.* 2006;**1**:e97.
72. Urso ML, Clarkson PM, Hittel D, Hoffman EP, & Thompson PD. Changes in ubiquitin proteasome pathway gene expression in skeletal muscle with exercise and statins. *Arterioscler Thromb Vasc Biol.* 2005;**25**:2560–6.
73. Johnson TE, Zhang XH, et al. Statins induce apoptosis in rat and human myotube cultures by inhibiting protein geranylgeranylation but not ubiquinone. *Toxicol Appl Pharmacol.* 2004;**200**:237–50.
74. Graaf MR, Richel DJ, van Noorden CJF, & Guchelaar HJ. Effects of statins and farnesyltransferase inhibitors on the development and progression of cancer. *Cancer Treat Rev.* 2004;**30**:609–41.
75. Kuoppala J, Lamminpää A, & Pukkala E. Statins and cancer: a systematic review and meta-analysis. *Eur J Cancer.* 2008;**44**:2122–32.
76. Capell BC & Collins FS. Human laminopathies: nuclei gone genetically awry. *Nat Rev Genet.* 2006;**7**:940–52.
77. Rankin J & Ellard S. The laminopathies: a clinical review [published correction appears in *Clin Genet.* 2007;**71**:293]. *Clin Genet.* 2006;**70**:261–74.
78. Vladutiu GD, Simmons Z, et al. Genetic risk factors associated with lipid-lowering drug-induced myopathies. *Muscle Nerve.* 2006;**34**:153–62.
79. Lennernas H. Clinical pharmacokinetics of atorvastatin. *Clin Pharmacokinet.* 2003;**42**:1141–60.
80. Mwinyi J, Johne A, Bauer S, Roots I, & Gerloff T. Evidence for inverse effects of OATP-C (SLC21A6) 5 and 1b haplotypes on pravastatin kinetics. *Clin Pharmacol Ther.* 2004;**75**:415–21.
81. Davidson MH. Controversy surrounding the safety of cerivastatin. *Expert Opin Drug Saf.* 2002;**1**:207–12.
82. Shitara Y, Hirano M, Sato H, & Sugiyama Y. Gemfibrozil and its glucuronide inhibit the organic anion transporting polypeptide 2 (OATP2/OATP1B1:SLC21A6)-mediated hepatic uptake and CYP2C8-mediated metabolism of cerivastatin: analysis of the mechanism of the clinically relevant drug-drug interaction between cerivastatin and gemfibrozil. *J Pharmacol Exp Ther.* 2004;**311**:228–36.
83. Schneck DW, Birmingham BK, et al. The effect of gemfibrozil on the pharmacokinetics of rosuvastatin. *Clin Pharmacol Ther.* 2004;**75**:455–63.
84. Pasanen MK, Neuvonen M, Neuvonen PJ, & Niemi M. SLCO1B1 polymorphism markedly affects the pharmacokinetics of simvastatin acid. *Pharmacogenet Genomics.* 2006;**16**:873–9.
85. Link E, Parish S, Armitage J, et al. SLCO1B1 variants and statin-induced myopathy – a genomewide study. *N Engl J Med.* 2008;**359**:789–99.
86. Caro J, Klittich W, McGuire A, Ford I, Pettitt D, Norrie J, & Shepherd J. International economic analysis of primary prevention of cardiovascular disease with pravastatin in WOSCOPS. West of Scotland Coronary Prevention Study. *Eur Heart J.* 1999;**20**:263–8.

13 Genomics of the Drug-Induced Long-QT Syndrome

Dan M. Roden, Prince J. Kannankeril, Stefan Kääb, and Dawood Darbar

The QT interval is a readily measured parameter on the surface electrocardiogram (ECG). It represents the time from the onset of ventricular depolarization to the end of ventricular repolarization, and is therefore a rough indicator of the duration of action potentials in individual ventricular cells (Figure 13.1). Marked prolongation of the QT interval on the surface ECG necessarily indicates prolongation of at least some action potentials within the ventricle (Figure 13.1). While there is some evidence that QT-interval prolongation can be antiarrhythmic, the change may also signal susceptibility to a mechanistically and morphologically distinct polymorphic ventricular tachycardia, torsades de pointes (TdP) (Figure 13.2, left).

A HISTORY OF TORSADES DE POINTES

In 1957, Jervell and Lange-Nielsen described a kindred with a high incidence of sudden death in childhood (4 of 6 children) with marked prolongation of the QT interval and congenital deafness in affected subjects. The parents were unaffected, so the assumption, which proved correct, was that the Jervell-Lange-Nielsen syndrome represented an autosomal recessive disorder. In the early 1960s, two groups independently reported autosomal dominant transmission of a syndrome of QT prolongation, normal hearing, and episodes of syncope and sudden death (the Romano-Ward syndrome). Interestingly, the initial descriptions of these syndromes did not identify ventricular arrhythmia as the mechanism of death.

The first drug to be associated with polymorphic ventricular tachycardia was quinidine. Quinidine was initially introduced in the 1920s by the Dutch cardiologist Wenckebach as an antiarrhythmic for patients with atrial fibrillation (AF). It was recognized early in the drug's use that occasional patients developed syncopal episodes, often within the first dose or two and attributed to the drug's vasodilator properties. In 1964, however, Selzer and Wray (1) described polymorphic ventricular tachycardia as the cause of quinidine syncope. Interestingly, in their initial report, they made no mention of the fact that the QT interval was markedly prolonged before the development of polymorphic ventricular tachycardia, although this is now obvious in the published tracings.

The term "torsades de pointes" to describe this long-QT-related arrhythmia was coined in 1966 by Dessertenne (2), who reported recurrent syncope in an elderly woman with advanced heart block. Surprisingly, the syncope was not due to very slow heart rates, but rather to occasional episodes of a very rapid polymorphic ventricular tachycardia. Dessertenne described the tachycardia as a slow twisting of the electrical axis, similar to other twists or "torsades" (Figure 13.2, right) seen in nature.

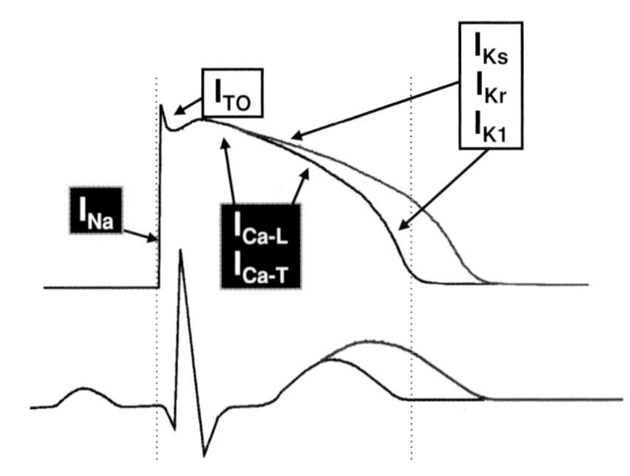

Figure 13.1. Correspondence between action potentials (*top*) and ECG (*bottom*). The vertical lines indicate the beginning and end of the QT interval, and correspond to action potential duration. The red tracings indicate an ECG with a long QT, and how this reflects prolongation of the action potential. Some key ion currents determining cellular depolarization (*black boxes*) and repolarization (*open boxes*) during various phases of the action potential are shown.

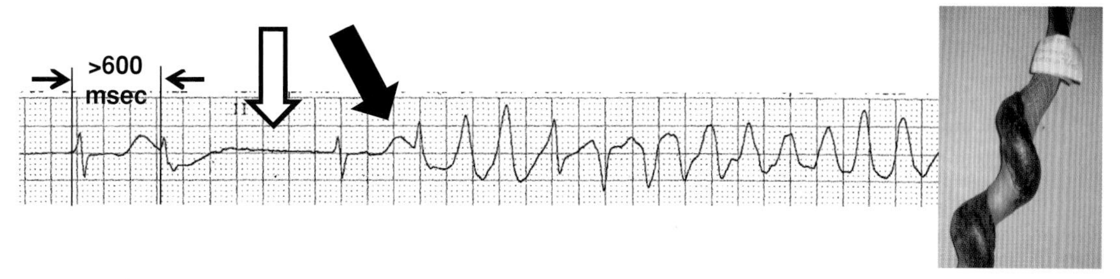

Figure 13.2. (*Left*) Typical episode of TdP showing typical features of diLQTS: very clear QT prolongation (*black arrow*) and a pause in cardiac rhythm before the development of the arrhythmia (*open arrow*). The thin lines represent the rough onset and end of the QT interval, here >600 ms; normal is <450–480. (*Right*) The label on this carved walking stick is "Reworked Torsades in Hazelnut Wood."

GENETICS OF QT CONTROL

Action potential duration is controlled by pore-forming protein complexes, termed "ion channels," through which specific ions flow often in a time- and voltage-dependent fashion. First principles in electrophysiology dictate that action potential prolongation reflects either reduced outward (repolarizing) current generally carried by potassium channels in the heart or increased inward current through sodium or calcium channels (Figure 13.1). Linkage analysis in large kindreds, followed by interrogation of (or occasionally cloning of) candidate genes in linked loci, has now demonstrated that the congenital long-QT syndrome (LQTS) is a disease of control of the action potential duration, and the most common mutations are those in genes encoding ion channels (3). The first three disease genes to be identified were the potassium channel genes *KCNQ1* and *KCNH2* (also known as *HERG*), where mutations reduce outward current, and the cardiac sodium channel gene *SCN5A*, where mutations increase inward current during repolarization. Together, mutations in these three genes are thought to account for more than 80 percent of all cases of congenital LQTS.

KCNQ1 and *KCNH2* encode protein subunits for pore-forming structures that underlie two important repolarizing potassium currents in the heart: I_{Ks} and I_{Kr}. In vitro, recapitulation of I_{Ks} requires transfection of both *KCNQ1* and *KCNE1*, which encodes a KCNQ1-associated protein that modifies its function. By contrast, expression of *KCNH2* alone recapitulates many the features of I_{Kr}, although function-modifying proteins have also been associated with this channel, and indeed virtually all ion channels.

As shown in Table 13.1, mutations in genes encoding ion channels and function-modifying associated proteins have now been described as rarer causes of congenital LQTS, and for each of these, the functional defect is loss of potassium current or increased inward current. The exception is *ANK2*, which encodes the protein ankyrin-B. The functional defect conferred by *ANK2* mutations is failure of appropriate subcellular distribution of molecules important in maintenance of normal intracellular calcium concentration (4). The clinical phenotype produced by *ANK2* mutations is broad and not limited to LQT; therefore, "Ankyrin-B syndrome" is now the preferred term (5). Interestingly, as also shown in Table 13.1, mutations in the LQTS disease genes may confer other phenotypes.

The identification of disease genes in the congenital LQTS has two important near-term implications for studies of the pharmacogenetics of the QT interval and its response to drugs. First, in any candidate gene survey, the congenital LQTS disease genes are a high priority. Second, with the increasing ability to identify disease-associated mutations in kindreds, it is becoming increasingly clear that penetrance in LQTS is highly incomplete (6); that is, there are many (perhaps a majority of) mutation carriers who do not have a manifest phenotype. The mechanism underlying this variable penetrance is under intensive study, but the finding raises the possibility that some patients with the drug-induced form of the disease may in fact be subclinical mutation carriers, a hypothesis whose testing is further discussed below.

THE DRUG-INDUCED LONG-QT SYNDROME

Drug-induced TdP remained an electrophysiologic curiosity until the late 1980s, when it was also recognized in certain patients receiving a widely used nonsedating antihistamine, terfenadine (7). Terfenadine is a very potent QT-prolonging drug, but undergoes near-complete presystemic metabolism to fexofenadine, a nonsedating antihistamine without QT-prolonging properties (8). When the presystemic metabolism was inhibited (by concomitant drug administration, overdose, or advanced liver disease), terfenadine accumulated in the systemic circulation, resulting in QT prolongation, TdP, and cases of sudden death. As a consequence of this episode, as well as a number of other cases of drugs with a

Table 13.1. Congenital LQTSs

	Gene	*Gene Product Function*	*Ion Current Change*	*Frequency*	*Other Manifestations*	*Other Syndromes Associated with Mutations*
LQT1	*KCNQ1*	Pore-forming subunit for the repolarizing cardiac potassium delayed rectifier current I_{Ks}	$\downarrow I_{Ks}$	45%	Two LQT1 or LQT5 alleles → deafness common (Jervell Lange-Nielsen Syndrome)	Short-QT syndrome (gain-of-function mutations; rare)
LQT2	*KCNH2*	Pore-forming subunit for the repolarizing cardiac potassium delayed rectifier current I_{Kr}	$\downarrow I_{Kr}$	45%		Short-QT syndrome (gain-of-function mutations; rare)
LQT3	*SCN5A*	Pore-forming subunit for the depolarizing cardiac sodium current I_{Na}	$\uparrow I_{Na}$	10%		Brugada syndrome (J-point elevation on ECG and sudden death) Conduction system disease AF Dilated cardiomyopathy
LQT4	*ANK2*	Targeting calcium-regulating proteins to correct subcellular location	Many	Rare	Other arrhythmias: sinus brady; AF	
LQT5	*KCNE1*	Function-modifying protein for I_{Ks}	$\downarrow I_{Ks}$	Rare		
LQT6	*KCNE2*	Function-modifying protein for I_{Kr}	$\downarrow I_{Kr}$ (?)	Rare		
LQT7	*KCNJ2*	Pore-forming subunit for the cardiac potassium inward rectifier current I_{K1}	$\downarrow I_{K1}$	Rare	"Andersen-Tawil syndrome"; neuromuscular manifestations, atypical facies	Short-QT syndrome (gain-of-function mutations; rare)
LQT8	*CACNA1C*	Pore-forming subunit for the L-type cardiac calcium current	$\uparrow I_{Ca\text{-}L}$	Rare	"Timothy syndrome" Syndactyly, retardation	Short-QT intervals and ST-segment elevation syndrome (loss-of-function mutations; rare)
LQT9	*SCN4B*	Function-modifying protein for I_{Na}	$\uparrow I_{Na}$	Rare		
LQT10	*CAV3*	Caveolins target membrane proteins for removal from cell surface; function-modifying protein for I_{Na}	$\uparrow I_{Na}$	Rare		
LQT11	*SNTA1*	α-1-Syntrophin: Cytoskeletal protein and function-modifying protein for I_{Na}	$\uparrow I_{Na}$	Rare		
LQT12	*AKAP9*	AKAP-docking protein kinase A to KCNQ1	$\downarrow I_{Ks}$	Rare		

very low, but clinically important, risk for TdP, this form of proarrhythmia has become one of the most common causes for drug relabeling or withdrawal over the past two decades; examples include the antihistamines terfenadine and astemizole, the antibiotic sparfloxacin, and the gastric prokinetic agent cisapride. In addition, because of this concern, screening new drug candidates for QT-prolonging potential in cellular and animal systems, as well as in human volunteers, has become a routine part of the drug development process (9). Although criteria have been established to identify drugs that may have a QT-prolonging potential, the ability of such approaches to precisely quantify risk in a large population remains uncertain. The regulatory and financial burdens involved in considering QT issues in the drug development process are substantial, and this concern may be one factor underlying the slowing in the rate of New Drug Applications.

Virtually all drugs that prolong the QT interval, whether used in the therapy of arrhythmias or for "noncardiovascular" indications (e.g., terfenadine), do so by blocking I_{Kr} (10). However, not every patient exposed to an I_{Kr} blocker develops marked QT prolongation, and indeed even with administration of very potent I_{Kr}-blocking antiarrhythmics, such as ibutilide or dofetilide, the incidence of TdP remains at less than 5 percent (11, 12). Thus, the question regarding the mechanisms underlying variability in QT prolongation by I_{Kr}-blocking drugs remains open.

I_{Kr} block results in marked prolongation of the action potentials, and this effect is highly heterogeneous among cell types within the ventricle (13). In some cells, there is minimal action potential prolongation, whereas in others action potential prolongation is dramatic and may result in secondary depolarizing upstrokes, termed early afterdepolarizations (EADs) (13, 14). TdP is thought to arise from the combination of the cellular heterogeneity, which creates an arrhythmia-prone heart, and arrhythmitriggers provided by EADs.

CLINICAL FEATURES OF DRUG-INDUCED TdP

Clinical studies have identified a number of consistent risk factors for drug-induced long-QT syndrome (diLQTS) (Table 13.1).

Risk factors for drug-induced TdP
Female gender
Hypokalemia, hypomagnesemia
Bradycardia
Recent conversion from AF
Other structural heart disease
Genetic variants
Subclinical LQTS
Common variants
SCN5A S1103Y

These are relevant to the present discussion because they are important covariates in any type of large-scale genomic analysis. Women are consistently overrepresented among cases of diLQTS, through mechanisms that remain uncertain (15). Some data suggest that sex hormones modulate potassium channel expression and the extent of QT-interval prolongation upon challenge with an I_{Kr} blocker (16). Bradycardia and hypokalemia both increase risk of TdP, and these increase the likelihood of EADs in vitro (14). Interestingly, hypokalemia potentiates QT prolongation by drugs both by directly reducing I_{Kr} and by increasing drug binding to the channel (17). The presence of structural heart disease also confers further risk. The mechanism is uncertain, but it is known that certain forms of structural heart disease are accompanied by downregulation of potassium channel expression, an acquired variant of the LQTS (18). Thus, the extent of I_{Kr} block required to generate TdP may be less in such situations, as discussed further below. Similarly, patients with longer-QT intervals predrug appear to be at increased risk for TdP during drug exposure. Some studies suggest that the period immediately after conversion of AF to sinus rhythm may be one of especially high risk of TdP (19). The mechanism for this increased risk is unknown, but some data indicate that AF itself results in shorter than expected QT intervals at any given heart rate, consistent with the observation that TdP is rare when AF is the underlying rhythm (20, 21). The explanation for increased susceptibility after conversion of AF to sinus rhythm remains uncertain.

Antiarrhythmic drugs with QT-prolonging potential (such as quinidine, dofetilide, sotalol, or ibutilide) produce TdP in 1 percent to 5 percent of exposed subjects (11, 12, 22). Other drugs with known torsades potential, such as terfenadine or cisapride, generate the arrhythmia much less commonly, and true incidence figures are actually difficult to estimate. A list of drugs associated with TdP, and a rough gauge of the risk, is maintained at www.torsades.org. The reason for the discrepancy in incidence between the two types of drugs is not clear. One possibility is that antiarrhythmic drugs are used in patients with structural heart disease, and often for AF, and both of these increase risk.

CHALLENGES IN EXECUTING STUDIES OF diLQTS GENETICS

The issues in applying advances in contemporary genetics to the problem of diLQTS are similar to those for other rare adverse drug effects (ADEs).

Case Ascertainment and Definition

Finding cases of rare ADEs generally requires databases accrued during drug development (by industry) or sets gathered by networks of investigators brought together for specific case and control identification. The Trans-Atlantic Network of Excellence "Alliance Against Sudden Cardiac Death" represents such an example (http://www.allianceagainstscd.org). Investigators in this network have accrued over 200 cases of diLQTS, and controls exposed to culprit drugs and who did not

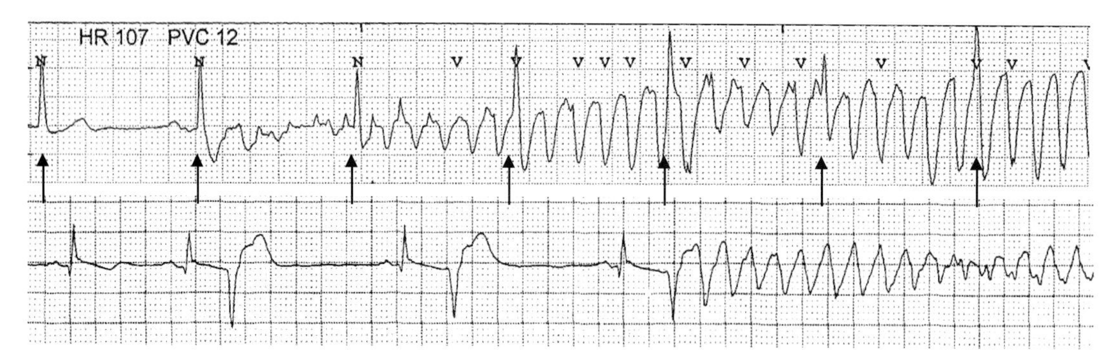

Figure 13.3. Polymorphic rhythm tracings misinterpreted as diLQTS. The top rhythm appears to be a polymorphic ventricular tachycardia, but is actually a recording artifact superimposed on a regular rhythm indicated by the arrows. The bottom rhythm shows no QT prolongation, and was observed in a patient with acute myocardial ischemia, another setting in which polymorphic ventricular tachycardia (but without QT prolongation) can be seen; this arrhythmia is clinically and mechanistically distinct from diLQTS.

develop marked QT prolongation. These cohorts represent an important starting point for studies evaluating clinical as well as genetic risk factors for the arrhythmia (23).

Precise curation of the phenotype is especially crucial in rare ADEs, because the numbers of cases may be small, and so the consequences of misclassification correspondingly large. Figures 13.2 and 13.3 contrast three patients referred for evaluation of TdP, and illustrate the potential for misclassification. The rhythm in Figure 13.2 shows typical features of diLQTS. The top rhythm in Figure 13.3 appears to be a polymorphic ventricular tachycardia, but is actually a recording artifact superimposed on a regular rhythm indicated by the arrows. The bottom rhythm in Figure 13.3 shows no QT prolongation, and was observed in a patient with acute myocardial ischemia, another setting in which polymorphic ventricular tachycardia (but without QT prolongation) can be seen; this arrhythmia is clinically and mechanistically distinct from diLQTS.

Cases may be those with the full-blown phenotype (TdP) or could also include those with no TdP but displaying marked QT-interval prolongation (e.g., recognized early to allow drug withdrawal). The decision about which case definition to use is analogous to deciding whether to include in studies of hepatotoxicity only clear cases (e.g., those progressing to transplant) or whether to also analyze cases with transiently elevated liver enzymes.

Identifying Controls

The ideal controls are similar patients exposed to drug who do not develop excessive QT prolongation or TdP. This is reasonable because most cases occur within days of drug initiation, but patients occasionally do develop the arrhythmia weeks or more later, often with the concomitant development of other risk factors like hypokalemia.

Because TdP is a rare event, large population controls may also be appropriate. Although these will necessarily include individuals at risk for TdP, this will be a small subset.

Identifying Candidate Genes

Analysis of genetic contributors to risk will focus on both candidate and unbiased approaches. Obvious candidate genes are congenital LQTS disease genes, their protein partners, pathways that modulate ion-channel activity and electrical propagation, and systems controlling intracellular calcium, a key modulator of both contractile and electrical activity in cardiac myocytes (Figure 13.4).

An interesting strategy to identifying further candidate genes has been the use of high-throughput screening in drug-sensitized embryonic zebra fish. Exposure to I_{Kr} blockers results in heart rate slowing in these animals (24). Screening large numbers of mutagenized zebra fish for resistance to heart slowing with a high dose of the potent and specific I_{Kr} blocker dofetilide or excessive heart rate slowing to low-dose dofetilide identified fifteen new genes as potential modulators of the drug response phenotype (25). These included valentine (*vtn*), previously implicated in the regulation of cardiomyocyte function and vascular morphogenesis during development (26), as well as vtn partners heart of glass (heg) and santa (san) (26, 27).

Another candidate gene for diLQTS emerged from a genome-wide analysis of variability in the normal-QT interval (28). This study identified a locus on chromosome 1 near a gene encoding an ancillary subunit of the neuronal nitric oxide synthase gene (*NOS1AP*). The encoded protein is expressed in the heart and affects

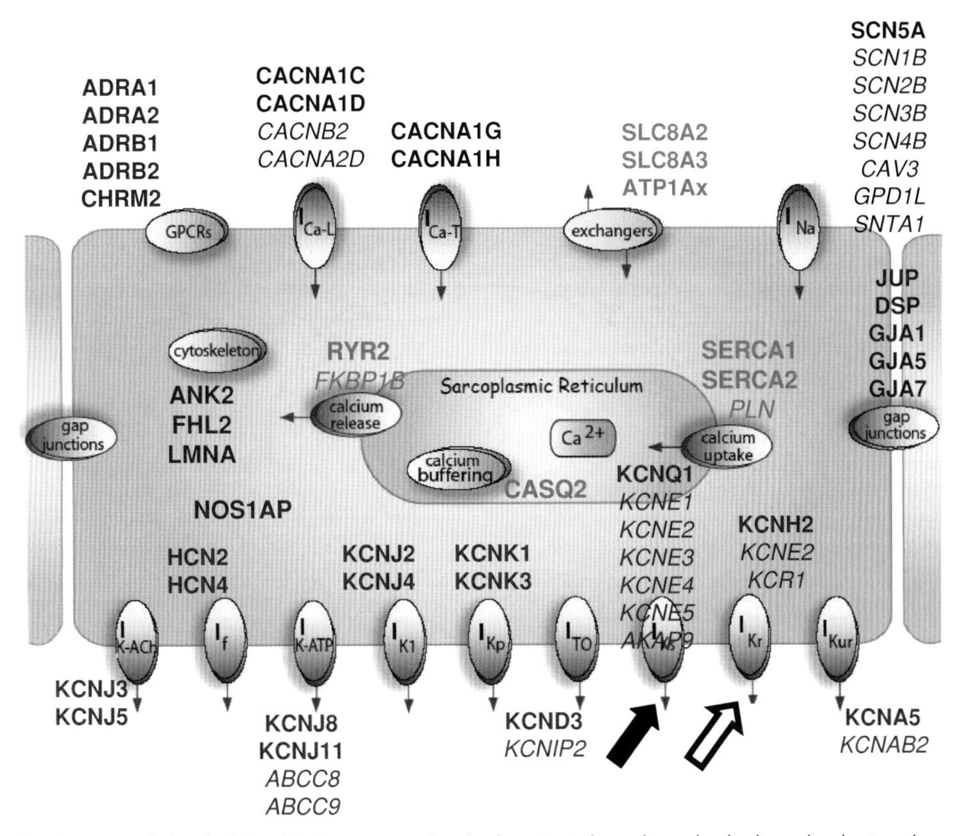

Figure 13.4. A cartoon of electrical signaling in a myocyte showing important channels conducting inward and outward currents as well as other protein complexes modulating cardiac electrical behavior. The genes whose expression results in these elements of electrical signaling are shown (major [pore-forming] subunits in bold; function-modifying genes in italics): inward currents (*black*); outward currents (*blue*); pumps/exchangers (*pink*); intracellular calcium control (*green*); and other components such as adrenergic receptors or gap junctions responsible for cell-cell communication (*gray*). The open arrow identifies the I_{Kr} complex that is the target of drugs causing diLQTS, and the solid arrow indicates the I_{Ks} complex, a major component of repolarization reserve (see text). Adapted from http://www.pharmgkb.org/do/serve?objId=PA2033&objCls=Pathway.

cardiac action potential duration probably by modulating ion-channel function (29).

A Systems View of Cardiac Repolarization – the Concept of "Repolarization Reserve"

As our understanding of the physiology of cardiac repolarization has become increasingly refined, it is possible to adopt a systems approach to considering why some patients are more susceptible to diLQTS than others. We have suggested the term "reduced repolarization reserve" to describe the situation in which the ability to execute normal action potential repolarization is compromised but clinically inapparent until drug challenge (30, 31). This framework applies equally to other ADEs, and to understanding the variability in drug action more generally, as well.

After the cloning of congenital LQTS disease genes, reports began to emerge of diLQTS in patients with minimal baseline QT prolongation, in other words, those with no manifest baseline phenotype (32, 33). Interestingly, the most commonly implicated gene is not *KCNH2*, whose expression generates I_{Kr}, but genes determining the activity of other ion channels that are not targets of drugs causing diLQTS, notably *KCNQ1* (32, 33), *SCN5A* (34), and *ANK2* (5).

We postulate that this result reflects the dependence of normal cardiac repolarization on physiologic function of multiple ion currents, "repolarization reserve." A variant in *KCNQ1* that decreases I_{Ks}, for example, might well remain clinically inapparent because a robust I_{Kr} is present to accomplish normal repolarization; supporting this idea is the finding that most patients with congenital LQTS due to *KCNQ1* mutations remain asymptomatic throughout life (35). In this scenario, however, studies both in human ventricular myocytes (36) and in computational models (37) indicate that administration of an I_{Kr} blocker might result in exaggerated QT prolongation, because of the absence of the usual repolarizing contribution of I_{Ks}. Although this formulation may apply to many proteins that contribute to normal repolarization, available data support an especially prominent role for

I_{Ks} in maintaining normal repolarization reserve (38). More recently, altered expression of microRNAs regulating I_{Ks} function has been implicated as an additional contributor to repolarization reserve (39).

THE GENETICS AND GENOMICS OF diLQTS

To date, studies to identify a potential role for genetic variants in modulating diLQTS risk have been confined to screening for occult congenital LQTS and to case-control studies examining the frequencies of common candidate nonsynonymous variants. Intensive candidate gene and genome-wide analyses are underway, but the results have not yet been reported. Following the case reports above, screening coding regions of common congenital LQTS disease genes has identified mutations in 5 percent to 20 percent of patients with diLQTS (40–42).

The nonsynonymous *SCN5A* variant resulting in S1103Y – identified almost exclusively in African Americans – generates a dysfunctional channel and has been associated with increased susceptibility to a range of arrhythmia phenotypes, including diLQTS (43, 44). The common *KCNH2* polymorphism resulting in K897T has been associated with both shorter (45) and longer (46, 47) QT intervals in large populations, but there is no reported association with diLQTS.

As described above, single-nucleotide polymorphisms (SNPs) in *NOS1AP* have been identified as modulators of the normal QT interval (28, 48). To date, there has not been an association between these SNPs and diLQTS. However, two studies have examined the association between *NOS1AP* SNPs and sudden cardiac death (SCD) in the broad population. The Rotterdam Heart Study replicated a strong association between *NOS1AP* SNPs and QT duration in 5,374 middle-aged and elderly subjects, but found no relationship to SCD (49). By contrast, an analysis of 14,737 European-Americans in the Atherosclerosis Risk In Communities study and the Cardiovascular Health Study identified *NOS1AP* alleles with an SCD relative risk of 1.3 (50). Interestingly, not all the *NOS1AP* SNPs associated with SCD predicted QT prolongation, and there was no association in 4,394 African American subjects. SCD is an enormous public health problem, accounting for approximately 15 percent of all adult deaths in the United States. Although diLQTS likely represents a tiny fraction of this problem, common cases may share some mechanisms with diLQTS. Indeed, baseline prolongation of the QT interval has been implicated as a risk factor for SCD (51–53).

SUMMARY AND FUTURE DIRECTIONS

Drug-induced QT prolongation represents an ongoing clinical problem and a special challenge to the drug development and regulatory communities. An in-depth understanding of the underlying pathophysiology has been instrumental not only in identifying candidate genes and pathways modulating risk, but also in formulating an overall framework with which to approach analysis of the problem. Difficulties in unraveling the role of genetic variants in risk are common to the study of other rare and serious ADEs. These include precise definitions as well as mechanisms to accumulate cases and controls. The development of multicenter and international networks represents one potential avenue to progress.

REFERENCES

1. Selzer A & Wray HW. Quinidine syncope, paroxysmal ventricular fibrillations occurring during treatment of chronic atrial arrhythmias. *Circulation.* 1964;**30**:17.
2. Dessertenne F. La tachycardie ventriculaire à deux foyers opposés variables. *Arch Mal Coeur.* 1966;**59**:263–72.
3. Keating MT & Sanguinetti MC. Molecular and cellular mechanisms of cardiac arrhythmias. *Cell.* 2001;**104**:569–80.
4. Mohler PJ, Schott JJ, Gramolini AO, Dilly KW, Guatimosim S, duBell WH, et al. Ankyrin-B mutation causes type 4 long-QT cardiac arrhythmia and sudden cardiac death. *Nature.* 2003;**421**:634–9.
5. Mohler PJ, Le Scouarnec S, Denjoy I, Lowe JS, Guicheney P, Caron L, et al. Defining the cellular phenotype of "ankyrin-B syndrome" variants: human ANK2 variants associated with clinical phenotypes display a spectrum of activities in cardiomyocytes. *Circulation.* 2007;**115**:432–41.
6. Priori SG, Napolitano C, & Schwartz PJ. Low penetrance in the long-QT syndrome: clinical impact. *Circulation.* 1999;**99**:529–33.
7. Monahan BP, Ferguson CL, Killeavy ES, Lloyd BK, Troy J, Cantilena LR, et al. Torsades de pointes occurring in association with terfenadine use. *JAMA.* 1990;**264**:2788–90.
8. Woosley RL, Chen Y, Freiman JP, & Gillis RA. Mechanism of the cardiotoxic actions of terfenadine. *JAMA.* 1993;**269**:1532–6.
9. Fenichel RR, Malik M, Antzelevitch C, Sanguinetti M, Roden DM, Priori SG, et al. Drug-induced torsades de pointes and implications for drug development. *J Cardiovasc Electrophysiol.* 2004;**15**:475–95.
10. Sanguinetti MC, Jiang C, Curran ME, & Keating MT. A mechanistic link between an inherited and an acquired cardiac arrhythmia: *HERG* encodes the IKr potassium channel. *Cell.* 1995;**81**:299–307.
11. Torp-Pedersen C, Moller M, Bloch-Thomsen PE, Kober L, Sandoe E, Egstrup K, et al. Dofetilide in patients with congestive heart failure and left ventricular dysfunction. Danish Investigations of Arrhythmia and Mortality on Dofetilide Study Group. *N Engl J Med.* 1999;**341**:857–65.
12. Murray KT. Ibutilide. *Circulation.* 1999;**97**:493–7.
13. Shimizu W & Antzelevitch C. Cellular basis for long QT, transmural dispersion of repolarization, and torsade de pointes in the long QT syndrome. *J Electrocardiol.* 1999;**32**(suppl):177–84.

14. Roden DM & Hoffman BF. Action potential prolongation and induction of abnormal automaticity by low quinidine concentrations in canine Purkinje fibers. Relationship to potassium and cycle length. *Circ Res.* 1985;**56**:857–67.
15. Makkar RR, Fromm BS, Steinman RT, Meissner MD, & Lehmann MH. Female gender as a risk factor for torsades de pointes associated with cardiovascular drugs. *JAMA.* 1993;**270**:2590–7.
16. Drici MD, Burklow TR, Haridasse V, Glazer RI, & Woosley RL. Sex hormones prolong the QT interval and down regulate potassium channel expression in the rabbit heart. *Circulation.* 1996;**94**:1471–4.
17. Yang T & Roden DM. Extracellular potassium modulation of drug block of IKr: implications for torsades de pointes and reverse use-dependence. *Circulation.* 1996;**93**:407–11.
18. Kääb S, Nuss HB, Chiamvimonvat N, O'Rourke B, Pak PH, Kass DA, et al. Ionic mechanism of action potential prolongation in ventricular myocytes from dogs with pacing-induced heart failure. *Circ Res.* 1996;**78**:262–73.
19. Choy AMJ, Darbar D, Dell'Orto S, & Roden DM. Increased sensitivity to QT prolonging drug therapy immediately after cardioversion to sinus rhythm. *J Am Coll Cardiol.* 1999;**34**:396–401.
20. Darbar D, Kimbrough J, Jawaid A, McCray R, Ritchie MD, & Roden DM. Persistent atrial fibrillation is associated with reduced risk of torsades de pointes in patients with drug-induced long QT syndrome. *J Am Coll Cardiol.* 2008;**51**:836–42.
21. Darbar D, Hardin B, Harris P, & Roden DM. A rate-independent method of assessing QT-RR slope following conversion of atrial fibrillation. *J Cardiovasc Electrophysiol.* 2007;**18**:636–41.
22. Soyka LF, Wirtz C, & Spangenberg RB. Clinical safety profile of sotalol in patients with arrhythmias. *Am J Cardiol.* 1990;**65**:74A–81A.
23. Kaab S, Pfeufer A, Hinterseer M, Nabauer M, Yalilzadeh S, George AL, Norris KJ, Wilde AA, Bezzina CR, Schulze-Bahr E, et al. Common gene variants associated with drug induced long-QT syndrome. *Circulation.* 2005;**112**:II-357.
24. Milan DJ, Peterson TA, Ruskin JN, Peterson RT, & Macrae CA. Drugs that induce repolarization abnormalities cause bradycardia in zebrafish. *Circulation.* 2003;**107**:1355–8.
25. Milan DJ, Jones IL, Amsterdam AH, Rosenbaum DS, Roden D, & MacRae CA. Abstract 637: A pharmacogenetic screen for modifiers of drug induced QT prolongation reveals 15 novel genes. *Circulation.* 2007;**116**:II.
26. Mably JD, Chuang LP, Serluca FC, Mohideen MA, Chen JN, & Fishman MC. Santa and valentine pattern concentric growth of cardiac myocardium in the zebrafish. *Development.* 2006;**133**:3139–46.
27. Mably JD, Mohideen MA, Burns CG, Chen JN, & Fishman MC. Heart of glass regulates the concentric growth of the heart in zebrafish. *Curr Biol.* 2003;**13**:2138–47.
28. Arking DE, Pfeufer A, Post W, Kao WH, Newton-Cheh C, Ikeda M, et al. A common genetic variant in the NOS1 regulator NOS1AP modulates cardiac repolarization. *Nat Genet.* 2006;**38**:644–51.
29. Chang KC, Barth AS, Sasano T, Kizana E, Kashiwakura Y, Zhang Y, et al. CAPON modulates cardiac repolarization via neuronal nitric oxide synthase signaling in the heart. *Proc Natl Acad Sci USA.* 2008;**105**:4477–82.
30. Roden DM. Taking the idio out of idiosyncratic – predicting torsades de pointes. *PACE.* 1998;**21**:1029–34.
31. Kaab S, Hinterseer M, Nabauer M, & Steinbeck G. Sotalol testing unmasks altered repolarization in patients with suspected acquired long-QT-syndrome – a case-control pilot study using i.v. sotalol. *Eur Heart J.* 2003;**24**:649–57.
32. Donger C, Denjoy I, Berthet M, Neyroud N, Cruaud C, Bennaceur M, et al. KVLQT1 C-terminal missense mutation causes a forme fruste long-QT syndrome. *Circulation.* 1997;**96**:2778–81.
33. Napolitano C, Schwartz PJ, Brown AM, Ronchetti E, Bianchi L, Pinnavaia A, et al. Evidence for a cardiac ion channel mutation underlying drug-induced QT prolongation and life-threatening arrhythmias. *J Cardiovasc Electrophysiol.* 2000;**11**:691–6.
34. Makita N, Horie M, Nakamura T, Ai T, Sasaki K, Yokoi H, et al. Drug-induced long-QT syndrome associated with a subclinical SCN5A mutation. *Circulation.* 2002;**106**:1269–74.
35. Priori SG, Schwartz PJ, Napolitano C, Bloise R, Ronchetti E, Grillo M, et al. Risk stratification in the long-QT syndrome. *N Engl J Med.* 2003;**348**:1866.
36. Jost N, Virag L, Bitay M, Takacs J, Lengyel C, Biliczki P, et al. Restricting excessive cardiac action potential and QT prolongation: a vital role for IKs in human ventricular muscle. *Circulation.* 2005;**112**:1392–9.
37. Silva J & Rudy Y. Subunit interaction determines IKs participation in cardiac repolarization and repolarization reserve. *Circulation.* 2005;**112**:1384–91.
38. Roden DM & Yang T. Protecting the heart against arrhythmias: potassium current physiology and repolarization reserve. *Circulation.* 2005;**112**:1376–8.
39. Xiao L, Xiao J, Luo X, Lin H, Wang Z, & Nattel S. Feedback remodeling of cardiac potassium current expression: a novel potential mechanism for control of repolarization reserve. *Circulation.* 2008.
40. Yang P, Kanki H, Drolet B, Yang T, Wei J, Viswanathan PC, et al. Allelic variants in long QT disease genes in patients with drug-associated Torsades de Pointes. *Circulation.* 2002;**105**:1943–8.
41. Lehtonen A, Fodstad H, Laitinen-Forsblom P, Toivonen L, Kontula K, & Swan H. Further evidence of inherited long QT syndrome gene mutations in antiarrhythmic drug-associated torsades de pointes. *Heart Rhythm.* 2007;**4**:603–7.
42. Paulussen AD, Gilissen RA, Armstrong M, Doevendans PA, Verhasselt P, Smeets HJ, et al. Genetic variations of KCNQ1, KCNH2, SCN5A, KCNE1, and KCNE2 in drug-induced long QT syndrome patients. *J Mol Med.* 2004;**82**:182–8.
43. Splawski I, Timothy KW, Tateyama M, Clancy CE, Malhotra A, Beggs AH, et al. Variant of SCN5A sodium channel implicated in risk of cardiac arrhythmia. *Science.* 2002;**297**:1333–6.
44. Plant LD, Bowers PN, Liu Q, Morgan T, Zhang T, State MW, et al. A common cardiac sodium channel variant associated with sudden infant death in African Americans, SCN5A S1103Y. *J Clin Invest.* 2006;**116**:430–5.
45. Pietila E, Fodstad H, Niskasaari E, Laitinen PP, Swan H, Savolainen M, et al. Association between HERG K897T polymorphism and QT interval in middle-aged Finnish women. *J Am Coll Cardiol.* 2002;**40**:511–14.

46. Bezzina CR, Verkerk AO, Busjahn A, Jeron A, Erdmann J, Koopmann TT, et al. A common polymorphism in KCNH2 (HERG) hastens cardiac repolarization. *Cardiovasc Res.* 2003;**59**:27–36.
47. Newton-Cheh C, Guo CY, Larson MG, Musone SL, Surti A, Camargo AL, et al. Common genetic variation in KCNH2 is associated with QT interval duration: the Framingham Heart Study. *Circulation.* 2007;**116**: 1128–36.
48. Post W, Shen H, Damcott C, Arking DE, Kao WH, Sack PA, et al. Associations between genetic variants in the NOS1AP (CAPON) gene and cardiac repolarization in the Old Order Amish. *Hum Hered.* 2007;**64**:214–19.
49. Aarnoudse AJ, Newton-Cheh C, de Bakker PIW, Straus SMJM, Kors JA, Hofman A, et al. Common NOS1AP variants are associated with a prolonged QTc interval in the Rotterdam Study. *Circulation.* 2007;**116**:10–16.
50. Kao WH, Arking DE, Post W, Rea TD, Sotoodehnia N, Prineas RJ, et al. Genetic variations in nitric oxide synthase 1 adaptor protein are associated with sudden cardiac death in US white community-based populations. *Circulation.* 2009;**119**:940–51.
51. Schwartz PJ & Wolf S. QT interval prolongation as a predictor of sudden death in patients with myocardial infarction. *Circulation.* 1978;**56**:1074–7.
52. Straus SM, Kors JA, De Bruin ML, Van Der Hooft CS, Hofman A, Heeringa J, et al. Prolonged QTc interval and risk of sudden cardiac death in a population of older adults. *J Am Coll Cardiol.* 2006;**47**:362–7.
53. Chugh SS, Reinier K, Singh T, Uy-Evanado A, Socoteanu C, Peters D, et al. Determinants of prolonged QT interval and their contribution to sudden death risk in coronary artery disease: the Oregon Sudden Unexpected Death Study. *Circulation.* 2009;**119**:663–70.

14 Pharmacogenetics of Diabetes

Mark C. H. de Groot and Olaf H. Klungel

INTRODUCTION

Health Impact

Diabetes mellitus has become a major public health epidemic, affecting more than 250 million individuals worldwide in 2008, increasing to 380 million in 2025. Each year 3.8 million deaths are attributable to diabetes. An even greater number die of cardiovascular disease made worse by diabetes-related lipid disorders and hypertension (1).

Type 2 diabetes is much more common than type 1 diabetes, and accounts for approximately 90 percent of all diabetes worldwide (2). Type 1 diabetes is characterized by a lack of production of insulin in the body, whereas type 2 diabetes is due to the body's diminished insulin secretion from pancreatic β-cells and the resistance of tissues to insulin (3).

Microvascular complications from type 2 diabetes are common and include retinopathy, leading to various degrees of visual impairment, including blindness; neuropathy, leading to pain and numbness; chronic and recurrent infected skin ulcers in the extremities, which can lead to amputation; and nephropathy, ultimately leading to renal failure and death. Macrovascular complications involve cardiovascular disease, stroke, and peripheral vascular disease (2, 4).

Although it is not within the scope of this chapter to present a full overview of the physiology and pharmacology of diabetes, a background will be given to discuss the genetic factors in the pharmacotherapeutic treatment of diabetes in "Etiology of Diabetes, Insulin Resistance, and Insulin Signaling." A short overview of pharmacotherapy options in diabetes ("Overview of Treatment Options") supports an understanding of the genetic modifications of pharmacotherapeutic effects discussed in "Genetic Influence on Effectiveness of Blood Glucose-Lowering Medication."

Etiology of Diabetes, Insulin Resistance, and Insulin Signaling

As mentioned under "Health Impact," two main forms of diabetes mellitus can be distinguished. Type 1 diabetes is caused by autoimmune processes that destroy the pancreatic β-cells resulting in an absolute deficiency of insulin secretion. Therefore, only insulin is used as treatment.

Type 2 diabetes results from a combination of insulin resistance and impaired insulin secretion by β-cell dysfunction. In healthy subjects. normoglycemia is maintained by the balanced interplay between insulin action and insulin secretion. When adaptation of β-cell function becomes insufficient to compensate for a decrease in insulin action, the subject progresses from normal to impaired glucose tolerance, and finally becomes type 2 diabetic (3). Insulin resistance is present when the biological effects of insulin are less than expected for both glucose disposal in skeletal muscle and the suppression of endogenous glucose production primarily in the liver (5). Treatment options for type 2 diabetes include all glucose-lowering drugs, but also lifestyle changes such as exercise and diet.

OVERVIEW OF TREATMENT OPTIONS

Goals of Treatment

The management of diabetes should be aimed at preventing hypoglycemia in the short term and at protecting patients from the long-term complications. Insulin resistance plays an essential role in the pathogenesis of type 2 diabetes and especially its adverse cardiovascular outcomes, and treatment should be focused on improvement of tissue insulin sensitivity. Lifestyle intervention, including modest exercise and diet for weight loss, clearly

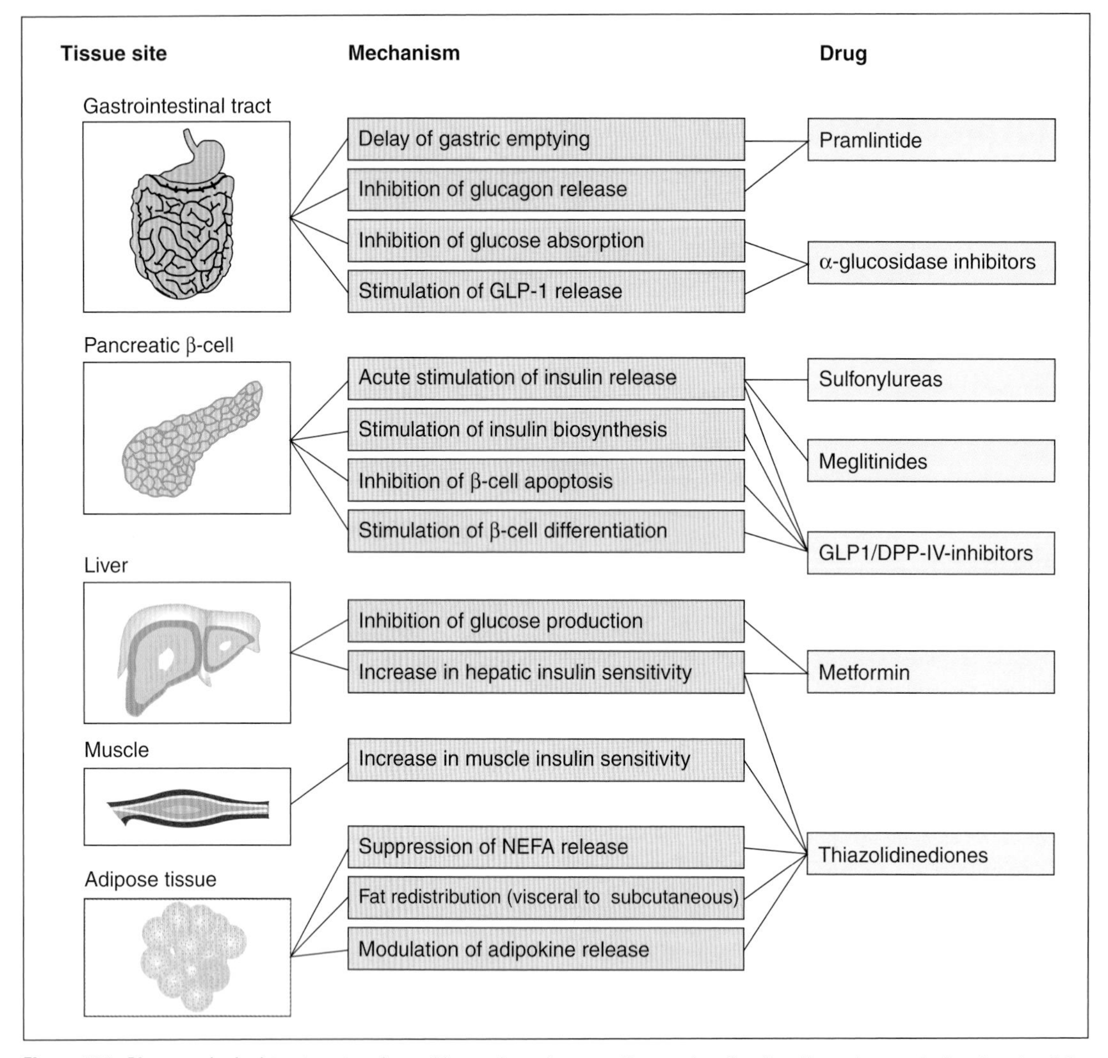

Figure 14.1. Pharmacological treatment options of hyperglycemia according to site of action. Drug classes depicted at the right side are connected through the mechanism of action to the corresponding tissue site. GLP-1 = glucagon-like-peptide 1, DPP-IV = dipeptidylpeptidase IV (3).

reduces the risk of the progression of impaired glucose tolerance to overt diabetes (6) and improves many of the cardiovascular risk parameters of the metabolic syndrome. A tight glycemic control is beneficial to reduce and delay the microvascular complications like retinopathy, nephropathy, and neuropathy (7), and macrovascular disease after a prolonged time of treatment, as well (8).

In addition to the treatment of hyperglycemia, therapy to reduce cardiovascular risk factors such as hypertension or hyperlipidemia is pivotal in patients with diabetes.

In the next sections pharmacological options will be discussed, and their site of action is shown in Figure 14.1.

Insulin Secretagogues: Sulfonylureas and Meglitinides

Secretion of insulin from pancreatic β-cells is stimulated by both blood glucose levels and activation of the sulfonylurea receptor (SUR) (Figure 14.2). When sulfonylureas bind to SUR, or by high intracellular energy levels, the SUR is activated and adjacent K_{ATP} channels (potassium inward rectifier [KIR]6.2 channel) close. Closure of the inward K^+ rectifier current leads to membrane depolarization and the opening of voltage-gated Ca^{2+} channels, which triggers release of preformed insulin-containing granules (9). A consequence of this insulin secretion is a suppression of endogenous

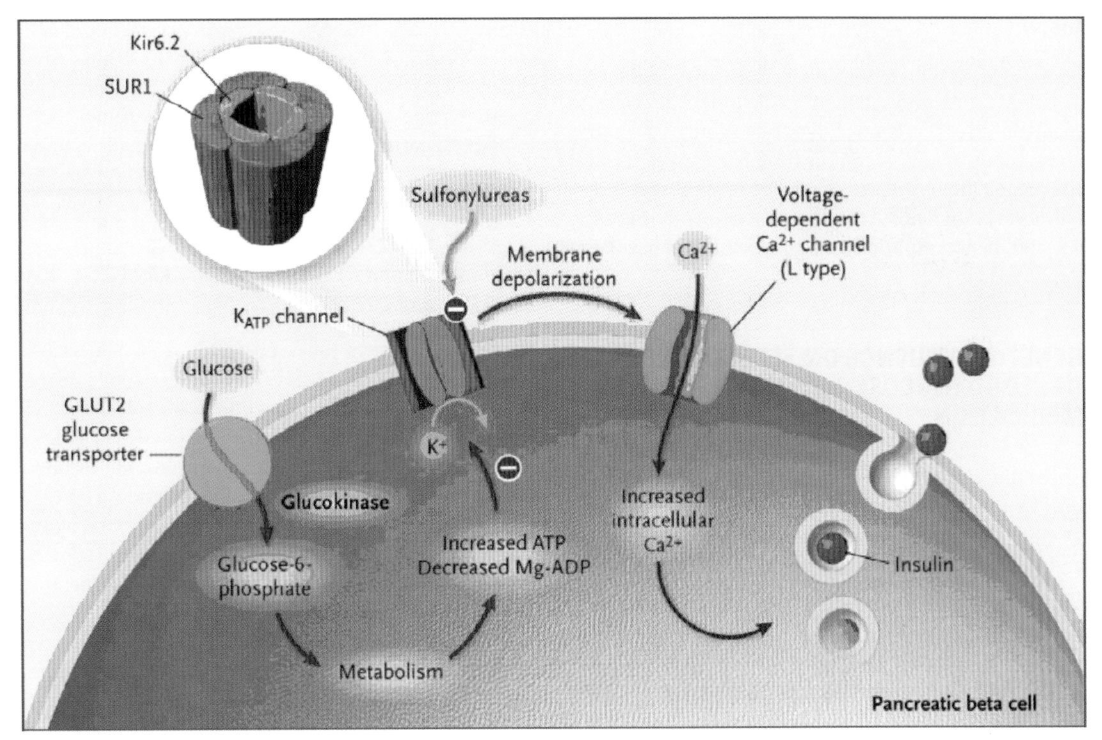

Figure 14.2. Schematic representation of the role of ATP-sensitive potassium channel in regulating insulin secretion in the pancreatic beta cell. Activation of the KATP channel, which consists of four sulfonylurea-receptor (SUR1) subunits and four Kir6.2 subunits in an octomeric structure, leads to closure of the channel. This activation can be induced either by an increased ATP/ADP ratio resulting from metabolism of glucose that entered the cell *or* by activation of the external side of the SUR. The resulting increase of potassium leads to membrane depolarization and hence activation of voltage-dependent calcium channels. Insulin exocytosis is then triggered by the increased intracellular Ca^{2+} concentration.

glucose production and the stimulation of glucose disappearance. Repaglinide and nateglinide are short-acting non-sulfonylurea insulin secretagogues collectively known as meglitinides. They also act on K_{ATP} channels, but bind on distinct sites (10). It is evident that the effectiveness of insulin secretagogues requires at least a basic ability of pancreatic islets to secrete insulin.

Glucagon-Like Peptide 1-Based Therapy

The gut-derived incretin hormone glucagon-like peptide 1 (GLP-1) has glucose-dependent insulinotropic properties, has trophic effects on β-cells, inhibits intestinal mobility, and decreases appetite, all of which reduce blood glucose levels (11). Half-life of GLP-1 in the circulation is only two to five minutes (12), mainly by degradation catalyzed by dipeptidyl peptidase-4 (DPP-4) and renal clearance. Therefore, degradation-resistant GLP-1 receptor agonists (incretin mimetics) and inhibitors of DPP-4 activity (incretin enhancers) are being developed (11).

Insulin Sensitizers: Metformin and Thiazolidinediones

Metformin and phenformin were discovered in the 1920s in a search for guanidine-containing compounds with antidiabetic activities. These biguanides were clinically introduced in Europe in the 1950s (13). Metformin can improve insulin action in hepatic and extrahepatic tissues. By inhibiting glycogenolysis and glyconeogenesis, it may decrease endogenous glucose production. It increases insulin receptor tyrosine kinase activity and increases both the number and activity of GLUT-4 transporters, responsible for glucose uptake in muscle cells. All these actions facilitate the disposal of glucose from the blood (14, 15).

Thiazolidinediones act as agonists on peroxisome proliferator-activated γ-receptors (PPARγ) and restore insulin sensitivity both directly and indirectly via lipid metabolism (16). The pleiotropic effects of these substances on inflammation, for instance, may additionally contribute to the effectiveness of these compounds.

Insulin

Exogenous insulin is indicated in patients with type 1 diabetes and some patients with type 2 diabetes to maintain glycemic control. Several regimens to mimic endogenous insulin secretion have been explored over the years. Different formulations of insulin with different absorption rates from the subcutaneous injection site and, hence, duration and onset of action have been developed.

GENETIC INFLUENCE ON EFFECTIVENESS OF BLOOD GLUCOSE-LOWERING MEDICATION

The influence of genetic variation on the effectiveness of pharmacotherapy can be categorized in three groups: pharmacokinetic effects, pharmacodynamics, and genetic drug effects where altered pathology, for example, metabolic pathways, influences drug effectiveness indirectly.

Candidate Genes on Pharmacokinetics

Drug response to oral hypoglycemic agents may be influenced by variation in the genes responsible for the metabolism or activation of a drug, transport of a drug to its active site, or excretion. Genes and polymorphisms therein associated with pharmacokinetics can accordingly be identified. Several studies on candidate genes show an association between genetic polymorphisms and pharmacokinetics of diabetes medication, as described in the next section.

Sulfonylureas and CYP2C9 Polymorphisms

Cytochrome P450 (CYP), for example, CYP3A5, genes are involved in drug metabolism. CYP3A5 is a principal catalyst of the biotransformation of repaglinide (17). Transporters include the organic anion transporters (OATs), for example, SLO1B1, and the organic cation transporters (OCTs), for example, SLC22A, that are involved in the uptake of many hydrophilic organic cations (18).

Major CYP enzymes responsible for metabolic clearance of a wide variety of drugs are CYP2C9, CYP2C8, and CYP2D6 (19). Several studies showed that healthy carriers of the Ile359Leu polymorphism of the CYP2C9 gene (also referred as CYP2C9*3) had decreased elimination of tolbutamide, glyburide, glimepiride, and chlorpropamide regardless of whether subjects were of white, Korean, or Chinese descent. Only in Chinese subjects was a clinical effect on blood glucose response and insulin response reported in CYP2C9*3 carriers (20), whereas other studies found no significant glycemic effects (21, 22). Interestingly, carriers of CYP2C9*3/*3 and CYP2C9*3/2* were overrepresented among diabetic patients admitted to the emergency department for severe hypoglycemia after sulfonylurea treatment (23). In diabetic patients carrying the CYP2C9*3 allele, lower dosages of tolbutamide were reported to be required for adequate glycemic control (24).

Meglitinides and CYP2C9, CYP2C8, CYP3A5, SLCO1B1, and CYP2D6

In healthy carriers of CYP2C9*3 (25), CYP2C8*3 (26, 27), and SLCO1B1 521T/C (28), metabolic degradation and clearance of meglitinides were decreased, which could theoretically lead to an increased risk of hypoglycemia in diabetic patients. Meglitinide elimination is unaffected by CYP2D6*4/*5 (25) and CYP3A5*3 (27) polymorphisms. In addition, the effects of repaglinide metabolism and clearance have been reported with two polymorphisms of the SLCO1B1 gene that codes for the hepatic OAT. Both the SLCO1B1 521T/C and the SLCO1B1-11187G/A variants resulted in vivo in increased peak plasma concentrations and areas under the plasma concentration-time curve (AUCs) (27, 29). This also led to lower blood glucose levels that were significantly associated with an enhanced effect of repaglinide in the case of the SLCO1B1-11187GA genotype (27).

Thiazolidinediones, CYP Enzymes, and SLO1B1

Thiazolidinediones are extensively metabolized in the liver by CYP isoenzymes. CYP2C8 is primarily involved in both rosiglitazone and pioglitazone, whereas CYP2C9 is more important in rosiglitazone (30) and CYP3A4 in pioglitazone (31) biodegradation (32).

Some in vivo studies show at least a tendency of a lower mean AUC of rosiglitazone in carriers of the CYP2C8*3 polymorphism (33–35), but no differences in blood glucose were observed (33).

Because gemfibrozil, an inhibitor of OATP1B1 in vitro, considerably increases the plasma concentrations of rosiglitazone and pioglitazone in vivo in humans, and these thiazolidinediones are competitive inhibitors of OATP1B1, polymorphisms of the encoding SLCO1B1 gene might influence pharmacokinetics. However, the SLCO1B1 521T/C polymorphism was not associated with changes in rosiglitazone or pioglitazone pharmacokinetics in healthy volunteers (36).

Biguanides and OCT Polymorphisms

Metformin is not metabolized and is mainly eliminated in urine. OCTs 1 and 2 are responsible for the hepatic (OCT1) and renal (OCT2) uptake and transport of metformin (13, 37). Several polymorphisms in encoding OCT1 genes, also known as SLC22A1, are associated with decreased metformin uptake (38, 39), and even different results have been found in oral glucose tolerance tests in healthy volunteers (39). Other studies have found only a small contribution of OCT1 and OCT2 polymorphisms to the variation of the clinical efficacy of metformin (40, 41). Recently, a multidrug and toxin extrusion (MATE) transporter protein family was identified, assigned as the SLC 47 family (42, 43). Metformin is a substrate for the MATE1 transporter located in the bile canicular membrane in the hepatocyte and the brush border of the renal epithelium and, hence, responsible for the final step of metformin excretion through bile and urine. MATE2-K is another member of this transporter family, also located in the brush border of renal epithelium and supposed to be involved in metformin excretion (44). The SLC47A1 and the SLC47A2 genes code these transporters, respectively. Preliminary results show a reduction of HbA1c level in metformin users, consistent with reduced MATE1 transporter activity associated with the rs2289669 G>A single-nucleotide polymorphism (SNP) in the SLC47A1 gene of these diabetic patients (45).

Candidate Genes Related to Pharmacodynamics

Genes related to the pharmacodynamics of glucose-lowering drugs may have an impact on drug responses. Gene products expressed as drug targets, such as receptors and signal transduction modulators, are relevant to the pharmacodynamics of diabetes drugs. Thus, polymorphisms in genes encoding for these drug targets may affect the response to a drug (46). In this section, the known polymorphisms will be discussed per drug or drug class and their drug target. Because most information is available on the effects of stimulation of insulin release by stimulation of the pancreatic β-cell and insulin sensitivity of target organs, we will focus on this.

Sulfonylureas and SUR1 and Kir6.2 Polymorphisms

As elaborated in "Insulin Secretagogues: Sulfonylureas and Meglitinides," the SUR has a key role in regulation of insulin release by the pancreatic β-cell. K_{ATP} channels are composed through association of the Kir6.2 pore with the SUR1 regulatory subunit (Figure 14.2). Polymorphisms in genes coding for the SUR1 receptor may affect binding affinity for sulfonylureas and, hence, explain part of the variation in response to these drugs.

Activating mutations in the ABCC8 gene-encoded SUR1 are associated with both transient and neonatal diabetes. In vitro data show decreased tolbutamide sensitivity in some of the identified polymorphisms (47).

Interestingly, *loss-of-function* gene defects in either KCNJ11 or ABCC8 resulting in inactivating K_{ATP} mutations have been implicated in hyperinsulinemic hypoglycemia of infancy (48). Effectiveness of treatment with the K_{ATP} channel opener diazoxide seems to depend on the genetic profile (20).

Activating mutations in the Kir6.2 pore-encoding gene (KCNJ11), favoring opening of the K_{ATP} channels, have been identified in both transient and permanent neonatal diabetes mellitus (49). Although the closure of the K_{ATP} channels regulated by high-energetic-status glucose or glucagon in the pancreatic cells is impaired or absent in these patients with one of the six studied mutations (R201H, R201C, V59M, V59G, Q52R, I296L), direct action on the SUR1 receptors via tolbutamide results in insulin release in these patients.

Meglitinides and Polymorphisms

Knowledge on the pharmacodynamics of repaglitinide with different genotypes is sparse. Kalliokoski et al. (36) investigated the effects of SLCO1B1 (encoding OATP1B1) c.521T→C (p.Val174Ala) SNP on the pharmacokinetics and pharmacodynamics of repaglinide. Although the pharmacokinetics was affected by genotype, no statistically significant effects on serum blood glucose levels were detected. For nateglinide, no differences were detected for these SNPs at all (36).

Sulfonylureas and Biguanides and Insulin Receptor Substrate Polymorphisms

The insulin receptor substrate family plays a role in the insulin-signaling pathway by phosphorylation of inositol 3-phosphate (50). An increased risk of nonresponse to sulfonylureas has been observed with polymorphisms of the Arg972 allele of IRS1 (51). The decreased effectiveness of metformin therapy in lowering fasting insulin levels and insulin resistance was seen in patients with polycystic ovary syndrome having this polymorphism in the IRS gene (52).

Sulfonylureas and Biguanides and KCNJ11 Polymorphisms

The common *KCNJ11* Glu23Lys or E23K polymorphism representing a glutamate → lysine (K) change at position 23 is consistently associated with type 2 diabetes (53, 54). Patients with impaired glucose tolerance or newly diagnosed type 2 diabetes and carriers of the lysine variant were associated with a more frequent secondary failure of metformin added to sulfonylurea therapy (55). The biological explanation for the differential response to metformin remains to be unraveled.

Sulfonylureas and Biguanides and Transcription Factor-7-Like 2 Polymorphisms

In the search for diabetes genes, transcription factor-7-like 2 (TCF7L2) was revealed as an unexpected suspect for a type 2 diabetes gene by the DECODE group in Iceland (56) in 2006. The mechanism(s) by which risk alleles in TCF7L2 increase the risk of type 2 diabetes most likely involve the impairment of insulin secretion (57). Besides an increased risk for type 2 diabetes (58), the reduced effectiveness of sulfonylurea therapy was reported in a Scottish cohort of diabetic patients. The effectiveness of metformin treatment was not changed (59).

Thiazolidiones, Acarbose, and PPARγ Polymorphisms

In key target tissues for insulin action, such as adipose tissue, skeletal muscle, and liver, PPARγs are found. These receptors are involved in regulation of adipocyte differentiation, lipid homeostasis, and insulin action (60). The thiazolidinedione compounds exert their insulin-sensitizing effect through these receptors in the treatment of type 2 diabetes (61). Among the polymorphisms studied, only the (Pro12Ala) polymorphism in the PPARG gene shows a consistent association with risk for type 2 diabetes. The less frequent alanine allele (16 percent) was associated with decreased diabetes risk (odds ratio, 0.78; $P < 0.045$ one-tailed) (62). Clinical experience shows that the considerable number of therapeutic nonresponders to thiazolidiones to be about one-third. The pharmacodynamic effects of PPARG 12A on the response to pioglitazone (63) and troglitazone (64) were generally not found (65). An exception is a small study where the response to rosiglitazone in patients with the alanine allele was better; a larger reduction occurred in their fasting glucose and HA1C values than in these values in Pro12Pro homozygotes (66).

Acarbose has been reported to prevent the development of diabetes; this effect is stronger among carriers of the Ser482 allele of the Gly482Ser polymorphism of the PPARGC1A gene (67) and in carriers of the PPARG Pro12Ala polymorphism (67).

Thiazolidiones, Acarbose, and Adiponectin Polymorphisms

Adiponectin (ACDC; ADIPOQ) is a protein secreted by adipocytes and is known to be a potent insulin sensitizer. A low fasting adiponectin concentration is associated with low insulin-stimulated skeletal muscle insulin receptor tyrosine phosphorylation. Although ADIPOQ gene expression in adipose tissue is associated with obesity, insulin resistance, and type 2 diabetes, hypoadiponectinemia is more strongly related to the degree of insulin resistance than the degree of adiposity or glucose tolerance (68). Genetic polymorphisms may be involved in the regulation of adiponectin (69).

Carriers of the GG genotype for ADIPOQ +45T/G polymorphism showed smaller reductions in fasting plasma glucose level and HbA1c value after rosiglitazone treatment than heterozygotes and homozygous wild-type carriers (70). No association was found between the ADIPOQ +45T/G polymorphism and the development of type 2 diabetes mellitus among acarbose-treated subjects with impaired glucose tolerance (71). However, the TT genotype of the ADIPOQ +276G/T polymorphism was associated with a higher risk of type 2 diabetes than the GG genotype in all subjects treated with acarbose.

Incretin Mimetics/DPP-4 Inhibitors

Incretins include gastrointestinal hormones such as GLP-1 and glucose-dependent insulinotropic peptide (GIP) that stimulate insulin response after oral glucose ingestion. In patients with type 2 diabetes this response is diminished and may be enhanced by incretin mimetics such as exenatide or by inhibiting GLP-1 degradation by DDP-4 inhibitors (72). In a small study among eight TCF7L2 TT/TC carriers and ten controls with the CC wild-type genotype, the incretin response was assessed from ratios of insulin secretory rates (IRS) during oral and isoglycemic glucose infusions (73). Among those with at least one variant TCF7L2 allele the incretin response was 30 percent lower in comparison with wild-type homozygous subjects. This effect was not due to reduced secretion of GLP-1 and GIP, but probably due to the reduced β-cell sensitivity to incretins.

Candidate Genes in the Causal Pathway

Finally, there is a growing interest in genes that are in the causal pathway of diseases and are able to influence the drug response. A complicating factor is that most diseases have a polygenetic origin and that different genetic pathways may therefore operate in patients with the same phenotype. These genetic differences may also lead to different responses to drug treatment. So far, no genes in this category have been identified.

CONCLUSION

Many pharmacogenetic investigations suggest that genetic variation plays a role in the response to glucose-lowering drugs. However, few genetic variants have been identified that consistently predict the response to specific glucose-lowering drugs.

So far, relevant pharmacogenetic interactions have been identified with regard to the CYP2C9*3 variant and sulfonylureas, resulting in decreased clearance and increased glucose-lowering response, increased glucose-lowering response to repaglinide in SLCO1B1 variant carriers, and increased HbA1c reduction in response to metformin in genes coding for the MATE1 transporter.

Most pharmacogenetic studies have focused on pharmacokinetic parameters, were performed in healthy volunteers, and have included incompletely assessed intragenic variation and relatively small sample sizes. Future studies should include patients with diabetes and measure clinically more important outcomes such as glucose-lowering response and eventually micro- and macrovascular complications. Furthermore, advancements in design (randomization and blinding of treatments, adequate sample size), analysis (taking multiple testing into account to prevent false discoveries), and conduct of pharmacogenetic studies (including replication studies) should contribute to a better understanding of the influence of genetic variation on the response to glucose-lowering drugs. SNPs within several candidate genes involved in pharmacokinetic and pharmacodynamic pathways of glucose-lowering drug response and genome-wide association studies to identify new genetic variants may help to consistently assess these pharmacogenetic interactions.

When important pharmacogenetic effects have been consistently demonstrated, trials that randomly assign patients to genotype-guided glucose-lowering drug choice versus standard glucose-lowering drug treatment algorithms are needed to demonstrate the cost-effectiveness of this individualized treatment approach.

REFERENCES

1. *International Diabetes Federation. Diabetes Atlas.* 3rd ed. 2007 [cited 2009 16–2–09]. http://www.idf.org.
2. World Health Organization. *Factsheet: Diabetes.* 2008 [cited 2009 16–2–09]; November 2008. http://www.who.int/mediacentre/factsheets/fs312/en/index.html.
3. Stumvoll M, Goldstein BJ, & van Haeften TW. Type 2 diabetes: principles of pathogenesis and therapy. *Lancet.* 2005;**365**:1333–46.
4. Campbell RK. Type 2 diabetes: where we are today: an overview of disease burden, current treatments, and treatment strategies. *J Am Pharm Assoc.* 2009;**49**(suppl 1):S3–9.
5. Dinneen S, Gerich J, & Rizza R. Carbohydrate metabolism in non-insulin-dependent diabetes mellitus. *N Engl J Med.* 1992;**327**:707–13.
6. Knowler WC, Barrett-Connor E, Fowler SE, Hamman RF, Lachin JM, Walker EA, et al. Reduction in the incidence of type 2 diabetes with lifestyle intervention or metformin. *N Engl J Med.* 2002;**346**:393–403.
7. Schellhase KG, Koepsell TD, & Weiss NS. Glycemic control and the risk of multiple microvascular diabetic complications. *Fam Med.* 2005;**37**:125–30.
8. Holman RR, Paul SK, Bethel MA, Matthews DR, & Neil HA. 10-year follow-up of intensive glucose control in type 2 diabetes. *N Engl J Med.* 2008;**359**:1577–89.
9. Ashcroft FM & Rorsman P. Electrophysiology of the pancreatic beta-cell. *Prog Biophys Mol Biol.* 1989;**54**: 87–143.
10. Fuhlendorff J, Rorsman P, Kofod H, Brand CL, Rolin B, MacKay P, et al. Stimulation of insulin release by repaglinide and glibenclamide involves both common and distinct processes. *Diabetes.* 1998;**47**:345–51.
11. Drucker DJ & Nauck MA. The incretin system: glucagon-like peptide-1 receptor agonists and dipeptidyl peptidase-4 inhibitors in type 2 diabetes. *Lancet.* 2006;**368**: 1696–705.
12. Orskov C, Wettergren A, & Holst JJ. Biological effects and metabolic rates of glucagonlike peptide-17–36 amide and glucagonlike peptide-17–37 in healthy subjects are indistinguishable. *Diabetes.* 1993;**42**:658–61.
13. Reitman ML & Schadt EE. Pharmacogenetics of metformin response: a step in the path toward personalized medicine. *J Clin Invest.* 2007;**117**:1226–9.
14. Wiernsperger NF & Bailey CJ. The antihyperglycaemic effect of metformin: therapeutic and cellular mechanisms. *Drugs.* 1999;**58**(suppl 1):31–9; discussion 75–82.
15. Klip A & Leiter LA. Cellular mechanism of action of metformin. *Diabetes Care.* 1990;**13**:696–704.
16. Staels B & Fruchart JC. Therapeutic roles of peroxisome proliferator-activated receptor agonists. *Diabetes.* 2005;**54**:2460–70.
17. Bidstrup TB, Bjornsdottir I, Sidelmann UG, Thomsen MS, & Hansen KT. CYP2C8 and CYP3A4 are the principal enzymes involved in the human in vitro biotransformation of the insulin secretagogue repaglinide. *Br J Clin Pharmacol.* 2003;**56**:305–14.
18. Jonker JW & Schinkel AH. Pharmacological and physiological functions of the polyspecific organic cation transporters: OCT1, 2, and 3 (SLC22A1–3). *J Pharmacol Exp Ther.* 2004;**308**:2–9.
19. Rettie AE & Jones JP. Clinical and toxicological relevance of CYP2C9: drug-drug interactions and pharmacogenetics. *Annu Rev Pharmacol Toxicol.* 2005;**45**:477–94.
20. Ashcroft FM. ATP-sensitive potassium channelopathies: focus on insulin secretion. *J Clin Invest.* 2005;**115**: 2047–58.
21. Kalliokoski A, Neuvonen M, Neuvonen PJ, & Niemi M. The effect of SLCO1B1 polymorphism on repaglinide pharmacokinetics persists over a wide dose range. *Br J Clin Pharmacol.* 2008;**66**:818–25.
22. Kirchheiner J, Brockmoller J, Meineke I, Bauer S, Rohde W, Meisel C, et al. Impact of CYP2C9 amino acid polymorphisms on glyburide kinetics and on the insulin and glucose response in healthy volunteers. *Clin Pharmacol Ther.* 2002;**71**:286–96.
23. Holstein A, Plaschke A, Ptak M, Egberts EH, El-Din J, Brockmoller J, et al. Association between CYP2C9 slow metabolizer genotypes and severe hypoglycaemia on medication with sulphonylurea hypoglycaemic agents. *Br J Clin Pharmacol.* 2005;**60**:103–6.

24. Becker ML, Visser LE, Trienekens PH, Hofman A, van Schaik RH, & Stricker BH. Cytochrome P450 2C9 *2 and *3 polymorphisms and the dose and effect of sulfonylurea in type II diabetes mellitus. *Clin Pharmacol Ther.* 2008;**83**:288–92.
25. Kirchheiner J, Meineke I, Muller G, Bauer S, Rohde W, Meisel C, et al. Influence of CYP2C9 and CYP2D6 polymorphisms on the pharmacokinetics of nateglinide in genotyped healthy volunteers. *Clin Pharmacokinet.* 2004;**43**:267–78.
26. Niemi M, Leathart JB, Neuvonen M, Backman JT, Daly AK, & Neuvonen PJ. Polymorphism in CYP2C8 is associated with reduced plasma concentrations of repaglinide. *Clin Pharmacol Ther.* 2003;**74**:380–7.
27. Niemi M, Backman JT, Kajosaari LI, Leathart JB, Neuvonen M, Daly AK, et al. Polymorphic organic anion transporting polypeptide 1B1 is a major determinant of repaglinide pharmacokinetics. *Clin Pharmacol Ther.* 2005;**77**:468–78.
28. Zhang W, He YJ, Han CT, Liu ZQ, Li Q, Fan L, et al. Effect of SLCO1B1 genetic polymorphism on the pharmacokinetics of nateglinide. *Br J Clin Pharmacol.* 2006;**62**:567–72.
29. Kalliokoski A, Neuvonen M, Neuvonen PJ, & Niemi M. No significant effect of SLCO1B1 polymorphism on the pharmacokinetics of rosiglitazone and pioglitazone. *Br J Clin Pharmacol.* 2008;**65**:78–86.
30. Baldwin CC. Characterization of the cytochrome P450 enzymes involved in the *in vitro* metabolism of rosiglitazone. *Br J Clin Pharmacol.* 1999;**48**:424–32.
31. Jaakkola T, Laitila J, Neuvonen PJ, & Backman JT. Pioglitazone is metabolised by CYP2C8 and CYP3A4 in vitro: potential for interactions with CYP2C8 inhibitors. *Basic Clin Pharmacol Toxicol.* 2006;**99**:44–51.
32. Aquilante CL. Pharmacogenetics of thiazolidinedione therapy. *Pharmacogenomics.* 2007;**8**:917–31.
33. Kirchheiner J, Thomas S, Bauer S, Tomalik-Scharte D, Hering U, Doroshyenko O, et al. Pharmacokinetics and pharmacodynamics of rosiglitazone in relation to CYP2C8 genotype. *Clin Pharmacol Ther.* 2006;**80**:657–67.
34. Pedersen RS, Damkier P, & Brosen K. The effects of human CYP2C8 genotype and fluvoxamine on the pharmacokinetics of rosiglitazone in healthy subjects. *Br J Clin Pharmacol.* 2006;**62**:682–9.
35. Hruska MW, Amico JA, Langaee TY, Ferrell RE, Fitzgerald SM, & Frye RF. The effect of trimethoprim on CYP2C8 mediated rosiglitazone metabolism in human liver microsomes and healthy subjects. *Clin Pharmacol Ther.* 2005; **59**(1):70–9.
36. Kalliokoski A, Neuvonen M, Neuvonen PJ, & Niemi M. Different effects of SLCO1B1 polymorphism on the pharmacokinetics and pharmacodynamics of repaglinide and nateglinide. *J Clin Pharmacol.* 2008;**48**:311–21.
37. Takane H, Shikata E, Otsubo K, Higuchi S, & Ieiri I. Polymorphism in human organic cation transporters and metformin action. *Pharmacogenomics.* 2008;**9**:415–22.
38. Tzvetkov MV, Vormfelde SV, Balen D, Meineke I, Schmidt T, Sehrt D, et al. The effects of genetic polymorphisms in the organic cation transporters OCT1, OCT2, and OCT3 on the renal clearance of metformin. *Clin Pharmacol Ther.* 2009;**86**:299–306.
39. Shu Y, Sheardown SA, Brown C, Owen RP, Zhang S, Castro RA, et al. Effect of genetic variation in the organic cation transporter 1 (OCT1) on metformin action. *J Clin Invest.* 2007;**117**:1422–31.
40. Shikata E, Yamamoto R, Takane H, Shigemasa C, Ikeda T, Otsubo K, et al. Human organic cation transporter (OCT1 and OCT2) gene polymorphisms and therapeutic effects of metformin. *J Hum Genet.* 2007;**52**:117–22.
41. Becker ML, Visser LE, van Schaik RH, Hofman A, Uitterlinden AG, & Stricker BH. Genetic variation in the organic cation transporter 1 is associated with metformin response in patients with diabetes mellitus. *Pharmacogenomics J.* 2009;**9**:242–7.
42. Otsuka M, Matsumoto T, Morimoto R, Arioka S, Omote H, & Moriyama Y. A human transporter protein that mediates the final excretion step for toxic organic cations. *Proc Natl Acad Sci USA.* 2005;**102**:17923–8.
43. Terada T & Inui K. Physiological and pharmacokinetic roles of H+/organic cation antiporters (MATE/SLC47A). *Biochem Pharmacol.* 2008;**75**:1689–96.
44. Tanihara Y, Masuda S, Sato T, Katsura T, Ogawa O, & Inui K. Substrate specificity of MATE1 and MATE2-K, human multidrug and toxin extrusions/H(+)-organic cation antiporters. *Biochem Pharmacol.* 2007;**74**: 359–71.
45. Becker ML, Visser LE, van Schaik RHN, Hofman A, Uitterlinden AG, & Stricker BHC. Genetic variation in the multidrug and toxin extrusion 1 transporter protein influences the glucose-lowering effect of metformin in patients with diabetes: a preliminary study. *Diabetes.* 2009;**58**: 745–9.
46. Mancinelli L, Cronin M, & Sadee W. Pharmacogenomics: the promise of personalized medicine. *AAPS Pharm Sci.* 2000;**2**:E4.
47. Sattiraju S, Reyes S, Kane GC, & Terzic A. K(ATP) channel pharmacogenomics: from bench to bedside. *Clin Pharmacol Ther.* 2008;**83**:354–7.
48. Dekelbab BH & Sperling MA. Recent advances in hyperinsulinemic hypoglycemia of infancy. *Acta Paediatr.* 2006;**95**:1157–64.
49. Gloyn AL, Pearson ER, Antcliff JF, Proks P, Bruining GJ, Slingerland AS, et al. Activating mutations in the gene encoding the ATP-sensitive potassium-channel subunit Kir6.2 and permanent neonatal diabetes. *N Engl J Med.* 2004;**350**:1838–49.
50. Mousavinasab F, Tahtinen T, Jokelainen J, Koskela P, Vanhala M, Oikarinen J, et al. Common polymorphisms in the PPARgamma2 and IRS-1 genes and their interaction influence serum adiponectin concentration in young Finnish men. *Mol Genet Metab.* 2005;**84**:344–8.
51. Sesti G, Marini MA, Cardellini M, Sciacqua A, Frontoni S, Andreozzi F, et al. The Arg972 variant in insulin receptor substrate-1 is associated with an increased risk of secondary failure to sulfonylurea in patients with type 2 diabetes. *Diabetes Care.* 2004;**27**:1394–8.
52. Ertunc D, Tok EC, Aktas A, Erdal EM, & Dilek S. The importance of IRS-1 Gly972Arg polymorphism in evaluating the response to metformin treatment in polycystic ovary syndrome. *Hum Reprod.* 2005;**20**:1207–12.
53. Florez JC, Jablonski KA, Kahn SE, Franks PW, Dabelea D, Hamman RF, et al. Type 2 diabetes-associated missense polymorphisms KCNJ11 E23K and ABCC8 A1369S influence progression to diabetes and response to

interventions in the Diabetes Prevention Program. *Diabetes.* 2007;**56**:531–6.
54. van Dam RM, Hoebee B, Seidell JC, Schaap MM, de Bruin TW, & Feskens EJ. Common variants in the ATP-sensitive K+ channel genes KCNJ11 (Kir6.2) and ABCC8 (SUR1) in relation to glucose intolerance: population-based studies and meta-analyses. *Diabet Med.* 2005;**22**:590–8.
55. Sesti G, Laratta E, Cardellini M, Andreozzi F, Del Guerra S, Irace C, et al. The E23K variant of KCNJ11 encoding the pancreatic beta-cell adenosine 5′-triphosphate-sensitive potassium channel subunit Kir6.2 is associated with an increased risk of secondary failure to sulfonylurea in patients with type 2 diabetes. *J Clin Endocrinol Metab.* 2006;**91**:2334–9.
56. Grant SF, Thorleifsson G, Reynisdottir I, Benediktsson R, Manolescu A, Sainz J, et al. Variant of transcription factor 7-like 2 (TCF7L2) gene confers risk of type 2 diabetes. *Nat Genet.* 2006;**38**:320–3.
57. Hattersley AT. Prime suspect: the TCF7L2 gene and type 2 diabetes risk. *J Clin Invest.* 2007;**117**:2077–9.
58. Lyssenko V, Lupi R, Marchetti P, Del Guerra S, Orho-Melander M, Almgren P, et al. Mechanisms by which common variants in the TCF7L2 gene increase risk of type 2 diabetes. *J Clin Invest.* 2007;**117**:2155–63.
59. Pearson ER, Donnelly LA, Kimber C, Whitley A, Doney AS, McCarthy MI, et al. Variation in TCF7L2 influences therapeutic response to sulfonylureas: a GoDARTs study. *Diabetes.* 2007;**56**:2178–82.
60. Mudaliar S & Henry RR. New oral therapies for type 2 diabetes mellitus: the glitazones or insulin sensitizers. *Annu Rev Med.* 2001;**52**:239–57.
61. Krentz AJ & Bailey CJ. Oral antidiabetic agents: current role in type 2 diabetes mellitus. *Drugs.* 2005;**65**:385–411.
62. Altshuler D, Hirschhorn JN, Klannemark M, Lindgren CM, Vohl MC, Nemesh J, et al. The common PPARgamma Pro12Ala polymorphism is associated with decreased risk of type 2 diabetes. *Nat Genet.* 2000;**26**:76–80.
63. Bluher M, Lubben G, & Paschke R. Analysis of the relationship between the Pro12Ala variant in the PPAR-gamma2 gene and the response rate to therapy with pioglitazone in patients with type 2 diabetes. *Diabetes Care.* 2003;**26**: 825–31.
64. Snitker S, Watanabe RM, Ani I, Xiang AH, Marroquin A, Ochoa C, et al. Changes in insulin sensitivity in response to troglitazone do not differ between subjects with and without the common, functional Pro12Ala peroxisome proliferator-activated receptor-gamma2 gene variant: results from the Troglitazone in Prevention of Diabetes (TRIPOD) study. *Diabetes Care.* 2004;**27**:1365–8.
65. Moore AF & Florez JC. Genetic susceptibility to type 2 diabetes and implications for antidiabetic therapy. *Annu Rev Med.* 2008;**59**:95–111.
66. Kang ES, Park SY, Kim HJ, Kim CS, Ahn CW, Cha BS, et al. Effects of Pro12Ala polymorphism of peroxisome proliferator-activated receptor gamma2 gene on rosiglitazone response in type 2 diabetes. *Clin Pharmacol Ther.* 2005;**78**:202–8.
67. Andrulionyte L, Zacharova J, Chiasson JL, & Laakso M. Common polymorphisms of the PPAR-gamma2 (Pro12Ala) and PGC-1alpha (Gly482Ser) genes are associated with the conversion from impaired glucose tolerance to type 2 diabetes in the STOP-NIDDM trial. *Diabetologia.* 2004;**47**:2176–84.
68. Weyer C, Funahashi T, Tanaka S, Hotta K, Matsuzawa Y, Pratley RE, et al. Hypoadiponectinemia in obesity and type 2 diabetes: close association with insulin resistance and hyperinsulinemia. *J Clin Endocrinol Metab.* 2001;**86**: 1930–5.
69. Hara K, Boutin P, Mori Y, Tobe K, Dina C, Yasuda K, et al. Genetic variation in the gene encoding adiponectin is associated with an increased risk of type 2 diabetes in the Japanese population. *Diabetes.* 2002;**51**:536–40.
70. Kang ES, Park SY, Kim HJ, Ahn CW, Nam M, Cha BS, et al. The influence of adiponectin gene polymorphism on the rosiglitazone response in patients with type 2 diabetes. *Diabetes Care.* 2005;**28**:1139–44.
71. Zacharova J, Chiasson JL, & Laakso M. The common polymorphisms (single nucleotide polymorphism [SNP] +45 and SNP +276) of the adiponectin gene predict the conversion from impaired glucose tolerance to type 2 diabetes: the STOP-NIDDM trial. *Diabetes.* 2005;**54**:893–9.
72. Gallwitz B. New therapeutic strategies for the treatment of type 2 diabetes mellitus based on incretins. *Rev Diabet Stud.* 2005;**2**:61–9.
73. Villareal DT, Robertson H, Bell GI, Patterson BW, Tran H, Wice B, & Polonsky KS. TCF7L2 variant rs7903146 affects the risk of type 2 diabetes by modulating incretin action. *Diabetes.* 2010;**59**:479–85.

15 Pharmacogenetics – Therapeutic Area – Respiratory

Kelan Tantisira and Scott Weiss

INTRODUCTION: THE BURDEN OF RESPIRATORY DISEASE

Respiratory disease refers to the broad category of illnesses affecting the upper and lower airways, the lung parenchyma, and the pulmonary vasculature. To date, however, studies focusing on the pharmacogenetics of respiratory disease have been largely concentrated in the areas of lung cancer and obstructive lung disease. This is primarily due to the relative rarity of many pulmonary diseases, providing insufficient sample size for pharmacogenetic studies. In addition, for many respiratory diseases, there is a relative paucity of available therapies, providing no current alternatives even if a priori prediction of poor response to therapies was available. For many of these diseases (e.g., interstitial pulmonary fibrosis, chronic hypersensitivity pneumonitis, and primary pulmonary hypertension), multicenter studies and/or industry collaborations are warranted. This chapter focuses on the genetics of drug treatment response in obstructive lung disease, with an emphasis on the pharmacogenetics of asthma. Although lung cancer is the leading cause of cancer deaths in both males and females in America and clearly of great importance, the pharmacogenetics of oncologic disorders has been covered separately in Chapter 10. We will review how the respiratory pharmacogenetics fits within the context of the four major categories of pharmacogenetic response and detail approaches to defining respiratory pharmacogenetic phenotypes. The chapter continues with a review of the literature on the pharmacogenetic associations that have been characterized in relation to asthma and chronic obstructive pulmonary disease (COPD). We conclude with some thoughts on the future of pharmacogenetics in these two diseases. However, we first provide an overview of the burden of obstructive respiratory disease.

Chronic obstructive respiratory disease, including COPD and asthma, is currently the fourth leading cause of death in the United States and is projected to move into third place nationwide by 2020 (1). Although mortality rates for the top two leading causes of death in the United States, heart disease and cancer, respectively, are decreasing, deaths from chronic lung disease continue to rise. According to the National Center for Health Statistics, more than 16 million people in the United States have been diagnosed with COPD, and it is estimated that another 16 million cases are undiagnosed (2). COPD cost the nation more than $30 billion in 2000, $14.7 billion in direct health care costs and $15.7 billion in indirect costs. In 2007, there were 22.9 million Americans with asthma, with an estimated annual cost of $19.7 billion ($14 billion in direct costs, of which the largest component was medication costs). Whereas COPD is a leading cause of morbidity and mortality in the elderly, asthma remains the leading cause of hospitalizations and school absences in children.

Pharmacogenetic Categories

Pharmacogenetics has traditionally been divided into four categories based on the effects of genetic variability on the pharmacologic properties of a drug. The categories include genetic variation related to pharmacokinetics, pharmacodynamics, idiosyncratic reactions, and disease pathogenesis. *Pharmacokinetics* studies the metabolic effects of an administered drug, including those related to the absorption, distribution, tissue localization, biotransformation, and excretion of the drug. The major areas of pharmacokinetic pharmacogenetics have been in the evaluation of genetic differences in drug-metabolizing enzymes and drug transporters; these are reviewed elsewhere in this text (Chapter 3 by Blaschke). One of the unique aspects of respiratory pharmacogenetics is that the most commonly prescribed medications for asthma and COPD are taken by the inhalational route, effectively delivering medication to the tracheobronchial tree while simultaneously limiting side effects. Therefore, although local (target tissue) transport and metabolism may play a role, *the role of first-pass metabolism and intestinal drug transport is negligible for these drugs.* Consequently, almost no pharmacokinetic studies related to asthma and COPD pharmacogenetics have been performed.

Pharmacodynamics is the study of the biochemical and physiological consequences of the administration of a drug and its mechanism of action (i.e., the effect of a drug at its therapeutic target). Genetic variation may lead to interindividual differences in the response to a drug despite the presence of appropriate concentrations of the drug at its intended target. In this category, the genetic variation is typically present at the site of the target or one of the downstream participants in the target's mechanistic pathway, thereby modulating the effects of the drug. The majority of studies related to the pharmacogenetics of asthma and COPD fall in this category.

Although pharmacogenetic predictors of efficacy at therapeutic target sites will likely become a primary basis for "individualized therapy," identification and replication of pharmacodynamic pharmacogenetic associations have proven difficult in comparison with those related to pharmacokinetics (3). At least three reasons may explain this discrepancy. The first is that the biology of drug metabolism and drug transport is comparatively straightforward. For instance, most compounds have one principle enzyme responsible for their metabolism. However, the physiology of most drug target pathways is fairly complex, providing multiple venues that require investigation within the pathway. The second is that differences in metabolism related to genetics can be as high as 10,000-fold, whereas differences in target binding related to genetics are generally less than 20-fold (4). The resultant statistical ability (power) to detect differences in the metabolism of a drug within a population for any given compound is clearly far greater than the ability to detect the variation in drug targets. Finally, the pattern of inheritance permits an easier identification of pharmacokinetic variants. Typically, the inheritance pattern underlying the pharmacogenetic effects related to differences in drug metabolism or transport can often be defined as Mendelian, such as the X-linked inheritance underlying glucose-6-phosphate dehydrogenase deficiency. In contrast, the inheritance patterns of pharmacodynamic pharmacogenetic relationships usually demonstrate relationships consistent with the complex traits underlying the purpose of the drug. Thus, these target responses tend to be determined by polygenic or gene-environment interactions, which are much more difficult to identify.

The *idiosyncratic* category of pharmacogenetic response to drugs includes the individuals that experience an adverse drug response (ADR) to a therapeutic agent that could not be anticipated on the basis of the known drug target and its pharmacokinetics. One example of the idiosyncratic category would be the genetic predisposition to the development of peripheral neuropathy in certain individuals taking the antituberculosis drug, isoniazid, which is metabolized by arylamine-*N*-acetyltransferase 2 (*NAT2*), a phase II drug-metabolizing enzyme.

The final pharmacogenetic category is that of genetic factors influencing *disease pathogenesis*. Genetic variants that affect properties related to the mechanistic basis of disease, including disease susceptibility or progression, can also affect the ability of a therapeutic agent to treat the disease.

Pulmonary Pharmacogenetic Phenotypes

Concurrent with advances in knowledge of the human genome, the need for better documentation and determination of phenotypic responses has been noted, with a call for the development of a "Human Phenome Project" (5). The rationale for these proposals is that, whereas enormous effort has been focused on the genotypic basis for phenotypic variation, but little is known about how to integrate this effort with the environmental influences on phenotypes, or even of how to accurately measure phenotypic features, in particular, for pharmacogenetics and genomics (5). Accurate phenotypic definition is crucial in genetic studies to avoid diagnostic misclassification (6). Two types of misclassification commonly occur in genetic studies: (1) nondifferential misclassification, in which the probability of error as to phenotype (in this case, a drug response or ADR) does not depend on exposure (genetic status), and (2) differential misclassification, in which it does. Nondifferential misclassification of phenotype reduces the observed genetic relative risk toward the null value, sometimes quite dramatically. An example of nondifferential misclassification is measurement error. For instance, let us assume that one were interested in studying the association of a candidate gene with bronchodilator response, with a significant response calculated as a 15 percent improvement from baseline. However, for each participant, the postbronchodilator measurement is read as 5 percent higher than the actual value. In this case, an increased number of individuals both with and without the genetic variant of interest would be placed into the "significant response" category. Therefore, there would be a decreased likelihood of detecting an association of the genetic variant with a significant bronchodilator response, if such an association actually existed.

Differential misclassification can bias the observed relative risk in either direction, depending on the different values of sensitivity and specificity among relatives of cases and controls. An example of differential misclassification would be an attempt to compare cytochrome P450 (CYP) genotypes (CYP metabolizes nicotine) in lifelong smoker cases older than age 50 with those of teenager smoking controls in relationship to response to smoking cessation therapy. In this case, variant CYP genotypes may actually be a predictor of true nicotine addiction and therefore would be expected to be more prevalent in the lifelong smokers in comparison with the teenagers. Therefore, the relative odds of response to smoking cessation therapy given a variant allele may appear to be falsely low in the lifelong smokers in comparison with the controls. The problem of phenotypic misclassification is especially prominent in complex disease traits and in studies of gene-environment interactions (6), including pharmacogenetics.

Pharmacogenetic analyses are inherently longitudinal in nature. That is, for a given individual genotype, one measures the phenotypic characteristic of interest before and after the administration of the therapeutic agent of interest. In the case of measuring the effects of genetic variation on a drug target, an intermediate phenotype or a measure of disease severity is obtained preceding and following the administration of the drug.

For the pharmacodynamic or disease-modifying categories of pharmacogenetic response common to asthma and COPD, examples of drug receptor/effector gene polymorphisms that lead to a complete lack of protein have not been identified to date (3). Moreover, instead of a bimodal or multimodal distribution common to drug metabolism, the functional response to these variations tends to be spread across a unimodal distribution. This suggests that multiple polymorphisms in multiple genes, and environmental factors, as well, may influence the overall response. In these cases, the effect of a single polymorphism is small, and the required sample sizes to detect significant pharmacogenetic associations are large. *Indeed, the single greatest problem with pulmonary pharmacogenetic studies published to date has been the relatively small sample sizes of the underlying clinical cohorts.* In addition, the measurement of drug receptor response or disease-modifying phenotypes is often imprecise, highly variable over time, or may be inappropriate for modeling pharmacologic response (3). Therefore, the identification of a valid drug response phenotype that is repeatable over time and is able to be measured accurately and precisely is especially important for the pharmacodynamic and disease-modifying pharmacogenetic studies commonly used in respiratory disease.

What measures best define a differential response to drugs in asthma and COPD? For greatest statistical power, this may be best accomplished by way of an easily measured, quantitative trait. For instance, the response to inhaled corticosteroids in asthma and COPD can be assessed by the change in forced expiratory volume at one second (FEV_1) over time (Figure 15.1). Other spirometric measures of lung function (e.g., forced vital capacity [FVC], FEV_1/FVC ratio, mid-maximal expiratory flow [FEF_{25-75}]) can be assessed in a similar fashion. Additional quantitative phenotypes commonly used to define

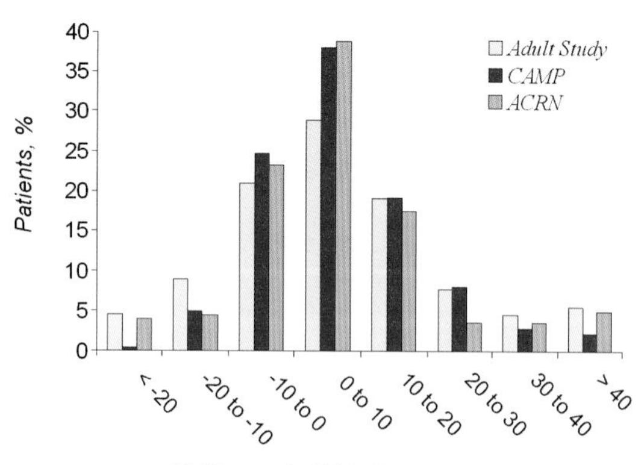

Figure 15.1. Relatively normal distribution of change in level of lung function (expressed as FEV_1) following inhaled corticosteroid administration for 6 to 8 weeks in three clinical trial populations. Although the average change in FEV_1 in each population is approximately 7 percent for each clinical trial, there is a significant proportion of subjects that did not improve while on treatment. ACRN, Asthma Clinical Research Network. (From Tantisira et al, *Hum Mol Genet.* 2004;**13:**1353–9.)

drug response in obstructive airways disease include the bronchodilator response (difference in lung function before and after inhaled β-agonist medications), provocative concentration of methacholine causing a 20 percent decrement in FEV_1 (PC_{20}, a measure of airways reactivity), and quantitative measures of atopy (e.g., eosinophil count, immunoglobulin E [IgE] levels) and inflammation (e.g., sputum neutrophil or eosinophil counts). The relatively normal distribution of many of these quantitative traits allows for the easy analysis of genetic effects with the use of standard regression models.

As knowledge of pulmonary pathophysiology increases, quantitative biomarkers will become increasingly appealing as phenotypic outcomes for pharmacogenetic studies. These biomarkers will strongly correlate with disease activity, outcomes, and response to therapy. Therefore, genetic predictors of biomarker response may serve as a proxy for overall drug treatment response. One example of these biomarkers is the measurement of exhaled breath condensates for nitric oxide, which have been strongly clinically correlated with asthma diagnosis and severity, eosinophilia, and airway responsiveness (7). The production of exhaled nitric oxide depends on the presence of arginine via nitric oxide synthase; however, arginase also competes for arginine as a substrate. Increased gene expression of arginase has been associated with the diagnosis of asthma (8), and variants of the arginase I gene have been associated with the differential response to inhaled bronchodilator medications (9). One additional quantitative trait biomarker that will probably play an increasingly important role in forthcoming pulmonary pharmacogenetic studies is that of gene expression. Although expression can clearly correlate with clinical outcome, the field of integrative genomics, that is, the combination of genetic and genomic (including gene expression) datasets, is in its relative infancy in relation to disease and therapeutic outcome prediction. By being associated with gene expression, a given genetic variant is more likely to be of functional significance. Moreover, because expression quantitative trait loci (eQTL) typically explain a large proportion of the variability of the expression trait, identified pharmacogenetic eQTL are more likely to replicate across multiple populations.

In evaluating a phenotype based on change in a quantitative trait over time, it is important to account for the baseline value in the definition of the phenotype or in the analytic phase. Subjects at the extremes of the response distribution at baseline may be there because they truly represent the extremes of the trait distribution for a given study. Alternatively, others at the extremes may be there only because they happened, through random variation, to have an extreme value when they were first measured. Thus, on subsequent measurements, these subjects would be expected to have less extreme values (i.e., closer to the population mean). This phenomenon is known as "regression to the mean" (10). Factors that are likely to increase potential for regression to the mean include (a) poor correlation of the trait of interest within an individual, also known as the intraclass correlation, and (b) initial extreme deviations of the observed value from the population mean. An example of accounting for the baseline value in the definition of a phenotype would be the definition of the change in lung function over time as a percentage of baseline, or the use of the baseline value as a covariate in a multivariate analysis.

In addition to quantitative traits, respiratory disorders lend themselves to qualitative measures. Although not as powerful from a statistical standpoint, these qualitative traits may be more clinically relevant because they are generally centered on the control of disease related to the initiation of medication. These qualitative traits can be either ordinal (ranked) or dichotomous (yes/no) in nature. As an example of an ordinal trait, symptom control in asthma is often designated on a 0 to 3 scale, in which 0 = asymptomatic, 1 = mild symptoms, 2 = moderate symptoms, and 3 = severe symptoms. A common dichotomous trait for both asthma and COPD is exacerbations, which can be defined as an acute worsening of the underlying disease requiring urgent or emergent medical care. Qualitative trait analysis has less statistical power to detect pharmacogenetic loci. In addition, the susceptibility to pulmonary symptoms and exacerbations may be related to environmental conditions and thus may not be fully indicative of a medication's success. Nonetheless, because of their direct clinical translation, given adequate sample sizes, these phenotypes should be increasingly interrogated in future pharmacogenetic studies.

Summary

As with any genetic study, the quality of the association is dependent on both the definition of a good response phenotype as well as sample size. Whereas sample sizes have limited the generalizability of many asthma pharmacogenetic studies to date, asthma does not suffer from the lack of high quality drug response phenotypes. Investigations into both quantitative and qualitative phenotypes will provide the most information with regard to biology of drug response, which will be crucial for pharmacogenetic modeling.

THE PHARMACOGENETICS OF ASTHMA

An estimated 300 million individuals worldwide are affected by asthma (11), a clinical disease characterized by inflammation, reversible airways obstruction, and bronchial responsiveness. Ninety percent of all asthma, including asthma in adults, has its origins in childhood. Asthma remains the leading disease that results in childhood school absences and hospitalizations.

Asthma is a genetic disease, noted to cluster in families for more than three centuries. Based on twin studies, the broad sense heritability estimates (proportion of the total variance of a trait due to genetic causes) of an asthma diagnosis ranges from ~36 percent to 75 percent. Asthma is a complex disease, in that no single gene is causal by itself. Instead, it likely results from the influence of multiple genetic, environmental, and developmental factors. There are large interindividual variations in each of the phenotypic responses to each of the three major classes of asthma treatment (12, 13) (e.g., Figure 15.1). Because the intraindividual response to treatment in patients with asthma is highly repeatable (14), and because phenotypes such as FEV_1 and PC_{20} are heritable traits, a genetic basis for the heterogeneity of this therapeutic response is plausible. Moreover, because asthma is a genetic disease, heritable factors that influence the natural history of asthma progression may also affect the ability of an individual to respond to therapy. Thus, given that the variation in the response to therapy and underlying disease pathogenesis may be largely due to genetic causes, the field of asthma is well suited to pharmacogenetic investigations.

Asthma Pharmacogenetic Phenotypes

In the asthma pharmacogenetics literature to date, most of the quantitative intermediate phenotypes used to assess pharmacodynamic responses have focused on measures of lung function or airway tone. For example, in a study by Szefler and colleagues (13), 38 percent of patients randomly assigned to inhaled budesonide or fluticasone demonstrated improvements in FEV_1 of less than 5 percent over the course of 24 weeks. These and other data have formed the basis for subsequent investigations into the genetic determinants of therapeutic response to both inhaled corticosteroids (15) and leukotriene antagonists (16) with the use of the change in FEV_1 as the primary outcome phenotype (Figure 15.1). Moreover, the FEV_1 correlates well with clinical asthma severity, degree of airway obstruction, and other physiologic markers of asthma severity, and with response to treatment (17). In addition, FEV_1 (17) and the therapeutic change in FEV_1 (14) are highly reproducible for an individual. Each of these suggests the potential for a genetic contribution to change in FEV_1 in response to medications.

In addition to the FEV_1, both PC_{20} and bronchodilator response (the change in FEV_1 immediately after the administration of a short-acting β-agonist medication) have been frequently cited as response phenotypes. These are logical choices, because they are easily interpretable, correlate well with clinical disease, and demonstrate far greater interindividual variability than intraindividual variablility (17). Other phenotypes, such as the peak expiratory flow rate (PEFR), have also been used. Clinically, increases in the variability of the PEFR have been used to suggest worsening asthma. However, although valid, PEFR variability as an outcome should be carefully scrutinized, because there is poor correlation of PEFR variability with symptoms or lung function (18), and PEFR measurements may have low reproducibility (19). Such poor repeatability for a phenotype may contribute to potential biases in pharmacogenetic study outcomes.

One qualitative asthma pharmacogenetic phenotype should be mentioned. Therapy in asthma is clearly not just directed toward the improvement of underlying lung function, but also toward the minimization of symptoms and the avoidance of exacerbations, including those related to emergency room visits and hospitalizations (20). Asthma pharmacogenetic studies evaluating genetic predictors of health care utilization while on asthma medications are underway.

A brief comment should be made with regard to phenotypes for the fourth category of pharmacogenetic response – genes that may influence the underlying pathogenesis or disease susceptibility to asthma. In general, these phenotypes should parallel the drug treatment response phenotypes (e.g., FEV_1 of PC_{20} over time). However, as the genetic and molecular basis for asthma becomes clearer over time, novel phenotypes will become increasingly prominent as pertains to both asthma pathogenesis and to drug treatment response. As noted previously, one example of this is the measurement of exhaled breath condensates for nitric oxide (7).

The Pharmacogenetics of Asthma – Overview of Studies to Date

In this section, we highlight the salient outcomes of the pharmacogenetic studies performed in asthma to date. These will be divided by therapeutic class of medication, focusing on β-agonists, leukotriene modifiers, and inhaled corticosteroids.

The β_2-Adrenergic Receptor Pathway

β-Adrenergic receptor agonists are the most commonly used class of medication in the treatment of asthma worldwide (21). Stimulation of the β_2-adrenergic receptor results in rapid and potent relaxation of airway smooth muscle. Interindividual variability in the response to these agents has been recognized for more than sixty years. For instance, the 1942 edition of Osler's *Principles and Practice of Medicine* noted, on the treatment of asthma, "Hypodermics of epinephrine, Mv to xv (0.3 to 1 cc) of a 1:1000 solution or of atropine, gr. 1/100 (0.6 mgm), may give prompt relief, but individual cases vary greatly" (22). Both short- and long-acting β-agonist agents are currently available and play distinct roles in the therapy of asthma (23).

The literature on β-agonists dominates the field of asthma pharmacogenetics. This is probably related to the relatively early identification (24) and resequencing (25) of the gene encoding the primary receptor for this class of medications, *ADRB2*. Two common nonsynonymous single-nucleotide polymorphisms (SNPs) have been identified and well characterized in the β_2-adrenergic receptor gene (*ADRB2*) (25). The first encodes for an amino acid change from arginine to glycine at position 16 (Arg16→Gly) and the second for an amino acid change from glutamic acid to glutamine at position 27 (Glu27→Gln). These SNPs are common, with minor allele frequencies between 40 percent and 50 percent in large cohort studies. In addition to potential roles influencing asthma risk and the level of baseline lung function, the pharmacogenetic effects of the *ADRB2* 16 and 27 loci have been evaluated in multiple studies. Despite the numerous studies of these loci, however, it should be noted that careful attempts to summarize these pharmacogenetic data have proven inconclusive because of the wide variability in study design and therapeutic end points (26). We review the most relevant studies in the following paragraphs.

Two primary phenotypic outcomes related to β-agonist medications can be commented upon. The acute pharmacogenetic response to the administration of β-agonist medications has focused on the change in FEV_1 from before to immediately after the administration of the therapy (typically for a medication such as albuterol [salbutamol], the second measurement is performed 15 to 30 minutes after the therapy). This change, known as the bronchodilator response, is a reasonable pharmacogenetic outcome, because these therapeutic agents are relied upon clinically to provide prompt relief to asthma symptoms. As noted, studies have differed substantially from one another with respect to study design and primary outcome assessed, resulting in a marked inconsistency in results. For example, in a longitudinal birth cohort of 269 children from Tucson, Arizona, Martinez et al. (27) examined the bronchodilator response at 30 minutes following albuterol administration (180 μg). Of those subjects, 39.7 percent were asthmatic, and only 11.5 percent were regular β_2-agonist users. After adjustment for asthma and wheeze, subjects who were Arg16 homozygotes were 5.3 times more likely to show a bronchodilator response than were Gly16 homozygotes (confidence interval [CI], 1.6–17.7). Heterozygotes showed an intermediate response. However, although the Arg16 allele was also associated with bronchodilator response in Mexican asthmatic patients (28), this was not replicated in family-based analyses of Puerto Rican (28) or white asthmatic patients (29).

Given the conflicting associations, it is likely that the bronchodilator response is, in fact, not specifically mediated by the Arg16Gly polymorphism. Rather, the *ADRB2* contribution to this response is probably mediated by a variant or variants in partial linkage disequilibrium with the Arg16Gly variant. This both supports the variability of associations within the literature and is supported by the literature focusing on haplotypic and other variants within *ADRB2*. Drysdale et al. (30), in a study of molecular haplotypes of *ADRB2*, demonstrated that acute bronchodilator response to inhaled albuterol was approximately twice as high in individuals homozygous for the most common haplotype, in comparison with those homozygous for the second most common haplotype ($P = 0.046$). Importantly, haplotypic analysis was more powerful in detecting associations in this population compared with single SNP analyses (30). Whereas a follow-up study of 176 whites failed to confirm the association between the Drysdale *ADRB2* haplotypes and acute bronchodilator response (31), the study did support the role of further investigations into other genetic variants.

In a large family-based study involving the probands of the Childhood Asthma Management Program (CAMP), Silverman et al. (29) genotyped a subset of eight of the Drysdale *ADRB2* SNPs, noting a novel association of a variant +523 base pairs after the ATG start site with bronchodilator response to albuterol. Given the proximity of this SNP to the true 3′-UTR, further work in better characterizing variation of the *ADRB2* 3′-UTR (and the promoter region) has occurred, with the description of several novel variants (32). By imputing haplotypic substructure, it is clear that the new variants subdivide the traditional *ADRB2* haplotypes even further. Cellular models support the 3′-UTR variants' role in *ADRB2* receptor expression (33). Clinical association studies are underway.

The vast majority of pharmacogenetic investigations of bronchodilator response to date have focused on *ADRB2*. Although some associations with bronchodilator response have been noted, predictability and reproducibility remain low. This is probably due to both an incomplete understanding of the haplotypic and functional structure of the *ADRB2* gene and the need to identify and characterize additional genes contributing to the variability in bronchodilator response. Investigations into other genes from the β-agonist pathway (Figure 15.2) have begun. For instance, variants in the 5′ region of the corticotropin-releasing hormone receptor 2 (*CRHR2*) gene have been associated with acute bronchodilator response in three independent cohorts (34).

Adenylate cyclase mediates signaling following activation of the β_2-adrenergic receptor (β_2-AR). A nonsynonymous polymorphism of the adenylate cyclase 9 gene (*ADCY9*) at amino acid 772 encoding an isoleucine to methionine substitution results in lower basal and β_2-AR-mediated adenylyl cyclase activities in cellular models (35). A recent evaluation of this nonsynonymous *ADCY9* SNP variant further demonstrated significantly greater albuterol-induced adenylyl cyclase activities in Met772-transfected lung cells compared with Ile772 (36). This effect was significantly enhanced upon the administration of exogenous corticosteroid. In the CAMP clinical trial of 1,041 childhood asthmatics, the association of the Met772 variant with increased bronchodilator response was noted, but only in those individuals randomly assigned to the inhaled corticosteroid arm of the trial (36).

Litonjua et al. (9) screened 844 SNPs in 111 candidate genes for association with bronchodilator response in 209 CAMP children and their parents, using a novel algorithm implemented in a family-based association test that ranked SNPs in order of statistical power. Genes that had SNPs with median power in the highest quartile were then taken for replication analyses in three other asthma cohorts. SNPs from one gene, *ARG1*, were associated with bronchodilator response in three independent cohorts following the CAMP screen. One *ARG1* SNP, rs2781659, survived Bonferroni correction for multiple testing (combined P value = 0.00048, adjusted P value = 0.047). Thus, several novel bronchodilator genes have now been identified. Although identifying a novel bronchodilator response gene, the *ARG1* article also has provided a novel method for efficiently screening and replicating pharmacogenetic loci (Figure 15.3).

Although perhaps the most commonly evaluated β-agonist phenotype is the acute bronchodilator response, because it directly assesses medication efficacy at its target site, a second phenotypic outcome evaluated for this class of medications has been the potential for downregulation of β-agonist receptor responsivity (tachyphylaxis)

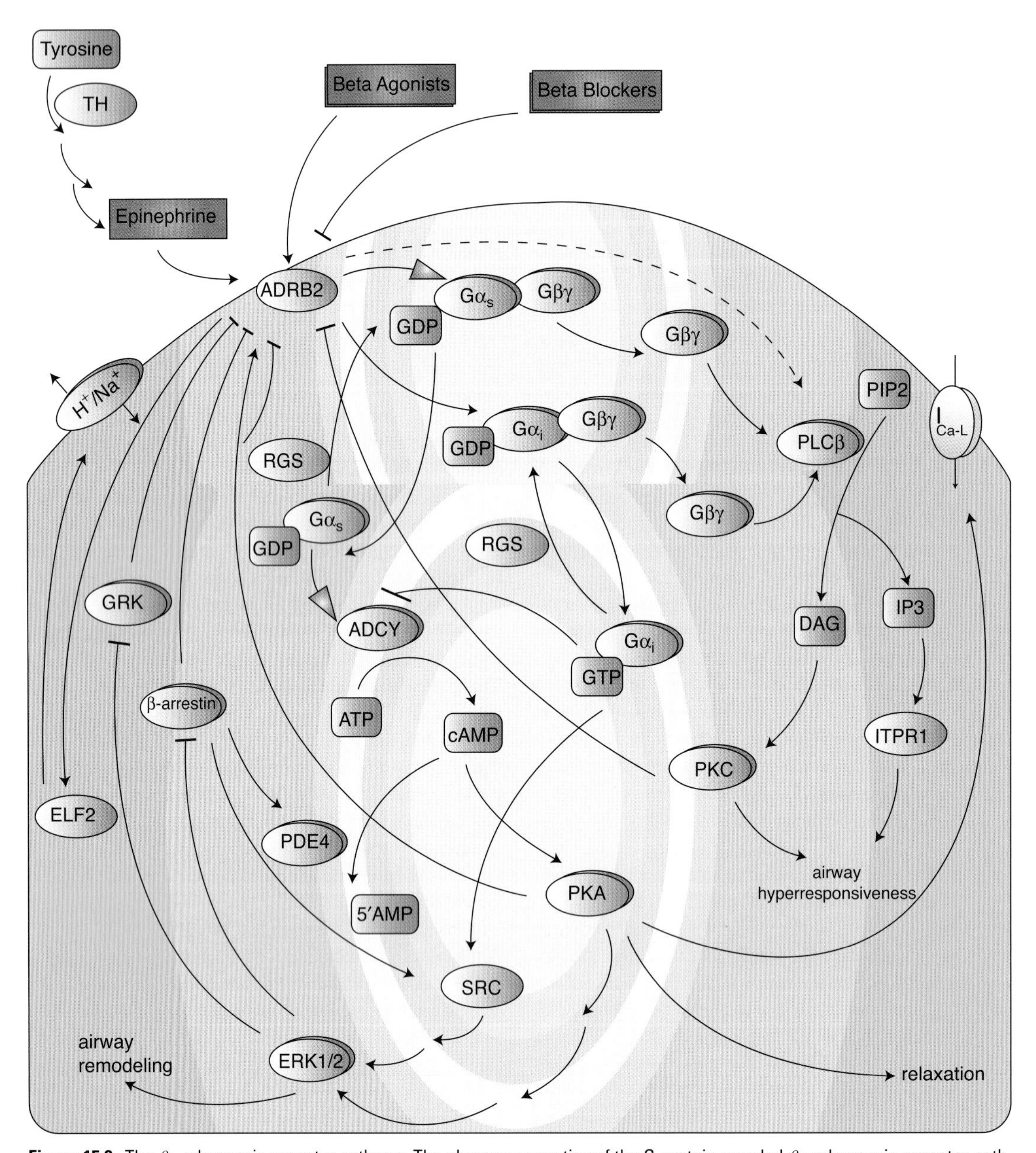

Figure 15.2. The β_2-adrenergic receptor pathway. The pharmacogenetics of the G-protein-coupled β_2-adrenergic receptor pathway to date has been primarily limited to investigations involving the primary receptor (*ADRB2*), although interrogations of other genes has begun. Further queries both within and outside this pathway, including epistatic interactions, will be necessary to get a comprehensive picture of the genetic response to this class of asthma medications. The complexity is as great, if not greater for the corticosteroid and leukotriene pathways. Details related to the specifics of this pathway, and the corticosteroid and leukotriene pathways, as well, are available on the PharmGKB website. (From PharmGKB, www.pharmgkb.org, accessed July 17, 2008.)

with the chronic administration of this class of therapy in association with *ADRB2* genetic variants. In a multicentered trial conducted by the Asthma Clinical Research Network, 255 asthmatics of mild severity were randomly assigned to either regular (180 μg four times daily) or as-needed albuterol use and were assessed over a 16-week period for evidence of tachyphylaxis and clinical deterioration (as measured by a fall in morning PEFR). No difference in PEFR variation was observed between treatment groups (37). However, in a post hoc analysis, Arg16 homozygotes significantly decreased their PEFRs with regular use of albuterol therapy in comparison with both Arg16 homozygotes receiving prn (as needed) albuterol and Gly16 homozygotes in either treatment group (38). Adding support to this association was an in vivo study of twenty-six healthy subjects, in which Arg16

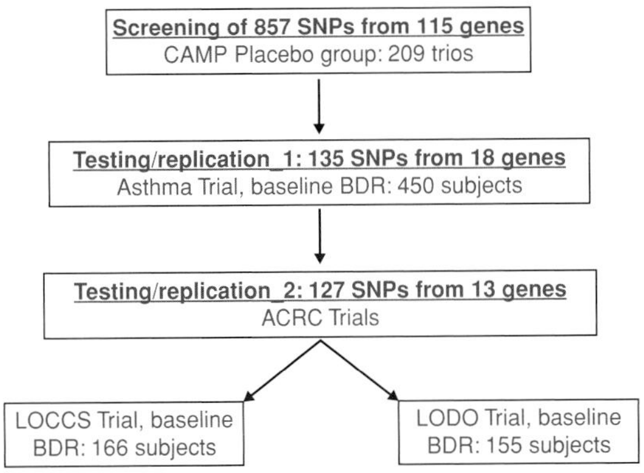

Screening of 857 SNPs from 115 genes
CAMP Placebo group: 209 trios

Testing/replication_1: 135 SNPs from 18 genes
Asthma Trial, baseline BDR: 450 subjects

Testing/replication_2: 127 SNPs from 13 genes
ACRC Trials

LOCCS Trial, baseline BDR: 166 subjects

LODO Trial, baseline BDR: 155 subjects

Figure 15.3. Screening and replication strategy as a way to rapidly and efficiently identify novel pharmacogenetic loci of interest. In the screening stage, SNPs with the greatest statistical power for association and replication are identified and ranked. Those with the highest relative power are genotyped and tested for association in independent trial populations. (From Litonjua et al., *Am J Respir Crit Care Med.* 2008;**178**:688–94.)

homozygotes demonstrated almost complete physiologic desensitization within 90 minutes of constant infusion of isoproterenol in comparison with individuals with homozygous Gly16 receptors, who showed only 50 percent desensitization as late as 120 minutes ($P = 0.006$) (39).

In the first prospective, genotype-stratified study of the effect of polymorphisms of the *ADRB2* gene on treatment-related changes in lung function, Israel et al. (40) matched asthmatic individuals homozygous for Arg/Arg ($n = 37$) at position 16 to those homozygous for Gly/Gly ($n = 41$), by level of FEV_1. The investigators of this trial (named BARGE for the β-Agonist Response by Genotype) then randomly assigned all of the genotype-stratified individuals in a double-blind, crossover study of regularly scheduled albuterol therapy (four times daily) versus placebo over two 16-week periods with an intervening 8-week washout period. Again, subjects homozygous for Arg/Arg had lower morning peak flow rates during regularly scheduled treatment versus placebo (−10 L/min, $P = 0.02$), whereas those homozygous for Gly/Gly had higher rates during those times (14 L/min, $P = 0.02$). The Arg/Arg minus Gly/Gly difference was very significant for morning peak flow (−24 L/min, $P = 0.0003$) and also reported to be significantly different while on treatment in regard to evening peak flow, FEV_1, morning symptom score, and need for rescue medication.

The accompanying editorial to this article pointed out three interesting data issues whose significance remains to be determined (41). First, the improvement in the morning peak flow with regularly scheduled albuterol in the Gly/Gly group continued during the washout period. Interestingly, there was an improvement in peak flow during the washout for those homozygous for Gly/Gly at position 16 in a previous study as well (42). Next, it was noted that no deterioration of the Arg/Arg individuals in BARGE occurred while on regularly scheduled albuterol (41). Instead, increases on placebo occurred after the run-in or washout periods. This is distinct from previously reported findings (42). Finally, the editorial noted that the Arg/Arg group experienced a large increase in peak flow during the initial run-in period, despite averaging only one puff of β-agonist daily before the run-in (41). Clearly, these findings need to be replicated. Additional limitations not mentioned in the editorial, including the relatively small sample size, the use of multiethnic populations without being able to investigate epistatic or haplotypic interactions, and the actual very small percentage of the phenotypic variance (in this case, morning peak flow interindividual variability) explained, would argue against making broad generalizations from these data at this time. Moreover, because the regularly scheduled use of short-acting β-agonist medications is not a normal part of asthma management guidelines, the clinical relevance of these data remains uncertain.

Although short-acting β-agonist medications are not clinically indicated for routine daily usage, long-acting β-agonist medications form a normal part of therapy for the moderate to severe asthmatic. Several analyses of the Arg16Gly polymorphism relationship to clinical outcomes in subjects taking long-acting β-agonists have been performed with conflicting results. One study supported worsened lung function (43), and another increased exacerbations given Arg/Arg homozygosity (44). However, two others with considerably larger sample sizes (i.e., hundreds of each genotypic variant) could not demonstrate any significant clinical differences in either outcome based on genotype (45, 46). It is likely, based on these results, that the pharmacogenetics of long-acting β-agonists is similar to that of short-acting β-agonists, with additional *ADRB2* loci and multiple other genes contributing to their response, rather than the response being mediated primarily via the Arg16Gly polymorphism. Until these additional loci are identified, there is currently no substantive evidence for genotypic prediction of adverse outcomes related to long-acting β-agonists.

In sum, studies on *ADRB2* provide evidence that the gene is probably associated with bronchodilator response, but that genetic prediction of this response may both vary by ethnic group and require the addition of loci from other genes, such as *CRHR2*, *AC9*, and *ARG1*, to be of sufficient magnitude for clinical significance. In addition, individuals homozygous for *ADRB2* Arg16 demonstrate significant β_2-AR desensitization, and these individuals (approximately 15 percent of the population) may be at risk of clinical deterioration with the regular use of short-, but not long-acting β-agonists. Clinical relevance remains uncertain, because regular use of

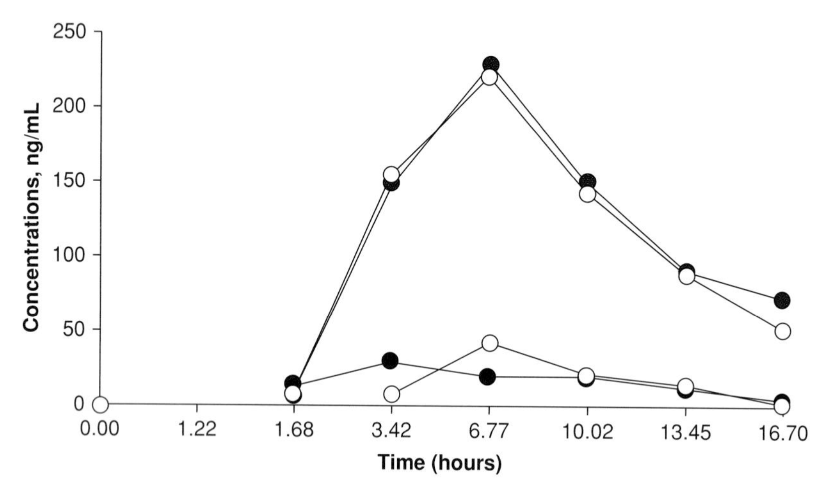

Figure 15.4. Pharmacokinetic variability in montelukast. Plasma concentration versus time profiles in two healthy volunteers after oral administration of single 10-mg doses of montelukast as a tablet (Singulair®; *filled circles*) and an encapsulated split-tablet formulation (*open circles*). (From Lima. *Mol Diag Ther.* 2007;**11:** 97–104.)

short-acting β-agonists is no longer part of normal asthma treatment guidelines. It is noteworthy, however, because it is the severe asthmatics who take these medications regularly on whom the potential impact may be large. Studies focusing on the role of *ADRB2* genotype within severe asthmatics are warranted. In addition, better studies with standardized phenotypes in large populations need to be performed, because many of the described associations have not been replicated. This work, along with better characterization of the other genes involved in the β-agonist pathway, their association with clinical outcomes, and their interactions with *ADRB2*, is underway.

The Leukotriene Antagonist Pathway

Leukotriene modifiers are the only orally administered class of the three most commonly prescribed classes of asthma medications and, thus, are generally preferred among patients (47). Currently, two classes of antileukotriene drugs have been approved for asthma treatment, the leukotriene Cys-LT_1 receptor antagonists (e.g., montelukast, zafirlukast, and pranlukast) and the 5-lipoxygenase (5-LO) inhibitor (zileuton). Despite the overall improvement versus placebo with antileukotriene drugs, the interindividual response is highly variable (12); genetic factors may be responsible for some of this heterogeneity (14).

The oral administration of the leukotriene modifiers makes them the only class of commonly prescribed asthma medications with the potential to undergo significant first-pass effects and, hence, be subject to their effects being modulated via genetic variants that control metabolism and/or drug transport. Metabolism for both the 5-LO inhibitors and for leukotriene receptor antagonists is largely mediated through the cytochrome P450 system, with *CYP2C9* and *CYP3A4* responsible for metabolism of both classes (*CYP1A2* also plays a role in the metabolism of 5-LO inhibitors) (48). Studies have yet to examine the role of variation in drug-metabolizing enzymes to leukotriene modifier response. Nonetheless, it has now been shown that montelukast demonstrates significant interindividual variability in plasma levels (Figure 15.4) (48). This variability appears to be mediated at least in part via the organic anion transporter, OATP2B1, which is encoded by the gene *SLCO2B1* (solute carrier organic anion transporter family, member 2B1) (49). The *SLCO2B1* SNP, rs12422149, has now been associated with variation in plasma montelukast levels, with heterozygotes demonstrating an ~30 percent reduction in levels versus wild-type genotypes (49). Association studies of clinical montelukast response phenotypes with *SLCO2B1* genotypes are underway.

Expression of 5-LO protein is regulated at a number of levels (50). At the transcriptional level, binding of different transcription factors to specific consensus binding sites in the 5-LO gene (*ALOX5*) promoter region has been shown to be important for expression in vitro. There are two well-studied *ALOX5* promoter-binding transcription factors, Sp1 and early growth response factor-1 (Egr-1). A family of *ALOX5* polymorphisms has been identified in this region that consists of a deletion of one or two SP1 -GGGCGG- consensus binding sites or the addition of one of these sites (51). Approximately 78 percent of subjects have the wild-type allele at this microsatellite locus. In a clinical drug trial of the zileuton-like 5-LO inhibitor, ABT-761, asthmatics with at least one wild-type allele had an average improvement in their FEV_1 of 19 percent, whereas patients with any two of the mutant alleles had an average decrement in their FEV_1 of 1 percent (52).

Given the previous associations, Lima et al. (16) genotyped on twenty-eight SNPs in five leukotriene pathway genes in subjects participating in a clinical trail comparing the efficacy of montelukast versus low-dose theophylline versus placebo as add-on therapy in mild to moderate asthma. DNA was obtained in 248 participants, and the primary analysis was performed on sixty-one whites taking montelukast. Outcomes assessed included change in FEV_1 and the presence of exacerbations during one, three, and six months time. At six months, two SNPs were associated with change in FEV_1, one in the *MRP1* gene (rs119774) and the second in *ALOX5* (rs2115819). Of potentially greater interest, variants from three genes were associated with the risk of exacerbations. Variation in one *LTA4H* SNP (rs2660845) was associated with a fourfold increase in the risk of at least one exacerbation over the six-month follow-up, whereas variation in the previously described *ALOX5* promoter repeat and the *LTC4S* A-444C SNP was associated with a 73 percent and a 76 percent reduction in exacerbations, respectively. The association of *ALOX5* promoter repeat with decreased exacerbations was reported in one additional study (53). Klotsman et al. (54) analyzed twenty-five SNPs in ten candidate genes in a population of 174 asthmatics taking montelukast over 12 weeks. Associations with differential change in peak flow rates were noted with two *ALOX5* and two *CYSLTR2* SNPs.

The primary limitation to the pharmacogenetic studies of leukotriene modifiers to date has been sample size. No study to date has reported association results of a single ethnic population in excess of 100 individuals (Klotsman et al. combined two separate clinical trials in their results with only 138 total whites analyzed). Additional, even smaller studies were excluded from this chapter. All in all, the data to date suggest that there are probably multiple leukotriene pathway pharmacogenetic loci modulating the therapeutic response to these agents and that *ALOX5* likely modulates at least part of the response. Nonetheless, these studies remain in their infancy and require both larger sample sizes and replication across multiple additional cohorts before predictive modeling can begin to be considered.

The Corticosteroid Pathway

Inhaled corticosteroids are the most effective and commonly used drugs for the chronic treatment of asthma, but they may be associated with serious adverse effects (20, 23). As noted previously, the intraindividual response to inhaled corticosteroid treatment in patients with asthma is highly repeatable (14); therefore, it is reasonable to postulate a genetic difference for the response to inhaled corticosteroids in asthma.

A number of family and twin studies have demonstrated consistent evidence that endogenous levels of glucocorticoids, usually measured as plasma cortisol levels at certain times of the day, are heritable (55). In turn, decrements in endogenous plasma cortisol levels at night (56) and during periods of stress (57) have been associated with nocturnal and stress-related asthma, respectively. Furthermore, cortisol levels in nocturnal asthma may be partially resistant to the effects of corticotropin (56). Overall, these studies suggest a potential role for genetic factors regulating endogenous cortisol production in the pathogenesis and long-term treatment response of asthma.

In a study of fourteen candidate genes selected for their biologic relevance to the entire corticosteroid pathway, including synthesis, binding, and metabolism, we found a significant association between the eight-week response to inhaled steroids and SNPs from the corticotropin-releasing hormone receptor 1 (*CRHR1*) gene in adult asthmatics (15). *CRHR1* was also significantly associated with the eight-week response to inhaled steroids in a clinical trial of children with asthma. Moreover, one particular SNP and one specific haplotype in the *CRHR1* gene that predicted good response to inhaled steroids in both populations were noted. The SNP, rs242941 (minor allele frequency ~30 percent) was associated with positive treatment response in both the Adult Study and the pediatric study, CAMP ($P =$ 0.025 and 0.006, respectively). In the Adult Study, the mean percentage of change in FEV_1 for those homozygous for the minor allele was 13.28 ± 3.11, compared with 5.49 ± 1.40 for those homozygous for the wild-type allele. Similarly, in CAMP, the percentage of change was 17.80 ± 6.77 versus 7.57 ± 1.50 for the variant and wild-type homozygotes, respectively. In CAMP, the evaluation of the placebo arm revealed no association of rs242941 or any of the other genotyped SNPs with change in lung function. Moreover, although inhaled corticosteroid usage was associated with improved FEV_1 at eight weeks ($P < 0.001$), variation in rs242941 significantly enhanced the improvement in lung function associated with this form of therapy (interaction $P = 0.02$).

The haplotype of interest had similar, but larger improvements in FEV_1 on inhaled steroids. On average, those imputed to have the haplotype of interest had two to three times the short-term response to inhaled steroids compared with those without the haplotype. The overall explained phenotypic variance was small (<5 percent in both populations), however, suggesting that multiple other factors (including additional genes) are responsible for the variability in the response to inhaled corticosteroids. Because *CRHR1* is the primary receptor for corticotropin-releasing factor in the brain and thereby modulates adrenocorticotropic hormone and endogenous cortisol levels, these results confirm the role of genetics in the asthma outcomes related to endogenous steroid levels.

TBX21 encodes for the transcription factor T-bet, which is crucial in the development of naïve T-lymphocyte production. The T-bet knockout mouse spontaneously develops airways inflammation and hyperresponsiveness suggestive of asthma (58). Only one common nonsynonymous SNP has been identified in the *TBX21* gene, encoding for a replacement of histidine by glutamine at amino acid position 33 (H33Q). In a study of 701 CAMP children, 4.5 percent were noted to be heterozygous for this variant. After limiting the analysis to the white children, each of the H33Q individuals on inhaled corticosteroids was noted to demonstrate a marked improvement in airways hyperresponsiveness as measured by the PC_{20}, compared with either the steroid H33H homozygotes or any individual not taking inhaled steroids (interaction $P = 0.0002$) (59). The average improvement in the level of PC_{20} in those H33Q individuals taking inhaled corticosteroids was to the level associated with nonasthmatics (Figure 15.5).

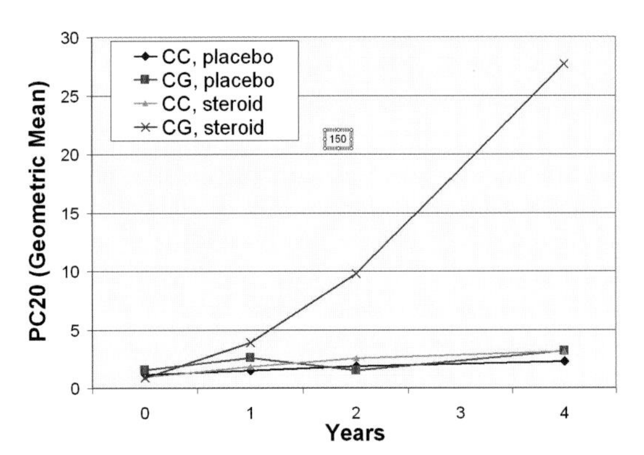

Figure 15.5. Geometric mean value of PC_{20} at yearly intervals over the four-year follow-up, stratified by treatment group assignment and H33Q genotype. The mean PC_{20} of individuals on corticosteroids who also possessed a copy of the 33Q variant improved significantly by the end of the first year of the study and continued to improve substantially with time and corticosteroid usage, resulting in a mean PC_{20} value at four years within the range normally ascribed to individuals without the diagnosis of asthma. (From Tantisira et al., *Proc Natl Acad Sci USA.* 2004;**101**:18099–104.)

In addition to the pharmacogenetic associations with lung function and airways responsiveness, a novel variant in the *FCER2* gene (which encodes for the low-affinity IgE receptor) has been associated with differential frequency of asthma exacerbations while on inhaled corticosteroids (60). The SNP, rs28364072, was associated with increased risk of exacerbations in asthmatic children taking inhaled corticosteroids, despite the generally protective effects of this class of medications (Figure 15.6). Relative risk, expressed as hazard ratios, for exacerbations in those homozygous for the variant allele was 3.95 (95% CI, 1.64–9.51) for white children and 3.08 (95%

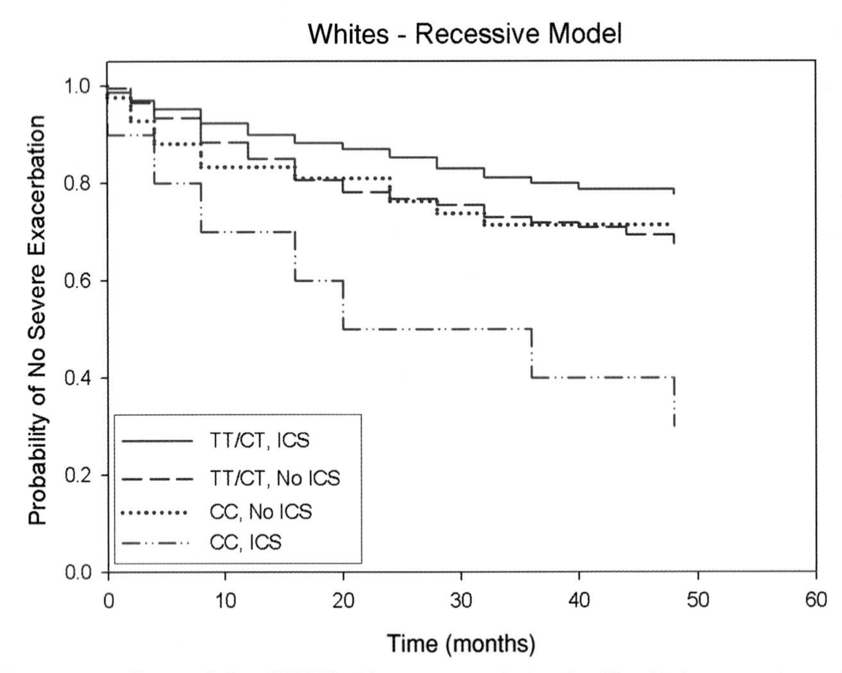

Figure 15.6. Treatment-specific association of T2206C with severe exacerbations in whites. Under a recessive model, the risk for severe exacerbations for subjects homozygous for the T2206C mutant allele while on inhaled corticosteroids (ICS) is significantly greater than the wild-type and heterozygotes for the T2206C allele on ICS. There is no difference in relative risk for any of the T2206C alleles in those subjects not taking ICS. There was a significant interaction between genotype and ICS use (interaction $P = 0.004$). (From Tantisira et al., *J Allergy Clin Immunol.* 2007;**120**:1285–91.)

CI, 1.00–9.47) for African American children. Of interest, this novel variant was also associated with both higher IgE levels and with differential expression of the *FCER2* gene, supporting the contention that variation in *FCER2* can adversely affect the normal negative feedback mechanism in the control of IgE synthesis and action.

As previously stated, first-pass drug metabolism and intestinal transport, in general, play minor roles in asthma pharmacogenetics. However, because systemic corticosteroids, administered via the oral or intravenous routes, play a major role in the management of acute exacerbations of asthma (and COPD), it should be mentioned that circulating corticosteroids are metabolized in the liver, primarily by *CYP3A4*, resulting in considerable interindividual variability (61). Pharmacokinetic characteristics at or near the target receptor site can affect therapeutic efficacy of corticosteroids. For instance, plasma protein binding, drug clearance, half-life, mean absorption time from the lungs to the systemic circulation, and lipid conjugation have all been related to target bioavailability in studies of inhaled corticosteroids (62). Of interest, local drug transport (via P-glycoprotein) and metabolism (via 11β-hydroxysteroid dehydrogenase) of corticosteroids specific to target tissues occurs; genetic variation within *MDR1* (which encodes for P-glycoprotein) has been associated with differential effects on cytokine production after dexamethasone administration (63). Nonetheless, although important, because these processes are target tissue specific, the variability in response mediated by local transport and metabolism phenotypically equates to pharmacodynamic differences.

As with the leukotrienes, pharmacogenetic investigations into the response to inhaled corticosteroid therapies, while promising, remain in their formative stages. Because the effects of corticosteroids are fairly pervasive, future studies may be best focused at a level above the traditional candidate gene studies. This is the basis of the following section.

Asthma Pharmacogenetics – The Future

Before beginning a discussion that optimistically describes the future of asthma pharmacogenetics, it is prudent to look back with a cautious eye upon the limitations of the studies performed to date. Active reading of the literature surrounding association studies related to this field should focus not only on the results of the association being evaluated within the study, but also the quality control of the study itself. In particular, many studies clearly suffer from problems related to inadequate sample size, improper matching of cases to controls, imprecise phenotypic definitions, potential population stratification, multiple comparisons, or the inability to replicate findings (64). As noted with regard to the *ADRB2* gene, these problems are beginning to gain the increasing attention that they deserve (65). One additional "limitation" to the asthma pharmacogenetics studies to date is the relatively small proportion of phenotypic variability that any one described genetic variant explains. No singular genetic marker or, for that matter, candidate gene, has been demonstrated to account for any more than a few percent of the variability for a given therapeutic outcome.

However, even given these limitations, the search for the genetic basis for asthma treatment response is becoming increasingly informative and closer to clinical applicability. In the near future, the field of asthma pharmacogenetics should continue to grow along three prominent lines. First, a better understanding of the molecular mechanisms of therapeutic agents will help to better elucidate exactly what constitutes a therapeutic "responder" from a "nonresponder." Although multiple working definitions have been proposed, no singular or optimal phenotypic definition of response to any of the three well-established asthma medication classes has been established.

In addition to improved asthma pharmacogenetic phenotypes, continued work into known and novel therapeutic pathways for identification of novel candidate genes and functional variants influencing drug treatment response will need to continue. Two high-throughput parallel approaches have sought to rapidly identify novel candidate genes and loci – genome-wide association studies and genome-wide expression studies. The promise of genome-wide association studies is evident. By interrogating hundreds of thousands of SNPs across the genome, novel loci contributing to a given pharmacogenetic effect can be rapidly and efficiently identified. Despite this promise, no genome-wide association study focusing on asthma pharmacogenetics has been published to date. Aside from the fact that genome-wide association studies remain costly, the major limitation to these studies as applied to asthma pharmacogenetics has been sample size. Most asthma clinical trials enroll subjects in the dozens to hundreds range. Because many of these trials have more than two treatment arms, sample size for a given treatment outcome is generally quite modest. This contrasts directly with the thousands of subjects projected for most currently available genotyping platforms as necessary for discovery of loci contributing modest effects (e.g., genotype relative risk <2) (66). Traditional analyses of datasets containing treatment response to a couple of hundred subjects would be filled with large numbers of both false-positive and false-negative associations. Approaches that seek to maximize statistical power (minimizing false negatives) and to perform distinct screening and testing stages (minimizing false positives) have been published (67). This has been practically implemented in the *ARG1* association with bronchodilator response described previously (9). Use of these types of approaches may thus allow for the successful application of genome-wide association studies in currently available asthma clinical trials.

An alternative approach to rapid identification of novel asthma pharmacogenetic loci is the identification of genes influencing drug treatment response via a differential expression approach. Despite the ready availability of expression microarray technology, few studies have been conducted specifically evaluating differences in drug response as they correlate with asthma treatment response. Harkonarson et al. (68) evaluated glucocorticoid-sensitive and -resistant asthmatics. Glucocorticoid sensitivity was defined as improvement in lung function, symptoms, or exacerbations with inhaled corticosteroids, and glucocorticoid resistance was defined as failure to improve lung function or clinical status even after a course of oral corticosteroids. Expression using high-density oligonucleotide microarrays (11,812 genes) was measured in peripheral blood mononuclear cells from eighty-seven patients at baseline, and following stimulation with interleukin-1β/tumor necrosis factor-α with or without dexamethasone pretreatment. Nine hundred twenty-three genes were differentially expressed following interleukin-1β/tumor necrosis factor-α stimulation and normalized via dexamethasone treatment. Fifteen genes predicted glucocorticoid response category with a reported accuracy of 84 percent. Given that this study focused only on phenotypic extremes and involved only one type of asthma therapy, additional studies of this type are warranted. Nonetheless, the promise of rapidly identifying a small subset of candidate genes by their expression profiles in this manner is clearly appealing.

The final near-future direction for asthma pharmacogenetics is that consideration must be given to predictive modeling incorporating both additive (two loci contributing equally to a given outcome) and epistatic (multiplicative) effects of multiple genes and loci. However, the concept of predictive modeling does not have to be limited to just SNP loci. Although genome-wide association studies and expression profiling will help to identify novel genotypic associations and might be sufficient by themselves to form a prognostic test for drug treatment response in asthma, the more likely scenario for both prognostic testing and the development of novel asthma therapies will be a combination of genetics, genomics, and, perhaps, animal studies, proteomics, and metabolomics, as well. These "integrative pharmacogenomics" and "systems biology" approaches have yet to be applied to asthma, but the recent application of an integrative approach to the response to metformin in diabetes (69) makes this approach promising.

In conclusion, despite significant progress, with associations of genetic variation with therapeutic response to each of the three major classes of asthma medications having now been reported, the field of asthma pharmacogenetics remains in its infancy. However, the outlook for this field remains promising. Combining these newer integrative approaches aimed at ascertaining function and association simultaneously will only serve to accelerate the possibility that a prognostic test which will serve to optimize therapeutic response in asthma, while minimizing the potential for side effects, will be developed in the near future.

THE PHARMACOGENETICS OF COPD

Worldwide projections note that COPD is likely to be fifth in burden of disease and third with respect to mortality by 2020 (1). COPD is a clinical disease consisting of cough, sputum production, and/or dyspnea. The diagnosis is confirmed by an obstructive defect on spirometry that does not fully reverse upon administration of a short-acting bronchodilator. A postbronchodilator FEV_1 <80 percent of predicted in combination with an FEV_1/FVC ratio of <70 percent is sufficient to confirm the diagnosis (70). In the United States in 1993, COPD accounted for an estimated \$23.9 billion in health care costs, of which \$14.7 billion were direct costs (70). After hospitalizations, medication costs are the largest component of direct medical expenditures for COPD (71). Medication usage in COPD has traditionally been to control symptoms and complications related to the disease, because no medications have been demonstrated to modify the long-term decline in lung function that is the hallmark of the disease (70). Given the higher prevalence of the disease and higher proportion of nonresponders to medication therapy in COPD relative to asthma, the ability to predict those individuals most likely to benefit from therapy would have far-reaching global implications.

Like asthma, COPD is a genetic disease (72) with α_1-antitrypsin deficiency providing the classic example of a Mendelian disorder resulting in emphysema. However, even in COPD that is not related to α_1-antitrypsin deficiency, there is a significant genetic component, with the heritability estimates of FEV_1, FEV_1/FVC, and FEF_{25-75}/FVC ratio between 0.31 and 0.45 in families of patients with COPD (73). Three major classes of therapeutic agents are currently available for the treatment of COPD, including β-agonists (both short- and long-acting), inhaled corticosteroids, and anticholinergic agents (also both short- and long-acting) (74). Like asthma, large interindividual variation exists in the treatment response to both corticosteroids (75) and to bronchodilator agents (76) (Figure 15.7). This variability in response may have prognostic significance; patients on the long-acting anticholinergic agent tiotropium with a large short-term improvement in FEV_1 had improved lung function and fewer exacerbations at the end of one year's therapy in comparison with those with a poor short-term response (77). Finally, there appears to be high repeatability of the response to therapy in

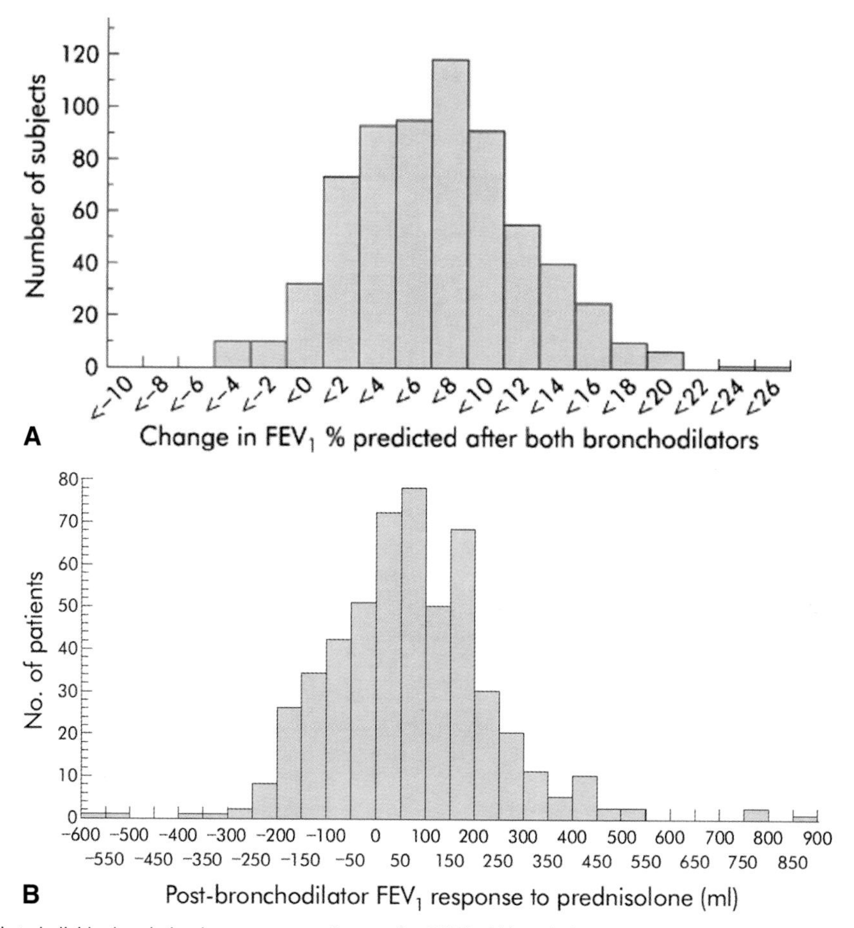

Figure 15.7. Interindividual variation in response to therapy for COPD. Although the average response to treatment for COPD is often cited as no different than placebo, there is significant interindividual variability in the response to both bronchodilators (**A**) and to corticosteroids (**B**). As with asthma, this suggests that factors other than the medications, including genetics, contribute to the therapeutic response. (From Calverley et al., *Thorax*. 2003;**58**:659–64 and Burge et al., *Thorax*. 2003;**58**:654–8.)

COPD (78). As with asthma, the genetic basis for the underlying disease along with the large interindividual variation and high repeatability to the response to therapy strongly suggests a significant role for pharmacogenetics in COPD.

Nevertheless, almost no pharmacogenetic studies of COPD have been performed to date. This may be due to the relative deemphasis on medical therapies, given the failure, on average, to pharmacologically modify lung function in COPD. However, one could argue that, given the burden of disease and the lack of therapeutic response, it would be even more important to be able to readily identify those genetically most likely to benefit from medications. In this section, after a brief review of potential COPD pharmacogenetic phenotypes, we will discuss the pharmacogenetics of smoking cessation. Although treatment to assist with smoking cessation is clearly not equivalent to direct pharmacologic therapy directed at the disease, smoking cessation remains the single most effective way to stop the progression of COPD (70). Therefore, understanding the genetics of smoking cessation are paramount to COPD management. Other COPD pharmacogenetic studies focusing on bronchodilator and corticosteroid medications are clearly warranted.

Potential COPD Pharmacogenetic Phenotypes

Despite the differences in pathophysiology, many of the phenotypes used in asthma pharmacogenetic studies would also be appropriate phenotypes for COPD. Because, like asthma, the majority of medications prescribed for COPD are inhaled, the role of pharmacokinetic studies is mitigated. However, given the increased severity and decreased overall response rates, more COPD patients will take oral medications. In the case of pharmacodynamic studies of COPD therapies, the intermediate phenotypes or measures of disease severity tend to parallel those of asthma, including measurements of lung function, airways responsiveness, and exacerbations (70, 74). Therefore, the methodologic approach toward the pharmacogenetic investigation of these intermediate phenotypes would also parallel those for asthma. Perhaps more so than asthma, the therapy of COPD is also often specifically directed toward the minimization of symptoms and the avoidance of exacerbations (70, 74). As such, genetic studies predicting relative therapeutic success of a given medication using exacerbations or health care utilization as an outcome would be of interest.

One other potentially interesting COPD pharmacogenetic phenotype may, in reality, actually represent a "surgicogenetic" phenotype instead. The treatment of COPD now includes various methods aimed at removing severely emphysematous portions of the lung, allowing fuller expansion of the underlying lung tissue. Lung-volume reduction surgery has been associated with increased exercise capacity, and with decreased mortality in patients with upper lobe predominance and low exercise capacity (79). In contrast, patients with homogeneous distribution of their underlying COPD and high-baseline exercise capacity are at increased risk of surgical mortality with negligible functional improvement (79). Investigation of the genetic factors influencing lobar predominance and exercise capacity in COPD is underway.

The Pharmacogenetics of Smoking Cessation

Cigarette smoking is the single most preventable cause of disease and death in the United States (80). Smoking cigarettes is also clearly the single greatest risk factor in the development of COPD (81) and its progression. It is well established that smoking cessation has substantial immediate and long-term health benefits, regardless of the age or relative health of the individual (82). There is evolving evidence that smoking behavior and response to pharmacologic therapy designed to assist with smoking cessation may be influenced by genetics. Nicotine as a drug is known to exert its addictive effects through activation of the mesolimbic dopaminergic pathway. Genetic differences influencing both nicotine metabolism and nicotine's effect at its target site have been investigated, primarily to identify genetic risk factors for smoking behavior and nicotine addiction.

There are two principal classes of medications for the assistance of smoking cessation efforts. Currently, the most common pharmacologic interventions for smoking cessation are alternative forms of nicotine replacement, including gums, inhalational sprays, lozenges, and transdermal patches. Interestingly, because the use of nicotine replacement medications is so prevalent as a therapeutic approach toward smoking cessation, the genetic factors influencing nicotine addiction are also likely to influence the efficacy of nicotine replacement medications. Bupropion is an antidepressant that inhibits reuptake of dopamine, noradrenaline, and serotonin in the central nervous system, is a noncompetitive nicotine receptor antagonist, and at high concentrations inhibits the firing of noradrenergic neurons in the locus coeruleus (83). It is not clear which of these effects accounts for the antismoking activity of the drug, but inhibition of the reductions in levels of dopamine and noradrenaline levels in the central nervous system that occur in nicotine withdrawal is likely to be important. The majority of individuals for all smoking cessation therapies relapse. In the paragraphs that follow, we review, in brief, the pharmacogenetic studies that have focused on both nicotine replacement therapy and bupropion.

Pharmacokinetic Paradigm

Nicotine is primarily metabolized by the cytochrome P450 system. Specifically, *CYP2A6* and *CYP2D6* mediate the conversion of nicotine to cotinine (84, 85). Recent heritability estimates derived from twin nicotine metabolism studies noted that 59.4 percent (95% CI, 44.7–70.7) of the weight-adjusted rate of total clearance of nicotine may be attributable to additive genetic influences (86). Similar figures (60.8 percent; 95% CI, 46.9–71.5) were noted for clearance of nicotine by the cotinine pathway. It was hypothesized more than a decade ago that nicotine dependence might be modified by variation in *CYP2D6* (85). Homozygotes for the *CYP2D6*5*, *CYP2D6*4*, and *CYP2D6*3* alleles are associated with poor metabolism (87), whereas those for *CYP2D6*2* and *CYP2D6*1* are associated with ultrarapid metabolism (88). Overall, studies in general support the hypothesis that the level of metabolism correlates with addiction to cigarettes. For instance, Saarikoski et al. examined 976 individuals (302 never smokers, 383 variable smokers, and 292 heavy smokers). The odds ratio (OR) for the presence of an ultrarapid CYP2D6 metabolism phenotype in heavy smokers versus never smokers was 4.2 (95% CI, 1.8–9.8) (89). The association of *CYP2A6* with smoking behavior has also been postulated, because *CYP2A6* may account for up to 90 percent of the nicotine → cotinine conversion (84). Because one of the primary therapeutic approaches toward smoking cessation is via the use of nicotine replacement medications, genetic variation in drug metabolism may be associated with the ability to respond to nicotine replacement therapy, with ultrarapid metabolizers having the greatest therapeutic need and, therefore, the least likelihood of responding to this form of treatment. Unfortunately, this hypothesis has not yet been formally tested.

Bupropion is metabolized by *CYP2B6*. A *CYP2B6* polymorphism (1459C>T) and its association with success of bupropion therapy for smoking cessation has been evaluated. Lerman et al. (90) examined the 1459C>T variant in a trial of 197 subjects taking bupropion along with 229 placebo controls. Overall, the minor allele (lower-activity variant) was associated with less abstinence at the end of therapy than the wild-type variant (OR, 1.56; 95% CI, 1.01–2.41 for relapse). At the end of six months, the men, but not the women, harboring the variant allele had a significantly higher relapse rate in those taking bupropion.

Pharmacodynamic Paradigm

Genetic variation at the site of the dopaminergic receptor in the central nervous system and its ability to influence the therapeutic response to smoking cessation efforts have also recently been investigated (91). The D2 dopamine receptor gene (*DRD2*) is a G-protein-coupled receptor located on postsynaptic dopaminergic neurons that is centrally involved in reward-mediating mesocorticolimbic pathways and is felt to modulate the effects of nicotine.

In a pharmacogenetic study of 755 smokers participating in a placebo-controlled trial of transdermal nicotine, nicotine replacement therapy was significantly more effective than placebo for carriers of the alternatively spliced A1 allele of *DRD2* but not for A2 homozygotes (92). The difference in the ORs for the treatment effect between the genotype groups was significant after the first week of treatment (patch/placebo OR, 2.8; 95% CI, 1.7–4.6 for any A1 allele vs. OR, 1.4; 95% CI, 0.9–2.1 for A2 homozygotes; P for difference in ORs = 0.04). This study also examined a polymorphism in the dopamine β-hydroxylase gene (*DBH*). Transdermal nicotine was more effective (OR, 3.6 for transdermal nicotine vs. placebo) in producing abstinence among smokers with both the *DRD2*A1* allele and the *DBH*A* allele in comparison with smokers with other genotypes. This study was continued longitudinally and suggested that abstinence at six and twelve months posttreatment was associated with the *DRD2*A1* variant. However, the effect was observed only among women (93). Two additional functional variants of *DRD2* have also been evaluated with respect to nicotine replacement therapy, an indel of the C allele at position –141 and a [C/T] SNP at position 957 (94). Both the deletion and the C957T variant alleles were associated with significantly higher quit rates on nicotine replacement therapy, compared with the wild-type alleles.

The μ-opioid receptor is the primary site of action for the rewarding effects of the endogenous opioid peptide, β-endorphin, which is released after acute and short-term nicotine administration (95). A common (25 percent to 30 percent of whites) Asn40Asp polymorphism of the μ-opioid receptor gene (*OPRM1*) increases the binding affinity of β-endorphin for this receptor threefold, relative to the wild-type Asn40 *OPRM1* (96). Among 320 white smokers, persons carrying the Asp40 variant were significantly more likely than those homozygous for the Asn40 variant to be abstinent at the end of nicotine replacement treatment (97). The treatment response differed by therapy type, and was significant among smokers who received transdermal nicotine (quit rates of 52 percent vs. 33 percent for the Asp40 and Asn40 groups, respectively; OR, 2.4), but not among smokers who received nicotine nasal spray. Further studies are warranted to investigate the reasons behind these differences.

Variants of the dopamine pathway have also been associated with differential response to smoking cessation with the use of bupropion. The *DRD2* –141C insertion (compared with the deletion's association with nicotine replacement therapy) was significantly associated with higher quit rates in a placebo-controlled trial of 414 smokers taking bupropion (94). In addition, the *DRD2* gene may interact with the dopamine transporter (*SLC6A3*) gene. In the same placebo-controlled study, smokers with both the *DRD2*A2* and *SLC6A3*9* variants had significantly higher abstinence rates at the end of treatment, and longer latency periods until relapse, as well. Although confirmatory studies are needed, the development of genetic prediction models of response to smoking cessation therapies will likely be forthcoming in the near future.

In a large candidate gene study, Conti et al. (98) genotyped 1,295 SNPs in genes from the nicotinic receptor and dopaminergic systems in 222 smokers receiving bupropion. They describe the association of two SNPs in *CHRNB2* (cholinergic receptor, nicotinic, beta 2) with decreased odds of smoking cessation (OR, 0.40; 95% CI, 0.25–0.67, P = 0.0004) (Figure 15.8). An SNP within *TDO2* (tryptophan 2,3-dioxygenase) was also associated with increased quit rates, but at a more nominal level of significance. Replication of these findings are necessary. Nonetheless, it can be concluded from this study that the pharmacogenetics of smoking cessation is modulated by multiple genes, and it is likely that genome-wide approaches will be useful in the future.

The Future of COPD Pharmacogenetics

The field of COPD pharmacogenetics is truly in its infancy. There is an increasing number of surgical therapeutic options, including lung-volume reduction surgery and lung transplantation, in the treatment of COPD. However, the potential to predict those patients with COPD who are most likely to continue to maintain their functional status from medications alone would have a huge impact on the potential timing and benefit of such interventions. Moreover, pharmacogenetic prediction of therapeutic response would also help to alleviate the potentially devastating side effects of long-term therapy in these patients. Finally, pharmacogenetic predictors related to the ability to minimize symptoms

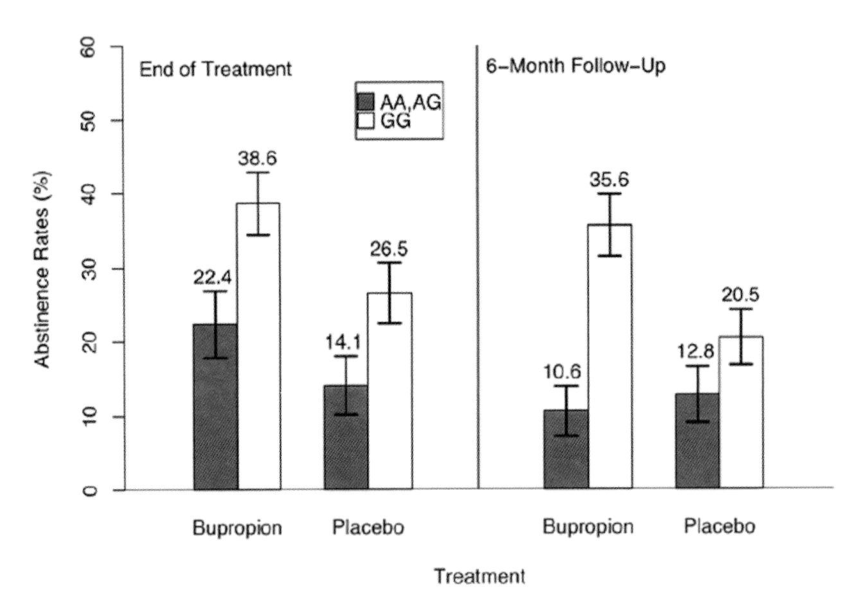

Figure 15.8. Abstinence rates by CHRNB2 rs2072661 and by bupropion treatment comparing individuals with at least one variant allele for CHRNB2 rs2072661 (*filled bars*) with those with both common alleles (*unfilled bars*) were stratified by treatment (bupropion or placebo) and estimated at each time point (end of treatment and six-month follow-up). (From Conti et al., *Hum Mol Genet.* 2008;**17**:2834–48.)

and/or health care use would be of tremendous benefit. Although we know of no large-scale COPD pharmacogenetic studies that are underway, a large genetic epidemiology study of COPD (COPD GENE – www.nationaljewish.org/copd-gene/index.aspx) is underway; ancillary pharmacogenetic studies are planned.

CONCLUSION

Asthma and COPD are both on the rise and contribute to significant morbidity, mortality, and cost in the United States and worldwide. Much of this cost is drug related. Variation in the response to drugs in these diseases has a significant heritable component. The pharmacogenetics of asthma and COPD is largely limited to pharmacodynamic (drug target) studies because of the inhalational nature of the most common medications. Although the time required to develop true prognostic testing for genetics of drug response for obstructive pulmonary diseases may still be years, clarification of the phenotypic characteristics and approaches to studying and analyzing pharmacogenetic data are in hand. By incorporating genome-wide, genomic, and systems biology approaches, the time line for "individualized medicine" as it relates to diseases of the lung can be accelerated.

DISCUSSION QUESTIONS

1. What are the factors indicating that the drug treatment response in pulmonary disease is mediated via genetics?
2. What are the main differences in pulmonary pharmacogenetics studies resulting from the fact that the majority of medications are administered via the inhalational route?
3. What has been the major limitation to the pulmonary pharmacogenetics studies to date? How can appropriate phenotyping help?
4. Why is the Arg16Gly locus of the *ADRB2* gene unlikely to be the major determinant of the treatment response to inhaled β-agonists?
5. Describe the approaches that are or will be taking place to rapidly identify novel candidate genes or loci in the study of pharmacogenetics in asthma and COPD?

REFERENCES

1. *Chronic Obstructive Pulmonary Disease.* Bethesda, MD: National Heart, Lung, and Blood Institute; 2003.
2. Petty TL. Definition, epidemiology, course, and prognosis of COPD. *Clin Cornerstone.* 2003;**5**:1–10.
3. Johnson JA & Lima JJ. Drug receptor/effector polymorphisms and pharmacogenetics: current status and challenges. *Pharmacogenetics.* 2003;**13**:525–34.
4. Nebert DW. Pharmacogenetics and pharmacogenomics: why is this relevant to the clinical geneticist? *Clin Genet.* 1999;**56**:247–58.
5. Freimer N & Sabatti C. The human genome project. *Nat Genet.* 2003;**34**:15–21.
6. Garcia-Closas M, Rothman N, & Lubin J. Misclassification in case-control studies of gene-environment interactions: assessment of bias and sample size. *Cancer Epidemiol Biomarkers Prev.* 1999;**8**:1043–50.

7. Smith AD & Taylor DR. Is exhaled nitric oxide measurement a useful clinical test in asthma? *Curr Opin Allergy Clin Immunol.* 2005;**5**:49–56.
8. Zimmermann N, King NE, Laporte J, Yang M, Mishra A, Pope SM, Muntel EE, Witte DP, Pegg AA, Foster PS, Hamid Q, & Rothenberg ME. Dissection of experimental asthma with DNA microarray analysis identifies arginase in asthma pathogenesis. *J Clin Invest.* 2003;**111**:1863–74.
9. Litonjua AA, Lasky-Su JA, Schneiter K, Tantisira KG, Lazarus R, Klanderman B, Lima JJ, Irvin CG, Peters SP, Hanrahan JP, Liggett SB, Hawkins GA, Meyers DA, Bleecker ER, Lange C, & Weiss ST. ARG1 is a novel bronchodilator response gene: screening and replication in four asthma cohorts. *Am J Respir Crit Care Med.* 2008;**178**:688–94.
10. Morton V & Torgerson DJ. Effect of regression to the mean on decision making in health care. *BMJ.* 2003;**326**:1083–4.
11. Masoli M, Fabian D, Holt S, & Beasley R. The global burden of asthma: executive summary of the GINA Dissemination Committee report. *Allergy.* 2004;**59**:469–78.
12. Malmstrom K, Rodriguez-Gomez G, Guerra J, Villaran C, Pineiro A, Wei LX, Seidenberg BC, & Reiss TF. Oral montelukast, inhaled beclomethasone, and placebo for chronic asthma. A randomized, controlled trial. Montelukast/Beclomethasone Study Group. *Ann Intern Med.* 1999;**130**:487–95.
13. Szefler SJ, Martin RJ, King TS, Boushey HA, Cherniack RM, Chinchilli VM, Craig TJ, Dolovich M, Drazen JM, Fagan JK, Fahy JV, Fish JE, Ford JG, Israel E, Kiley J, Kraft M, Lazarus SC, Lemanske RF Jr, Mauger E, Peters SP, & Sorkness CA. Significant variability in response to inhaled corticosteroids for persistent asthma. *J Allergy Clin Immunol.* 2002;**109**:410–18.
14. Drazen JM, Silverman EK, & Lee TH. Heterogeneity of therapeutic responses in asthma. *Br Med Bull.* 2000;**56**:1054–70.
15. Tantisira KG, Lake S, Silverman ES, Palmer LJ, Lazarus R, Silverman EK, Liggett SB, Gelfand EW, Richter BG, Israel E, Gabriel S, Altshuler D, Lander ES, Drazen JM, & Weiss ST. Corticosteroid pharmacogenetics: association of sequence variants in CRHR1 with improved lung function in asthmatics treated with inhaled corticosteroids. *Hum Mol Genet.* 2004;**13**:1353–9.
16. Lima JJ, Zhang S, Grant A, Shao L, Tantisira KG, Allayee H, Wang J, Sylvester J, Holbrook J, Wise R, Weiss ST, & Barnes K. Influence of leukotriene pathway polymorphisms on response to montelukast in asthma. *Am J Respir Crit Care Med.* 2005;**173**:379–85.
17. Enright PL, Lebowitz MD, & Cockroft DW. Physiologic measures: pulmonary function tests. Asthma outcome. *Am J Respir Crit Care Med.* 1994;**149**(2 pt 2):S9–18; discussion S19–20.
18. Brand PL, Duiverman EJ, Waalkens HJ, van Essen-Zandvliet EE, & Kerrebijn KF. Peak flow variation in childhood asthma: correlation with symptoms, airways obstruction, and hyperresponsiveness during long-term treatment with inhaled corticosteroids. Dutch CNSLD Study Group. *Thorax.* 1999;**54**:103–7.
19. Frischer T, Meinert R, Urbanek R, & Kuehr J. Variability of peak expiratory flow rate in children: short and long term reproducibility. *Thorax.* 1995;**50**:35–9.
20. National Heart, Lung, and Blood Institute. *Highlights of the Expert Panel Report 2: Guidelines for the Diagnosis and Management of Asthma.* Bethesda, MD: NIH Publications; 1997:97–4051A.
21. Raby BA & Weiss ST. Beta2-adrenergic receptor genetics. *Curr Opin Mol Ther.* 2001;**3**:554–66.
22. *Osler's Principles and Practice of Medicine.* 14 ed. New York: D. Appleton-Century Company; 1942.
23. Chung KF & O'Byrne P. Pharmacological agents used to treat asthma. In: Chung KF & Fabbri LM, eds. *Asthma.* Vol 8. Sheffield: European Respiratory Society Journals Ltd.; 2003:458.
24. Kobilka BK, Dixon RA, Frielle T, Dohlman HG, Bolanowski MA, Sigal IS, Yang-Feng TL, Francke U, Caron MG, & Lefkowitz RJ. cDNA for the human beta 2-adrenergic receptor: a protein with multiple membrane-spanning domains and encoded by a gene whose chromosomal location is shared with that of the receptor for platelet-derived growth factor. *Proc Natl Acad Sci USA.* 1987;**84**:46–50.
25. Reihsaus E, Innis M, MacIntyre N, & Liggett SB. Mutations in the gene encoding for the beta 2-adrenergic receptor in normal and asthmatic subjects. *Am J Respir Cell Mol Biol.* 1993;**8**:334–9.
26. Contopoulos-Ioannidis DG, Alexiou GA, Gouvias TC, & Ioannidis JP. An empirical evaluation of multifarious outcomes in pharmacogenetics: beta-2 adrenoceptor gene polymorphisms in asthma treatment. *Pharmacogenet Genomics.* 2006;**16**:705–11.
27. Martinez FD, Graves PE, Baldini M, Solomon S, & Erickson R. Association between genetic polymorphisms of the beta2-adrenoceptor and response to albuterol in children with and without a history of wheezing. *J Clin Invest.* 1997;**100**:3184–8.
28. Choudhry S, Ung N, Avila PC, Nazario S, Casal J, Torres A, Rodriguez-Santana JR, Fagan JK, Lilly C, Salas J, Selman M, Chapela R, Sheppard D, Weiss ST, Ford JG, Boushey HA, Drazen JM, Rodriguez-Cintron W, Silverman EK, & Gonzalez Burchard E. Pharmacogenetic differences in response to albuterol between Puerto Rican and Mexican asthmatics. *Am J Respir Crit Care Med.* 2005;**171**:563–70.
29. Silverman EK, Kwiatkowski DJ, Sylvia JS, Lazarus R, Drazen JM, Lange C, Laird NM, & Weiss ST. Family-based association analysis of beta2-adrenergic receptor polymorphisms in the childhood asthma management program. *J Allergy Clin Immunol.* 2003;**112**:870–6.
30. Drysdale CM, McGraw DW, Stack CB, Stephens JC, Judson RS, Nandabalan K, Arnold K, Ruano G, & Liggett SB. Complex promoter and coding region beta 2-adrenergic receptor haplotypes alter receptor expression and predict in vivo responsiveness. *Proc Natl Acad Sci USA.* 2000;**97**:10483–8.
31. Taylor DR, Epton MJ, Kennedy MA, Smith AD, Iles S, Miller AL, Littlejohn MD, Cowan JO, Hewitt T, Swanney MP, Brassett KP, & Herbison GP. Bronchodilator response in relation to beta2-adrenoceptor haplotype in patients with asthma. *Am J Respir Crit Care Med.* 2005;**172**:700–3.
32. Hawkins GA, Tantisira K, Meyers DA, Ampleford EJ, Moore WC, Klanderman B, Liggett SB, Peters SP, Weiss ST, & Bleecker ER. Sequence, haplotype, and association analysis of ADRbeta2 in a multiethnic asthma case-control study. *Am J Respir Crit Care Med.* 2006;**174**:1101–9.

33. Panebra A, Schwarb MR, Swift SM, Weiss ST, Bleecker ER, Hawkins GA, & Liggett SB. Variable-length poly-C tract polymorphisms of the beta2-adrenergic receptor 3′-UTR alter expression and agonist regulation. *Am J Physiol Lung Cell Mol Physiol.* 2008;**294**:L190–5.
34. Poon AH, Tantisira KG, Litonjua AA, Lazarus R, Xu J, Lasky-Su J, Lima JJ, Irvin CG, Hanrahan JP, Lange C, & Weiss ST. Association of corticotropin-releasing hormone receptor-2 genetic variants with acute bronchodilator response in asthma. *Pharmacogenet Genomics.* 2008;**18**:373–82.
35. Small KM, Brown KM, Theiss CT, Seman CA, Weiss ST, & Liggett SB. An Ile to Met polymorphism in the catalytic domain of adenylyl cyclase type 9 confers reduced beta2-adrenergic receptor stimulation. *Pharmacogenetics.* 2003;**13**:535–41.
36. Tantisira KG, Small KM, Litonjua AA, Weiss ST, & Liggett SB. Molecular properties and pharmacogenetics of a polymorphism of adenylyl cyclase type 9 in asthma: interaction between {beta}-agonist and corticosteroid pathways. *Hum Mol Genet.* 2005;**14**:1671–7.
37. Drazen JM, Israel E, Boushey HA, Chinchilli VM, Fahy JV, Fish JE, Lazarus SC, Lemanske RF, Martin RJ, Peters SP, Sorkness C, & Szefler SJ. Comparison of regularly scheduled with as-needed use of albuterol in mild asthma. Asthma Clinical Research Network. *N Engl J Med.* 1996;**335**:841–7.
38. Israel E, Drazen JM, Liggett SB, Boushey HA, Cherniack RM, Chinchilli VM, Cooper DM, Fahy JV, Fish JE, Ford JG, Kraft M, Kunselman S, Lazarus SC, Lemanske RF Jr, Martin RJ, McLean DE, Peters SP, Silverman EK, Sorkness CA, Szefler SJ, Weiss ST, & Yandava CN. Effect of polymorphism of the beta(2)-adrenergic receptor on response to regular use of albuterol in asthma. *Int Arch Allergy Immunol.* 2001;**124**:183–6.
39. Dishy V, Sofowora GG, Xie HG, Kim RB, Byrne DW, Stein CM, & Wood AJ. The effect of common polymorphisms of the beta2-adrenergic receptor on agonist-mediated vascular desensitization. *N Engl J Med.* 2001;**345**:1030–5.
40. Israel E, Chinchilli VM, Ford JG, Boushey HA, Cherniack R, Craig TJ, Deykin A, Fagan JK, Fahy JV, Fish J, Kraft M, Kunselman SJ, Lazarus SC, Lemanske RF, Jr, Liggett SB, Martin RJ, Mitra N, Peters SP, Silverman E, Sorkness CA, Szefler SJ, Wechsler ME, Weiss ST, & Drazen JM. Use of regularly scheduled albuterol treatment in asthma: genotype-stratified, randomised, placebo-controlled cross-over trial. *Lancet.* 2004;**364**:1505–12.
41. Tattersfield AE & Hall IP. Are beta2-adrenoceptor polymorphisms important in asthma – an unravelling story. *Lancet.* 2004;**364**:1464–6.
42. Israel E, Drazen JM, Liggett SB, Boushey HA, Cherniack RM, Chinchilli VM, Cooper DM, Fahy JV, Fish JE, Ford JG, Kraft M, Kunselman S, Lazarus SC, Lemanske RF, Martin RJ, McLean DE, Peters SP, Silverman EK, Sorkness CA, Szefler SJ, Weiss ST, & Yandava CN. The effect of polymorphisms of the beta(2)-adrenergic receptor on the response to regular use of albuterol in asthma. *Am J Respir Crit Care Med.* 2000;**162**:75–80.
43. Wechsler ME, Lehman E, Lazarus SC, Lemanske RF Jr, Boushey HA, Deykin A, Fahy JV, Sorkness CA, Chinchilli VM, Craig TJ, DiMango E, Kraft M, Leone F, Martin RJ, Peters SP, Szefler SJ, Liu W, & Israel E. beta-Adrenergic receptor polymorphisms and response to salmeterol. *Am J Respir Crit Care Med.* 2006;**173**:519–26.
44. Palmer CN, Lipworth BJ, Lee S, Ismail T, Macgregor DF, & Mukhopadhyay S. Arginine-16 beta2 adrenoceptor genotype predisposes to exacerbations in young asthmatics taking regular salmeterol. *Thorax.* 2006;**61**:940–4.
45. Bleecker ER, Postma DS, Lawrance RM, Meyers DA, Ambrose HJ, & Goldman M. Effect of ADRB2 polymorphisms on response to longacting beta2-agonist therapy: a pharmacogenetic analysis of two randomised studies. *Lancet.* 2007;**370**:2118–25.
46. Bleecker ER, Yancey SW, Baitinger LA, Edwards LD, Klotsman M, Anderson WH, & Dorinsky PM. Salmeterol response is not affected by beta2-adrenergic receptor genotype in subjects with persistent asthma. *J Allergy Clin Immunol.* 2006;**118**:809–16.
47. Bukstein DA, Bratton DL, Firriolo KM, Estojak J, Bird SR, Hustad CM, & Edelman JM. Evaluation of parental preference for the treatment of asthmatic children aged 6 to 11 years with oral montelukast or inhaled cromolyn: a randomized, open-label, crossover study. *J Asthma.* 2003;**40**:475–85.
48. Lima JJ. Treatment heterogeneity in asthma: genetics of response to leukotriene modifiers. *Mol Diagn Ther.* 2007;**11**:97–104.
49. Mougey EB, Feng H, Castro M, Irvin CG, & Lima JJ. Montelukast plasma concentration associates with a common SNP in SLCO2B1. *Clin Pharmacol Ther.* 2008;**83**(S1): S67.
50. Silverman ES & Drazen JM. The biology of 5-lipoxygenase: function, structure, and regulatory mechanisms. *Proc Assoc Am Physicians.* 1999;**111**:525–36.
51. In KH, Asano K, Beier D, Grobholz J, Finn PW, Silverman EK, Silverman ES, Collins T, Fischer AR, Keith TP, Serino K, Kim SW, De Sanctis GT, Yandava C, Pillari A, Rubin P, Kemp J, Israel E, Busse W, Ledford D, Murray JJ, Segal A, Tinkleman D, & Drazen JM. Naturally occurring mutations in the human 5-lipoxygenase gene promoter that modify transcription factor binding and reporter gene transcription. *J Clin Invest.* 1997;**99**:1130–7.
52. Drazen JM, Yandava CN, Dube L, Szczerback N, Hippensteel R, Pillari A, Israel E, Schork N, Silverman ES, Katz DA, & Drajesk J. Pharmacogenetic association between ALOX5 promoter genotype and the response to anti-asthma treatment. *Nat Genet.* 1999;**22**:168–70.
53. Telleria JJ, Blanco-Quiros A, Varillas D, Armentia A, Fernandez-Carvajal I, Jesus Alonso M, & Diez I. ALOX5 promoter genotype and response to montelukast in moderate persistent asthma. *Respir Med.* 2008;**102**:857–61.
54. Klotsman M, York TP, Pillai SG, Vargas-Irwin C, Sharma SS, Van Den Oord EJ, & Anderson WH. Pharmacogenetics of the 5-lipoxygenase biosynthetic pathway and variable clinical response to montelukast. *Pharmacogenet Genomics.* 2007;**17**:189–96.
55. Bartels M, Van Den Berg M, Sluyter F, Boomsma DI, & de Geus EJ. Heritability of cortisol levels: review and simultaneous analysis of twin studies. *Psychoneuroendocrinology.* 2003;**28**:121–37.
56. Sutherland ER, Ellison MC, Kraft M, & Martin RJ. Altered pituitary-adrenal interaction in nocturnal asthma. *J Allergy Clin Immunol.* 2003;**112**:52–7.

57. Laube BL, Curbow BA, Costello RW, & Fitzgerald ST. A pilot study examining the relationship between stress and serum cortisol concentrations in women with asthma. *Respir Med.* 2002;**96**:823–8.
58. Finotto S, Neurath MF, Glickman JN, Qin S, Lehr HA, Green FH, Ackerman K, Haley K, Galle PR, Szabo SJ, Drazen JM, De Sanctis GT, & Glimcher LH. Development of spontaneous airway changes consistent with human asthma in mice lacking T-bet. *Science.* 2002;**295**:336–8.
59. Tantisira KG, Hwang ES, Raby BA, Silverman ES, Lake SL, Richter BG, Peng SL, Drazen JM, Glimcher LH, & Weiss ST. TBX21: a functional variant predicts improvement in asthma with the use of inhaled corticosteroids. *Proc Natl Acad Sci USA.* 2004;**101**:18099–104.
60. Tantisira KG, Silverman ES, Mariani TJ, Xu J, Richter BG, Klanderman BJ, Litonjua AA, Lazarus R, Rosenwasser LJ, Fuhlbrigge AL, & Weiss ST. FCER2: a pharmacogenetic basis for severe exacerbations in children with asthma. *J Allergy Clin Immunol.* 2007;**120**:1285–91.
61. Abel SM, Maggs JL, Back DJ, & Park BK. Cortisol metabolism by human liver in vitro–I. Metabolite identification and inter-individual variability. *J Steroid Biochem Mol Biol.* 1992;**43**:713–19.
62. Winkler J, Hochhaus G, & Derendorf H. How the lung handles drugs: pharmacokinetics and pharmacodynamics of inhaled corticosteroids. *Proc Am Thorac Soc.* 2004;**1**:356–63.
63. Pawlik A, Baskiewicz-Masiuk M, Machalinski B, Kurzawski M, & Gawronska-Szklarz B. Involvement of C3435T and G2677T multidrug resistance gene polymorphisms in release of cytokines from peripheral blood mononuclear cells treated with methotrexate and dexamethasone. *Eur J Pharmacol.* 2005;**528**:27–36.
64. Weiss ST, Silverman EK, & Palmer LJ. Case-control association studies in pharmacogenetics. *Pharmacogenomics J.* 2001;**1**:157–8.
65. Contopoulos-Ioannidis DG, Alexiou G, Gouvias T, Ioannidis JPA. An empirical evaluation of multifarious outcomes in pharmacogenetics: beta2 adrenoreceptor polymorphisms and asthma treatment. *Pharmacogenet Genomics.* 2005;**16**:705–11.
66. Klein RJ. Power analysis for genome-wide association studies. *BMC Genet.* 2007;**8**:58.
67. Van Steen K, McQueen MB, Herbert A, Raby B, Lyon H, Demeo DL, Murphy A, Su J, Datta S, Rosenow C, Christman M, Silverman EK, Laird NM, Weiss ST, & Lange C. Genomic screening and replication using the same data set in family-based association testing. *Nat Genet.* 2005;**37**:683–91.
68. Hakonarson H, Bjornsdottir US, Halapi E, Bradfield J, Zink F, Mouy M, Helgadottir H, Gudmundsdottir AS, Andrason H, Adalsteinsdottir AE, Kristjansson K, Birkisson I, Arnason T, Andresdottir M, Gislason D, Gislason T, Gulcher JR, & Stefansson K. Profiling of genes expressed in peripheral blood mononuclear cells predicts glucocorticoid sensitivity in asthma patients. *Proc Natl Acad Sci USA.* 2005;**102**:14789–94.
69. Shu Y, Sheardown SA, Brown C, Owen RP, Zhang S, Castro RA, Ianculescu AG, Yue L, Lo JC, Burchard EG, Brett CM, & Giacomini KM. Effect of genetic variation in the organic cation transporter 1 (OCT1) on metformin action. *J Clin Invest.* 2007;**117**:1422–31.
70. Pauwels RA, Buist AS, Calverley PM, Jenkins CR, & Hurd SS. Global strategy for the diagnosis, management, and prevention of chronic obstructive pulmonary disease. NHLBI/WHO Global Initiative for Chronic Obstructive Lung Disease (GOLD) Workshop summary. *Am J Respir Crit Care Med.* 2001;**163**:1256–76.
71. Wilson L, Devine EB, & So K. Direct medical costs of chronic obstructive pulmonary disease: chronic bronchitis and emphysema. *Respir Med.* 2000;**94**:204–13.
72. DeMeo DL & Silverman EK. Genetics of chronic obstructive pulmonary disease. *Semin Respir Crit Care Med.* 2003;**24**:151–60.
73. DeMeo DL, Carey VJ, Chapman HA, Reilly JJ, Ginns LC, Speizer FE, Weiss ST, & Silverman EK. Familial aggregation of FEF(25–75) and FEF(25–75)/FVC in families with severe, early onset COPD. *Thorax.* 2004;**59**:396–400.
74. Rodriguez-Roisin R. The airway pathophysiology of COPD: implications for treatment. *J COPD.* 2005;2:253–62.
75. Burge PS, Calverley PM, Jones PW, Spencer S, & Anderson JA. Prednisolone response in patients with chronic obstructive pulmonary disease: results from the ISOLDE study. *Thorax.* 2003;**58**:654–8.
76. Calverley PM, Burge PS, Spencer S, Anderson JA, & Jones PW. Bronchodilator reversibility testing in chronic obstructive pulmonary disease. *Thorax.* 2003;**58**:659–64.
77. Tashkin D & Kesten S. Long-term treatment benefits with tiotropium in COPD patients with and without short-term bronchodilator responses. *Chest.* 2003;**123**:1441–9.
78. Ihre E & Larsson K. Airways responses to ipratropium bromide do not vary with time in asthmatic subjects. Studies of interindividual and intraindividual variation of bronchodilatation and protection against histamine-induced bronchoconstriction. *Chest.* 1990;**97**:46–51.
79. Fishman A, Martinez F, Naunheim K, Piantadosi S, Wise R, Ries A, Weinmann G, & Wood DE. A randomized trial comparing lung-volume-reduction surgery with medical therapy for severe emphysema. *N Engl J Med.* 2003;**348**:2059–73.
80. *U.S. Department of Health and Human Services. Healthy People 2010: Understanding and Improving Health.* 2nd ed. Washington, DC: U.S. Government Printing Office; 2000.
81. Mannino DM. Epidemiology and global impact of chronic obstructive pulmonary disease. *Semin Respir Crit Care Med.* 2005;**26**:204–10.
82. Edwards R. The problem of tobacco smoking. *BMJ.* 2004;**328**:217–19.
83. Roddy E. Bupropion and other non-nicotine pharmacotherapies. *BMJ.* 2004;**328**:509–11.
84. Malaiyandi V, Sellers EM, & Tyndale RF. Implications of CYP2A6 genetic variation for smoking behaviors and nicotine dependence. *Clin Pharmacol Ther.* 2005;**77**:145–58.
85. Cholerton S, Arpanahi A, McCracken N, Boustead C, Taber H, Johnstone E, Leathart J, Daly AK, & Idle JR. Poor metabolisers of nicotine and CYP2D6 polymorphism. *Lancet.* 1994;**343**:62–3.
86. Swan GE, Benowitz NL, Lessov CN, Jacob P 3rd, Tyndale RF, & Wilhelmsen K. Nicotine metabolism: the impact of CYP2A6 on estimates of additive genetic influence. *Pharmacogenet Genomics.* 2005;**15**:115–25.

87. Daly AK, Brockmoller J, Broly F, Eichelbaum M, Evans WE, Gonzalez FJ, Huang JD, Idle JR, Ingelman-Sundberg M, Ishizaki T, Jacqz-Aigrain E, Meyer UA, Nebert DW, Steen VM, Wolf CR, & Zanger UM. Nomenclature for human CYP2D6 alleles. *Pharmacogenetics.* 1996;**6**:193–201.
88. Johansson I, Lundqvist E, Bertilsson L, Dahl ML, Sjoqvist F, & Ingelman-Sundberg M. Inherited amplification of an active gene in the cytochrome P450 CYP2D locus as a cause of ultrarapid metabolism of debrisoquine. *Proc Natl Acad Sci USA.* 1993;**90**:11825–9.
89. Saarikoski ST, Sata F, Husgafvel-Pursiainen K, Rautalahti M, Haukka J, Impivaara O, Jarvisalo J, Vainio H, & Hirvonen A. CYP2D6 ultrarapid metabolizer genotype as a potential modifier of smoking behaviour. *Pharmacogenetics.* 2000;**10**:5–10.
90. Lerman C, Shields PG, Wileyto EP, Audrain J, Pinto A, Hawk L, Krishnan S, Niaura R, & Epstein L. Pharmacogenetic investigation of smoking cessation treatment. *Pharmacogenetics.* 2002;**12**:627–34.
91. Berrettini WH & Lerman CE. Pharmacotherapy and pharmacogenetics of nicotine dependence. *Am J Psychiatry.* 2005;**162**:1441–51.
92. Johnstone EC, Yudkin PL, Hey K, Roberts SJ, Welch SJ, Murphy MF, Griffiths SE, & Walton RT. Genetic variation in dopaminergic pathways and short-term effectiveness of the nicotine patch. *Pharmacogenetics.* 2004;**14**(2):83–90.
93. Yudkin P, Munafo M, Hey K, Roberts S, Welch S, Johnstone E, Murphy M, Griffiths S, & Walton R. Effectiveness of nicotine patches in relation to genotype in women versus men: randomised controlled trial. *BMJ.* 2004;**328**:989–90.
94. Lerman C, Jepson C, Wileyto EP, Epstein LH, Rukstalis M, Patterson F, Kaufmann V, Restine S, Hawk L, Niaura R, & Berrettini W. Role of functional genetic variation in the dopamine D2 receptor (DRD2) in response to bupropion and nicotine replacement therapy for tobacco dependence: results of two randomized clinical trials. *Neuropsychopharmacology.* 2006;**31**:231–42.
95. Davenport KE, Houdi AA, & Van Loon GR. Nicotine protects against mu-opioid receptor antagonism by beta-funaltrexamine: evidence for nicotine-induced release of endogenous opioids in brain. *Neurosci Lett.* 1990;**113**: 40–6.
96. Bond C, LaForge KS, Tian M, Melia D, Zhang S, Borg L, Gong J, Schluger J, Strong JA, Leal SM, Tischfield JA, Kreek MJ, & Yu L. Single-nucleotide polymorphism in the human mu opioid receptor gene alters beta-endorphin binding and activity: possible implications for opiate addiction. *Proc Natl Acad Sci USA.* 1998;**95**:9608–13.
97. Lerman C, Wileyto EP, Patterson F, Rukstalis M, Audrain-McGovern J, Restine S, Shields PG, Kaufmann V, Redden D, Benowitz N, & Berrettini WH. The functional mu opioid receptor (OPRM1) Asn40Asp variant predicts short-term response to nicotine replacement therapy in a clinical trial. *Pharmacogenomics J.* 2004;**4**:184–92.
98. Conti DV, Lee W, Li D, Liu J, Van Den Berg D, Thomas PD, Bergen AW, Swan GE, Tyndale RF, Benowitz NL, & Lerman C. Nicotinic acetylcholine receptor (beta)2 subunit gene implicated in a systems-based candidate gene study of smoking cessation. *Hum Mol Genet.* 2008;**17**:2834–48.

Pharmacogenomics Associated with Therapy for Acid-Related Disorders

16

Takahisa Furuta

Peptic ulcer and gastroesophageal reflux disease (GERD) are common benign upper gastrointestinal disorders. The major causes of peptic ulcer are *Helicobacter pylori* (*H. pylori*) infection and nonsteroidal anti-inflammatory drugs (NSAIDs), including aspirin. No matter whether caused by *H. pylori*, NSAIDs, or both, however, gastric acid plays a markedly important role in the pathogenesis of peptic ulcer. Furthermore, the major cause of GERD is the reflux of gastric acid from the stomach to the esophagus. Gastric acid therefore acts as the most important pathogenic factor of upper gastrointestinal disorders, and acid inhibition with a proton pump inhibitor (PPI) is the major strategy against them. PPIs in combination with one or two antibiotics are also used for the eradication of *H. pylori.*

PPIs are substitutes of benzimidazole and are mainly metabolized by the cytochrome P450 (CYP) system in the liver (1). The principal enzyme in this metabolism is CYP2C19 (1), although CYP3A4 also plays a role (2–6). Given the presence of interindividual differences in CYP2C19 activity, the pharmacokinetics (PKs) and pharmacodynamics (PDs) of PPIs largely depend on polymorphisms in CYP2C19 (7).

Similarly, NSAIDs such as ibuprofen and naproxen are metabolized by CYP2C9, and so the PKs and PDs of NSAIDs largely depend on polymorphisms in this gene (8).

Here, we describe the influence of genetic polymorphisms in CYP2C19 on the PKs and PDs of PPIs and on clinical outcomes of PPI-based therapies for *H. pylori* infection and GERD. We also discuss CYP2C19 genotype-based personalized treatment of *H. pylori* infection with PPI-based regimens, and recent knowledge on the pharmacogenomics of NSAID ulcer, as well.

EFFECTS OF CYP2C19 POLYMORPHISM ON THE PKs AND PDs OF PPIs

Genetic Differences in the PPI-Metabolizing Enzyme, CYP2C19

PPIs are mainly metabolized by CYP2C19 (Figure 16.1), for which interindividual differences in activity have been identified. Polymorphisms of this enzyme are classified into the three genotype groups of rapid metabolizers (RM: *1/*1), intermediate metabolizers (IM: *1/*X), and poor metabolizers (PM: *X/*X), for which *1 = wild-type allele and *X = the mutated allele (*2 or *3) (9).

PMs of this enzyme are also subject to substantial interethnic differences in frequency, at 2.5 percent in white Americans, 3.7 percent in African Americans, 2.8 percent in white Europeans, 4.8 percent in Shona Zimbabweans, 19.8 percent in Han Chinese, 13.4 percent in Bai Chinese, 12.6 percent in Koreans, and 18.0 percent to 22.5 percent in Japanese (10–18).

Various genetic mutations in CYP2C19 polymorphism from ethnically different populations have been discovered (http://www.imm.ki.se/CYPalleles/cyp2c19.htm). The PM-related CYP2C19 polymorphism in Japanese can be explained by the combination of two point mutations, *CYP2C19*2* of exon 5 and *CYP2C19*3* of exon 4 (15, 19, 20). *CYP2C19*2* is a single base pair mutation (from guanine to adenine) at position 681 of exon 5 of *CYP*2C19, which creates a truncated nonfunctional protein. This defect accounts for approximately 75 percent to 83 percent of the PM allele in both Japanese and white subjects (20). de Morais et al. (19) also detected another mutation, *CYP*2C19*3, which consists of a guanine to adenine mutation at position 636 of exon 4 of *CYP*2C19, and which creates a premature stop codon. Recently, the *CYP*2C19*17 allele has been identified as an ultrarapid extensive metabolizer genotype of

Table 16.1. Frequencies of CYP2C19 Phenotypes and Alleles in Different Populations

	Phenotype Frequency			Allele Frequency			
Ethnicity	*RM*	*IM*	*PM*	**1*	**2*	**3*	*Reference*
Africans	0.68	0.28	0.04	0.823	0.173	0.004	(12)
Asians	0.35	0.46	0.19	0.58	0.29	0.13	(16)
Whites	0.69	0.28	0.03	0.83	0.17	0	(18)

RM = rapid metabolizer of *CYP*2C19 (*1/*1); IM = intermediate metabolizer of *CYP*2C19 (*1/*2 or *1/*3); PM = poor metabolizer of *CYP*2C19 (*2/*2, *2/*3 or *3/*3). Modified from reference (23).

CYP2C19 (21), although the impact of this polymorphism appears to differ among different ethnic groups (22).

Representative frequencies of phenotypes and alleles in different ethnic groups are summarized in Table 16.1 (23).

Effect of CYP2C19 Polymorphism on the PKs of PPIs

Omeprazole, the representative and first clinically available PPI, is mainly metabolized by CYP2C19 to 5-hydroxyomeprazole, which is then metabolized by CYP3A4 to 5-hydroxyomeprazole sulfone. Omeprazole is partially first metabolized by CYP3A4 to omeprazole sulfone, which is then metabolized by CYP2C19 to 5-hydrozyomeprazole sulfone (Figure 16.1). On single administration of omeprazole at 20 mg, plasma omeprazole concentrations differ among the three different CYP2C19 genotype groups (RM, IM, and PM) (Figure 16.2) (7), with levels in the PM group sustained for an extended period after dosing, and mean values for the area under the plasma concentration-time curve (AUC) about 13 times those of the RM group (Table 16.2).

Several articles describing the PKs of PPIs in relation to CYP2C19 genotypes are summarized in Table 16.2 (23).

Effect of CYP2C19 Polymorphism on Gastric Acid Inhibition with PPIs

On administration of a PPI (i.e., omeprazole 20 mg), the intragastric pH profile also differs among the three different genotype groups (Figure 16.3). Mean 24-hour intragastric pH level is lowest in the RM group, higher in the IM group, and highest in the PM group (7). The acid inhibition attained by omeprazole in the RM group under the so-called standardized dosing scheme, in fact, seems insufficient (7, 24).

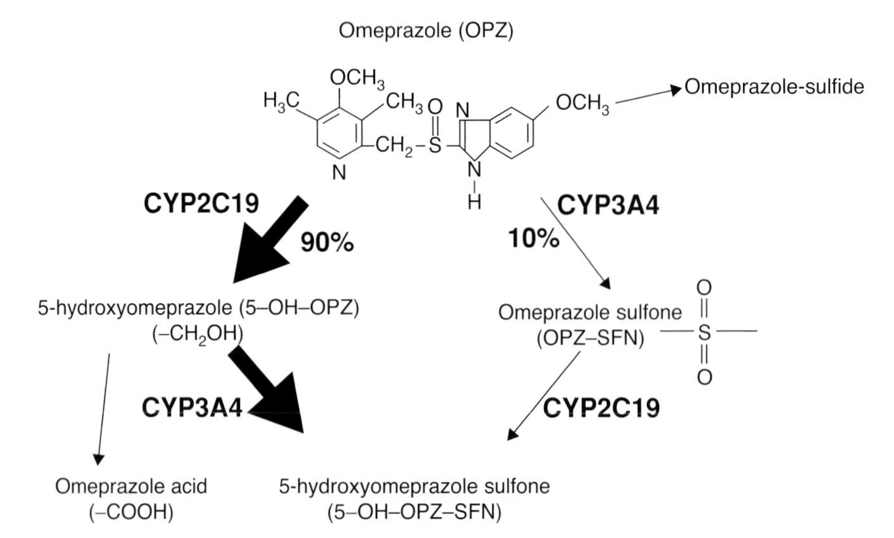

Figure 16.1. Metabolism of omeprazole (OPZ) in relation to cytochrome P450 (CYP) isoenzymes. Weight of arrows indicates the relative contribution of the different enzyme pathways. OPZ is mainly metabolized by CYP2C19 to 5-hydroxyomeprazole (5-OH-OPZ), which is then metabolized to 5-hydroxyomeprazole sulfone (5-OH-OPZ-SFN). OPZ is also metabolized by CYP3A4 to omeprazole sulfone (OPZ-SFN), which is then metabolized by CYP2C19 to 5-OH-OPZ-SFN. Modified from reference (23).

Table 16.2. CYP2C19 Genotype Group and AUC_{0-24h} (ng·h/mL) of PPIs

Regimen	*RM*	*IM*	*PM*	*RM:IM:PM Ratio*	*Reference*
OPZ 20 mg day 1	421	1,403	5,109	1:3.3:12.2	(7)
OPZ 20 mg day 1	524	1,096	5,607	1:2.1:10.7	(25)
OPZ 20 mg day 8	1,057	2,418	7,153	1:2.3:6.8	(25)
RPZ 20 mg day 1	696	1,609	2,329	1:2.3:3.3	(25)
RPZ 20 mg day 8	464	1,398	2,437	1:3.0:5.3	(25)
LPZ 30 mg day 8	1,980	4,775	10,663	1:2.4:5.4	(41)

RM = rapid metabolizer of *CYP*2C19 (*1/*1); IM = intermediate metabolizer of *CYP*2C19 (*1/*2 or *1/*3); PM = poor metabolizer of *CYP*2C19 (*2/*2, *2/*3, *3/*3); OPZ = omeprazole; LPZ = lansoprazole; RPZ = rabeprazole. Modified from reference (23).

Several reports on intragastric pH profiles after PPI dosing are summarized in Table 16.3 (23). The differences in acid inhibition by a PPI among the different CYP2C19 genotype groups are considered to be due to their different plasma concentrations: the profound inhibition in PMs is derived from their high plasma PPI concentrations, and the low acid inhibition in RMs is similarly ascribable to their low plasma PPI concentrations (25, 26).

Intragastric pH on Frequent PPI Dosing

Administration of lansoprazole at 30 mg four times daily ensures sustained plasma levels throughout the day (Figure 16.4A), achieving complete acid inhibition even in those with the RM genotype of CYP2C19 (Figure 16.4B). Interestingly, the C_{max} (peak plasma concentration level) of lansoprazole under this regimen in RMs is no higher than that obtained with a single daily dose of 30 mg (Figure 16.4A), and not as high as that observed in PMs, albeit that inhibition is still sufficient. This indicates that the determinant of the acid-inhibitory effect of PPIs is not C_{max} or AUC, but rather the duration of time above a certain threshold level (27). The administration regimen of a PPI (i.e., dose as well as dosing schedule) is thus a key point in attaining appropriate intragastric pH levels, and should be determined on the basis of sufficient acid inhibition for the CYP2C19 genotype status of the individual patient.

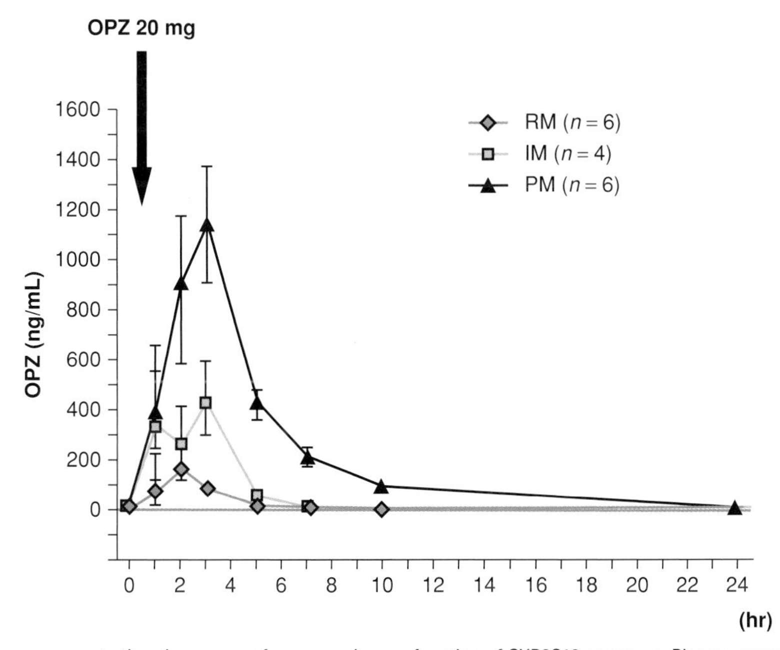

Figure 16.2. Plasma concentration-time curve of omeprazole as a function of CYP2C19 genotype. Plasma concentration of OPZ is highest in the PM group, intermediate in the IM group, and lowest in the RM group (7). RM = rapid metabolizer of CYP2C19; IM = intermediate metabolizer of CYP2C19; PM = poor metabolizer of CYP2C19; PP = per-protocol; and ITT = intention-to-treat.

Table 16.3. Intragastric pH Attained by a PPI as a Function of *CYP*2C19 Genotype Status

Regimen	*RM*	*IM*	*PM*	*Reference*
OPZ 20 mg day 1	2.1	3.3	4.5	(7)
OPZ 20 mg day 1	2.3	3.3	4.1	(25)
OPZ 20 mg day 8	4.1	4.7	5.9	(25)
LPZ 30 mg day 8	4.4	4.9	5.4	(26)
RPZ 20 mg day 1	3.3	4.2	5.3	(25)
RPZ 20 mg day 8	4.8	5.0	6.0	(25)

RM = rapid metabolizer of CYP2C19 (*1/*1); IM = intermediate metabolizer of CYP2C19 (*1/*2 or *1/*3); PM = poor metabolizer of CYP2C19 (*2/*2, *2/*3, *3/*3); OPZ = omeprazole; LPZ = lansoprazole; RPZ = rabeprazole. Modified from reference (23).

Influence of CYP2C19 Genotype Status on the PKs and PDs of Different PPIs

The influence of CYP2C19 genotype status on the PKs and PDs of PPIs differs among the different PPIs (Tables 16.2 and 16.3). The ratio of PM to RM in AUC is highest for omeprazole, followed by lansoprazole, and is lowest for rabeprazole. A similar tendency is observed in intragastric pH levels during PPI treatment. Esomeprazole was reported to be less affected by CYP2C19 genotype status (28), but a recent report by Kirsch et al. (29) indicated that it is, in fact, considerably affected by CYP2C19 genotype status. The impact of CYP2C19 polymorphisms on PKs and PDs of the newly generated PPIs should be tested and compared with those of omeprazole and lansoprazole in the same subjects to clarify the impact of CYP2C19 polymorphism on PKs and PDs of different PPIs.

EFFECT OF CYP2C19 POLYMORPHISM ON GERD TREATMENT BY PPI

GERD, a common disorder estimated to affect 35 percent to 40 percent of adults in Western countries (30), is now the major indication for PPIs. Healing rates of GERD achieved with PPIs are usually high, although some patients appear refractory to standard doses (e.g., 20 mg of omeprazole or 30 mg of lansoprazole) (31–33). One reason for this refractoriness was recently identified with regard to the metabolism of a PPI (i.e., lansoprazole).

When lansoprazole was given by single daily administration at 30 mg for eight weeks to GERD patients positive for mucosal breaks (grades A to D in Los Angeles Classification), the healing rate of mucosal breaks was lowest in the RM group, followed in order by the IM and PM groups (Figure 16.5). In particular, the healing rate of grade C or D GERD in patients with the RM genotype of CYP2C19 was dramatically low (1/6 = 16.7%; confidence interval [CI], 0.4%–64.1%) (34). Similarly, Kawamura et al. (35) reported that, in the administration of 30 mg to patients with erosive reflux esophagitis (low-grade GERD), the healing rate of mucosal breaks was again lowest in the RM patients, indicating that the degree of acid inhibition achieved by single daily

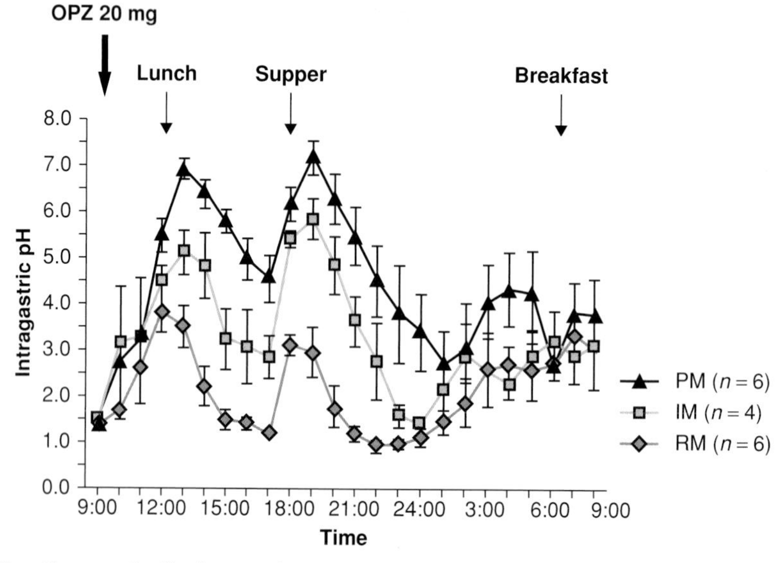

Figure 16.3. Profiles of intragastric pH values as a function of CYP2C19 genotype status for administration of omeprazole (OPZ) at 20 mg (7). RM = rapid metabolizer of CYP2C19; IM = intermediate metabolizer of CYP2C19; PM = poor metabolizer of CYP2C19; PP = per-protocol; and ITT = intention-to-treat.

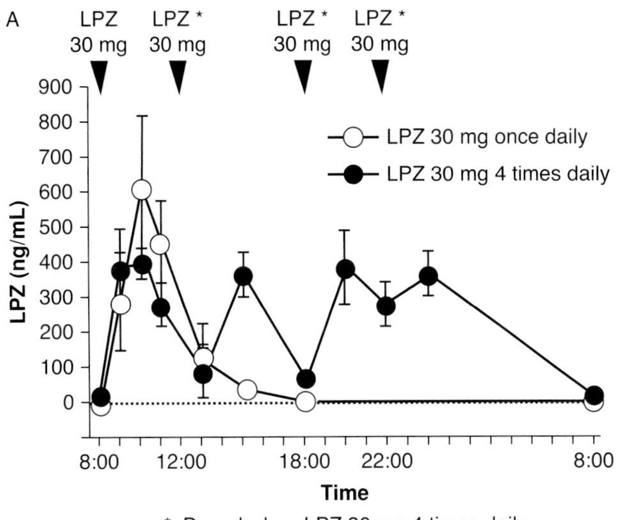

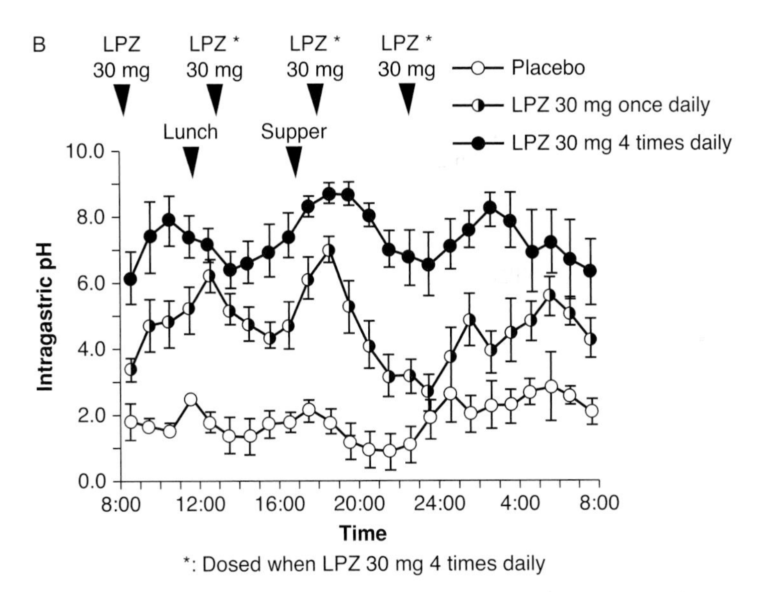

Figure 16.4. Mean (±SE) plasma concentration-time curves (**A**) and mean (±SE) intragastric pH values versus time course (**B**) for lansoprazole (LPZ) after the final administration of 30 mg of LPZ given once daily and 30 mg of lansoprazole given four times daily for eight days in five subjects with the RM genotype. With LPZ at 30 mg four times daily, plasma levels of LPZ were sustained during all dosing intervals, providing complete acid inhibition (i.e., intragastric pH approximately 7.0) (41).

administration of 30 mg is clinically sufficient for GERD treatment in the PM genotype group, but possibly insufficient in some IM and RM patients. Kawamura et al. (36) also reported that the recurrence rate of GERD under maintenance therapy with lansoprazole at 15 mg depended on CYP2C19 genotype status, indicating that the CYP2C19 genotype status of GERD patients must be considered not only in initial therapy with a PPI, but also in maintenance therapy. The CYP2C19 RM genotype should be considered a risk factor of GERD refractoriness to usual PPI treatment.

Nocturnal acid breakthrough (NAB), defined as an intragastric pH lower than 4 lasting for more than one hour during the overnight period, is another factor associated with the success or failure of treatment of GERD with PPIs (37–40). Interestingly, the frequency of NAB depends on the CYP2C19 genotype status, and is most frequently seen in subjects with the RM genotype (26, 41), which is likely associated with the lowest healing rate of GERD among the three genotype groups (34, 35). On the basis of this discussion, one therapeutic strategy for the treatment of GERD refractory to the usual dose of a

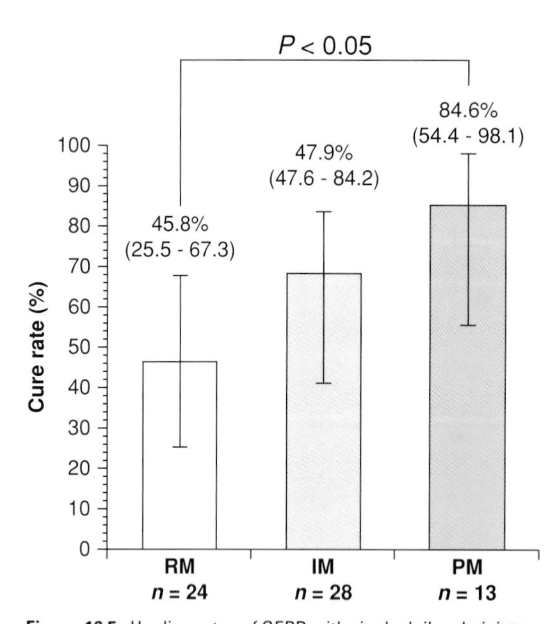

Figure 16.5. Healing rates of GERD with single daily administration of lansoprazole at 30 mg for eight weeks in the different CYP2C19 genotype groups. Bars indicate 95% confidence intervals (95% CI). A significant difference in healing rates among the three different CYP2C19 genotype groups was seen (34). RM = rapid metabolizer of CYP2C19; IM = intermediate metabolizer of CYP2C19; PM = poor metabolizer of CYP2C19; PP = per-protocol; ITT = intention-to-treat.

PPI may be an increased dose of PPI, in patients with the RM genotype of CYP2C19 at least.

It has been reported that the acid inhibitory effect of histamine 2 (H2) receptor antagonists, such as famotidine, is not affected by CYP2C19 genotype status and that these drugs are superior to lansoprazole in RM patients during the night (26). The addition of an evening dose of an H2 receptor antagonist to a morning dose of a PPI is effective in controlling NAB in individuals who are resistant to a usual PPI treatment (42–44). Concomitant treatment with a PPI plus an H2 receptor antagonist might therefore be another therapeutic strategy for RM patients with GERD refractory to treatment with a usual constant dose of a PPI alone. Overall, CYP2C19 genotype status is considered a predictable determinant of outcome in PPI-based GERD therapy, and individualized therapeutic strategies based on individual CYP2C19 genotype status are expected to increase healing rates with initial treatment for GERD.

EFFECT OF CYP2C19 POLYMORPHISM ON PEPTIC ULCER HEALING WITH A PPI

Acid inhibition is required for peptic ulcer healing. As with GERD, it has recently been reported that ulcer healing by a PPI depends on CYP2C19 genotype status (45). On single daily administration of omeprazole at 20 mg, gastric ulcer healing in those with the RM genotype was slower than in those with the IM and PM genotype, and improvement ratios in the ulcer area in RMs, IMs, and PMs on treatment for one week were 46.3 percent, 61.7 percent, and 63.2 percent, respectively; this difference between the RMs and IMs was significant. Interestingly, corresponding values for rabeprazole at 10 mg daily for one week were 60.8 percent, 65.0 percent, and 55.3 percent, respectively, without significant difference. Further, improvement ratios with rabeprazole in RMs and IMs were significantly greater than that with omeprazole in RMs, indicating the usefulness of CYP2C19 genotype-guided selection of a PPI in the treatment of peptic ulcer (45).

EFFECTS OF CYP2C19 POLYMORPHISM ON PPI-BASED ERADICATION THERAPY OF *H. pylori*

CYP2C19 Genotype Status as the Therapeutic Determinant of PPI-Based Eradication Therapy for *H. pylori*

H. pylori infection is associated with a variety of upper gastrointestinal disorders, such as peptic ulcer, mucosa-associated lymphoid tissue lymphoma, and gastric cancer (46–49). Eradication of this pathogen is useful for the treatment of these diseases. Current regimens for eradication consist of a PPI plus one or two antibacterial agents, such as amoxicillin, clarithromycin, and metronidazole (50).

The roles of PPIs in *H. pylori* eradication therapy are classified as follows: first, to increase the stability and bioavailability of antibiotics in the stomach by raising intragastric pH to neutral levels (51); second, to neutralize intragastric pH levels to allow *H. pylori* to reach the growth phase, and thus become more sensitive to antibiotics, such as amoxicillin (52, 53); third, to suppress acid secretion and thereby increase the concentration of antibiotic in the stomach, such as amoxicillin (54); and fourth, to exert their own inherent anti-*H. pylori* effect (55). These four characteristics make PPIs key drugs in *H. pylori* eradication therapy.

The first report on the influence of CYP2C19 polymorphism on *H. pylori* eradication rates was with regard to dual PPI/amoxicillin therapy: eradication rates by dual therapy with a single daily dose of omeprazole at 20 mg and amoxicillin at 500 mg four times daily for two weeks were approximately 30 percent in RMs, 60 percent in IMs, and 100 percent in PMs (Table 16.3) (56). These different eradication rates are assumed to reflect the differences in plasma omeprazole levels among the different CYP2C19 genotype groups. Other reports on the influence of CYP2C19 polymorphism on dual PPI/amoxicillin therapy are summarized in Table 16.4.

Table 16.4. Eradication Rates of *H. pylori* Infection by PPI-Based Regimens as a Function of CYP2C19 Genotype Status (%)

Regimen	*RM*	*IM*	*PM*	*Reference*
OPZ 20 mg 4 times daily + AMPC 500 mg 4 times daily for 2 weeks	29	60	100	(56)
OPZ 20 mg twice daily + AMPC 500 mg 3 times daily for 1 week	40	42	100	(77)
RPZ 10 mg twice daily + AMPC 500 mg 3 times daily for 2 weeks	61	92	94	(71)
OPZ 20 mg or LPZ 30 mg twice daily + AMPC 500 mg 3 times daily + CAM 200 mg 3 times daily for 1 week	73	92	98	(61)
OPZ 20 mg twice daily + AMPC 1,000 mg twice daily + CAM 500 mg twice daily for 1 week	60	84	100	(78)
LPZ 30 mg twice daily + AMPC 750 mg twice daily; CAM 200 mg twice daily for 1 week	63	87	100	(79)
OPZ 20 mg, LPZ 30 mg or RPZ 10 mg + AMPC 1,500 mg; CAM 600 mg for 1 week	69	77	82	(80)

OPZ = omeprazole; LPZ = lansoprazole; RPZ = rabeprazole; AMPC = amoxicillin; CAM = clarithromycin. Modified from reference (23).

One current regimen for the eradication of *H. pylori* is triple therapy with a PPI, amoxicillin, and clarithromycin (50, 57). Interestingly, clarithromycin is not only metabolized by CYP3A4, but also is a potent inhibitor of CYP3A4. CYP3A4 is involved in the sulfoxidation of PPIs (58). Clarithromycin also affects the activity of CYP2C19 (59). A small part of clarithromycin is metabolized by CYP2C19. On coadministration of a PPI and clarithromycin, a drug-drug interaction can therefore occur, which can in turn lead to a difference in plasma clarithromycin level among the different CYP2C19 genotype groups (60).

Differences in plasma clarithromycin and PPI levels among the different CYP2C19 genotype groups do indeed produce differences in eradication rates for *H. pylori* infection. In our previous study (61), eradication rates by triple therapy with single daily doses of omeprazole 40 mg or lansoprazole 60 mg, amoxicillin 1,500 mg, and clarithromycin 600 mg for one week were 73 percent in RMs, 92 percent in IMs, and 98 percent in PMs (Table 16.4). Incidence of the RM genotype in the group without eradication was high, whereas that of the PM genotype was markedly low. Other reports on the influence of CYP2C19 polymorphism on the eradication rates by triple therapy are summarized in Table 16.4. Although a few studies found no statistical significance in eradication rates among the different CYP2C19 genotype groups (62, 63), most did indeed report such a tendency. Taken together, these findings indicate that one reason for the failure of *H. pylori* eradication by triple PPI/amoxicillin/clarithromycin therapies is an insufficient dose of the PPI (omeprazole or lansoprazole) in RMs.

Another important determinant of the success or failure of *H. pylori* eradication by a triple PPI/amoxicillin/clarithromycin therapy is bacterial resistance to clarithromycin. Resistance is due to a mutation in 23S rRNA, which can be detected by genetic testing (64–67). We previously reported a dramatically low eradication rate in RM patients infected with clarithromycin-resistant strains of *H. pylori* of 7.1 percent (61). We therefore conclude that the major factors associated with the success or failure of *H. pylori* eradication by a triple PPI/amoxicillin/clarithromycin therapy are not only the CYP2C19 genotype status of the patients, but also bacterial resistance of *H. pylori* strains to clarithromycin.

Pharmacogenomics-Based Rescue Regimens after Eradication Failure with Standard PPI/Amoxicillin/Clarithromycin Therapy at Usual Doses

The majority of patients experiencing *H. pylori* infection eradication failure with PPI/amoxicillin/clarithromycin therapy at the usual doses have the RM genotype of CYP2C19, a clarithromycin-resistant strain, or both. In contrast, however, amoxicillin-resistant strains of *H. pylori* are quite rare (68), and treatment with an increased dose of a PPI and amoxicillin is expected to succeed when initial triple therapy with a PPI, amoxicillin, and clarithromycin at the usual doses fails. Bayerdorffer et al. (69) reported that eradication rates of approximately 90 percent could be achieved by dual therapy with omeprazole 40 mg three times daily plus amoxicillin 750 mg three times daily for two weeks, indicating the feasibility of eradication by dual therapy with a PPI and amoxicillin at high doses with frequent dosings. In our institution, we have selected dual therapy with high doses of PPI plus amoxicillin as a retreatment strategy, in which lansoprazole 30 mg or rabeprazole 10 mg plus 500 mg of amoxicillin are given four times daily for two weeks. Thanks to the above-noted acid inhibition (intragastric pH is kept at approximately 7 for 24 hours) provided by PPI given four times daily (see "Effect of CYP2C19 Polymorphism on Gastric Acid Inhibition with PPIs") and consequent maximization of the stability and bioavailability of amoxicillin obtained at

Table 16.5. Cure Rates of *H. pylori* Infection by Treatment with High Doses of a PPI Plus Amoxicillin (PP Analysis)

Regimen	*Eradication Rates % (PP)*	*Reference*
OPZ 40 mg 3 times daily + AMPC 750 mg 3 times daily for 2 weeks	91	(69)
LPZ 30 mg 4 times daily + AMPC 500 mg 4 times daily for 2 weeks	96.7	(61)
RPZ 10 mg 4 times daily + AMPC 500 mg 4 times daily for 2 weeks	100.0	(70, 71)
RPZ 10 mg 4 times daily + AMPC 500 mg 4 times daily for 2 weeks	93.8	(81)
OPZ 40 mg 4 times daily + AMPC 750 mg 4 times daily for 2 weeks	83.8	(72)
RPZ 10 mg 4 times daily + AMPC 500 mg 4 times daily for 2 weeks	93.8	(82)

PP = per protocol analysis; OPZ = omeprazole; LPZ = lansoprazole; RPZ = rabeprazole; AMPC = amoxicillin.
Modified from reference (23).

approximately pH 7.0, eradication rates have been sufficient, at higher than 90 percent (61, 70, 71). Miehlke et al. (72) also reported that dual rescue therapy with markedly high doses of omeprazole and amoxicillin (omeprazole 40 mg four times daily plus amoxicillin 750 mg four times daily) provided the same efficacy as a quadruple therapy. Reported eradication rates by treatment with high doses of a PPI plus amoxicillin are summarized in Table 16.5.

Tailored Strategy for *H. pylori* Infection Based on CYP2C19 Genotype and *H. pylori* 23S rRNA

The principles underlying the strategy for *H. pylori* eradication can be summarized as follows: first, select an antibacterial agent to which the *H. pylori* is sensitive; and second, optimize environmental conditions in the stomach to maximize the stability and bioavailability of the selected antibacterial agent by coadministration with a PPI at a dose consistent with the individual CYP2C19 genotype status.

We recently succeeded in developing a tailored strategy for *H. pylori* infection, as follows (73). Patients infected with clarithromycin-sensitive strains are treated with triple therapy with a PPI, amoxicillin, and clarithromycin for one week. Patients infected with clarithromycin-resistant strains are treated with dual therapy with a PPI and amoxicillin four times daily for two weeks. As a preliminary study, we first optimized dosing schedules of lansoprazole in relation to CYP2C19 genotypes in a 24-hour intragastric pH-monitoring study in healthy volunteers, which showed that lansoprazole provided sufficient intragastric pH for triple therapy at 30 mg three times daily for RMs, 15 mg three times daily for IMs, and 15 mg twice daily for PMs; and for dual therapy at 30 mg four times daily for RMs, 15 mg four times daily for IMs, and 15 mg twice daily for PMs (73). We then applied these dosing schemes for lansoprazole to the tailored regimen below.

Susceptibility of *H. pylori* strains to clarithromycin is determined by the measurement of point mutations of 23S rRNA at positions 2142 and 2143, as previously reported (67). Patients infected with clarithromycin-sensitive strains are treated with 200 mg of clarithromycin three times daily, 500 mg of amoxicillin three times daily, and personalized doses of lansoprazole (i.e., 30 mg three times daily in RMs, 15 mg three times daily in IMs, and 15 mg twice daily in PMs) for one week; and those with clarithromycin-resistant strains with 500 mg of amoxicillin four times daily and personalized dose of lansoprazole (i.e., 30 mg four times daily in RMs, 15 mg four times daily in IMs, and 15 mg twice daily in PMs) for two weeks. This tailored strategy achieved the satisfactory eradication rate of 96 percent (Figure 16.6).

PHARMACOGENOMIC RISK FACTORS FOR NSAID/ASPIRIN ULCER

NSAIDs, such as ibuprofen, are metabolized by CYP2C9. In subjects with the PM genotype of CYP2C9 (i.e., *3/*3), plasma ibuprofen levels are high, which results in the slower recovery of plasma prostaglandin levels decreased by ibuprofen than in subjects with wild-type alleles (8). Recently, CYP2C9 genotype status has been found to be associated with the risk of NSAID ulcer (74); because plasma NSAID levels are sustained for a longer time, subjects with the *CYP2C9*2* or *CYP2C9*3* allele (or both) are at higher risk of NSAID-induced gastroduodenal lesions (Figure 16.7).

The activity of UGT1A6, which is involved in the metabolism of aspirin, is polymorphic. Shiotani et al. (75) reported that patients with the slow metabolizer genotype of UGT1A6 are at higher risk of aspirin-induced peptic ulcer, whereas van Oijen et al. (76)

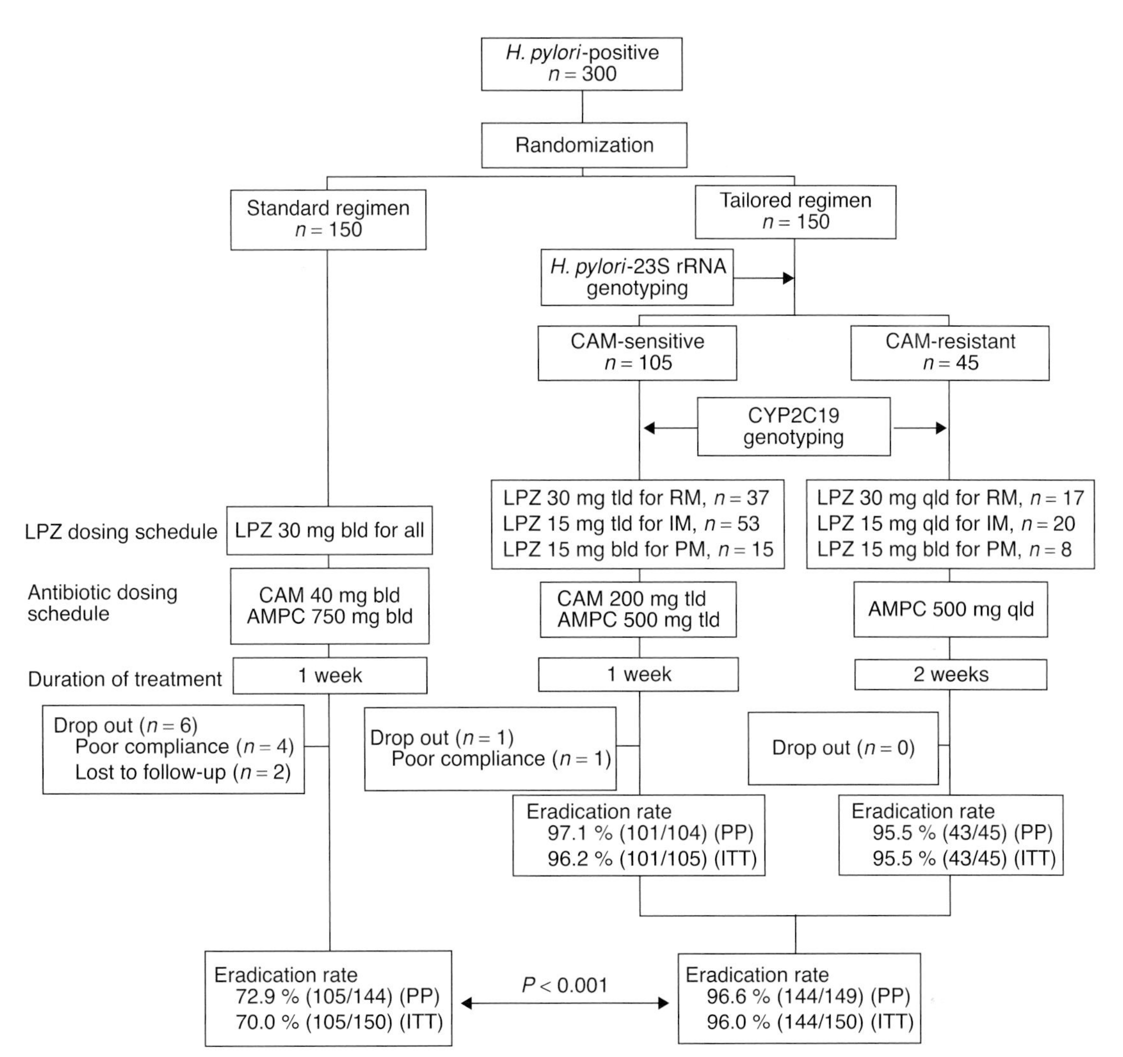

Figure 16.6. Schema of allocation of *H. pylori*-positive patients to standard and tailored regimen groups (73). AMPC = amoxicillin; LPZ = lansoprazole; RM = rapid metabolizer of CYP2C19; RM = rapid metabolizer of CYP2C19; IM = intermediate metabolizer of CYP2C19; PM = poor metabolizer of CYP2C19; PP = per-protocol; ITT = intention-to-treat.

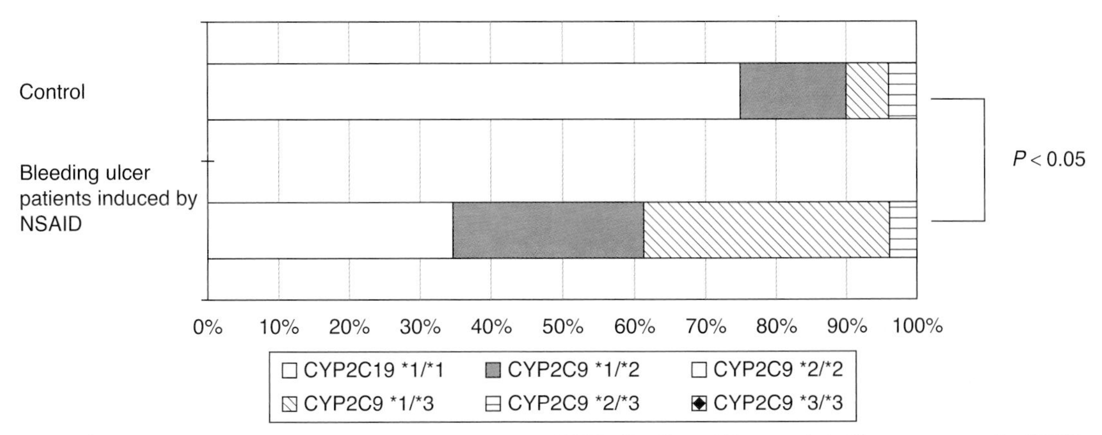

Figure 16.7. CYP2C9 genotype status and risk of bleeding ulcer by NSAID. Significant differences in incidence were seen by CYP2C9 genotype between controls and case subjects with NSAID-induced bleeding ulcer (74).

reported that there was no association between UGT1A6 polymorphisms and the prevalence of gastric complaints in cardiovascular patients treated with aspirin.

CONCLUSION

Because the clinical effects of PPIs depend on the CYP2C19 genotype status of the patient, adjustment of the dose and dosing schedule of these agents in accordance with the CYP2C19 genotype will optimize the efficacy of treatment with them. Adverse NSAID/aspirin-induced gastroduodenal complications are also related to the polymorphisms of NSAID/aspirin-metabolizing enzymes. Patients with a slow-metabolizer genotype for NSAIDs/aspirin should be concomitantly treated with a PPI to prevent NSAID/aspirin-induced mucosal damage. Pharmacogenomics-based strategies require genotyping tests in advance and therefore appear costly, but we anticipate that their costs will be offset, at least in part, by the higher eradication rates and lower incidence of adverse events thereby provided. Indications for pharmacogenomics will expand in the near future.

REFERENCES

1. Andersson T. Pharmacokinetics, metabolism and interactions of acid pump inhibitors. Focus on omeprazole, lansoprazole and pantoprazole. *Clin Pharmacokinet.* 1996; **31**(1):9–28.
2. Andersson T, Regardh CG, Dahl-Puustinen ML, & Bertilsson L. Slow omeprazole metabolizers are also poor S-mephenytoin hydroxylators. *Ther Drug Monit.* 1990;**12**: 415–16.
3. Andersson T, Regardh CG, Lou YC, Zhang Y, Dahl ML, & Bertilsson L. Polymorphic hydroxylation of S-mephenytoin and omeprazole metabolism in Caucasian and Chinese subjects. *Pharmacogenetics.* 1992;**2**:25–31.
4. Sohn DR, Kobayashi K, Chiba K, Lee KH, Shin SG, & Ishizaki T. Disposition kinetics and metabolism of omeprazole in extensive and poor metabolizers of S-mephenytoin 4′-hydroxylation recruited from an Oriental population. *J Pharmacol Exp Ther.* 1992;**262**:1195–202.
5. Pearce RE, Rodrigues AD, Goldstein JA, & Parkinson A. Identification of the human P450 enzymes involved in lansoprazole metabolism. *J Pharmacol Exp Ther.* 1996; **277**:805–16.
6. Yamazaki H, Inoue K, Shaw PM, Checovich WJ, Guengerich FP, & Shimada T. Different contributions of cytochrome P450 2C19 and 3A4 in the oxidation of omeprazole by human liver microsomes: effects of contents of these two forms in individual human samples. *J Pharmacol Exp Ther.* 1997;**283**:434–42.
7. Furuta T, Ohashi K, Kosuge K, Zhao XJ, Takashima M, Kimura M, et al. CYP2C19 genotype status and effect of omeprazole on intragastric pH in humans. *Clin Pharmacol Ther.* 1999;**65**:552–61.
8. Kirchheiner J, Meineke I, Freytag G, Meisel C, Roots I, & Brockmoller J. Enantiospecific effects of cytochrome P450 2C9 amino acid variants on ibuprofen pharmacokinetics and on the inhibition of cyclooxygenases 1 and 2. *Clin Pharmacol Ther.* 2002;**72**:62–75.
9. Furuta T, Shirai N, Sugimoto M, Ohashi K, & Ishizaki T. Pharmacogenomics of proton pump inhibitors. *Pharmacogenomics.* 2004;**5**:181–202.
10. Ishizaki T, Sohn DR, Kobayashi K, Chiba K, Lee KH, Shin SG, et al. Interethnic differences in omeprazole metabolism in the two S-mephenytoin hydroxylation phenotypes studied in Caucasians and Orientals. *Ther Drug Monit.* 1994;**16**: 214–15.
11. Xiao ZS, Goldstein JA, Xie HG, Blaisdell J, Wang W, Jiang CH, et al. Differences in the incidence of the CYP2C19 polymorphism affecting the S-mephenytoin phenotype in Chinese Han and Bai populations and identification of a new rare CYP2C19 mutant allele. *J Pharmacol Exp Ther.* 1997;**281**:604–9.
12. Xie HG, Kim RB, Stein CM, Wilkinson GR, & Wood AJ. Genetic polymorphism of (S)-mephenytoin 4′-hydroxylation in populations of African descent. *Br J Clin Pharmacol.* 1999;**48**:402–8.
13. Marinac JS, Balian JD, Foxworth JW, Willsie SK, Daus JC, Owen R, et al. Determination of CYP2C19 phenotype in black Americans with omeprazole: correlation with genotype. *Clin Pharmacol Ther.* 1996;**60**:138–44.
14. Masimirembwa C, Bertilsson L, Johansson I, Hasler JA, & Ingelman-Sundberg M. Phenotyping and genotyping of S-mephenytoin hydroxylase (cytochrome P450 2C19) in a Shona population of Zimbabwe. *Clin Pharmacol Ther.* 1995;**57**:656–61.
15. de Morais SM, Goldstein JA, Xie HG, Huang SL, Lu YQ, Xia H, et al. Genetic analysis of the S-mephenytoin polymorphism in a Chinese population. *Clin Pharmacol Ther.* 1995;**58**:404–11.
16. Kubota T, Chiba K, & Ishizaki T. Genotyping of S-mephenytoin 4′-hydroxylation in an extended Japanese population. *Clin Pharmacol Ther.* 1996;**60**:661–6.
17. Roh HK, Dahl ML, Tybring G, Yamada H, Cha YN, & Bertilsson L. CYP2C19 genotype and phenotype determined by omeprazole in a Korean population. *Pharmacogenetics.* 1996;**6**:547–51.
18. Xie HG, Stein CM, Kim RB, Wilkinson GR, Flockhart DA, & Wood AJ. Allelic, genotypic and phenotypic distributions of S-mephenytoin 4′-hydroxylase (CYP2C19) in healthy Caucasian populations of European descent throughout the world. *Pharmacogenetics.* 1999;**9**:539–49.
19. De Morais SM, Wilkinson GR, Blaisdell J, Meyer UA, Nakamura K, & Goldstein JA. Identification of a new genetic defect responsible for the polymorphism of (S)-mephenytoin metabolism in Japanese. *Mol Pharmacol.* 1994;**46**:594–8.
20. de Morais SM, Wilkinson GR, Blaisdell J, Nakamura K, Meyer UA, & Goldstein JA. The major genetic defect responsible for the polymorphism of S-mephenytoin metabolism in humans. *J Biol Chem.* 1994;**269**:15419–22.
21. Sim SC, Risinger C, Dahl ML, Aklillu E, Christensen M, Bertilsson L, et al. A common novel CYP2C19 gene variant causes ultrarapid drug metabolism relevant for the drug

response to proton pump inhibitors and antidepressants. *Clin Pharmacol Ther.* 2006;**79**:103–13.

22. Kurzawski M, Gawronska-Szklarz B, Wrzesniewska J, Siuda A, Starzynska T, & Drozdzik M. Effect of CYP2C19*17 gene variant on *Helicobacter pylori* eradication in peptic ulcer patients. *Eur J Clin Pharmacol.* 2006;**62**:877–80.
23. Furuta T, Sugimoto M, Shirai N, & Ishizaki T. CYP2C19 pharmacogenomics associated with therapy of *Helicobacter pylori* infection and gastro-esophageal reflux diseases with a proton pump inhibitor. *Pharmacogenomics.* 2007;**8**:1199–210.
24. Saitoh T, Fukushima Y, Otsuka H, Hirakawa J, Mori H, Asano T, et al. Effects of rabeprazole, lansoprazole and omeprazole on intragastric pH in CYP2C19 extensive metabolizers. *Aliment Pharmacol Ther.* 2002;**16**:1811–17.
25. Shirai N, Furuta T, Moriyama Y, Okochi H, Kobayashi K, Takashima M, et al. Effects of CYP2C19 genotypic differences in the metabolism of omeprazole and rabeprazole on intragastric pH. *Aliment Pharmacol Ther.* 2001;**15**:1929–37.
26. Shirai N, Furuta T, Xiao F, Kajimura M, Hanai H, Ohashi K, et al. Comparison of lansoprazole and famotidine for gastric acid inhibition during the daytime and night-time in different CYP2C19 genotype groups. *Aliment Pharmacol Ther.* 2002;**16**:837–46.
27. Sugimoto M, Furuta T, Shirai N, Kajimura M, Hishida A, Sakurai M, et al. Different dosage regimens of rabeprazole for nocturnal gastric acid inhibition in relation to cytochrome P450 2C19 genotype status. *Clin Pharmacol Ther.* 2004;**76**:290–301.
28. Kale-Pradhan PB, Landry HK, & Sypula WT. Esomeprazole for acid peptic disorders. *Ann Pharmacother.* 2002;**36**:655–63.
29. Kirsch C, Kuhlisch E, Lindner F, Grossman D, Morger A, Madisch A, et al. Influence of cytochrome P450 2c19 polymorphism on intragastric acidity during treatment with high-dose esomeprazole and pantoprazole. *Gastroenterology.* 2007;**132**(suppl 2):A-607.
30. Nebel OT, Fornes MF, & Castell DO. Symptomatic gastroesophageal reflux: incidence and precipitating factors. *Am J Dig Dis.* 1976;**21**:953–6.
31. Bardhan KD, Hawkey CJ, Long RG, Morgan AG, Wormsley KG, Moules IK, et al. Lansoprazole versus ranitidine for the treatment of reflux oesophagitis. UK Lansoprazole Clinical Research Group. *Aliment Pharmacol Ther.* 1995;**9**:145–51.
32. Bardhan KD. The role of proton pump inhibitors in the treatment of gastro-oesophageal reflux disease. *Aliment Pharmacol Ther.* 1995;**9**(suppl 1):15–25.
33. Kirchgatterer A, Aschl G, Hinterreiter M, Stadler B, & Knoflach P. [Current concepts in therapy of reflux disease]. *Wien Med Wochenschr.* 2001;**151**:266–9.
34. Furuta T, Shirai N, Watanabe F, Honda S, Takeuchi K, Iida T, et al. Effect of cytochrome P4502C19 genotypic differences on cure rates for gastroesophageal reflux disease by lansoprazole. *Clin Pharmacol Ther.* 2002;**72**:453–60.
35. Kawamura M, Ohara S, Koike T, Iijima K, Suzuki J, Kayaba S, et al. The effects of lansoprazole on erosive reflux oesophagitis are influenced by CYP2C19 polymorphism. *Aliment Pharmacol Ther.* 2003;**17**:965–73.
36. Kawamura M, Ohara S, Koike T, Iijima K, Suzuki H, Kayaba S, et al. Cytochrome P450 2C19 polymorphism influences the preventive effect of lansoprazole on the recurrence of erosive reflux esophagitis. *J Gastroenterol Hepatol.* 2007;**22**:222–6.
37. Peghini PL, Katz PO, Bracy NA, & Castell DO. Nocturnal recovery of gastric acid secretion with twice-daily dosing of proton pump inhibitors. *Am J Gastroenterol.* 1998;**93**:763–7.
38. Adachi K, Fujishiro H, Katsube T, Yuki M, Ono M, Kawamura A, et al. Predominant nocturnal acid reflux in patients with Los Angeles grade C and D reflux esophagitis. *J Gastroenterol Hepatol.* 2001;**16**:1191–6.
39. Klinkenberg-Knol EC & Meuwissen SG. Combined gastric and oesophageal 24-hour pH monitoring and oesophageal manometry in patients with reflux disease, resistant to treatment with omeprazole. *Aliment Pharmacol Ther.* 1990;**4**:485–95.
40. Ours TM, Fackler WK, Richter JE, & Vaezi MF. Nocturnal acid breakthrough: clinical significance and correlation with esophageal acid exposure. *Am J Gastroenterol.* 2003;**98**:545–50.
41. Furuta T, Shirai N, Xiao F, Ohashi K, & Ishizaki T. Effect of high-dose lansoprazole on intragastic pH in subjects who are homozygous extensive metabolizers of cytochrome P4502C19. *Clin Pharmacol Ther.* 2001;**70**:484–92.
42. Xue S, Katz PO, Banerjee P, Tutuian R, & Castell DO. Bedtime H2 blockers improve nocturnal gastric acid control in GERD patients on proton pump inhibitors. *Aliment Pharmacol Ther.* 2001;**15**:1351–6.
43. Peghini PL, Katz PO, & Castell DO. Ranitidine controls nocturnal gastric acid breakthrough on omeprazole: a controlled study in normal subjects. *Gastroenterology.* 1998;**115**:1335–9.
44. Kinoshita Y, Adachi K, & Fujishiro H. Therapeutic approaches to reflux disease, focusing on acid secretion. *J Gastroenterol.* 2003;**38**(suppl 15):13–19.
45. Ando T, Ishikawa T, Kokura S, Naito Y, Yoshida N, & Yoshikawa T. Endoscopic analysis of gastric ulcer after one week's treatment with omeprazole and rabeprazole in relation to CYP2C19 genotype. *Dig Dis Sci.* 2008;**53**:933–7.
46. Blaser MJ. Hypotheses on the pathogenesis and natural history of *Helicobacter pylori*-induced inflammation. *Gastroenterology.* 1992;**102**:720–7.
47. Parsonnet J, Blaser MJ, Perez-Perez GI, Hargrett-Bean N, & Tauxe RV. Symptoms and risk factors of *Helicobacter pylori* infection in a cohort of epidemiologists. *Gastroenterology.* 1992;**102**:41–6.
48. Wotherspoon AC. *Helicobacter pylori* infection and gastric lymphoma. *Br Med Bull.* 1998;**54**:79–85.
49. Uemura N, Okamoto S, Yamamoto S, Matsumura N, Yamaguchi S, Yamakido M, et al. *Helicobacter pylori* infection and the development of gastric cancer. *N Engl J Med.* 2001;**345**:784–9.
50. Malfertheiner P, Megraud F, O'Morain C, Bazzoli F, El-Omar E, Graham D, et al. Current concepts in the management of *Helicobacter pylori* infection: the Maastricht III Consensus Report. *Gut.* 2007;**56**:772–81.
51. Grayson ML, Eliopoulos GM, Ferraro MJ, & Moellering RC Jr. Effect of varying pH on the susceptibility of Campylobacter pylori to antimicrobial agents. *Eur J Clin Microbiol Infect Dis.* 1989;**8**:888–9.
52. Scott D, Weeks D, Melchers K, & Sachs G. The life and death of *Helicobacter pylori*. *Gut.* 1998;**43**(suppl 1):S56–60.

53. Scott DR, Weeks D, Hong C, Postius S, Melchers K, & Sachs G. The role of internal urease in acid resistance of *Helicobacter pylori. Gastroenterology.* 1998;**114**:58–70.
54. Goddard AF, Jessa MJ, Barrett DA, Shaw PN, Idstrom JP, Cederberg C, et al. Effect of omeprazole on the distribution of metronidazole, amoxicillin, and clarithromycin in human gastric juice. *Gastroenterology.* 1996;**111**:358–67.
55. Midolo PD, Turnidge JD, Lambert JR, & Bell JM. Oxygen concentration influences proton pump inhibitor activity against *Helicobacter pylori* in vitro. *Antimicrob Agents Chemother.* 1996;**40**:1531–3.
56. Furuta T, Ohashi K, Kamata T, Takashima M, Kosuge K, Kawasaki T, et al. Effect of genetic differences in omeprazole metabolism on cure rates for *Helicobacter pylori* infection and peptic ulcer. *Ann Intern Med.* 1998;**129**:1027–30.
57. Asaka M, Sugiyama T, Kato M, Satoh K, Kuwayama H, Fukuda Y, et al. A multicenter, double-blind study on triple therapy with lansoprazole, amoxicillin and clarithromycin for eradication of *Helicobacter pylori* in Japanese peptic ulcer patients. *Helicobacter.* 2001;**6**:254–61.
58. Andersson T, Miners JO, Veronese ME, & Birkett DJ. Identification of human liver cytochrome P450 isoforms mediating secondary omeprazole metabolism. *Br J Clin Pharmacol.* 1994;**37**:597–604.
59. Rodrigues AD, Roberts EM, Mulford DJ, Yao Y, & Ouellet D. Oxidative metabolism of clarithromycin in the presence of human liver microsomes. Major role for the cytochrome P4503A (CYP3A) subfamily. *Drug Metab Dispos.* 1997;**25**:623–30.
60. Furuta T, Ohashi K, Kobayashi K, Iida I, Yoshida H, Shirai N, et al. Effects of clarithromycin on the metabolism of omeprazole in relation to CYP2C19 genotype status in humans. *Clin Pharmacol Ther.* 1999;**66**:265–74.
61. Furuta T, Shirai N, Takashima M, Xiao F, Hanai H, Sugimura H, et al. Effect of genotypic differences in CYP2C19 on cure rates for *Helicobacter pylori* infection by triple therapy with a proton pump inhibitor, amoxicillin, and clarithromycin. *Clin Pharmacol Ther.* 2001;**69**:158–68.
62. Dojo M, Azuma T, Saito T, Ohtani M, Muramatsu A, & Kuriyama M. Effects of CYP2C19 gene polymorphism on cure rates for *Helicobacter pylori* infection by triple therapy with proton pump inhibitor (omeprazole or rabeprazole), amoxicillin and clarithromycin in Japan. *Dig Liver Dis.* 2001;**33**:671–5.
63. Inaba T, Mizuno M, Kawai K, Yokota K, Oguma K, Miyoshi M, et al. Randomized open trial for comparison of proton pump inhibitors in triple therapy for *Helicobacter pylori* infection in relation to CYP2C19 genotype. *J Gastroenterol Hepatol.* 2002;**17**:748–53.
64. Versalovic J, Osato MS, Spakovsky K, Dore MP, Reddy R, Stone GG, et al. Point mutations in the 23S rRNA gene of *Helicobacter pylori* associated with different levels of clarithromycin resistance. *J Antimicrob Chemother.* 1997;**40**:283–6.
65. Stone GG, Shortridge D, Versalovic J, Beyer J, Flamm RK, Graham DY, et al. A PCR-oligonucleotide ligation assay to determine the prevalence of 23S rRNA gene mutations in clarithromycin-resistant *Helicobacter pylori. Antimicrob Agents Chemother.* 1997;**41**:712–14.
66. Menard A, Santos A, Megraud F, & Oleastro M. PCR-restriction fragment length polymorphism can also detect point mutation A2142C in the 23S rRNA gene, associated with *Helicobacter pylori* resistance to clarithromycin. *Antimicrob Agents Chemother.* 2002;**46**:1156–7.
67. Furuta T, Sagehashi Y, Shirai N, Sugimoto M, Nakamura A, Kodaira M, et al. Influence of CYP2C19 polymorphism and *Helicobacter pylori* genotype determined from gastric tissue samples on response to triple therapy for *H. pylori* infection. *Clin Gastroenterol Hepatol.* 2005;**3**:564–73.
68. Adamek RJ, Suerbaum S, Pfaffenbach B, & Opferkuch W. Primary and acquired *Helicobacter pylori* resistance to clarithromycin, metronidazole, and amoxicillin – influence on treatment outcome. *Am J Gastroenterol.* 1998;**93**:386–9.
69. Bayerdorffer E, Miehlke S, Mannes GA, Sommer A, Hochter W, Weingart J, et al. Double-blind trial of omeprazole and amoxicillin to cure *Helicobacter pylori* infection in patients with duodenal ulcers. *Gastroenterology.* 1995;**108**:1412–17.
70. Furuta T, Shirai N, Xiao F, Takashima M, Sugimoto M, Kajimura M, et al. High-dose rabeprazole/amoxicillin therapy as the second-line regimen after failure to eradicate *H. pylori* by triple therapy with the usual doses of a proton pump inhibitor, clarithromycin and amoxicillin. *Hepatogastroenterology.* 2003;**50**:2274–8.
71. Furuta T, Shirai N, Takashima M, Xiao F, Hanai H, Nakagawa K, et al. Effects of genotypic differences in CYP2C19 status on cure rates for *Helicobacter pylori* infection by dual therapy with rabeprazole plus amoxicillin. *Pharmacogenetics.* 2001;**11**:341–8.
72. Miehlke S, Kirsch C, Schneider-Brachert W, Haferland C, Neumeyer M, Bastlein E, et al. A prospective, randomized study of quadruple therapy and high-dose dual therapy for treatment of *Helicobacter pylori* resistant to both metronidazole and clarithromycin. *Helicobacter.* 2003;**8**:310–19.
73. Furuta T, Shirai N, Kodaira M, Sugimoto M, Nogaki A, Kuriyama S, et al. Pharmacogenomics-based tailored versus standard therapeutic regimen for eradication of *H. pylori. Clin Pharmacol Ther.* 2007;**81**:521–8.
74. Pilotto A, Seripa D, Franceschi M, Scarcelli C, Colaizzo D, Grandone E, et al. Genetic susceptibility to nonsteroidal anti-inflammatory drug-related gastroduodenal bleeding: role of cytochrome P450 2C9 polymorphisms. *Gastroenterology.* 2007;**133**:465–71.
75. Shiotani A, Sakakibara T, Takami M, Yamanaka Y, Imamura H, Tarumi K, et al. Upper gastrointestinal ulcer in Japanese patients taking low dose aspirin. *Gastroenterology.* 2008;**134**(suppl 1):A736.
76. van Oijen MG, Huybers S, Peters WH, Drenth JP, Laheij RJ, Verheugt FW, et al. Polymorphisms in genes encoding acetylsalicylic acid metabolizing enzymes are unrelated to upper gastrointestinal health in cardiovascular patients on acetylsalicylic acid. *Br J Clin Pharmacol.* 2005;**60**:623–8.
77. Tanigawara Y, Aoyama N, Kita T, Shirakawa K, Komada F, Kasuga M, et al. CYP2C19 genotype-related efficacy of omeprazole for the treatment of infection caused by *Helicobacter pylori. Clin Pharmacol Ther.* 1999;**66**:528–34.
78. Sapone A, Vaira D, Trespidi S, Perna F, Gatta L, Tampieri A, et al. The clinical role of cytochrome p450 genotypes in *Helicobacter pylori* management. *Am J Gastroenterol.* 2003;**98**:1010–15.

79. Okudaira K, Furuta T, Shirai N, Sugimoto M, & Miura S. Concomitant dosing of famotidine with a triple therapy increases the cure rates of *Helicobacter pylori* infections in patients with the homozygous extensive metabolizer genotype of CYP2C19. *Aliment Pharmacol Ther.* 2005;**21**:491–7.
80. Kawai T, Kawakami K, Mikinori K, Takei K, Itoi T, Moriyasu F, et al. Efficacy of low-dose proton pump inhibitor (PPI) in the eradication of *Helicobacter pylori* following combination PPI/AC therapy in Japan. *Hepatogastroenterology.* 2007;**54**:649–54.
81. Furuta T, Baba S, Takashima M, Shirai N, Xiao F, Futami H, et al. H+/K+-adenosine triphosphatase mRNA in gastric fundic gland mucosa in patients infected with *Helicobacter pylori. Scand J Gastroenterol.* 1999;**34**:384–90.
82. Shirai N, Sugimoto M, Kodaira C, Nishino M, Ikuma M, Kajimura M, et al. Dual therapy with high doses of rabeprazole and amoxicillin versus triple therapy with rabeprazole, amoxicillin, and metronidazole as a rescue regimen for *Helicobacter pylori* infection after the standard triple therapy. *Eur J Clin Pharmacol.* 2007;**63**:743–9.

17 Pharmacogenetics of Rheumatology: Focus on Rheumatoid Arthritis

Robert M. Plenge, Yvonne C. Lee, Soumya Raychaudhuri, and Daniel H. Solomon

Rheumatoid arthritis (RA) is the most common inflammatory rheumatic disease, with a population prevalence of 2 percent to 3 percent and an annual incidence rate of 0.5 percent to 1.0 percent. Like many autoimmune conditions, it affects women more commonly than men (4:1). The prevalence of RA increases from the third decade through the eighth. Because it is the most common inflammatory rheumatic disease, most pharmacogenetic studies in adult rheumatology have concentrated on treatments for RA, and this will be the focus of the chapter.

Clinically, RA is a painful condition associated with significant disability. It primarily affects the joints, resulting in pain, swelling, and loss of musculoskeletal function. Because of its chronic nature, small changes accumulate and many patients experience substantial disability over the decades that they live with the condition. Although classical extraarticular manifestations (e.g., vasculitis, Felty's syndrome, spine disease) have diminished in frequency, there is an increasing awareness that the inflammation of RA places patients at increased risk for common conditions, such as cardiovascular disease, osteoporosis, and insulin resistance. These chronic comorbidities also contribute to the disability associated with RA.

The costs of RA are high, including both direct medical costs and the indirect costs of work loss and disability. The annual direct medical costs of RA were recently estimated at $9,600. More than half of these costs were attributed to drug costs. This figure is likely to increase because several of the new biologic drugs cost more than $20,000 per year (see Table 17.1). Compared with the direct costs, the indirect costs of RA are even higher. Forty percent of patients with RA will reduce or stop their work outside of the home within 10 years of developing the disease.

TREATMENT PARADIGMS

In considering how pharmacogenetic information may influence clinical decisions, it is first useful to understand the basic treatment paradigm in RA. Key decision points are whether to administer a disease-modifying antirheumatic drug (DMARD) or non-DMARD therapy, and how to advance to more aggressive DMARDs in patients for whom initial therapy fails.

Non-DMARD therapies, principally glucocorticoids and nonsteroidal anti-inflammatory drugs (NSAIDs), are often used as adjuncts to DMARDs. The NSAIDs, selective and nonselective, have proven efficacy in reducing the pain of RA and thus are used for their analgesic effects for residual pain. Glucocorticoids have a more controversial role in treating RA. They are used in one-quarter to one-half of all patients with RA, typically at very low dosages (i.e., 2 to 10 mg per day). However, there are data suggesting that large dosages early in the treatment regimen may modify the disease course and improve outcomes (1).

Recent developments in DMARDs have propelled the treatment of RA forward. Up until the late 1990s, only a handful of treatments existed for RA that demonstrated substantial benefit without commonly causing adverse events. In the past decade, at least six new DMARDs have become available in clinical practice, and now 60 percent to 70 percent of patients can achieve low disease activity or remission. Many of these treatments are biologically derived DMARDs, such as monoclonal antibodies targeting specific effector functions of the immune system involved in the pathophysiology of RA. These biologic DMARDs provide many more treatment options for patients with refractory disease. However, they may be associated with substantial toxicity for some patients (malignancy and infection), and they are

Table 17.1. Selected Antirheumatic Drugs

Agent (Trade Name)	*Presumed Mechanism of Action*	*Selected Side Effects*	*Annual Cost*
Abatacept (Orencia)	Selective costimulatory modulator through interactions with CTLA-4	Infection	\$19,440[a]
Adalimumab (Humira)	TNF-α blockade	Infection, lymphoma	\$17,945
Azathioprine (Imuran)	Disruption of purine metabolism	Leukopenia	\$565
Etanercept (Enbrel)	TNF-α blockade	Infection, lymphoma	\$16,842
Infliximab (Remicade)	TNF-α blockade	Infection, lymphoma	\$17,580[b]
Methotrexate	Disruption of purine metabolism	Infection, lymphoma, hepatitis, interstitial pneumonitis	\$246
Rituximab (Rituxan)	B-cell depletion through binding CD20	Infection	\$12,180

TNF, tumor necrosis factor; TPMT, thiopurine methyltransferase.
Note: Annual costs are based on drugstore.com and Medicare list prices. Methotrexate was originally approved for use in cancer.
[a]Infliximab costs are based on 420 mg every 8 weeks.
[b]Abatacept costs are based on 750 mg every 4 weeks.

extremely expensive at costs of \$15,000 to \$30,000 per year (see Table 17.1).

Current treatment paradigms rely on inadequate responses with standard therapies before consideration of biologic DMARDs (see Figure 17.1). These paradigms are based on a few good studies. Only a small number of treatment trials compare realistic treatment options, and even fewer compare treatment strategies, such as combinations of therapy and/or sequences of treatment. Rather, most trials enroll patients in whom a standard therapy has failed, such as methotrexate (MTX), and randomly assign them to placebo versus a new therapy. In the trials that have examined treatment strategies, response to a given medication guides treatment decisions.

Importantly, these strategy trials have demonstrated the short-term and long-term benefits of early treatment combined with prompt therapeutic changes when therapy is failing. Prompt step-ups in treatment to gain control of inflammation are important in minimizing joint damage. Thus, the ability to predict which therapeutic strategy will control a patient's inflammation – for example, genetic predictors of treatment response – would have great value in reducing the long-term damage associated with RA while at the same time controlling costs of unnecessary treatment.

Patient factors are potential candidates for treatment decision making. These include variables that are prognostic markers for RA severity: serologic status (i.e., rheumatoid factor [RF], anticitrullinated cyclic peptide antibody), early development of focal joint erosions, and early functional loss. Although these variables may help clinicians to identify patients likely to develop severe progressive RA, they do not specifically identify patients who are more or less likely to respond to therapy.

Candidate Genes in Pharmacogenetic Studies

In theory, genetic markers may define patient subsets that respond differently to different DMARDs (both toxicity and efficacy). In broad terms, these genes can be placed into three categories: genes that influence RA susceptibility, genes that influence drug metabolism, and genes that influence the biological pathway of the drug. The majority of pharmacogenetic studies have focused on candidate genes in these categories. As will be discussed at the end of the chapter, advances in human genetics have led to the ability to screen the human genome in an unbiased manner, alleviating the need to pick specific genes as part of a pharmacogenetic study. Nonetheless, this framework is useful to understand pharmacogenetic studies to date.

Genetic markers of RA susceptibility may define patient subsets that predict response to DMARDs. The strongest known genetic risk factor for RA is a collection of alleles at the HLA-DRB1 gene known as "shared epitope" (SE) alleles. These alleles encode QKRAA, QRRAA, or RRRAA amino acid sequences in positions 70 to 74 of the DR-β chain (see Table 17.2). Recently, several other genetic risk factors have been identified and validated, including a missense allele at the *PTPN22* gene (2) and noncoding alleles near the *STAT4* gene (3), *TNFAIP3* gene (also known as *A20*) (4, 5), and the *TRAF1-C5* genes (6, 7). Intriguingly, several of these genes reside in the tumor necrosis factor (TNF)-signaling pathway (*TNF-alpha induced protein 3* [*TNFAIP3*] and *TNF receptor-associated factor 1* [*TRAF1*]). Other genetic risk factors that have not yet been unequivocally validated include alleles near *PADI4* (8, 9), *CTLA4* (9), and *IRF5* (10).

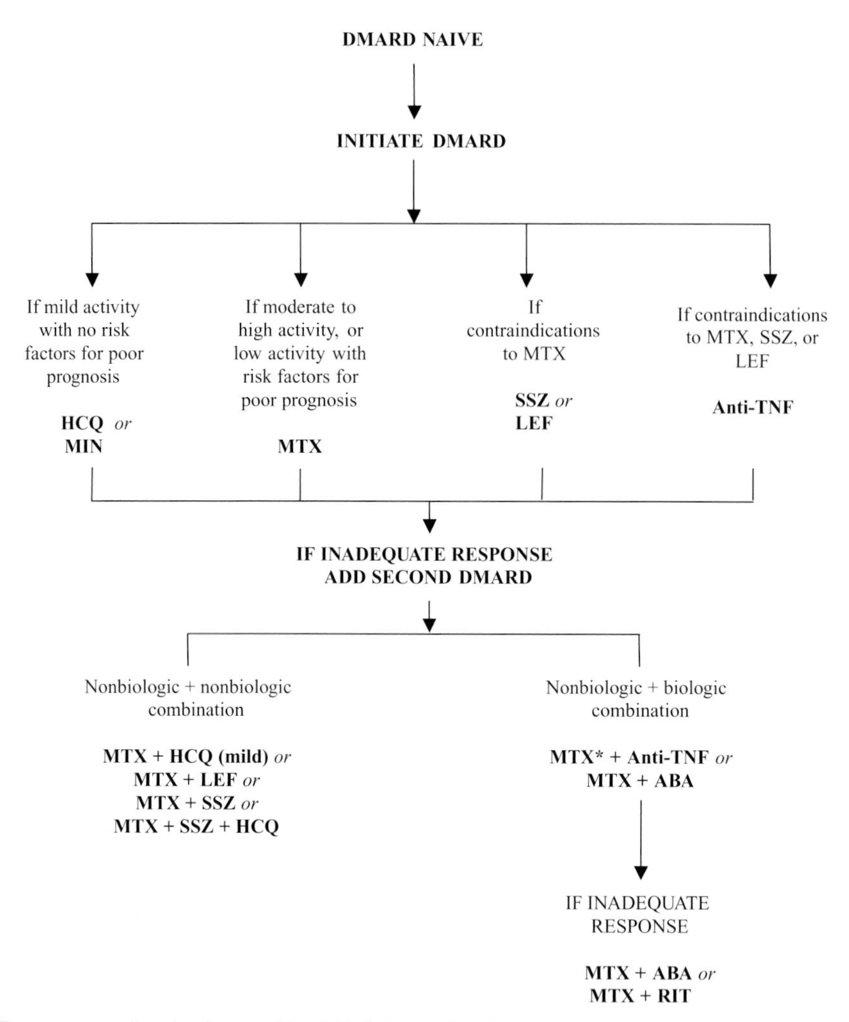

Figure 17.1. Treatment paradigm for rheumatoid arthritis (adapted from Bathon et al., *Hochberg's Textbook of Rheumatology.* M. C. Hochberg, A. J. Silman, J. S. Smolen, M. E. Weinblatt, M. H. Weisman, Rheumatology Edition, 4th edn, Mosby-Elsevier, 2008. Hardcover, 2056 p, text with continually updated online reference, 2-volume set, ISBN 9780323044295).

The genetics of response to DMARD therapy may differ from RA susceptibility variants. It is possible that genes in specific biological pathways or genes involved in drug metabolism may influence treatment response. For each of the drugs we describe here, we provide a basic overview of the relevant biological pathways.

Challenges to Pharmacogenetics in Rheumatic Diseases

The promise of pharmacogenetics in rheumatic disease is real; however, there are a number of difficult methodologic issues. First, defining pharmacologic response is not straightforward. For a disease like RA, there are a number of different response measures (e.g., American College of Rheumatology RA Core Data Set, the Disease Activity Score (DAS), and several simpler disease activity indices), and they can be measured at different time points. Thus, debate persists about when and how to measure response and its required duration. Second, pharmacologic response can be assessed in patients starting a given therapy (new users) or those on therapy already (persistent users). Because patients stop treatments when they are not working, measuring response in persistent users is more likely to demonstrate benefits than measuring response among new users. Third, the statistical power to detect a significant pharmacogenetic interaction is usually limited and depends on a strong association between a treatment response and the genetic allele. Thus, unless the pharmacogenetic associations are very large, substantial samples of patients beginning therapy with careful follow-up are required.

Table 17.2. HLA-DRB1 "Shared Epitope" Susceptibility Alleles

		Amino Acid Location				
		70	*71*	*72*	*73*	*74*
DRB1 Alleles	*Low-Resolution*	*Q*	*R*	*R*	*A*	*A*
DRB1*0101	DR1	Q	R	R	A	A
DRB1*0102	DR1	–	–	–	–	–
DRB1*0103	DR1	D	E	–	–	–
DRB1*03	DR3	–	K	–	G	R
DRB1*0401	DR4	–	K	–	–	–
DRB1*0402	DR4	D	E	–	–	–
DRB1*0403	DR4	–	–	–	–	E
DRB1*0404	DR4	–	–	–	–	–
DRB1*0405	DR4	–	–	–	–	–
DRB1*0407	DR4	–	–	–	–	E
DRB1*0408	DR4	–	–	–	–	–
DRB1*0411	DR4	–	–	–	–	E
DRB1*07	DR7	D	–	–	G	Q
DRB1*08	DR8	D	–	–	–	L
DRB1*0901	DR9	R	–	–	–	E
DRB1*1001	DR10	R	–	–	–	–
DRB1*1101	DR11	D	–	–	–	–
DRB1*1102	DR11	D	E	–	–	–
DRB1*1103	DR11	D	E	–	–	–
DRB1*1104	DR11	D	–	–	–	–
DRB1*12	DR12	D	–	–	–	–
DRB1*1301	DR13	D	E	–	–	–
DRB1*1302	DR13	D	E	–	–	–
DRB1*1303	DR13	D	K	–	–	–
DRB1*1323	DR13	D	E	–	–	–
DRB1*1401	DR14	R	–	–	–	E
DRB1*1402	DR14	–	–	–	–	–
DRB1*1404	DR14	R	–	–	–	E
DRB1*15	DR2	–	A	–	–	–
DRB1*16	DR16	D	–	–	–	–

The issue of statistical power is displayed in Figure 17.2. Each curve represents the required sample size of cases to achieve 80 percent power for a *P* value $<0.5 \times 10^{-7}$ (a standard cutoff in genetic studies for statistical significance) for a range of odds ratios (ORs) (*y*-axis) and specific allele frequencies (*x*-axis). For example, with 100 cases (orange line) and an allele frequency of 10 percent, an OR of 4.9 would be required to achieve 80 percent power for a *P* value $<0.5 \times 10^{-7}$. These curves suggest that very large sample sizes (approximately 2,000 cases) are required for ORs that we might anticipate in the pharmacogenetics of RA (2.0). However, it is important to emphasize that we do not yet understand the genetic architecture underlying treatment response. It is possible that certain genetic loci predict response with a high OR and that these alleles are common in the general population. It is also possible that the genetic architecture will mirror what is now being observed for non-major histocompatibility complex (MHC) associations in RA susceptibility, where common alleles have modest effect sizes (OR < 1.50).

SPECIFIC PHARMACOGENETIC ASSOCIATIONS

Different methods can be used to identify drug-gene interactions in pharmacogenetic studies. To date, most of these interactions have been explored by using a candidate gene approach. Candidate genes typically code for proteins known to be involved in a given drug's metabolism or a given disease's pathophysiologic pathway. Future genome-wide association studies will likely lead to new candidate genes for pharmacogenetic study.

Methotrexate

Methotrexate is a highly effective DMARD. It enters the cell via reduced folate carrier-1 (RFC-1) and inhibits folate-dependent enzymes such as dihydrofolate reductase (DHFR), thymidylate synthase (TSER), and 5-aminoamidazole-4-carboxamide ribonucleotide transformylase (ATIC) (11). These enzymes have interrelated roles in pathways involving folate metabolism, homocysteine-methionine production, and anti-inflammatory adenosine production (see Figure 17.3). The exact mechanism by which MTX exerts its anti-inflammatory effects is not clear.

More than 65 percent of RA patients on MTX have at least 50 percent improvement in pain and joint swelling (12). As with all medications, however, MTX can have toxic side effects, and adverse effects are the most common cause of MTX discontinuation (13). Severe hepatotoxicity requiring discontinuation of MTX can occur in up to 26 percent of patients. Folate supplementation lowers the risk of hepatotoxicity to approximately 4 percent (14).

Response to MTX is affected by both clinical and genetic factors. Hoekstra et al. (14) examined clinical predictors of MTX response in a randomized clinical trial of 411 RA patients treated with MTX ± placebo. Low baseline disease activity, male sex, low creatinine clearance, and NSAID use were associated with MTX efficacy. High body mass index and lack of folate supplementation were associated with MTX withdrawal due to hepatotoxicity. MTX-related diarrhea was associated with prior gastrointestinal (GI) events and young age (14).

Studies of genetic predictors of treatment response have varied in terms of design, study population, and treatment outcomes, but all studies have used a candidate gene approach. In this review, we highlight four large

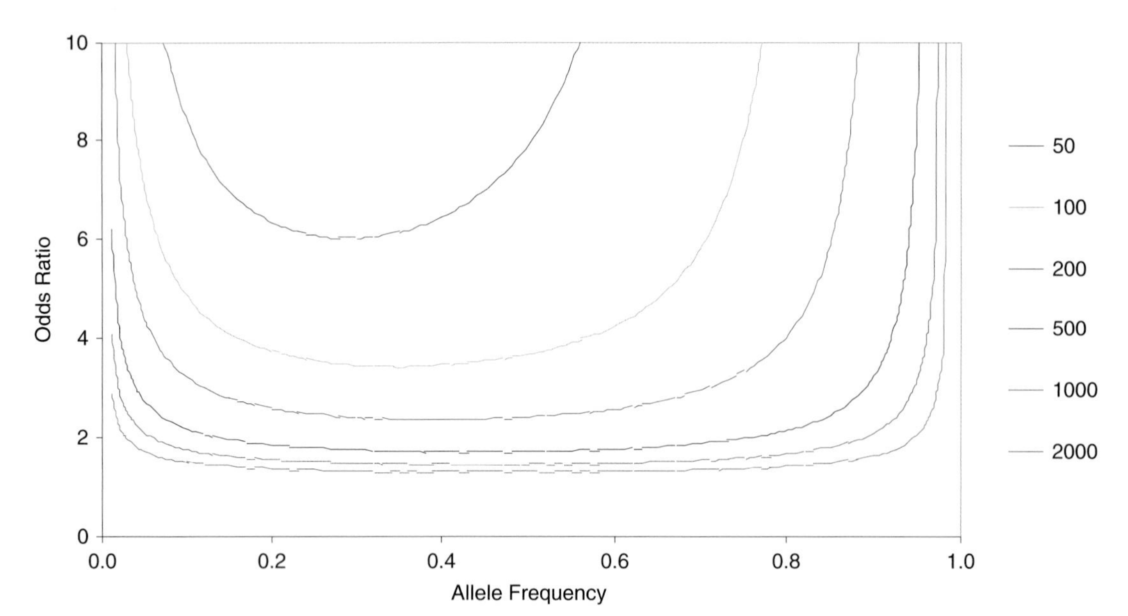

Figure 17.2. Sample size requirements based on a range of allele frequencies and odds ratios. Each curve represents the required sample size of cases to achieve 80 percent power for a *P* value $<0.5 \times 10^{-7}$ for a range of odds ratios (*y*-axis) and specific allele frequencies (*x*-axis). For example, with 100 cases (orange line) and an allele frequency of 10 percent, an odds ratio of 4.9 would be required to achieve 80 percent power for a *P* value $<0.5 \times 10^{-7}$.

($n > 200$) studies that provide insight into the genetics of MTX treatment response (see Table 17.3). We focus on studies involving *MTHFR*, the gene encoding methylene tetrahydrofolate reductase, an enzyme involved in the regeneration of methionine from homocysteine (see Figure 17.3). We conclude this section with a discussion of how genetic information can be incorporated into clinical prediction models.

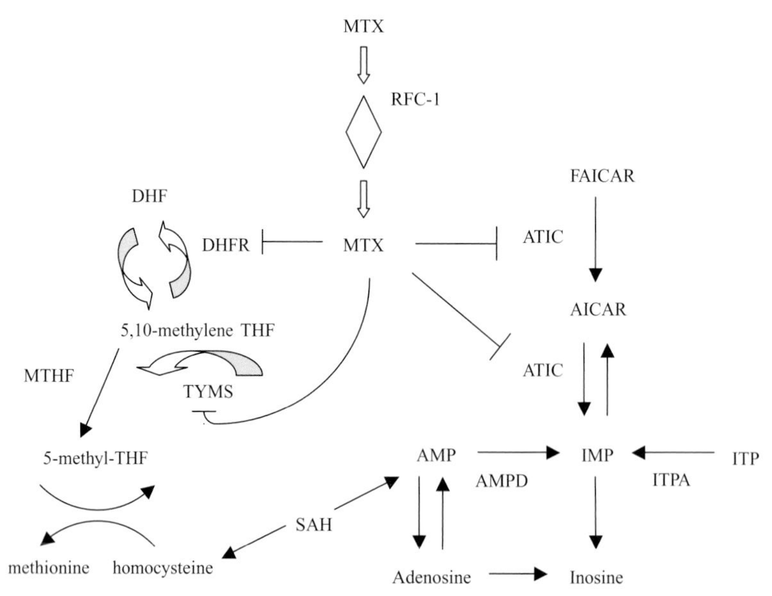

Figure 17.3. MTX enters the cell via RFC-1 and inhibits enzymes such as DHFR, TS, and ATIC. By inhibiting these enzymes, MTX has many downstream effects, including alteration of the conversion of homocysteine to methionine and increasing the levels of adenosine. MTX, methotrexate; RFC-1, reduced folate carrier-1; DHFR, dihydrofolate reductase; DHF, dihydrofolate; THF, tetrahydrofolate; TYMS, thymidylate synthase; ATIC, 5-aminoimidazole-4-carboxamide ribonucleotide transformylase; FAICAR, formyl-5-aminoimidazole-4-carboxamide ribonucleotide; AICAR, aminoimidazole carboxamide ribonucleotide; IMP, inosine monophosphate; ITP, inosine triphosphate; ITPA, inosine triphosphate pyrophosphatase; AMP, adenosine monophosphate; AMPD, adenosine monophosphate deaminase; SAH, *S*-adenosylhomocysteine.

Table 17.3. Three Genetic Studies of Response to MTX Therapy Selected Because of Size (>200 Samples) and Evaluation of *MTHFR* Variants

				Response		*Baseline Characteristics*					
Study	*Size*	*Dose*	*Outcome Criteria*	*Interval Assessment*	*Final Assessment*	*Disease Duration*	*Prior MTX*	*Concurrent DMARD*	*Trial Design*	*Genes*	*Major Conclusions*
van Ede et al.	$n = 228$	7.5 mg qwk, increased to maximum 25 mg qwk	Efficacy: DAS Toxicity: withdrawal due to adverse events	Efficacy q6wks Toxicity q3wks	48 weeks	79 months	No	Yes	Randomized clinical trial	*MTHFR* C677T	*MTHFR* 677CT and *MTHFR* TT genotypes are associated with increased risk of discontinuing MTX treatment due to adverse events. *MTHFR* C677T was not associated with MTX efficacy.
Wessels et al.	$n = 205$	7.5 mg qwk, increased to 15 mg qwk after 4 wks; then increased by 5 mg every 2 wks if insufficient clinical response	Efficacy: good clinical response, good clinical improvement, moderate clinical improvement Toxicity: reported adverse drug events	3 and 6 months	6 months	2 weeks	No	No	Randomized clinical trial	*MTHFR* C677T, *MTHFR* C1298A, *DHFR* G-437A, *DHFR* G35289A, *RFC* G80A	The *MTHFR* 1298AA genotype was associated with better clinical improvement than the *MTHFR* 1298CA and 1298CC genotypes. The *MTHFR* 1298CA and 1298CC genotypes were associated with an increased risk of overall adverse events compared with the *MTHFR* 1298 AA genotype. No association between MTHFR C677T, *DHFR* and *RFC* polymorphisms and MTX efficacy and toxicity.
Weisman et al.	$n = 214$	Mean 12.5 mg qwk	CNS side effects, GI side effects, alopecia, overall side effects	Single study visit	Single study visit	8 years	Yes	Yes	Cross-sectional observational study	*MTHFR* C677T, *ATIC* C347G, *SHMT1* C1420T, *TSER* *2/*2	*MTHFR* 677TT and *SHMT1* 1420CC are associated with CNS toxicity. *ATIC* 347 GG is associated with GI side effects. *TSER* *2/*2 and *SHMT1* 1420 CC are associated with alopecia. Participants with a toxicogenetic index >3 are seven times more likely to have a side effect than participants with a toxicogenetic index of 0.
Kim et al.	$n = 385$	Mean 10.8 mg qwk	MTX-related toxicity as assessed by medical record review and confirmed by patient interview	Single study visit	Single study visit	13.9 years	Yes	Yes	Retrospective	*MTHFR*	*MTHFR* 677CT and *MTHFR* 677TT were associated with an increased risk of MTX toxicity compared with *MTHFR* 677CC. Screening for *MTHFR* C677T may result in slightly lower annual direct medical costs.

qwk, once a week; q6wks, every 6 weeks; q3wks, every 3 weeks; CNS, central nervous system.

van Ede et al. (15) were the first to examine the association between *MTHFR* polymorphisms and MTX efficacy and toxicity. This study included 236 RA patients in a randomized, double-blind, placebo-controlled trial. Patients had long-standing disease and were not on any other DMARDs during the study. The primary efficacy outcome was the DAS. The primary toxicity end point was discontinuation of MTX because of an adverse effect. Only one single-nucleotide polymorphism (SNP), *MTHFR* C677T, was examined. The 677CT and 677TT genotypes were associated with an increased risk of MTX discontinuation because of adverse events (relative risk 2.01, 95% confidence interval [CI] 1.09–3.70) (15). There were no statistically significant differences between the genotypes with regard to MTX efficacy, but the study may not have had sufficient power to detect a small to moderate effect.

Wessels et al. (16) examined the association between polymorphisms in *MTHFR* and two other folate-dependent enzymes, *DHFR* and the RFC. The study population included participants in a randomized, multi-center, single-blind trial comparing the efficacy of different treatment strategies. In contrast to the previous study, this study focused on patients with early RA. Patients previously treated with DMARDs were excluded; only patients randomly assigned to MTX monotherapy were included. No a priori primary efficacy outcomes were defined; instead, the authors assessed multiple efficacy measures, including good clinical response (DAS $\leq$ 2.4), good clinical improvement (change in DAS $>$ 1.2), and moderate clinical improvement (change in DAS $>$ 0.6) at three and six months. The primary toxicity outcome was all reported adverse drug events (16).

As in the previous study, *MTHFR* 677CC was not significantly associated with MTX efficacy. In contrast to the previous study, there was no association between *MTHFR* 677CC and MTX toxicity. The *MTHFR* 1298AA genotype was associated with better clinical improvement than the *MTHFR* 1298CA and 1298CC genotypes, but no ORs or *P* values were provided for the association between this genotype and good clinical response. The *MTHFR* 1298CA and 1298CC genotypes were associated with an increased risk of overall adverse events compared to the *MTHFR* 1298AA genotype. *DHFR* G-437A, *DHFR* G35289A and *RFC* G80A polymorphisms were not associated with MTX efficacy or toxicity (16).

Weisman et al.'s (17) study examined the association between MTX toxicity, *MTHFR* C677T, and three other polymorphisms in folate-dependent enzymes (*ATIC*, *TSER*, and serine hydroxymethyltransferase [*SHMT1*]). In contrast to the studies by van Ede et al. and Wessels et al., this study was a cross-sectional analysis. The study population included 214 patients with established RA. Twenty-nine percent were treated concurrently with other DMARDs. The outcome included different measures of MTX toxicity, including overall side effects, GI side effects, central nervous system (CNS) side effects, alopecia, and cough/dyspnea. *MTHFR* 677TT was associated with an increased risk of CNS side effects but not other side effects. There was no significant association between *MTHFR* 677TT and the risk for overall side effects. *SHMT1* 1420CC was associated with an increased risk for CNS side effects and an increased risk of alopecia. *TSER* *2/*2 was associated with an increased risk of alopecia, and *ATIC* 347GG was associated with an increased risk for GI side effects (17).

Kim et al. (18) examined the association between *MTHFR* C677T and MTX toxicity. These investigators also performed a cost-effectiveness analysis of *MTHFR* polymorphism screening. This study included 385 Korean patients with long-standing RA, treated with MTX. Concurrent DMARD treatment was allowed. The primary outcome measure was MTX toxicity determined by a medical record review and confirmed by a patient interview. *MTHFR* 677CT and *MTHFR* 677TT were associated with an increased risk of MTX toxicity in comparison with *MTHFR* 677CC. According to a decision-tree model, the total expected cost was $710 per year for conventional dosing versus $658 per year for genotype-based dosing. The probability of continuing MTX at one year was 94 percent for conventional dosing and 96 percent for genotype-based dosing (18).

These four examples reflect the variation in study design, source population, and outcome measures that exists across studies, contributing multiple possible reasons for the lack of replication. In addition to the studies mentioned, several other studies have also examined the association between *MTHFR* C677T and MTX toxicity, but these studies have been smaller and have had conflicting results (19–23). Despite many differences, the two largest studies, van Ede et al. and Kim et al., both reported a statistically significant increased risk for toxicity in patients with the *MTHFR* 677CT and 677TT genotypes (15, 18). The third largest study showed a significant association between *MTHFR* 677TT and CNS toxicity (17). These observations suggest that smaller studies may have been underpowered.

Clinical Prediction Models

The ultimate goal of pharmacogenetics is to predict patient response to medical treatment. However, most genetic effects are small to modest, and individual SNPs may not adequately predict response. To surmount this barrier, investigators have developed the concept of a pharmacogenetic index, a system that assigns a point for each homozygous variant genotype carried by an individual. A higher score is associated with a greater likelihood of response to MTX. In a study of 108 RA patients, Dervieux et al. (24) showed that a pharmacogenetic index composed of the homozygous variant genotypes of *ATIC*, *TSER*, and *RFC-1* was able to predict physician assessment of patient response to MTX. Replication was not attempted, but a larger analysis of 226 RA patients, which included the 108 patients in the first study, also showed a significant association between a higher pharmacogenetic index score and lower disease activity (25).

Building on the concept of a pharmacogenetic index, Wessels et al. combined clinical and genetic predictors of MTX efficacy to develop a clinical prediction model that included sex, RF status, smoking status, baseline DAS, and SNPs in *ATIC*, *ITPA*, *AMPD1*, and *MTHFD1* (26). This model yielded a score ranging from 0 (high probability of response to MTX) to 11.5 (low probability of response to MTX). For scores $\leq$3.5, the positive response rate was 95 percent. For scores $\geq$6, the negative response rate was 86 percent. This model was replicated in a small cohort of thirty-eight patients. In this population, the positive response rate was 70 percent for scores $\leq$3.5, and the negative response rate was 72 percent for scores $\geq$6 (26). This study is interesting because it leads the field toward the goal of pharmacogenetics – the ability to predict an individual patient's response to treatment, but larger replication studies are needed before this prediction rule can be used in a clinical setting.

TNF-α Antagonists

A remarkable advance in the treatment of RA occurred with the introduction of monoclonal antibodies that block the inflammatory cytokine, TNF. This class of biologic agents, which currently includes three Food and Drug Administration-approved drugs (see below), is most often prescribed following the failure of one or several DMARDs such as MTX. Although anti-TNF therapy holds great promise for many patients, a substantial percentage of patients (~40 percent to 60 percent) do not respond to either DMARD or biologic therapy (27–29). Moreover, anti-TNF therapy is expensive (30), and can be associated with significant side effects, such as increased risk of infection (31) and malignancy (32). Thus, a better understanding of who responds and who develops serious toxicity to anti-TNF therapy would be a significant advance in patient care.

There are two classes of anti-TNF therapies: TNF-soluble receptors and neutralizing anti-TNF antibodies. *Etanercept* (Enbrel) is a fusion protein consisting of two p75 TNF receptors linked to the Fc part of human IgG1. It acts as a soluble receptor to bind TNF (as well as a related molecule, lymphotoxin-alpha [LTA]), thereby preventing TNF from binding to its cell surface receptor. It is most often given at a dose of 50 mg per week via subcutaneous injection, regardless of body mass index. *Infliximab* (Remicade; chimeric mouse-human antibody) and *adalimumab* (Humira; fully humanized antibody) are antibodies directed against TNF, leading to neutralization of TNF (and downstream signaling). Infliximab is administered intravenously, at a dose ranging from 3 mg/kg to 10 mg/kg per infusion. Adalimumab is generally administered as a subcutaneous injection at a dose of 40 mg every two weeks.

There are undoubtedly shared mechanisms between the two drug classes (e.g., downstream signaling factors), as illustrated by similar effects on the change in inflammatory cytokines, complement activation, lymphocyte trafficking, and apoptosis (33–35). However, different biological factors may influence the response to TNF-soluble receptors and neutralizing anti-TNF antibodies. Infliximab and adalimumab bind to transmembrane TNF on the surface of activated immune cells, whereas etanercept only binds soluble TNF (36); etanercept also binds a related molecule, LTA, whereas infliximab and adalimumab do not (37). These differences may account for different rates of granulomatous infections between the two drug classes (38). They may also underlie the reasons why infliximab and adalimumab are approved for treatment of Crohn's disease, but etanercept is not.

In considering the response to anti-TNF therapy, it is important to know whether a patient has previously received anti-TNF therapy (and thus presumably failed one TNF class). In clinical practice and in most studies, patients who do not respond to one TNF drug have some response to another TNF agent. However, the response rate is not as high, and nonresponse to the first agent is strongly correlated with nonresponse to the second agent (39). It is also important to classify patients as *primary* nonresponders (fail to respond upon initial administration of anti-TNF agents) or *secondary* nonresponders (initially respond, but efficacy diminishes with time), because the molecular mechanisms leading to both may be different.

There are clinical factors that predict response to anti-TNF therapy, and these must be taken into consideration when evaluating pharmacogenetic studies. Hyrich et al. (39), in a study of patients followed as part of the British Society for Rheumatology Biologics Register, found clinical predictors of response to anti-TNF therapy among RA patients. After receiving either infliximab or etanercept for six months, concurrent NSAIDs and/or MTX predicted better outcome. In addition, a low baseline score on the self-reported Health Assessment Questionnaire – a lower score means better functional status – predicted improved outcome. It is interesting that patient age, disease duration, and baseline disease status activity as measured by the DAS did not predict extent of response.

The primary purpose of our review of published association studies is to compare and contrast the characteristics across studies (e.g., clinical characteristics and response criteria), because (a) no single variant consistently replicates across all studies, and (b) the study characteristics may partially explain inconsistencies across the studies. Although we cite all studies (40–51), we highlight only three studies with >100 patients that have genotyped multiple variants across the MHC, because other studies are underpowered to detect anything but extremely large effect sizes (see Table 17.4). A recent review has given more attention to each individual study (52).

Criswell et al. (40)

This was a randomized control trial (RCT), and included 150 patients on low-dose etanercept (10 mg twice weekly) and 151 patients on standard-dose etanercept (25 mg twice weekly). This is the largest study to date. No other DMARDs were administered. Patients had new-onset disease and had not previously received anti-TNF therapy. The study used 50 percent improvement in disease activity according to the criteria of the American College of Rheumatology (ACR50 response) at 12 months to define response, and all patients were categorized as either responders or nonresponders. Univariate and multivariate analyses were performed, where covariates included sex, ethnicity, age, disease duration, and baseline values for RF and the tender and swollen joint counts. The major finding was that the presence of two HLA-DRB1 SE alleles was associated with better response to etanercept (25 mg twice weekly), with OR 4.3 (95% CI 1.8–10.3). Another finding was that an extended haplotype of specific SE alleles (*0101 and *0404) and DNA variants in the *LTA-TNF* region also predicted treatment response.

Marotte et al. (41)

This was a prospective, longitudinal, open-label trial with a 3 mg/kg infusion of infliximab in patients with long-standing disease. All patients were on concurrent MTX and had long-standing disease; the number of patients in whom prior TNF therapy had failed was not specified. The study investigated predictors of *receiving* infliximab treatment (in a collection of 930 RA patients), in addition to predictors of *responding* to infliximab treatment (in the subset of 198 patients who received infliximab). A stepwise multiple logistic regression analysis was performed with identified independent predictors of selection for infliximab treatment (e.g., younger age at onset, the presence of RF, and SE homozygosity). The study used ACR20 at thirty weeks to define response and assessed significance by comparing the percentage of patients meeting ACR20 in each of the three genotype classes (carriers of no, one, and two copies of SE alleles). A major finding was that patients who *received* infliximab treatment were more likely to carry SE alleles than were those who did not receive infliximab treatment. This result is consistent with SE alleles predicting more severe disease, as previously reported by others (53). However, there was no significant association between SE alleles (nor SNPs in other genes) and response to infliximab. The authors cautioned that, because the SE was associated with treatment selection, one might not anticipate observing an additional effect on response.

Padyukov et al. (42)

This was a study using a national surveillance program in Sweden that included 123 patients receiving standard-dose etanercept (25 mg twice weekly). Approximately half of the patients were on etanercept alone, and half were on another DMARD. Patients had long-standing disease, and the number in whom prior TNF therapy had failed was not specified (although all patients had failed at least one DMARD). Four SNPs in four different genes were investigated. The major finding was that no single SNP was associated with treatment response. The authors did not find an association between treatment response and SE alleles. No adjustment was made for clinical characteristics that might predict treatment response.

Azathioprine

Azathioprine is a DMARD commonly used in rheumatic diseases such as systemic lupus erythematosus and inflammatory myositis. It is less commonly used to treat RA. Azathioprine acts as an immunosuppressive through disruption of purine metabolism. Its metabolic pathway (shown in Figure 17.4) involves conversion to the metabolically active 6-mercaptopurine and then subsequent conversion to 6-thioguanine nucleotides, which disrupt the purine metabolism through interacting with DNA.

As with most immunosuppressive agents, there is a relatively narrow therapeutic index where azathioprine provides benefit without excess risk. Azathioprine's major adverse effect is myelosuppression with subsequent leucopenia. Although this side effect can be observed in approximately 10 percent of patients, it becomes severe and clinically significant in only 2 percent to 3 percent.

Thiopurine methyltransferase (TPMT) is a key enzymatic regulator of azathioprine metabolism. Reduced activity of TPMT results in excess accumulation of 6-thioguanine and an increase in the risk of azathioprine-induced leucopenia. The gene for TPMT maps to chromosome 6p22.3. Most variant alleles involve SNPs resulting in heterozygosity and reduced enzymatic activity or, less commonly, homozygosity with no TPMT activity. Approximately 1 in 10 are relatively deficient with heterozygous allelic patterns, and 1 in 300 whites are homozygous for TPMT deficiency.

Table 17.4. Selected Pharmacogenetic Studies of Response to TNF Antagonists

				Response		*Baseline Characteristics*					
Study	*Size*	*Anti-TNF Agent (Dose)*	*Primary Outcome Criteria*	*Interval Assessment*	*Final Assessment*	*Disease Duration*	*Prior TNF*	*Concurrent DMARDs*	*Trial Design*	*Genes*	*Major Conclusions*
Criswell et al.	$n = 150$ (10 mg BIW); $n = 151$ (25 mg BIW)	Etanercept 10 mg or 25 mg BIW	ACR50	0.5, 1, 6, 8, 10, 12 months	12 months	<3 years	New start	None	RCT of MTX vs. etanercept alone	HLA-DRB1 (SE alleles), TNF, LTA, TNFRSF1A, TNFRSF1B, FCGR2A, FCGR3A, FCGR3B	Presence of two SE alleles associated with better response to etanercept (25 mg BIW), with OR 4.3 (95% CI 1.8–10.3)
Marotte et al.	$n = 198$	Infliximab 3 mg/kg (at 0, 2, 6, and every 8 weeks)	ACR20	Baseline and week 30	Week 30	10.4 years	Not specified	All on MTX	Prospective longitudinal open-label trial	HLA-DRB1 (SE alleles), TNF, IL1B, IL1-RN	No significant effects of single DNA variants on treatment response
Padyukov et al.	$n = 123$	Etanercept 25 mg BIW	ACR20 and DAS28	3 months	3 months	14 years	Not specified	63 patients treated with etanercept alone, 53 with etanercept + MTX, and 7 with etanercept + other DMARDs	Nationwide surveillance program	HLA-DRB1 (SE alleles; data not shown), TNF, IL10, TGFB, IL1-RN	No significant effects of single DNA variants on treatment response

BIW, twice weekly.

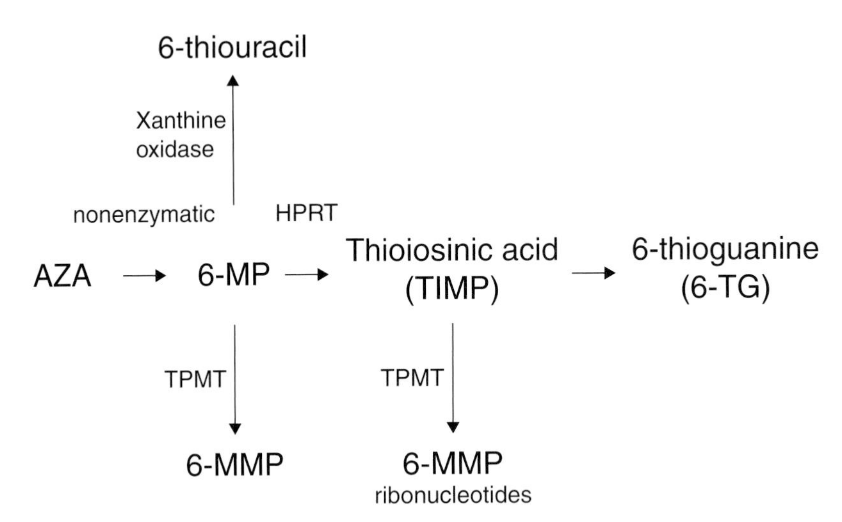

Figure 17.4. Azathioprine mechanism.

The relationship between TPMT genotypes and azathioprine-induced leucopenia has been noted in multiple case studies without any large-scale pharmacogenetic studies ever having been conducted (54). A fairly certain relationship without large epidemiologic studies is possible for several reasons. First, the pharmacologic mechanism of action of azathioprine is relatively well understood, and the importance of TPMT in its metabolism well recognized. Second, the buildup of 6-thioguanine nucleotides has clearly been related to severe leucopenia among patients using azathioprine (55). Finally, the relationship between TPMT variants and leukopenia is very strong.

Because the relationship between TPMT genotypes and azathioprine-induced leucopenia is fairly certain, several studies have examined the economics and utilization of this testing. Two studies have examined the cost-effectiveness of TPMT testing before initiating azathioprine (56, 57). The studies used slightly different methodologies and were conducted in the context of very different health care systems (Canada and South Korea). However, both studies found that a strategy that included TPMT testing was cost saving.

Despite these conclusions, complete adoption of routine TPMT testing has not occurred. One large study from the United Kingdom found that approximately one-third of consulting physicians reported not using TPMT testing before initiating azathioprine (58). Thus, despite scientific and financial utility, genetic testing will need to gain greater acceptance in the community before it can have a significant impact on medical care.

CONCLUSION

Pharmacogenetics has great potential to have an impact on the practice of rheumatology. For a disease like RA, clinicians and patients have many treatment options but few scientifically driven methods for selecting therapy. Current methods are based on empirics and opinion, and pharmacogenetic studies may help to create a more evidence-based approach. However, there are huge challenges to overcome before pharmacogenetics becomes part of daily practice. Methodologic barriers include small samples for study, few candidate genes, and concurrent treatments. In addition, a major challenge for incorporating pharmacogenetics into clinical practice will be teaching physicians how to use these tests in a rational manner that improves care and outcomes rather than only increasing costs.

REFERENCES

1. Landewe RB, Boers M, Verhoeven AC, Westhovens R, van de Laar MA, Markusse HM, et al. COBRA combination therapy in patients with early rheumatoid arthritis: long-term structural benefits of a brief intervention. *Arthritis Rheum.* 2002;**46**:347–56.
2. Begovich AB, Carlton VE, Honigberg LA, Schrodi SJ, Chokkalingam AP, Alexander HC, et al. A missense single-nucleotide polymorphism in a gene encoding a protein tyrosine phosphatase (PTPN22) is associated with rheumatoid arthritis. *Am J Hum Genet.* 2004;**75**:330–7. Epub 2004 Jun 18.
3. Remmers EF, Plenge RM, Lee AT, Graham RR, Hom G, Behrens TW, et al. STAT4 and the risk of rheumatoid arthritis and systemic lupus erythematosus. *N Engl J Med.* 2007;**357**:977–86.
4. Plenge RM, Cotsapas C, Davies L, Price AL, de Bakker PI, Maller J, et al. Two independent alleles at 6q23 associated with risk of rheumatoid arthritis. *Nat Genet.* 2007;**39**:1477–82. Epub 2007 Nov 4.
5. Thomson W, Barton A, Ke X, Eyre S, Hinks A, Bowes J, et al. Rheumatoid arthritis association at 6q23. *Nat Genet.* 2007;**39**:1431–3. Epub 2007 Nov 4.

6. Plenge RM, Seielstad M, Padyukov L, Lee AT, Remmers EF, Ding B, et al. TRAF1-C5 as a risk locus for rheumatoid arthritis – a genomewide study. *N Engl J Med.* 2007;**357**:1199–209. Epub 2007 Sep 5.
7. Kurreeman FA, Padyukov L, Marques RB, Schrodi SJ, Seddighzadeh M, Stoeken-Rijsbergen G, et al. A candidate gene approach identifies the TRAF1/C5 region as a risk factor for rheumatoid arthritis. *PLoS Med.* 2007;**4**:e278.
8. Suzuki A, Yamada R, Chang X, Tokuhiro S, Sawada T, Suzuki M, et al. Functional haplotypes of PADI4, encoding citrullinating enzyme peptidylarginine deiminase 4, are associated with rheumatoid arthritis. *Nat Genet.* 2003;**34**:395–402.
9. Plenge RM, Padyukov L, Remmers EF, Purcell S, Lee AT, Karlson EW, et al. Replication of putative candidate-gene associations with rheumatoid arthritis in >4,000 samples from North America and Sweden: association of susceptibility with PTPN22, CTLA4, and PADI4. *Am J Hum Genet.* 2005;**77**:1044–60. Epub 2005 Nov 1.
10. Sigurdsson S, Padyukov L, Kurreeman FA, Liljedahl U, Wiman AC, Alfredsson L, et al. Association of a haplotype in the promoter region of the interferon regulatory factor 5 gene with rheumatoid arthritis. *Arthritis Rheum.* 2007;**56**:2202–10.
11. Cronstein BN. Low-dose methotrexate: a mainstay in the treatment of rheumatoid arthritis. *Pharmacol Rev.* 2005;**57**:163–72.
12. Weinblatt ME, Maier AL, Fraser PA, & Coblyn JS. Longterm prospective study of methotrexate in rheumatoid arthritis: conclusion after 132 months of therapy. *J Rheumatol.* 1998;**25**:238–42.
13. Yazici Y, Sokka T, Kautiainen H, Swearingen C, Kulman I, & Pincus T. Long term safety of methotrexate in routine clinical care: discontinuation is unusual and rarely the result of laboratory abnormalities. *Ann Rheum Dis.* 2005;**64**:207–11. Epub 2004 Jun 18.
14. Hoekstra M, van Ede AE, Haagsma CJ, van de Laar MA, Huizinga TW, Kruijsen MW, et al. Factors associated with toxicity, final dose, and efficacy of methotrexate in patients with rheumatoid arthritis. *Ann Rheum Dis.* 2003;**62**:423–6.
15. van Ede AE, Laan RF, Blom HJ, Huizinga TW, Haagsma CJ, Giesendorf BA, et al. The C677T mutation in the methylenetetrahydrofolate reductase gene: a genetic risk factor for methotrexate-related elevation of liver enzymes in rheumatoid arthritis patients. *Arthritis Rheum.* 2001;**44**:2525–30.
16. Wessels JA, de Vries-Bouwstra JK, Heijmans BT, Slagboom PE, Goekoop-Ruiterman YP, Allaart CF, et al. Efficacy and toxicity of methotrexate in early rheumatoid arthritis are associated with single-nucleotide polymorphisms in genes coding for folate pathway enzymes. *Arthritis Rheum.* 2006;**54**:1087–95.
17. Weisman MH, Furst DE, Park GS, Kremer JM, Smith KM, Wallace DJ, et al. Risk genotypes in folate-dependent enzymes and their association with methotrexate-related side effects in rheumatoid arthritis. *Arthritis Rheum.* 2006;**54**:607–12.
18. Kim SK, Jun JB, El-Sohemy A, & Bae SC. Cost-effectiveness analysis of MTHFR polymorphism screening by polymerase chain reaction in Korean patients with rheumatoid arthritis receiving methotrexate. *J Rheumatol.* 2006;**33**:1266–74. Epub 2006 Jun 1.
19. Dervieux T, Greenstein N, & Kremer J. Pharmacogenomic and metabolic biomarkers in the folate pathway and their association with methotrexate effects during dosage escalation in rheumatoid arthritis. *Arthritis Rheum.* 2006;**54**:3095–103.
20. Hughes LB, Beasley TM, Patel H, Tiwari HK, Morgan SL, Baggott JE, et al. Racial or ethnic differences in allele frequencies of single-nucleotide polymorphisms in the methylenetetrahydrofolate reductase gene and their influence on response to methotrexate in rheumatoid arthritis. *Ann Rheum Dis.* 2006;**65**:1213–18. Epub 2006 Jan 26.
21. Berkun Y, Levartovsky D, Rubinow A, Orbach H, Aamar S, Grenader T, et al. Methotrexate related adverse effects in patients with rheumatoid arthritis are associated with the A1298C polymorphism of the MTHFR gene. *Ann Rheum Dis.* 2004;**63**:1227–31.
22. Kumagai K, Hiyama K, Oyama T, Maeda H, & Kohno N. Polymorphisms in the thymidylate synthase and methylenetetrahydrofolate reductase genes and sensitivity to the low-dose methotrexate therapy in patients with rheumatoid arthritis. *Int J Mol Med.* 2003;**11**:593–600.
23. Urano W, Taniguchi A, Yamanaka H, Tanaka E, Nakajima H, Matsuda Y, et al. Polymorphisms in the methylenetetrahydrofolate reductase gene were associated with both the efficacy and the toxicity of methotrexate used for the treatment of rheumatoid arthritis, as evidenced by single locus and haplotype analyses. *Pharmacogenetics.* 2002;**12**:183–90.
24. Dervieux T, Furst D, Lein DO, Capps R, Smith K, Walsh M, et al. Polyglutamation of methotrexate with common polymorphisms in reduced folate carrier, aminoimidazole carboxamide ribonucleotide transformylase, and thymidylate synthase are associated with methotrexate effects in rheumatoid arthritis. *Arthritis Rheum.* 2004;**50**:2766–74.
25. Dervieux T, Furst D, Lein DO, Capps R, Smith K, Caldwell J, et al. Pharmacogenetic and metabolite measurements are associated with clinical status in patients with rheumatoid arthritis treated with methotrexate: results of a multicentred cross sectional observational study. *Ann Rheum Dis.* 2005;**64**:1180–5. Epub 2005 Jan 27.
26. Wessels JA, van der Kooij SM, le Cessie S, Kievit W, Barerra P, Allaart CF, et al. A clinical pharmacogenetic model to predict the efficacy of methotrexate monotherapy in recent-onset rheumatoid arthritis. *Arthritis Rheum.* 2007;**56**:1765–75.
27. Maini R, St Clair EW, Breedveld F, Furst D, Kalden J, Weisman M, et al. Infliximab (chimeric anti-tumour necrosis factor alpha monoclonal antibody) versus placebo in rheumatoid arthritis patients receiving concomitant methotrexate: a randomised phase III trial. ATTRACT Study Group. *Lancet.* 1999;**354**:1932–9.
28. Weinblatt ME, Kremer JM, Bankhurst AD, Bulpitt KJ, Fleischmann RM, Fox RI, et al. A trial of etanercept, a recombinant tumor necrosis factor receptor:Fc fusion protein, in patients with rheumatoid arthritis receiving methotrexate. *N Engl J Med.* 1999;**340**:253–9.
29. Klareskog L, van der Heijde D, de Jager JP, Gough A, Kalden J, Malaise M, et al. Therapeutic effect of the

combination of etanercept and methotrexate compared with each treatment alone in patients with rheumatoid arthritis: double-blind randomised controlled trial. *Lancet.* 2004;**363**:675–81.
30. Chen YF, Jobanputra P, Barton P, Jowett S, Bryan S, Clark W, et al. A systematic review of the effectiveness of adalimumab, etanercept and infliximab for the treatment of rheumatoid arthritis in adults and an economic evaluation of their cost-effectiveness. *Health Technol Assess.* 2006;**10**:iii–iv.
31. Dixon WG, Watson K, Lunt M, Hyrich KL, Silman AJ, & Symmons DP. Rates of serious infection, including site-specific and bacterial intracellular infection, in rheumatoid arthritis patients receiving anti-tumor necrosis factor therapy: results from the British Society for Rheumatology Biologics Register. *Arthritis Rheum.* 2006;**54**:2368–76.
32. Chakravarty EF & Genovese MC. Associations between rheumatoid arthritis and malignancy. *Rheum Dis Clin North Am.* 2004;**30**:271–84.
33. Aeberli D, Seitz M, Juni P, & Villiger PM. Increase of peripheral CXCR3 positive T lymphocytes upon treatment of RA patients with TNF-alpha inhibitors. *Rheumatology (Oxford).* 2005;**44**:172–5. Epub 2004 Oct 27.
34. Agnholt J, Dahlerup JF, & Kaltoft K. The effect of etanercept and infliximab on the production of tumour necrosis factor alpha, interferon-gamma and GM-CSF in in vivo activated intestinal T lymphocyte cultures. *Cytokine.* 2003;**23**:76–85.
35. Catrina AI, Trollmo C, af Klint E, Engstrom M, Lampa J, Hermansson Y, et al. Evidence that anti-tumor necrosis factor therapy with both etanercept and infliximab induces apoptosis in macrophages, but not lymphocytes, in rheumatoid arthritis joints: extended report. *Arthritis Rheum.* 2005;**52**:61–72.
36. Scallon BJ, Moore MA, Trinh H, Knight DM, & Ghrayeb J. Chimeric anti-TNF-alpha monoclonal antibody cA2 binds recombinant transmembrane TNF-alpha and activates immune effector functions. *Cytokine.* 1995;**7**:251–9.
37. Gudbrandsdottir S, Larsen R, Sorensen LK, Nielsen S, Hansen MB, Svenson M, et al. TNF and LT binding capacities in the plasma of arthritis patients: effect of etanercept treatment in juvenile idiopathic arthritis. *Clin Exp Rheumatol.* 2004;**22**:118–24.
38. Furst DE, Wallis R, Broder M, & Beenhouwer DO. Tumor necrosis factor antagonists: different kinetics and/or mechanisms of action may explain differences in the risk for developing granulomatous infection. *Semin Arthritis Rheum.* 2006;**36**:159–67. Epub 2006 Jul 3.
39. Hyrich KL, Watson KD, Silman AJ, & Symmons DP. Predictors of response to anti-TNF-alpha therapy among patients with rheumatoid arthritis: results from the British Society for Rheumatology Biologics Register. *Rheumatology (Oxford).* 2006;**45**:1558–65. Epub 2006 May 16.
40. Criswell LA, Lum RF, Turner KN, Woehl B, Zhu Y, Wang J, et al. The influence of genetic variation in the HLA-DRB1 and LTA-TNF regions on the response to treatment of early rheumatoid arthritis with methotrexate or etanercept. *Arthritis Rheum.* 2004;**50**:2750–6.
41. Marotte H, Pallot-Prades B, Grange L, Tebib J, Gaudin P, Alexandre C, et al. The shared epitope is a marker of severity associated with selection for, but not with response to, infliximab in a large rheumatoid arthritis population. *Ann Rheum Dis.* 2006;**65**:342–7. Epub 2005 Aug 11.
42. Padyukov L, Lampa J, Heimburger M, Ernestam S, Cederholm T, Lundkvist I, et al. Genetic markers for the efficacy of tumour necrosis factor blocking therapy in rheumatoid arthritis. *Ann Rheum Dis.* 2003;**62**:526–9.
43. Martinez A, Salido M, Bonilla G, Pascual-Salcedo D, Fernandez-Arquero M, de Miguel S, et al. Association of the major histocompatibility complex with response to infliximab therapy in rheumatoid arthritis patients. *Arthritis Rheum.* 2004;**50**:1077–82.
44. Fabris M, Tolusso B, Di Poi E, Assaloni R, Sinigaglia L, & Ferraccioli G. Tumor necrosis factor-alpha receptor II polymorphism in patients from southern Europe with mild-moderate and severe rheumatoid arthritis. *J Rheumatol.* 2002;**29**:1847–50.
45. Kang CP, Lee KW, Yoo DH, Kang C, & Bae SC. The influence of a polymorphism at position -857 of the tumour necrosis factor alpha gene on clinical response to etanercept therapy in rheumatoid arthritis. *Rheumatology (Oxford).* 2005;**44**:547–52. Epub 2005 Feb 3.
46. Mugnier B, Balandraud N, Darque A, Roudier C, Roudier J, & Reviron D. Polymorphism at position -308 of the tumor necrosis factor alpha gene influences outcome of infliximab therapy in rheumatoid arthritis. *Arthritis Rheum.* 2003;**48**:1849–52.
47. Seitz M, Wirthmuller U, Moller B, & Villiger PM. The -308 tumour necrosis factor-alpha gene polymorphism predicts therapeutic response to TNFalpha-blockers in rheumatoid arthritis and spondyloarthritis patients. *Rheumatology (Oxford).* 2007;**46**:93–6. Epub 2006 May 23.
48. Cuchacovich M, Ferreira L, Aliste M, Soto L, Cuenca J, Cruzat A, et al. Tumour necrosis factor-alpha (TNF-alpha) levels and influence of -308 TNF-alpha promoter polymorphism on the responsiveness to infliximab in patients with rheumatoid arthritis. *Scand J Rheumatol.* 2004;**33**:228–32.
49. Tutuncu Z, Kavanaugh A, Zvaifler N, Corr M, Deutsch R, & Boyle D. Fcgamma receptor type IIIA polymorphisms influence treatment outcomes in patients with inflammatory arthritis treated with tumor necrosis factor alpha-blocking agents. *Arthritis Rheum.* 2005;**52**:2693–6.
50. Kastbom A, Bratt J, Ernestam S, Lampa J, Padyukov L, Soderkvist P, et al. Fcgamma receptor type IIIA genotype and response to tumor necrosis factor alpha-blocking agents in patients with rheumatoid arthritis. *Arthritis Rheum.* 2007;**56**:448–52.
51. Radstake TR, Fransen J, Toonen EJ, Coenen MJ, Eijsbouts AE, Donn R, et al. Macrophage migration inhibitory factor polymorphisms do not predict therapeutic response to glucocorticoids or to tumour necrosis factor alpha-neutralising treatments in rheumatoid arthritis. *Ann Rheum Dis.* 2007;**66**:1525–30. Epub 2007 Apr 24.
52. Coenen MJ, Toonen EJ, Scheffer H, Radstake TR, Barrera P, & Franke B. Pharmacogenetics of anti-TNF treatment in patients with rheumatoid arthritis. *Pharmacogenomics.* 2007;**8**:761–73.

53. Gorman JD, Lum RF, Chen JJ, Suarez-Almazor ME, Thomson G, & Criswell LA. Impact of shared epitope genotype and ethnicity on erosive disease: a meta-analysis of 3,240 rheumatoid arthritis patients. *Arthritis Rheum.* 2004;**50**:400–12.
54. Black AJ, McLeod HL, Capell HA, Powrie RH, Matowe LK, Pritchard SC, et al. Thiopurine methyltransferase genotype predicts therapy-limiting severe toxicity from azathioprine. *Ann Intern Med.* 1998;**129**:716–18.
55. Yates CR, Krynetski EY, Loennechen T, Fessing MY, Tai HL, Pui CH, et al. Molecular diagnosis of thiopurine S-methyltransferase deficiency: genetic basis for azathioprine and mercaptopurine intolerance. *Ann Intern Med.* 1997;**126**:608–14.
56. Marra CA, Esdaile JM, & Anis AH. Practical pharmacogenetics: the cost effectiveness of screening for thiopurine S-methyltransferase polymorphisms in patients with rheumatological conditions treated with azathioprine. *J Rheumatol.* 2002;**29**:2507–12.
57. Oh KT, Anis AH, & Bae SC. Pharmacoeconomic analysis of thiopurine methyltransferase polymorphism screening by polymerase chain reaction for treatment with azathioprine in Korea. *Rheumatology (Oxford).* 2004;**43**:156–63. Epub 2003 Aug 15.
58. Payne K, Newman W, Fargher E, Tricker K, Bruce IN, & Ollier WE. TPMT testing in rheumatology: any better than routine monitoring? *Rheumatology (Oxford).* 2007;**46**(5):727–9. Epub 2007 Jan 25.

18 Pharmacogenetics of Obstetric Therapeutics

David Haas and Jamie Renbarger

Obstetric pharmacology, in many respects, is still in its infancy. Because of valid concerns about the safety of drugs to both the mother and developing fetus, not as much data exist for drugs in pregnancy. There is also a distinct paucity of data regarding pharmacogenetics in pregnancy. At this time, many of the potential pharmacogenetic applications to pregnancy drug therapy are theoretical. This chapter summarizes the drug therapy for five of the major pregnancy conditions/complications, along with a brief discussion of potential pharmacogenetic or individualized pharmacotherapy applications for each. These conditions are preterm labor, depression, diabetes, the nausea and vomiting of pregnancy (NVP), and hypertension.

PRETERM LABOR

Preterm birth is the leading cause of morbidity and mortality in newborns in the United States. More than 12.5 percent of all live births are preterm, accounting for a large amount of health care spending for infants. The number of preterm births in the United States, as well as in many other industrialized countries, continues to rise. Preterm labor is the most common cause of hospitalization of pregnant women (1). Although the incidence and burden of preterm birth are relatively clear, the causes of preterm labor are not well understood (2).

Preterm labor, defined as delivery that occurs at less than thirty-seven and more than twenty weeks gestational age, can be spontaneous or due to other factors such as premature rupture of membranes, multiple gestation, maternal infection, or indicated preterm delivery for maternal or fetal indications. There have been several theories regarding the pathogenesis of preterm labor; however, it is not clear whether preterm labor represents an early activation of normal labor or results from a pathologic mechanism (3). As such, treatment of preterm labor has not always led to successful outcomes.

Why Do We Treat Preterm Labor?

As stated, preterm birth is a major cause of neonatal morbidity and mortality. In fact, preterm infants have a relative risk of developing cerebral palsy nearly forty times higher than term infants. With this in mind, stopping preterm birth becomes an important intervention to reduce pregnancy complications. However, many studies have demonstrated that various drugs used to treat preterm contractions do not reduce the rate of births before thirty-seven weeks gestation.

One of the greatest therapies that obstetricians possess to improve outcomes for preterm babies is antenatal corticosteroids. Antenatal corticosteroids, either betamethasone or dexamethasone, given to the mother when a preterm delivery is anticipated, have clearly been shown to decrease the incidence of neonatal mortality and morbidities including respiratory distress syndrome and intraventricular hemorrhage (4). In light of this, the main goal of tocolytic therapy (drugs used to stop contractions) is to prolong pregnancy at least long enough for the administration and effect of antenatal corticosteroids. Even delaying delivery by as little as twenty-four to forty-eight hours can improve neonatal outcomes. Antenatal corticosteroids are usually given to women in preterm labor at between twenty-four and thirty-four weeks gestational age.

Treatment of preterm labor contractions with tocolytic drugs is not universal. Some institutions, such as Parkland in Texas, do not treat preterm contractions (5). However, most obstetricians, when presented with a pregnant preterm labor patient at between twenty-four and thirty-four weeks gestation with intact membranes would opt to treat her with a tocolytic drug.

Which drug obstetricians should choose first is a difficult question to answer. The American College of Obstetricians and Gynecologists (ACOG) confirms that there is "no clear first-line tocolytic agent" (2). Comparison studies of the effectiveness of different tocolytic agents

Table 18.1. Comparison of Commonly Used Tocolytic Therapies

Drug Class	*Common Drugs*	*Administration*	*Common Side Effects*
Betamimetics	Terbutaline, Ritodrine	IV or SQ	Tachycardia, pulmonary edema, arrhythmia, tachycardia, hyperglycemia
Calcium channel blocker	Nifedipine	PO or sublingual	Hypotension, flushing, headache
Magnesium sulfate	–	IV or IM	Flushing, headache, dizziness, respiratory depression, pulmonary edema, cardiac arrhythmia
Nonsteroidal anti-inflammatories	Indomethacin, Ketorolac, Sulindac, Rofecoxib	PO, PR, IM, IV	Nausea, heartburn, fetal ductal constriction, oligohydramnios
Oxytocin antagonists	Atosiban	IV	Nausea

show conflicting results. Prolonged use of tocolytic drugs also has not been clearly demonstrated to show a benefit to the mother or baby. Opinions regarding which tocolytic agent is their first-line drug of choice vary widely, but the common classes of drugs that are used include (in alphabetical order by class): betamimetics, calcium channel blockers, magnesium sulfate, nonsteroidal anti-inflammatory drugs, and oxytocin antagonists.

What Drugs Are Used for Preterm Labor?

It must be noted that only one drug has ever been approved by the U.S. Food and Drug Administration for tocolysis. That drug is the betamimetic ritodrine. Ritodrine, however, is not currently available commercially. All other drugs are used off-label in their treatment of preterm labor. A summary of the classes of tocolytics, common administration, and side effects is provided in Table 18.1. Nonpharmacologic therapies such as bed rest have not been shown to decrease preterm birth, even in high-risk populations. Although a few studies have recently been published in which nitric oxide donors are used as tocolytics, the data are limited and this treatment is still confined to research protocols.(3)

Betamimetics

The most commonly used drug in this class is terbutaline. Other agents studied have been hexoprenaline and isoxsuprine. These drugs work on the β_2-adrenergic receptors that are found in the uterus. Stimulation of this receptor triggers a relaxation of the uterine muscle, thus stopping contractions. Because of the presence of these receptors in blood vessels, bronchioles, and the liver and the relative nonselectivity of these drugs, side effects are common to this class of medications. Tachycardia (maternal and fetal), hypotension, hyperglycemia, palpitations, nervousness, tremors, and other side effects are relatively common with these drugs (2, 3, 6). Although betamimetics have not demonstrated improvement in neonatal mortality or respiratory distress syndrome over placebo alone, they have demonstrated a reduction in delivery within forty-eight hours (7). These drugs can be given intravenously or subcutaneously, because oral therapy has not been shown to be beneficial for acute tocolysis. These drugs are usually not given to women with cardiac disease, uncontrolled diabetes, or cardiac arrhythmias (2). Genetic variations have been noted in the β-receptor that may have an impact on drug response. It is possible that some of the variation in response to betamimetics is due to receptor genetic variation. However, this has not been systematically studied.

Calcium Channel Blockers

The most commonly used drug in this class is nifedipine. Another agent in this class used as a tocolytic is nicardipine. As the name implies, these drugs block calcium channels in the uterine muscle that control the ability of the uterine muscle to contract. There are no placebo-controlled trials evaluating nifedipine's use as a tocolytic. Instead the literature supporting nifedipine is in direct comparison studies with other tocolytic agents. Because nifedipine is also used as an antihypertensive medication, one of the most common side effects seen with its use in tocolysis is hypotension (2, 3, 6). A systematic review showed that calcium channel blockers demonstrated superior effectiveness at delaying deliveries by seven days in comparison with other tocolytics (mainly betamimetics) (8), but other randomized trials have had mixed results. Nifedipine is most commonly given orally, but has been given sublingually for a loading dose in several studies and protocols. These drugs are not usually given to women with cardiac disease or preexisting hypotension. In addition, use at the same time as magnesium sulfate is usually avoided (2). Nifedipine is a CYP3A4/5 substrate. Currently, there are no data about genetic influences on calcium channel blocker response for preterm labor. However, it is reasonable to hypothesize that differences in drug-metabolizing activity may be responsible for some of the differences in side effects and efficacy seen with nifedipine.

Magnesium Sulfate

Magnesium sulfate is theorized to have its tocolytic effect by blocking calcium channels in the uterine muscle. Common side effects of magnesium sulfate infusions are flushing, nausea, drowsiness, and blurred vision. Side effects of magnesium sulfate are usually dose related (3). More severe side effects include pulmonary edema and cardiac arrhythmia. Concern has been raised about the adverse fetal effects such as lethargy, hypotonia, and others. The results regarding tocolytic efficacy for magnesium sulfate have been mixed, particularly in comparison with other tocolytic agents (9). Magnesium sulfate is given intravenously as a continuous infusion after a loading dose or intramuscularly if there is no intravenous access. Magnesium sulfate is contraindicated in women with myasthenia gravis.

Nonsteroidal Anti-Inflammatory Drugs

The most commonly studied drug for tocolysis in this class is indomethacin. Other agents that have been studied are ketorolac, sulindac, and rofecoxib. Prostaglandins stimulate uterine muscle contractions and, thus, these drugs act by inhibiting the production of prostaglandins, limiting the procontractile signals to the uterine muscle. These medications have few side effects, some of which are nausea and heartburn. There is concern for constriction of the fetal ductus arteriosus or the development of oligohydramnios with the prolonged use of these drugs in the third trimester. Constriction of the fetal ductus has not been seen within less than forty-eight hours of the use of indomethacin, so most authorities feel that short-term use of indomethacin under thirty-two weeks gestation is effective and safe (3). A systematic review found that indomethacin reduced delivery within forty-eight hours, seven days, and at less than thirty-seven weeks compared with placebo (10). Indomethacin is usually given rectally as a loading dose followed by oral therapy. Other drugs in this class, such as ketorolac, are given intramuscularly. These drugs are withheld from women with active renal or hepatic disease or active ulcer disease (2). Several drugs in this class are CYP2C9 and 2C19 substrates. As evidence of pharmacogenetic influences of these two CYPs emerges, there may be potential applications for assessing whether genetic polymorphisms may explain some of the differences in efficacy, side effects, or neonatal effects. This type of analysis has not been performed for these tocolytics, however.

Oxytocin Antagonists

These drugs are not available in the United States, but they are used effectively in other countries. The main drug in this class is atosiban. Oxytocin is a natural hormone that stimulates contractions in the uterus. Atosiban acts by competitively blocking the oxytocin receptor. Side effects are infrequent and include nausea. Studies comparing atosiban with other tocolytics have had mixed results, but have shown fewer side effects (3). Atosiban is given intravenously.

Summary

Tocolytic drugs may prolong pregnancy for two to seven days, which may allow for the administration of antenatal corticosteroids to improve neonatal outcomes. There is no clear-cut first-line therapy. Thus, clinical circumstances and physician preference dictate treatment (2). Because many of these drugs are substrates of polymorphic enzymes, the potential exists for pharmacogenetic applications in preterm labor therapy in the future.

PERINATAL DEPRESSION

Depressive disorders are common in the female population. In fact, the lifetime rate of major depressive disorder among women is twice that of men. Furthermore, vulnerability is not distributed evenly across the reproductive years; prevalence actually peaks during the childbearing years (11–14), making depression a frequent finding in pregnant women. Historically, pregnancy is considered a time of emotional well-being providing protection from depression; however, this notion is not supported by data in the literature (15, 16). Approximately 16 percent of pregnant women meet the diagnostic criteria for major depressive disorder, and up to 70 percent report symptoms of depression (17, 18). Although pharmacologic treatment for perinatal depression is helpful, it is not completely protective. One small study actually found that nearly 60 percent of the women who continue to take antidepressant medications during pregnancy experience a worsening of depressive symptoms that require an increase in medication dose to maintain euthymia (normal mood) (19). Hostetter et al. monitored pregnant women who either maintained or discontinued antidepressant medications after learning of their conception and found that 68 percent of those who discontinued their antidepressant medications had a relapse of depressive symptoms compared with 26 percent of those who continued their antidepressant (19). The challenge is that many symptoms of pregnancy overlap with the symptoms of depression, and, as such, many pregnant women with major depressive disorder are overlooked in clinical settings. Symptoms associated with depression according to the Diagnostic and Statistical Manual of Mental Disorders, fourth version (DSM-IV) (20) include depressed mood or the loss of interest or pleasure in usual activities combined with five or more of the following: substantial weight change, insomnia or hypersomnia, psychomotor agitation or retardation, fatigue, feelings of worthlessness or guilt, difficulty concentrating, and, in more severe cases, thoughts of suicide.

Why Do We Treat Depression during Pregnancy?

Untreated maternal depression during pregnancy is associated with an increased risk of obstetric complications such as preeclampsia (21) and adverse pregnancy outcomes, including a high rate of recurrent miscarriages (22), premature birth (23–27), low-birthweight infants (28–31), fetal growth restriction, and postnatal complications (32). Furthermore, the newborns of women with untreated depression during pregnancy cry more and are more difficult to console (30, 33–35). Maternal depression is also associated with increased life stress, decreased social support, poor maternal weight gain, smoking, and alcohol and drug use (34, 35), all of which can adversely affect infant outcomes. Later in life, children of mothers with untreated depression may be prone to behavior problems and emotional instability (35–37).

What Drugs Are Used to Treat Depression during Pregnancy?

The overall goal in treating perinatal depression is to achieve euthymia in the mother without causing harm to the fetus (38, 39) – thus requiring an evaluation of the potential benefits of treatment on the mother and the fetus against the potential risks of treatment (40–42). Treatment options for depression include pharmacotherapy and psychotherapy. Limited data show that structured psychotherapy may be an effective treatment of mild to moderate depression during pregnancy (43); however, antidepressant medication is the mainstay of treatment for depression. Many classes of antidepressant medications are available (Table 18.2); however, most data in the literature related to antidepressants in pregnancy are derived from the use of selective serotonin reuptake inhibitors (SSRIs).

SSRIs

Epidemiologic studies have been used to evaluate the safety of drug use in pregnancy. Cohort studies of multiple commonly used SSRIs (38) have reported no association between the use of those medications during pregnancy and major fetal malformations. There are two reports from the drug company GlaxoSmithKline published on their Web site (www.gskus.com/news/paroxetine/paxil_letter_e3.pdf) that infants exposed to paroxetine during early pregnancy had an increased risk of cardiovascular defects. However, the increase in risk was quite small (1.5 percent compared with 1 percent in the general population). This risk of cardiovascular defects related to paroxetine was also documented in two other studies (44, 45). This resulted in the manufacturer's changing paroxetine's pregnancy risk category from C to D (www.fda.gov/cder/drug/advisory/paroxetine20512.htm). In contrast, a meta-analysis and two other studies of antidepressants during pregnancy (46–48) conducted during the same period of time evaluated a combined total of 4,500 pregnancy outcomes and found no increased risk of major malformations. More recently, two articles published in the *New England Journal of Medicine* report findings of large case-control studies from multisite surveillance programs of the teratogenic effects of SSRIs (35, 49, 50). Despite conducting multiple statistical tests on the data – which increases the likelihood of finding statistically significant results of no clinical significance – the studies found very few, if any, teratogenic effects of these drugs. Many of the SSRIs are CYP2D6 substrates and inhibitors. Because CYP2D6 is known to be highly polymorphic, research is currently underway to determine whether these polymorphisms may account for perinatal depression treatment differences and the impact of SSRIs on the neonate.

Tricyclic Antidepressants

Tricyclic antidepressants (TCAs) have been available in the United States since the early 1960s and were commonly used by pregnant women before the introduction of SSRIs. Epidemiologic studies and reviews of early studies to evaluate the safety of these drugs in pregnancy reveal no association with major fetal malformations (51–55).

Atypical Antidepressants

These medications are non-SSRI and non-TCA antidepressants that work via pharmacodynamic mechanisms (35). The atypical antidepressants include such medications as bupropion, duloxetine, mirtazapine, nefazodone, and venlafaxine. The data in the literature are limited regarding the risks of fetal exposure to these medications; however, they do not suggest any increased risk of fetal malformations or adverse pregnancy outcomes associated with use of the atypical antidepressants (35, 56–61). One small study reported a significantly increased risk of spontaneous abortion in pregnant women exposed to bupropion with no increase in the risk of major malformations (35, 62); however, the manufacturer's registry of adverse events associated with this medication has not identified any increase in risk of spontaneous abortion (35). Many of these drugs also may be affected by several polymorphic enzymes. Research is needed to determine the role of individualized pharmacogenetic approaches to therapy for this group of drugs in pregnancy.

Table 18.2. Antidepressant Drugs

Class of Drug	*Antidepressant Medication*	*Teratogenicity*	*Notes*
SSRI	Citalopram	No increased risk of major malformations (PMID 17381382, 10501819, 12634662, 16325604, 16319254).	Recommended as second-line treatment for depression during pregnancy by the Canadian Psychiatric Association and Canadian Network for Mood and Anxiety Treatments (CANMAT) (11441773)
	Escitalopram	Minimal data available.	Risk remains to be established.
	Fluoxetine	No increased risk of major malformations (PMID 17381382 8793924, 10501819, 12450956, 12712058, 12634662, 16319254). No neurodevelopmental effects in childhood (8995088).	Recommended as first-line treatment for depression during pregnancy by CANMAT (11441773)
	Fluvoxamine	No increased risk of major malformations (PMID 17381382 9486756, 12712058, 12634662, 16319254).	Recommended as second-line treatment for depression during pregnancy by CANMAT (11441773)
	Paroxetine	Possible small increase in risk of major malformations and cardiac malformations (17381382 GlaxoSmithKline, Inc. Epidemiology study: updated preliminary report on bupropion and other antidepressants, including paroxetine, in pregnancy and the occurrence of cardiovascular and major congenital malformation. Available from http://ctr.gsk.co.uk/Summary/paroxetine/studylist.asp. Accessed October 9, 2006.).	ACOG recommends avoiding paroxetine use during pregnancy if possible.
	Sertraline	No increased risk of major malformations (PMID 17381382 9486756, 10501819, 12450956, 12712058, 12634662, 16319254).	Recommended as second-line treatment for depression during pregnancy by CANMAT (11441773)
TCA	Desipramine	No risk of major malformation identified (PMID 8829251, 11800539, 10517430, 9830392, 15990522).	Recommended as third-line treatment for depression during pregnancy by CANMAT (11441773)
	Nortriptyline	No risk of major malformation identified (PMID 8829251, 11800539, 10517430, 9830392, 15990522).	Recommended as third-line treatment for depression during pregnancy by CANMAT (11441773)
Atypical antidepressants	Bupropion	No risk of major malformation identified.	Limited data available. May increase the risk of spontaneous abortion (15746694).
	Duloxetine	No human data available.	
	Mirtazapine	Limited data reveal no risk of major malformations (PMID 17381382 16965209).	Limited data available. May increase risk of preterm birth and spontaneous abortion (17381382 16965209).
	Nefazodone	No risk of major malformation identified (19164848).	Limited data available. No increased risk of spontaneous abortion identified.
	Trazodone	No risk of major malformation identified (19164848).	Limited data available. No increased risk of spontaneous abortion identified.
	Venlafaxine	No risk of major malformation identified (19164848).	Limited data available. No increased risk of spontaneous abortion identified.

Effects of Antidepressants in Pregnancy

Exposure to tricyclics, SSRIs, or serotonin-norepinephrine reuptake inhibitors late in pregnancy has been associated with a transient neonatal adaptation syndrome (57, 63–67). This syndrome, present in 10 percent to 30 percent (57, 63, 64, 67–69) of exposed infants, may include such clinical findings as jitteriness, mild respiratory distress, transient tachypnea of the newborn, weak cry, poor tone, and neonatal intensive care unit admission. However, the effects seem to resolve within a few days without the need for specific treatment. In addition, prospective studies of the effect of antidepressant exposure in utero on childhood neurodevelopment reveal no effect on global IQ, language, temperament, or behavioral development (70–73).

There is a growing body of evidence to support the relative safety of SSRIs and TCAs during pregnancy (38). In making a treatment decision, one must consider the very small potential risk of antidepressant use during pregnancy against the risk of relapse of depression if treatment is discontinued. Factors associated with relapse during pregnancy include a greater than five-year history of depressive illness and a history of more than four relapses (74). As such, the most appropriate pharmacologic treatment of depression during pregnancy should be an individualized decision made by the physician and the patient together. In 2000 the American Academy of Pediatrics (AAP) published guidelines for pharmacologic treatment of pregnant women with a psychiatric disease (75) based on available data. With regard to antidepressant drug selection, the AAP recommended the SSRIs and certain TCAs, such as nortriptyline and desipramine, and advocates conducting prospective studies to evaluate the maternal and fetal effects of these drugs (18). In addition, given the data presented here, use of paroxetine should be avoided, if possible, in pregnant women and in women planning pregnancy (35). Evaluation with fetal echocardiography may be considered in women taking paroxetine during early pregnancy (35).

DIABETES IN PREGNANCY

Diabetes is becoming epidemic in the United States and other countries. It is estimated that the lifetime risk of diabetes in individuals born in 2000 is 33 percent for males and 39 percent for females (76). Diabetes is the most common medical complication in pregnancy (77). Pregestational diabetes is defined as diabetes that is present before conception. Gestational diabetes (GDM) presents later in pregnancy as a result of carbohydrate intolerance that develops during pregnancy, usually in the late second and third trimester. Gestational diabetes complicates between 2 percent and 5 percent of pregnancies in the United States (77). Most pregnant women are screened for GDM at the start of the third trimester.

Why Do We Treat Diabetes in Pregnancy?

Diabetes that exists before pregnancy carries tremendous risks to the developing fetus. Uncontrolled diabetes at conception can lead to congenital malformations with rates of up to 15 percent being documented (78). There is also an increased miscarriage rate, preterm delivery rate, and unexplained fetal death rate. The neonate born to a pregestational diabetic mother has increased risks of respiratory distress, electrolyte abnormalities, and altered growth (79). It is also believed that diabetes in pregnancy can cause early imprinting on the fetus and have effects later in life (80). Some of these include obesity, insulin resistance, and, potentially, cardiovascular disease. Achieving normal glycemic control pregestation is important to prevent early pregnancy loss, and congenital malformations, as well (79). Unfortunately, most diabetic women of reproductive age are unaware that they are diabetic before conception.

Gestational diabetes also affects the developing fetus through the final trimester of pregnancy. GDM is associated with preterm delivery, macrosomia, and hyperbilirubinemia (77). The babies of women with GDM are also at increased risk of operative delivery and birth trauma (77). The goal of treatment of GDM is to lower the likelihood of macrosomia and the consequences of neonatal hypoglycemia. Although evidence that treating GDM can prevent maternal and fetal complications is inconclusive, universal screening and treatment are widely practiced.

What Drugs Are Used to Treat Diabetes in Pregnancy?

Similar medications are used to treat both pregestational diabetes and GDM. Thus, they will be covered together here by class of medication. Dietary modifications are the first step in the management of GDM and are also part of the management plan for women who become pregnant with preexisting diabetes. Although the evidence supporting these modifications is generally inconclusive, they are standard practice in most centers. Women with GDM are also encouraged to perform physical activity in pregnancy, because exercise has been shown to reduce the need for insulin therapy.

Insulin

Trials have demonstrated that insulin treatment of women with GDM can reduce the likelihood of delivering a macrosomic baby (77). Insulin is usually added when dietary modifications fail to maintain fasting and postprandial glucose levels in a normal range. No particular insulin regimen or insulin dose has been demonstrated to be superior for GDM (77). Experts differ in their approach to insulin therapy, but usually begin with a simple regimen (once or twice per day) and work up to a more complex regimen as needed. Combinations of long-, intermediate-, and short-acting insulin are used. Free insulin apparently does not cross the placenta and appears to be safe in pregnancy.

Oral Hypoglycemic Agents

Traditionally, oral sulfonylureas were contraindicated in pregnancy. The early generation of these drugs crossed the placenta and had the potential to stimulate the fetal pancreas. However, the second-generation sulfonylurea, glyburide, has been used with success in women who did not achieve adequate glycemic control with diet alone (81). There were no apparent neonatal complications with glyburide therapy seen in the large trial. Because glyburide is given orally in doses ranging from 2.5 mg to 20 mg once or twice daily, it is often preferred by patients instead of the injections of insulin. More long-term follow-up data are needed for glyburide, but, at the current time, it is widely used by obstetricians when diet therapy alone has failed. Currently, no other oral agent has been shown to be safe and effective in GDM (77). Glyburide is a CYP2C9 substrate. As such, there may be polymorphic influences on the drug's efficacy to treat GDM. There are currently no reports correlating concentrations of oral hypoglycemic agents in pregnancy to CYP genotypes.

SUMMARY

Diabetes in pregnancy, both preexisting and developing during pregnancy, is common and can carry significant maternal and fetal risk. When diet therapy alone fails to control blood glucose, insulin and glyburide are the current therapies of choice for normalizing maternal glycemic control.

NAUSEA AND VOMITING OF PREGNANCY

Nausea and vomiting of pregnancy affects up to 85 percent of all pregnant women (82–84) and affects the health of both the woman and her fetus. Although NVP is most common during the first trimester of pregnancy, up to 20 percent of pregnant women are affected beyond twelve weeks of gestation and throughout the day (vs. isolated to the morning hours). Of the group of women affected, 1 percent to 3 percent experience a severe form of NVP called hyperemesis gravidarum (HG) which includes weight loss, dehydration, and electrolyte imbalances (85). Because "morning sickness" is so common in early pregnancy, the presence of NVP may be minimized by health care providers and pregnant women and, consequently, undertreated (86, 87).

Why Do We Treat Nausea and Vomiting of Pregnancy?

In mild cases, NVP causes discomfort and inconvenience; however, HG can pose significant risks to the health of the pregnant woman and the fetus – often necessitating hospitalization (85). Even the milder cases can have a significant impact on the quality of a woman's life and contribute significantly to health care costs and time lost from work (83, 84). Termination of otherwise wanted pregnancies has been reported in women experiencing severe, prolonged NVP (88). Because NVP is rarely life threatening, the goal of pharmacologic treatment is improvement in the quality of life for the pregnant women affected (89). Although this may ameliorate symptoms in many women, some women improve dramatically with pharmacologic intervention, and others continue to experience severe NVP despite treatment. In addition, there is evidence in support of a genetic predisposition to NVP, such as family history as a risk factor and variation in frequency between ethnic groups (90).

What Drugs Are Used to Treat Nausea and Vomiting of Pregnancy?

Pyridoxine (vitamin B_6) may be used as a single agent or in combination with the *antihistamine H_1 antagonist* doxylamine (Unisom). Two randomized controlled trials have evaluated pyridoxine alone for treatment of varying degrees of NVP. One small study compared pyridoxine 25 mg every eight hours with placebo and found a significant reduction in severe vomiting, but only a minimal effect on more mild disease (91). A second larger study evaluated pyridoxine 10 mg every eight hours versus placebo and found a reduction in both nausea and vomiting in the pyridoxine group (92). Pharmacologic doses of pyridoxine have not been found to be teratogenic. Furthermore, randomized, placebo-controlled trials of pyridoxine in combination with doxylamine have shown a 70 percent decrease in NVP (86, 93–95). The formulation of a delayed-release combination of doxylamine in combination with pyridoxine is not currently available in the United States; however, it is the drug of choice for treatment of NVP in Canada (96). Analysis of hospital admissions for NVP reveals a significant increase in hospital admissions for HG after the combination product was taken off the market in the United States in 1983 compared with the time when it was available (97). Although multiple studies show no teratogenic effects of the product (98), the manufacturer voluntarily removed it from the market because of litigation (99).

The *antihistamine H_1 antagonist* dimenhydrinate is also used for management of NVP. The safety of antihistamines during pregnancy is supported by a review of more than 200,000 first-trimester exposures (100). One randomized, double-blind, placebo-controlled trial of seventy-seven pregnant women evaluated the efficacy of dimenhydrinate in treatment of NVP and found this agent to be effective in reducing the symptoms of NVP (101).

A number of *dopamine antagonists* are used in the treatment of NVP, including phenothiazines, such as prochlorperazine, chlorpromazine, and promethazine, and metoclopramide. These drugs are not associated with an increased risk of major malformations (89). Three trials of various phenothiazines for treatment of NVP have been conducted, and all have shown an overall reduction in vomiting; however, the magnitude of responsiveness is quite variable (102–104). Metoclopramide for treatment of NVP has not been extensively studied, although it is commonly used in clinical practice (89, 105). It has not been associated with any increase in rates of malformation; however, there is limited literature addressing this (76, 106–108). One study in the literature compared the efficacy of metoclopramide in combination with pyridoxine with monotherapy with either prochlorperazine or promethazine. The study found that combination therapy with metoclopramide appears to be superior to either monotherapy regimen (109). Several drug choices in this class are CYP2D6 metabolites. As such, there may be genetic influences on the effectiveness of these antiemetics. Current research is underway to look for genetic associations with drug response for women with HG.

The *5-hydroxytryptamine 3 (5-HT_3) antagonist* ondansetron is commonly used as an antiemetic with cancer chemotherapy and is increasingly being used in the treatment of women with NVP. Einarson et al. (110) report no increased risk of major fetal malformations associated with ondansetron use by 176 pregnant women in their first trimester. In addition, no malformations were reported in a randomized controlled trial of fifteen patients exposed during the first trimester (111). This randomized, double-blind, controlled trial compared intravenous ondansetron with intravenous promethazine for treatment of severe NVP in hospitalized patients. The results of the trial revealed that ondansetron was equivalent to promethazine with respect to the following outcome measures: relief of nausea, daily weight gain, days of hospitalization, and treatment failures.

The herbal product *ginger* has also been shown to be safe for use during pregnancy to manage nausea and vomiting (112–114). One study comparing ginger capsules with placebo in twenty-seven women with HG revealed a reduction in episodes of vomiting in the treatment group (115). A second study of seventy women with NVP of varying severity found significant improvement in both nausea and vomiting (116). Because ondansetron is metabolized by several enzymes, the potential for one polymorphism to have a major impact on its effectiveness is limited. However, there may be 5-HT_3 receptor variants that could influence its effectiveness. This research is currently underway.

Corticosteroid treatment may also be beneficial for management of NVP. A randomized, double-blind, controlled trial compared methylprednisone with promethazine for treatment of NVP and found equal rates of improvement among the treatment groups; however, hospital readmission rates within two weeks of discharge were significantly lower in the steroid-treated group (117). A subsequent randomized trial found no reduction in rehospitalization rates (118). Although the use of corticosteroids in pregnancy is generally considered to be safe, three studies have confirmed a weak teratogenic effect of methylprednisolone use in the first trimester, with development of one to two cases of oral clefts per 1,000 treated women (119–122). As such, the use of corticosteroids for management of HG should be done with caution and should be avoided before ten weeks gestation (86). Corticosteroids are usually reserved for severe HG that does not respond to other medications. Prednisone has been demonstrated to be an inducer of CYP2C19. Thus, its coadministration with other drugs may have an impact on the effectiveness of the other drug.

The *anticholinergic* agent scopolamine is increasingly used for management of NVP in the United States. However, there are no data in the literature evaluating the safety or efficacy of this drug in the management of NVP.

Figure 18.1 (adapted from references 86 and 96) depicts a hierarchy of therapeutic interventions that balance safety and efficacy in the management of NVP.

HYPERTENSION IN PREGNANCY

The blood pressure (BP) criteria used to define high BP are a systolic BP $\geq$140 mmHg or a diastolic BP $\geq$90 mmHg. Hypertension in pregnancy can be classified as severe if the systolic BP is $\geq$180 mmHg or the diastolic BP is $\geq$110 mmHg (123, 124). Chronic hypertension is defined as hypertension that is present before pregnancy begins, the use of any antihypertensive medications before pregnancy, or hypertension that begins before the twentieth week of gestation. Chronic hypertension is estimated to have a prevalence of 3 percent of pregnant women (124).

In contrast, gestational hypertension is hypertension that develops after the twentieth week of gestation. When combined with significant proteinuria (>300 mg in a twenty-four-hour collection of urine), it is defined as preeclampsia (125, 126). Gestational hypertension is the most frequent cause of hypertension during pregnancy.

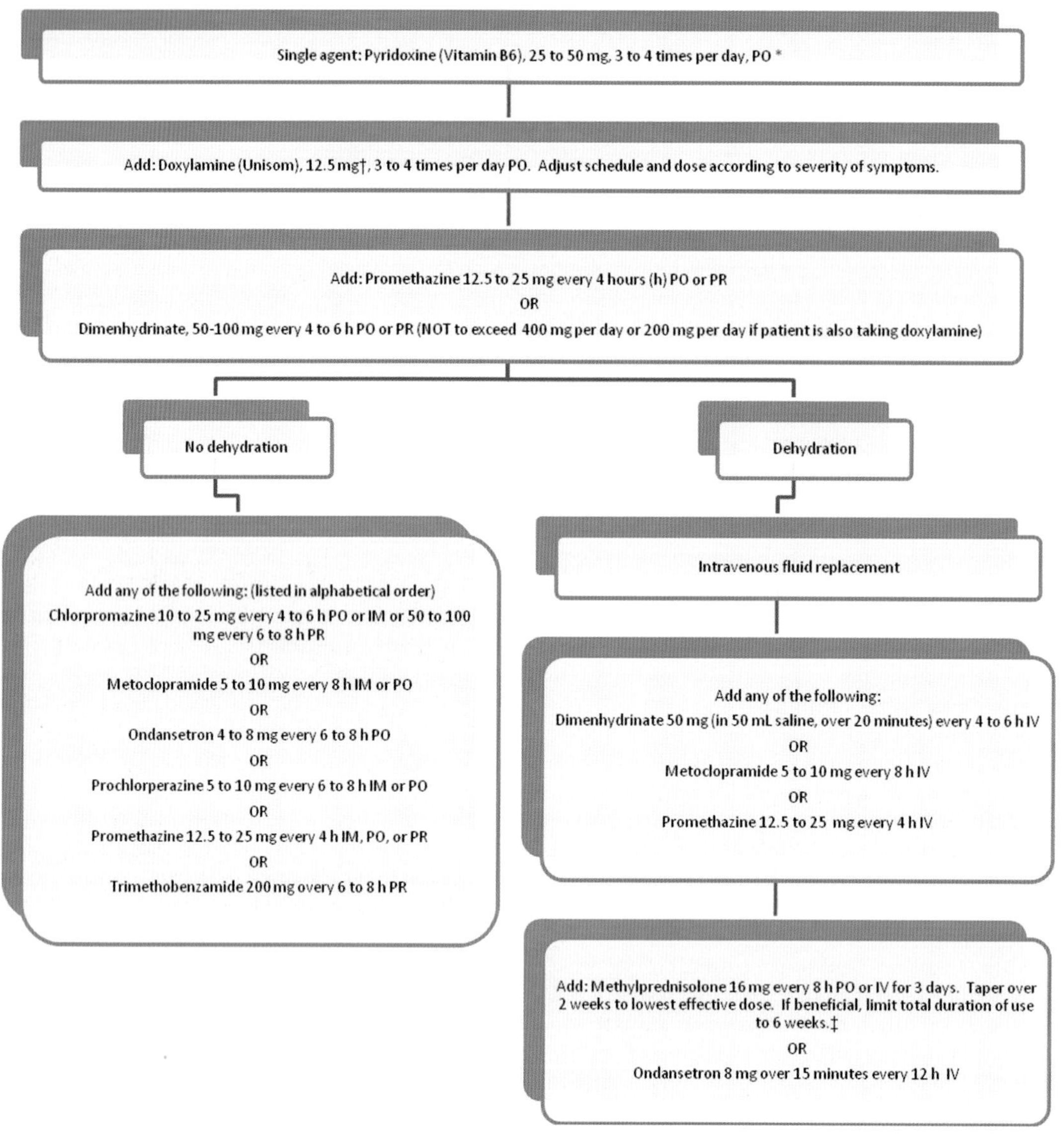

Figure 18.1. Treatment of nausea and vomiting of pregnancy (NVP) (if no improvement, proceed to the next step), *assuming all other causes of NVP have been ruled out.* (Adapted from Einarson A, Maltepe C, Boskovic R, & Koren G. Treatment of nausea and vomiting of pregnancy. An updated algorithm. *Can Fam Physician.* 2007;**53**:2109–11).

The rate ranges between 2 percent and 17 percent (126). The majority of cases of mild gestational hypertension develop at or beyond the thirty-seventh week of gestation.

Why Do We Treat Hypertension in Pregnancy?

Chronic hypertension is associated with several adverse outcomes. These include premature birth, intrauterine growth restriction, fetal death, placental abruption, cesarean delivery, and superimposed preeclampsia (123, 124). Perinatal mortality is also increased in comparison with the general obstetric population. These risks are significant for the health of the developing baby and the mother. Although treatment of nonpregnant women reduces the rate of stroke and cardiovascular morbidity and mortality, there is currently no compelling evidence that short-term antihypertensive therapy is beneficial for the mother in women with low-risk hypertension (123, 124). Women with mild hypertension can often be managed closely without antihypertensive

Table 18.3. Comparison of Commonly Used Antihypertensive Therapies in Pregnancy

Drug class	*Common Drugs*	*Administration*	*Common Side Effects*
β-Blocker	Atenolol, Metoprolol, Propranolol, Labetolol	25–100 mg PO daily 50–200 mg PO BID 40–240 mg PO BID 100 mg PO BID to start (can increase dose and frequency up to 2,400 mg per day) for acute hypertension: 20–40 mg IV every 10–15 min	Postural hypotension, fatigue
Central α-blocker	Methyldopa	250 mg PO daily (can increase dose and frequency up to 3 g per day)	Headache, sedation, dry mouth
Calcium channel blockers	Nifedipine	10 mg PO BID to TID (up to 120 mg per day) (Can use extended release 30–90 mg daily) (For acute treatment 10 mg oral or sublingual every 30 min)	Flushing, headache, nausea
Diuretic	Hydrochlorothiazide, Furosemide	12.5–50 mg PO daily 10–20 mg PO daily	Electrolyte disturbances
Vasodilator	Hydralazine	5–10 mg IV every 15–20 min for acute treatment (if not improved after 30 mg change drug)	Headache, tachycardia

PO = orally; BID = twice daily; TID = three times daily; IV = intravenously.

medications. Some authorities, however, advocate aggressive treatment of both mild and severe hypertension as a means to avoid pregnancy complications (127). Although the treatment of mild chronic hypertension in pregnancy may be inconsistent, most agree that women with more severe hypertension deserve drug therapy. In these women, antihypertensive therapy is needed to reduce the risk of stroke, congestive heart failure, or renal failure (124). In addition, the control of BP in these women may allow for the prolongation of the pregnancy and therefore improved perinatal outcomes.

Gestational hypertension later in pregnancy is treated in much the same way, with the exception that, if the woman is at term, delivery of the baby is usually considered. In addition, any pregnant woman in the third trimester with newly elevated BP will have an evaluation for preeclampsia. Elevated BP associated with both gestational hypertension and preeclampsia usually resolves after delivery of the baby.

What Drugs Are Used to Treat Hypertension in Pregnancy?

Many classes of medications have been used to treat hypertension in pregnancy. The safety information regarding the use of these drugs is not clearly established (124). The information on teratogenicity is limited, with the exception of animal data. Registries provide much of the safety information. A small amount of long-term infant follow-up information with the use of methyldopa and nifedipine is available. A summary of the medications commonly used to treat hypertension in pregnancy is given in Table 18.3. Experts recommend using pharmacotherapy to keep the BP below 140 mmHg systolic and 90 mmHg diastolic.

Angiotensin-converting enzyme (ACE) inhibitors and angiotensin receptor blockers block the conversion of angiotensin-I to the vasoconstrictive angiotensin-II. The use of ACE inhibitors in pregnancy has been associated with skull and renal defects in the fetus. Because of these effects, both ACE inhibitors and angiotensin receptor blockers are contraindicated in pregnancy. Women who become pregnant while taking one of these medications are counseled to stop the drug and to begin a different antihypertensive therapy.

β-Blockers

These drugs act by blocking the β-adrenergic receptors and decreasing sympathetic tone. Propranolol, atenolol, and metoprolol are commonly used. Labetolol, which has a combined α- and β-blockade, is also commonly used and has gained popularity as a first-line agent (124). A systematic review of β-blockers noted an increase in small-for-gestational age infants with the use of β-blockers for chronic hypertension, although this was seen for atenolol and not in the one trial included using labetolol (128). β-Blockers such as labetolol are used in both acute control of hypertension and in long-term treatment of hypertension in pregnancy. Postural hypotension can be seen as a side effect of these medications. Recent evidence in the cardiovascular literature has demonstrated that drug response to β-blockers may be influenced by polymorphic receptors or enzymes. We are unaware of ongoing studies of genetic influences on β-blocker therapy in pregnancy.

Central α-Blocker

One of the most commonly used drugs for treating hypertension in pregnancy is methyldopa, which acts centrally to reduce the sympathetic outflow, decreasing vascular tone. It appears to be relatively safe (123). It is rarely used any more in nonpregnant women and thus is not practical when needing to switch a woman to another antihypertensive drug in the postpartum period (124). Methyldopa may need to be dosed up to four times per day to achieve adequate effectiveness. Its side effects include dry mouth and drowsiness (124). Clonidine is a drug with both oral and transdermal formulations that also acts in this manner but is less commonly used in pregnancy.

Calcium Channel Blockers

Although many calcium channel blockers are currently on the market, nifedipine seems to be the most studied in pregnancy. Other drugs in this class that have been used are nicardipine, verapamil, and diltiazem. These agents have negative inotropic effects and act by blocking the entry of calcium into cells. Common side effects of nifedipine include flushing and headache. Nifedipine can be used both acutely and for long-term hypertension control, but the published experience is limited (129). Although no improvements in pregnancy outcomes over other antihypertensives have been established, no adverse fetal effects have been demonstrated. As discussed in the section on tocolytics, there are potential pharmacogenetic implications for nifedipine as a CYP3A4/5 substrate, but those implications have not been studied in pregnancy.

Diuretics

Diuretics are well established in the treatment of nonpregnant hypertensive patients. Concern has been raised, however, about the safety of diuretics, because they may counteract the normal blood volume expansion associated with pregnancy. However, a systematic review of trials with diuretics failed to demonstrate any adverse perinatal effects (130). Diuretics are options for treating hypertension in pregnancy but may be discontinued if the fetus develops growth restriction or the woman develops preeclampsia (131). Both thiazide diuretics and furosemide have been used with success in pregnancy.

Vasodilators

The main vasodilator used in pregnancy is hydralazine. Nitroprusside is another drug in this category. Hydralazine directly relaxes arterial smooth muscle. This drug has been used for a long time to treat hypertension in pregnancy. It is usually used intravenously in the acute setting, because it has only a weak antihypertensive effect as oral therapy (124, 132). The main side effects of hydralazine are headache and tachycardia.

SUMMARY

Hypertension is common before, during, and after pregnancy. Many different medications can be used for treating hypertension. An individualized approach to antihypertensive therapy in pregnancy is needed. More than one drug is often needed for adequate BP control in pregnancy. Optimizing BP in pregnancy can lead to improved maternal and neonatal outcomes.

REFERENCES

1. Savitz DA, Blackmore CA, & Thorp JM. Epidemiologic characteristics of preterm delivery: etiologic heterogeneity. *Am J Obstet Gynecol.* 1991;**164**:467–71.
2. ACOG Practice Bulletin No. 43, May 2003. Management of preterm labor. *Int J Gynaecol Obstet.* 2003;**82**:127–35.
3. Goldenberg RL. The management of preterm labor. *Obstet Gynecol.* 2002;**100**:1020–37.
4. Crowley P. Prophylactic corticosteroids for preterm birth. *Cochrane Database Syst Rev.* 2000:CD000065.
5. Chapter 36: Preterm birth. In: Cunningham FG, Leveno KJ, Bloom SL, Hauth JC, Rouse DJ, and Spong CY, eds. *Williams Obstetrics.* 23rd ed. New York: McGraw-Hill; 2010:804–831.
6. Haas DM. Preterm birth. *Clin Evid.* 2006:1966–85.
7. Anotayanonth S, Subhedar NV, Garner P, Neilson JP, & Harigopal S. Betamimetics for inhibiting preterm labour. *Cochrane Database Syst Rev.* 2004:CD004352.
8. King JF, Flenady VJ, Papatsonis DN, Dekker GA, & Carbonne B. Calcium channel blockers for inhibiting preterm labour. *Cochrane Database Syst Rev.* 2003: CD002255.
9. Crowther CA, Hiller JE, & Doyle LW. Magnesium sulphate for preventing preterm birth in threatened preterm labour [Systematic Review]. *Cochrane Database Syst Rev.* 2002:CD001060.
10. King J, Flenady V, Cole S, & Thornton S. Cyclo-oxygenase (COX) inhibitors for treating preterm labour. *Cochrane Database Syst Rev.* 2005:CD001992.
11. Kessler RC, Berglund P, Demler O, Jin R, Merikangas KR, & Walters EE. Lifetime prevalence and age-of-onset distributions of DSM-IV disorders in the National Comorbidity Survey Replication. *Arch Gen Psychiatry.* 2005;**62**:593–602.
12. Kessler RC, McGonagle KA, Zhao S, Nelson CB, Hughes M, Eshleman S, et al. Lifetime and 12-month prevalence of DSM-III-R psychiatric disorders in the United States. Results from the National Comorbidity Survey. *Arch Gen Psychiatry.* 1994;**51**:8–19.

13. Weissman MM, Bland RC, Canino GJ, Faravelli C, Greenwald S, Hwu HG, et al. Cross-national epidemiology of major depression and bipolar disorder. *JAMA.* 1996;**276**:293–9.
14. Weissman MM & Olfson M. Depression in women: implications for health care research. *Science.* 1995;**269**:799–801.
15. Kumar R & Robson KM. A prospective study of emotional disorders in childbearing women. *Br J Psychiatry.* 1984;**144**:35–47.
16. Emery J, Watson E, Rose P, & Andermann A. A systematic review of the literature exploring the role of primary care in genetic services. *Fam Pract.* 1999;**16**:426–45.
17. Affonso DD, Lovett S, Paul SM, & Sheptak S. A standardized interview that differentiates pregnancy and postpartum symptoms from perinatal clinical depression. *Birth.* 1990;**17**:121–30.
18. Evans BJ. What will it take to reap the clinical benefits of pharmacogenomics? *Food Drug Law J.* 2006;**61**:753–94.
19. Hostetter A, Stowe ZN, Strader JR Jr, McLaughlin E, & Llewellyn A. Dose of selective serotonin uptake inhibitors across pregnancy: clinical implications. *Depress Anxiety.* 2000;**11**:51–7.
20. American Psychiatric Association. *Diagnostic and Statistical Manual of Mental Disorders.* 4th ed. Washington, DC: American Psychiatric Association; 1994.
21. Kurki T, Eronen M, Lumme R, & Ylikorkala O. A randomized double-dummy comparison between indomethacin and nylidrin in threatened preterm labor. *Obstet Gynecol.* 1991;**78**:1093–7.
22. Sugiura-Ogasawara M, Furukawa TA, Nakano Y, Hori S, Aoki K, & Kitamura T. Depression as a potential causal factor in subsequent miscarriage in recurrent spontaneous aborters. *Hum Reprod.* 2002;**17**:2580–4.
23. Korebrits C, Ramirez MM, Watson L, Brinkman E, Bocking AD, & Challis JR. Maternal corticotropin-releasing hormone is increased with impending preterm birth. *J Clin Endocrinol Metab.* 1998;**83**:1585–91.
24. Orr ST, James SA, & Blackmore Prince C. Maternal prenatal depressive symptoms and spontaneous preterm births among African-American women in Baltimore, Maryland. *Am J Epidemiol.* 2002;**156**:797–802.
25. Orr ST & Miller CA. Maternal depressive symptoms and the risk of poor pregnancy outcome. Review of the literature and preliminary findings. *Epidemiol Rev.* 1995;**17**:165–71.
26. Perkins C, Balma D, & Garcia R. Why current breast pathology practices must be evaluated. A Susan G. Komen for the Cure white paper: June 2006. *Breast J.* 2007;**13**:443–7.
27. Steer CM & Petrie RH. A comparison of magnesium sulfate and alcohol for the prevention of premature labor. *Am J Obstet Gynecol.* 1977;**129**:1–4.
28. Gennaro S, York R, & Brooten D. Anxiety and depression in mothers of low birthweight and very low birthweight infants: birth through 5 months. *Issues Compr Pediatr Nurs.* 1990;**13**:97–109.
29. Hedegaard M, Henriksen TB, Sabroe S, & Secher NJ. The relationship between psychological distress during pregnancy and birth weight for gestational age. *Acta Obstet Gynecol Scand.* 1996;**75**:32–9.
30. Hoffman S & Hatch MC. Depressive symptomatology during pregnancy: evidence for an association with decreased fetal growth in pregnancies of lower social class women. *Health Psychol.* 2000;**19**:535–43.
31. Lobel M, DeVincent CJ, Kaminer A, & Meyer BA. The impact of prenatal maternal stress and optimistic disposition on birth outcomes in medically high-risk women. *Health Psychol.* 2000;**19**:544–53.
32. Bisits A, Madsen G, Knox M, Gill A, Smith R, Yeo G, et al. The Randomized Nitric Oxide Tocolysis Trial (RNOTT) for the treatment of preterm labor. *Am J Obstet Gynecol.* 2004;**191**:683–90.
33. Sondergaard C, Olsen J, Friis-Hasche E, Dirdal M, Thrane N, & Sorensen HT. Psychosocial distress during pregnancy and the risk of infantile colic: a follow-up study. *Acta Paediatr.* 2003;**92**:811–16.
34. Zuckerman H, Shalev E, Gilad G, & Katzuni E. Further study of the inhibition of premature labor by indomethacin. Part II double-blind study. *J Perinat Med.* 1984;**12**:25–9.
35. ACOG Practice Bulletin No. 87, November 2007. Use of psychiatric medications during pregnancy and lactation. *Obstet Gynecol.* 2007;**110**:1179–98.
36. Lyons-Ruth K, Wolfe R, & Lyubchik A. Depression and the parenting of young children: making the case for early preventive mental health services. *Harv Rev Psychiatry.* 2000;**8**:148–53.
37. Weissman MM, Prusoff BA, Gammon GD, Merikangas KR, Leckman JF, & Kidd KK. Psychopathology in the children (ages 6–18) of depressed and normal parents. *J Am Acad Child Psychiatry.* 1984;**23**:78–84.
38. Sawdy RJ, Lye S, Fisk NM, & Bennett PR. A double-blind randomized study of fetal side effects during and after the short-term maternal administration of indomethacin, sulindac, and nimesulide for the treatment of preterm labor. *Am J Obstet Gynecol.* 2003;**188**:1046–51.
39. Hendrick V & Altshuler L. Management of major depression during pregnancy. *Am J Psychiatry.* 2002;**159**:1667–73.
40. Buist A. Managing depression in pregnancy. *Aust Fam Physician.* 2000;**29**:663–7.
41. Coverdale JH, Chervenak FA, McCullough LB, & Bayer T. Ethically justified clinically comprehensive guidelines for the management of the depressed pregnant patient. *Am J Obstet Gynecol.* 1996;**174**:169–73.
42. Robert E. Treatment depression in pregnancy. *N Engl J Med.* 1996;**335**:1056–8.
43. Spinelli MG & Endicott J. Controlled clinical trial of interpersonal psychotherapy versus parenting education program for depressed pregnant women. *Am J Psychiatry.* 2003;**160**:555–62.
44. Einarson A & Koren G. Prescribing antidepressants to pregnant women: what is a family physician to do? *Can Fam Physician.* 2007;**53**:1412–14, 23–5.
45. Kallen B & Otterblad Olausson P. Antidepressant drugs during pregnancy and infant congenital heart defect. *Reprod Toxicol.* 2006;**21**:221–2.
46. Einarson TR & Einarson A. Newer antidepressants in pregnancy and rates of major malformations: a meta-analysis of prospective comparative studies. *Pharmacoepidemiol Drug Saf.* 2005;**14**:823–7.

47. Malm H, Klaukka T, & Neuvonen PJ. Risks associated with selective serotonin reuptake inhibitors in pregnancy. *Obstet Gynecol.* 2005;**106**:1289–96.
48. Wen SW, Yang Q, Garner P, Fraser W, Olatunbosun O, Nimrod C, et al. Selective serotonin reuptake inhibitors and adverse pregnancy outcomes. *Am J Obstet Gynecol.* 2006;**194**:961–6.
49. Alwan S, Reefhuis J, Rasmussen SA, Olney RS, & Friedman JM. Use of selective serotonin-reuptake inhibitors in pregnancy and the risk of birth defects. *N Engl J Med.* 2007;**356**:2684–92.
50. Louik C, Lin AE, Werler MM, Hernandez-Diaz S, & Mitchell AA. First-trimester use of selective serotonin-reuptake inhibitors and the risk of birth defects. *N Engl J Med.* 2007;**356**:2675–83.
51. Tantisira KG, Lake S, Silverman ES, Palmer LJ, Lazarus R, Silverman EK, et al. Corticosteroid pharmacogenetics: association of sequence variants in CRHR1 with improved lung function in asthmatics treated with inhaled corticosteroids. *Hum Mol Genet.* 2004;**13**:1353–9.
52. Austin MP & Mitchell PB. Psychotropic medications in pregnant women: treatment dilemmas. *Med J Aust.* 1998;**169**:428–31.
53. McElhatton PR, Garbis HM, Elefant E, Vial T, Bellemin B, Mastroiacovo P, et al. The outcome of pregnancy in 689 women exposed to therapeutic doses of antidepressants. A collaborative study of the European Network of Teratology Information Services (ENTIS). *Reprod Toxicol.* 1996;**10**:285–94.
54. Wisner KL, Gelenberg AJ, Leonard H, Zarin D, & Frank E. Pharmacologic treatment of depression during pregnancy. *JAMA.* 1999;**282**:1264–9.
55. Wisner KL, Perel JM, & Wheeler SB. Tricyclic dose requirements across pregnancy. *Am J Psychiatry.* 1993;**150**:1541–2.
56. Einarson A, Bonari L, Voyer-Lavigne S, Addis A, Matsui D, Johnson Y, et al. A multicentre prospective controlled study to determine the safety of trazodone and nefazodone use during pregnancy. *Can J Psychiatry.* 2003;**48**:106–10.
57. Kallen B. Neonate characteristics after maternal use of antidepressants in late pregnancy. *Arch Pediatr Adolesc Med.* 2004;**158**:312–16.
58. Kesim M, Yaris F, Kadioglu M, Yaris E, Kalyoncu NI, & Ulku C. Mirtazapine use in two pregnant women: is it safe? *Teratology.* 2002;**66**:204.
59. Rohde A, Dembinski J, & Dorn C. Mirtazapine (Remergil) for treatment resistant hyperemesis gravidarum: rescue of a twin pregnancy. *Arch Gynecol Obstet.* 2003;**268**:219–21.
60. Yaris F, Kadioglu M, Kesim M, Ulku C, Yaris E, Kalyoncu NI, et al. Newer antidepressants in pregnancy: prospective outcome of a case series. *Reprod Toxicol.* 2004;**19**:235–8.
61. Yaris F, Ulku C, Kesim M, Kadioglu M, Unsal M, Dikici MF, et al. Psychotropic drugs in pregnancy: a case-control study. *Prog Neuropsychopharmacol Biol Psychiatry.* 2005;**29**:333–8.
62. Chun-Fai-Chan B, Koren G, Fayez I, Kalra S, Voyer-Lavigne S, Boshier A, et al. Pregnancy outcome of women exposed to bupropion during pregnancy: a prospective comparative study. *Am J Obstet Gynecol.* 2005;**192**:932–6.
63. Chambers CD, Johnson KA, Dick LM, Felix RJ, & Jones KL. Birth outcomes in pregnant women taking fluoxetine. *N Engl J Med.* 1996;**335**:1010–15.
64. Costei AM, Kozer E, Ho T, Ito S, & Koren G. Perinatal outcome following third trimester exposure to paroxetine. *Arch Pediatr Adolesc Med.* 2002;**156**:1129–32.
65. Moses-Kolko EL, Bogen D, Perel J, Bregar A, Uhl K, Levin B, et al. Neonatal signs after late in utero exposure to serotonin reuptake inhibitors: literature review and implications for clinical applications. *JAMA.* 2005;**293**:2372–83.
66. Zeskind PS & Stephens LE. Maternal selective serotonin reuptake inhibitor use during pregnancy and newborn neurobehavior. *Pediatrics.* 2004;**113**:368–75.
67. Kalra S, Einarson A, & Koren G. Taking antidepressants during late pregnancy. How should we advise women? *Can Fam Physician.* 2005;**51**:1077–8.
68. Hendrick V, Smith LM, Suri R, Hwang S, Haynes D, & Altshuler L. Birth outcomes after prenatal exposure to antidepressant medication. *Am J Obstet Gynecol.* 2003;**188**:812–15.
69. Oberlander TF, Misri S, Fitzgerald CE, Kostaras X, Rurak D, & Riggs W. Pharmacologic factors associated with transient neonatal symptoms following prenatal psychotropic medication exposure. *J Clin Psychiatry.* 2004;**65**:230–7.
70. Casper RC, Fleisher BE, Lee-Ancajas JC, Gilles A, Gaylor E, DeBattista A, et al. Follow-up of children of depressed mothers exposed or not exposed to antidepressant drugs during pregnancy. *J Pediatr.* 2003;**142**:402–8.
71. Nulman I, Rovet J, Stewart DE, Wolpin J, Gardner HA, Theis JG, et al. Neurodevelopment of children exposed in utero to antidepressant drugs. *N Engl J Med.* 1997;**336**:258–62.
72. Nulman I, Rovet J, Stewart DE, Wolpin J, Pace-Asciak P, Shuhaiber S, et al. Child development following exposure to tricyclic antidepressants or fluoxetine throughout fetal life: a prospective, controlled study. *Am J Psychiatry.* 2002;**159**:1889–95.
73. Simon GE, Cunningham ML, & Davis RL. Outcomes of prenatal antidepressant exposure. *Am J Psychiatry.* 2002;**159**:2055–61.
74. Cohen LS, Altshuler LL, Harlow BL, Nonacs R, Newport DJ, Viguera AC, et al. Relapse of major depression during pregnancy in women who maintain or discontinue antidepressant treatment. *JAMA.* 2006;**295**:499–507.
75. Use of psychoactive medication during pregnancy and possible effects on the fetus and newborn. Committee on Drugs. American Academy of Pediatrics. *Pediatrics.* 2000;**105**:880–7.
76. Narayan KM, Boyle JP, Thompson TJ, Sorensen SW, & Williamson DF. Lifetime risk for diabetes mellitus in the United States. *JAMA.* 2003;**290**:1884–90.
77. ACOG Practice Bulletin No. 30, September 2001. Clinical management guidelines for obstetrician-gynecologists (replaces Technical Bulletin No. 200, December 1994). Gestational diabetes. *Obstet Gynecol.* 2001;**98**:525–38.
78. Cheung NW, McElduff A, & Ross GP. Type 2 diabetes in pregnancy: a wolf in sheep's clothing. *Aust N Z J Obstet Gynaecol.* 2005;**45**:479–83.
79. Chapter 52: Diabetes. In: Cunningham FG, Leveno KJ, Bloom SL, Hauth JC, Rouse DJ, and Spong CY, eds. *Williams Obstetrics.* 23rd ed. New York: McGraw-Hill; 2010:1104–25.

80. Feig DS & Palda VA. Type 2 diabetes in pregnancy: a growing concern. *Lancet.* 2002;**359**:1690–2.
81. Langer O, Conway DL, Berkus MD, Xenakis EM, & Gonzales O. A comparison of glyburide and insulin in women with gestational diabetes mellitus. *N Engl J Med.* 2000;**343**:1134–8.
82. Emelianova S, Mazzotta P, Einarson A, & Koren G. Prevalence and severity of nausea and vomiting of pregnancy and effect of vitamin supplementation. *Clin Invest Med.* 1999;**22**:106–10.
83. Gadsby R, Barnie-Adshead A, & Jagger C. A prospective study of nausea and vomiting during pregnancy. *Br J Gen Pract.* 1993;**43**:245–8.
84. Mazzotta P, Stewart D, Atanackovic G, Koren G, & Magee LA. Psychosocial morbidity among women with nausea and vomiting of pregnancy: prevalence and association with anti-emetic therapy. *J Psychosom Obstet Gynaecol.* 2000;**21**:129–36.
85. Miller F. Nausea and vomiting of pregnancy: the problem of perception–is it really a disease? *Am J Obstet Gynecol.* 2002;**186**:S182–3.
86. ACOG Practice Bulletin. Nausea and vomiting of pregnancy. *Obstet Gynecol.* 2004;**103**:803–14.
87. Attard C, Kohli M, Coleman S, Bradley C, Hux M, Atanackovic G, et al. The burden of illness of severe nausea and vomiting of pregnancy in the United States. *Am J Obstet Gynecol.* 2002;**186**:S220–7.
88. Mazzotta P, Magee L, & Koren G. Therapeutic abortions due to severe morning sickness. Unacceptable combination. *Can Fam Physician.* 1997;**43**:1055–7.
89. Magee LA, Chandra K, Mazzotta P, Stewart D, Koren G, & Guyatt GH. Development of a health-related quality of life instrument for nausea and vomiting of pregnancy. *Am J Obstet Gynecol.* 2002;**186**:S232–8.
90. Goodwin TM. Nausea and vomiting of pregnancy: an obstetric syndrome. *Am J Obstet Gynecol.* 2002;**186**:S184–9.
91. Sahakian V, Rouse D, Sipes S, Rose N, & Niebyl J. Vitamin B6 is effective therapy for nausea and vomiting of pregnancy: a randomized, double-blind placebo-controlled study. *Obstet Gynecol.* 1991;**78**:33–6.
92. Vutyavanich T, Wongtra-ngan S, & Ruangsri R. Pyridoxine for nausea and vomiting of pregnancy: a randomized, double-blind, placebo-controlled trial. *Am J Obstet Gynecol.* 1995;**173**:881–4.
93. Geiger CJ, Fahrenbach DM, & Healey FJ. Bendectin in the treatment of nausea and vomiting in pregnancy. *Obstet Gynecol.* 1959;**14**:688–90.
94. McGuinness BW & Binns DT. 'Debendox' in pregnancy sickness. *J R Coll Gen Pract.* 1971;**21**:500–3.
95. Wheatley D. Treatment of pregnancy sickness. *Br J Obstet Gynaecol.* 1977;**84**:444–7.
96. Einarson A, Maltepe C, Boskovic R, & Koren G. Treatment of nausea and vomiting in pregnancy: an updated algorithm. *Can Fam Physician.* 2007;**53**:2109–11.
97. Neutel CI & Johansen HL. Measuring drug effectiveness by default: the case of Bendectin. *Can J Public Health.* 1995;**86**:66–70.
98. McKeigue PM, Lamm SH, Linn S, & Kutcher JS. Bendectin and birth defects: I. A meta-analysis of the epidemiologic studies. *Teratology.* 1994;**50**:27–37.
99. Quinlan JD & Hill DA. Nausea and vomiting of pregnancy. *Am Fam Physician.* 2003;**68**:121–8.
100. Seto A, Einarson T, & Koren G. Pregnancy outcome following first trimester exposure to antihistamines: meta-analysis. *Am J Perinatol.* 1997;**14**:119–24.
101. Cartwright EW. Dramamine in nausea and vomiting of pregnancy. *West J Surg Obstet Gynecol.* 1951;**59**:216–34.
102. Newlinds JS. Nausea and vomiting in pregnancy: a trial of thiethylperazine. *Med J Aust.* 1964;**14**:234–6.
103. Fitzgerald JP. The effect of promethazine in nausea and vomiting of pregnancy. *N Z Med J.* 1955;**54**:215–18.
104. Lask S. Treatment of nausea and vomiting of pregnancy with antihistamines. *Br Med J.* 1953;**1**:652–3.
105. Einarson A, Koren G, & Bergman U. Nausea and vomiting in pregnancy: a comparative European study. *Eur J Obstet Gynecol Reprod Biol.* 1998;**76**:1–3.
106. Pinder RM, Brogden RN, Sawyer PR, Speight TM, & Avery GS. Metoclopramide: a review of its pharmacological properties and clinical use. *Drugs.* 1976;**12**:81–131.
107. Vaknin Z, Halperin R, Schneider D, Teitler J, Dar P, Herman A, et al. Hyperemesis gravidarum and nonspecific abnormal EEG findings: a preliminary report. *J Reprod Med.* 2006;**51**:623–7.
108. Buttino L Jr, Coleman SK, Bergauer NK, Gambon C, & Stanziano GJ. Home subcutaneous metoclopramide therapy for hyperemesis gravidarum. *J Perinatol.* 2000;**20**:359–62.
109. Bsat FA, Hoffman DE, & Seubert DE. Comparison of three outpatient regimens in the management of nausea and vomiting in pregnancy. *J Perinatol.* 2003;**23**:531–5.
110. Einarson A, Maltepe C, Navioz Y, Kennedy D, Tan MP, & Koren G. The safety of ondansetron for nausea and vomiting of pregnancy: a prospective comparative study. *BJOG.* 2004;**111**:940–3.
111. Sullivan CA, Johnson CA, Roach H, Martin RW, Stewart DK, & Morrison JC. A pilot study of intravenous ondansetron for hyperemesis gravidarum. *Am J Obstet Gynecol.* 1996;**174**:1565–8.
112. Borrelli F, Capasso R, Aviello G, Pittler MH, & Izzo AA. Effectiveness and safety of ginger in the treatment of pregnancy-induced nausea and vomiting. *Obstet Gynecol.* 2005;**105**:849–56.
113. Nordeng H & Havnen GC. Use of herbal drugs in pregnancy: a survey among 400 Norwegian women. *Pharmacoepidemiol Drug Saf.* 2004;**13**:371–80.
114. Portnoi G, Chng LA, Karimi-Tabesh L, Koren G, Tan MP, & Einarson A. Prospective comparative study of the safety and effectiveness of ginger for the treatment of nausea and vomiting in pregnancy. *Am J Obstet Gynecol.* 2003;**189**:1374–7.
115. Fischer-Rasmussen W, Kjaer SK, Dahl C, & Asping U. Ginger treatment of hyperemesis gravidarum. *Eur J Obstet Gynecol Reprod Biol.* 1991;**38**:19–24.
116. Vutyavanich T, Kraisarin T, & Ruangsri R. Ginger for nausea and vomiting in pregnancy: randomized, double-masked, placebo-controlled trial. *Obstet Gynecol.* 2001;**97**:577–82.
117. Safari HR, Fassett MJ, Souter IC, Alsulyman OM, & Goodwin TM. The efficacy of methylprednisolone in the

treatment of hyperemesis gravidarum: a randomized, double-blind, controlled study. *Am J Obstet Gynecol.* 1998;**179**:921–4.
118. Yost NP, McIntire DD, Wians FH Jr, Ramin SM, Balko JA, & Leveno KJ. A randomized, placebo-controlled trial of corticosteroids for hyperemesis due to pregnancy. *Obstet Gynecol.* 2003;**102**:1250–4.
119. Carmichael SL & Shaw GM. Maternal corticosteroid use and risk of selected congenital anomalies. *Am J Med Genet.* 1999;**86**:242–4.
120. Park-Wyllie L, Mazzotta P, Pastuszak A, Moretti ME, Beique L, Hunnisett L, et al. Birth defects after maternal exposure to corticosteroids: prospective cohort study and meta-analysis of epidemiological studies. *Teratology.* 2000;**62**:385–92.
121. Rodriguez-Pinilla E & Martinez-Frias ML. Corticosteroids during pregnancy and oral clefts: a case-control study. *Teratology.* 1998;**58**:2–5.
122. Shepard TH, ed. *Catalog of Teratogenic Agents.* 10th ed. Baltimore, MD: Johns Hopkins University Press; 2001.
123. ACOG Practice Bulletin No.29. Chronic hypertension in pregnancy. ACOG Committee on Practice Bulletins. *Obstet Gynecol.* 2001;**98**(suppl):177–85.
124. Sibai BM. Chronic hypertension in pregnancy. *Obstet Gynecol.* 2002;**100**:369–77.
125. ACOG Practice Bulletin No. 33, January 2002. Diagnosis and management of preeclampsia and eclampsia. *Obstet Gynecol.* 2002;**99**:159–67.
126. Sibai BM. Diagnosis and management of gestational hypertension and preeclampsia. *Obstet Gynecol.* 2003;**102**:181–92.
127. Easterling TR, Carr DB, Brateng D, Diederichs C, & Schmucker B. Treatment of hypertension in pregnancy: effect of atenolol on maternal disease, preterm delivery, and fetal growth. *Obstet Gynecol.* 2001;**98**:427–33.
128. Magee LA, Elran E, Bull SB, Logan A, & Koren G. Risks and benefits of beta-receptor blockers for pregnancy hypertension: overview of the randomized trials. *Eur J Obstet Gynecol Reprod Biol.* 2000;**88**:15–26.
129. Smith P, Anthony J, & Johanson R. Nifedipine in pregnancy. *BJOG.* 2000;**107**:299–307.
130. Collins R, Yusuf S, & Peto R. Overview of randomised trials of diuretics in pregnancy. *Br Med J (Clin Res Ed).* 1985;**290**:17–23.
131. Report of the National High Blood Pressure Education Program Working Group on High Blood Pressure in Pregnancy. *Am J Obstet Gynecol.* 2000;**183**:S1–22.
132. Chapter 45: Chronic hypertension. In: Cunningham FG, Leveno KJ, Bloom SL, Hauth JC, Rouse DJ, and Spong CY, eds. *Williams Obstetrics.* 23rd ed. New York: McGraw-Hill; 2010:983–995.

19

Pharmacogenomics of Psychiatric Drugs

David Mrazek

The primary objective of this chapter is to provide an overview of the clinical applications of psychiatric pharmacogenomic testing. Essentially, two current primary objectives of testing exist. The first is to identify medications for which a patient is at increased risk for developing side effects or adverse events. The second is to identify medications that are less likely to be effective. Ultimately, the objective of psychiatric pharmacogenomics will be to identify psychotropic medications that have a high probability of achieving a therapeutic response for an individual patient based on the identification of specific genetic variations (1).

Although clinical psychiatric pharmacogenomic testing was being provided at some academic centers before 2004, the introduction of the AmpliChip and its approval by the Food and Drug Administration in 2004 was an important milestone in the clinical adoption of genetic testing that was designed to provide guidance in the selection and dosing of psychiatric medications. Over the next 5 years, many academic medical centers introduced clinical pharmacogenomic testing of drug-metabolizing enzyme genes as a component of the clinical evaluation of patients with atypical responses to medication. As the cost of psychiatric pharmacogenomic testing decreases, there is a growing expectation that genotyping patients before initiating treatment with psychotropic medication will be a cost-effective clinical strategy to minimize adverse events.

CYTOCHROME P450 DRUG-METABOLIZING ENZYME GENES

The cytochrome P450 enzymes have been extensively studied and classified. These enzymes are involved in the metabolism of several hundred currently available medications. Although cytochrome P450 2D6 (CYP2D6) and cytochrome P450 2C19 (CYP2C19) are currently the most widely tested drug-metabolizing enzyme genes, both cytochrome P450 1A2 (CYP1A2) and cytochrome P450 2C9 (CYP2C9) have specific indications for genotyping to improve the outcomes of psychiatric patients. In the near future, the genotyping of other cytochrome P450 enzyme genes such as cytochrome P450 3A4 (CYP3A4), cytochrome P450 3A5 (CYP3A5), and cytochrome P450 2B6 (CYP2B6) is expected to become more widely available.

CYP2D6

CYP2D6 is involved in the metabolism of more than seventy currently available medications. These include many important psychotropic drugs. CYP2D6 is located on chromosome 22 and is highly variable. Currently, there are more than seventy-five formally recognized alleles of CYP2D6, and an additional fifty-five variants that are closely related to the primary alleles, as well. Many of these alleles produce products that have no enzymatic activity. In addition, there are alleles that produce the 2D6 enzyme with decreased activity.

There is considerable variation in allele frequencies across different geographic ancestral populations. For example, the completely inactive *3 allele and the partially active *9 allele are found only in European populations. In contrast, the *10 allele is the most frequent allele found in Japanese populations, but the allele frequency of the *10 allele is only 3 percent in Europe. Similarly, the frequency of the *17 allele is more common in African populations and very rare in either Europeans or Japanese.

The traditional objective of genotyping CYP2D6 is to be able to predict the capacity of an individual patient to metabolize 2D6 substrate medications. By tradition, four phenotypic categories have been established. These are ultrarapid metabolizers, extensive metabolizers, intermediate metabolizers, and poor metabolizers (2). Phenotypic categorization is derived from specified CYP2D6 genotypes. Although different strategies have been used to assign these phenotypes, all methodologies consider individuals with two active copies of CYP2D6 to be extensive or normal metabolizers. Similarly, individuals with three or more active alleles are universally categorized as having ultrarapid metabolism. Individuals who have one active and one inactive allele are generally referred to as intermediate metabolizers. Finally, those with two inactive alleles or one inactive and one deficient allele are generally referred to as poor metabolizers. However, there are categorization methodologies that divide patients into seven levels of metabolic activity in which patients with one inactive and one deficient allele are referred to as decreased intermediate metabolizers.

The CYP2D6 enzyme plays an important role in the metabolism of many tricyclic antidepressants and serotonin selective reuptake inhibitors, as well as venlafaxine. Serious side effects have occurred when patients who have poor metabolic capacity have been treated with 2D6 substrate medications. Exacerbations of most side effects have been reported in patients who were poor metabolizers and treated with venlafaxine (3). Even fatal outcomes have been reported in relatively rare cases of patients treated with 2D6 substrate psychotropic medications (4).

CYP2D6 also plays an important role in the metabolism of many typical and some atypical antipsychotic medications. Haloperidol and perphenazine are typical antipsychotic medications that are widely used and are primarily metabolized by the 2D6 enzyme. Risperidone is an atypical antipsychotic medication that is also primarily metabolized by the 2D6 enzyme. Aripiprazole and olanzapine are atypical antipsychotic medications that are substrates of 2D6. However, aripiprazole and olanzapine each have alternative enzymes that play a role in their metabolism.

CYP2D6 is involved in the metabolism of a number of important analgesics, including codeine and tramadol. These medications are frequently used in patients with pain syndromes who also may have a comorbid psychiatric illness.

There are two additional 2D6 substrate medications widely used that may present specific psychiatric clinical challenges. The first substrate medication is dextromethorphan, which is a cough suppressant. However, dextromethorphan is also a drug of abuse. Dextromethorphan is almost exclusively metabolized by the 2D6 enzyme. The second substrate medication is tamoxifen, which is used in the treatment of breast cancer. Tamoxifen is a prodrug and requires the 2D6 enzyme for its metabolism to endoxifen, which is the active metabolite of tamoxifen. Patients who are poor metabolizers of 2D6 substrate medications have been shown to have less successful clinical outcomes when treated with tamoxifen (5, 6). Given that many serotonin selective reuptake inhibitors are strong inhibitors of the 2D6 enzyme, treatment of breast cancer patients with medications such as paroxetine or fluoxetine can effectively create a phenocopy of a poor metabolizer and decrease the therapeutic effectiveness of tamoxifen.

The first major goal of psychiatric pharmacogenomic testing of CYP2D6 is to identify poor metabolizers who are at increased risk for the development of side effects or adverse events. The second major objective is to identify ultrarapid metabolizers who are unlikely to have a therapeutic response to standard treatment with 2D6 substrate psychotropic medications.

CYP2C19

CYP2C19 is located on chromosome 10 and is involved in the metabolism of several psychotropic medications. Like CYP2D6, CYP2C19 is highly variable, and there are a number of inactive alleles, as well as decreased activity alleles. Recently, the *17 allele has been demonstrated to produce more 2C19 enzyme (7). Consequently, it is now possible to identify patients who have an ultrarapid metabolic 2C19 phenotype on the basis of being homozygous for the *17 allele.

There is also considerable variation in the CYP2C19 allele frequency of different geographic ancestral populations. Both the *2 inactive allele and the inactive *3 allele are more common in Chinese populations than in European populations. However, the newly described enhanced *17 allele is more common in European and African populations than in Chinese populations.

CYP2C19 plays an important role in the metabolism of amitriptyline and clomipramine, and citalopram and escitalopram, as well. It is also the primary enzyme that is involved in the metabolism of diazepam.

CYP1A2

CYP1A2 is located on chromosome 15 and is involved in the metabolism of approximately forty currently available medications. It is somewhat unique among cytochrome P450 genes in that there is one common allele that is upregulated by exposure to environmental inducers. This results in some patients who are functioning as extensive metabolizers when they are not exposed to inducers, but are functioning like ultrarapid metabolizers when they are exposed to inducers. Nicotine and modafinil are important inducers. However, dietary substances such as cruciferous vegetables can also result in induction.

CYP1A2 produces the primary enzyme that is responsible for the metabolism of fluvoxamine. CYP1A2 also plays a substantial role in the primary metabolism of duloxetine and imipramine. The CYP1A2 enzyme is primarily responsible for the metabolism of clozapine and olanzapine. Because some individuals experience induction of CYP1A2 by smoking, the management of these antipsychotic medications in patients who episodically smoke can be clinically challenging.

CYP2C9

CYP2C9 is located on chromosome 10, and variability in CYP2C9 has less impact on the management of psychotropic medications. However, CYP2C9 plays a major role in the metabolism of warfarin and phenytoin.

One reason to genotype patients before prescribing fluoxetine is that patients who have poor metabolic capacity for CYP2D6 and CYP2C9 substrates are at a very increased risk of experiencing adverse events if they are treated with standard doses of fluoxetine. From a practical perspective, these two genotypes can be determined sequentially. The CYP2D6 genotype can be genotyped first. If the patient has adequate CYP2D6 metabolic capacity, the determination of the CYP2C9 metabolic capacity contributes little additional information that would be useful in managing treatment with fluoxetine. However, if the patient has poor metabolic capacity for CYP2D6, the determination of the CYP2C9 metabolic capacity can be very important, because treatment with fluoxetine is contraindicated in patients who are both poor 2D6 metabolizers and poor 2C9 metabolizers.

CATECHOLAMINE-*O*-METHYLTRANSFERASE GENE

The catecholamine-*O*-methyltransferase gene (COMT) produces an enzyme that is responsible for *O*-methylation of catecholamines. COMT codes for two similar enzymes. One is soluble and referred to as S-COMT. The other is membrane bound and is referred to as MB-COMT. COMT is located on chromosome 22.

Although many COMT variants have been identified, the most well-studied single-nucleotide polymorphism (SNP) is rs4680, which for many years has been referred to as the Val158Met polymorphism (8). The valine allele codes for the more active form of the enzyme, and patients who are homozygous for the valine allele have been considered to be at increased risk for the development of schizophrenia.

The allele frequency of rs4680 has been determined for many geographic ancestral populations. The less active methionine allele has an allele frequency of about 47 percent in many European populations. In contrast, the methionine allele has been documented to have an allele frequency of as low as 18 percent in some African and South American populations.

The lower-activity methionine allele has been associated with better treatment response to bupropion (9). Patients with high COMT activity are more likely to respond to typical antipsychotic medication in comparison with patients with lower activity (10). In contrast, patients with low COMT activity did better when treated with atypical antipsychotic olanzapine than patients with high COMT activity (11). In examining methylphenidate response, patients with the more active valine allele have been reported to be more likely to respond to methylphenidate than patients who are homozygous for the methionine allele (12).

NEUROTRANSMITTER TRANSPORTER GENES

Norepinephrine Transporter Gene

The norepinephrine transporter gene (SLC6A2) codes for the norepinephrine transporter protein. SLC6A2 is also referred to as NET or NET1. SLC6A2 is located on chromosome 16.

An association between variation in an SNP of SLC6A2, rs5569, and response to the antidepressant, nortriptyline, has been reported in a Korean sample. Patients who were homozygous for the guanine allele were reported to have a response rate of 83 percent to nortriptyline, compared with patients who had one or two copies of the adenine allele whose response rate was 43 percent (13). Variation of the rs5569 allele of SLC6A2 was also shown to be associated with response to risperidone and olanzapine in French patients. Specifically, patients who were homozygous for the adenine allele were reported to have more improvement of their positive symptoms than those who had a copy of the guanine allele. In this study, a similar association was reported for a second SLC6A2 variation, rs2242446. Patients who were homozygous for the thymine allele, or heterozygous for the cytosine and thymine allele of rs2242446, experienced greater improvement in their positive symptoms when treated with risperidone and olanzapine than did patients who were homozygous for the cytosine allele (14).

A variation in rs5569 of SLC6A2 has also been linked to methylphenidate response. In a Han Chinese sample, patients who had one or two copies of the guanine allele experienced greater symptom reduction than did patients who were homozygous for the adenine allele (47).

Dopamine Transporter Gene

The dopamine transporter gene (SLC6A3) codes for the dopamine transporter protein. SLC6A3 is also referred to as DAT or DAT1. SLC6A3 is located on chromosome 5.

The most widely studied genetic variation of SLC6A3 is rs28363170. This variation has also been referred to as the 40-base pair variable number tandem repeat (VNTR) located in the 3′-untranslated region. The alleles of the VNTR are defined by the number of 40-base pair repeats that are identified. There have been ten reported repeat variants. The shortest is the three-repeat allele. The longest is the thirteen-repeat allele (15). The most common rs28363170 alleles are the nine-repeat allele and the ten-repeat allele. The ten-repeat allele is the most common allele in all populations with the exception of some

Middle Eastern populations where the nine-repeat allele is the most common.

An association between rs28363170 and antidepressant response has been reported. Specifically, individuals who are homozygous for the ten-repeat allele experienced a higher response rate compared with individuals who were homozygous for the nine-repeat allele (16).

Research has also been conducted to identify variations in rs28363170 and response to methylphenidate. Although there does appear to be an association between the ten-repeat allele and a greater risk for developing attention deficit hyperactivity disorder in some populations, the association between variations of rs28363170 and methylphenidate response has been somewhat inconsistent (17–20).

Serotonin Transporter Gene

The serotonin transporter gene (SLC6A4) codes for the serotonin transporter protein. It is also referred to as 5HTT and SERT. SLC6A4 is located on chromosome 17.

The most widely studied variation of the serotonin transporter gene is rs4795541. This variation is also referred to as the indel promoter polymorphism. It has also been referred to as 5HTTLPR and SERTPR. rs4795541 is usually described as having two alleles, which differ in length by 43 or 44 base pairs. However, there are actually many variations of this indel polymorphism despite the convention of dichotomizing individuals as having either the long allele or the short allele (21). There is also an associated SNP, rs25531, that has been reported to influence the overall expression of SLC6A4 in individuals with the long form of the indel promoter (22).

Another important variation of SLC6A4 is usually referred to as the second intron VNTR. The most common alleles of this VNTR variation are the ten-repeat allele and the twelve-repeat allele. However, a nine-repeat and an eleven-repeat allele have also been identified, and the nine-repeat allele may be associated with increased gene expression.

There is considerable variation in the allele frequencies of these SLC6A4 variations in different geographic ancestral populations. The allele frequency of the long allele of the indel promoter polymorphism has been reported to be as high as 83 percent in some African populations and as low as 20 percent in Japanese populations. The allele frequency of the second intron VNTR alleles also varies by geographic ancestry. For example, the allele frequency of the ten-repeat allele has been reported to be as high as 47 percent in some European populations and as low as 2 percent in some Japanese populations.

The most extensive pharmacogenomic studies of variation in SLC6A4 have been linked to antidepressant response. A meta-analysis examined the results of fifteen studies and concluded that, for patients of European geographic ancestry, individuals who were homozygous for the long allele were more likely to have a positive response to selective serotonin reuptake inhibitors than those who were homozygous for the short allele (23). In contrast, some Asian studies have reported a better association with patients who were homozygous for the short allele (13).

In an analysis of the variation in SLC6A4 in a large North American sample, variations in both rs4795541 and the second intron VNTR were analyzed. In patients who identified themselves as being white, but not being Hispanic, an association between both the indel promoter polymorphism and the second intron VNTR variation was demonstrated. In those patients who were homozygous for the long allele of rs4795541 and who were not homozygous for the twelve-repeat allele of the VNTR in intron 2, approximately 76 percent experienced a complete remission when treated with citalopram (24).

The indel promoter variant of SLC6A4 has also been associated with response to clozapine. A variation in rs4795541, when considered within the context of five other gene variants, identified 77 percent of patients who responded to clozapine (25).

SEROTONIN RECEPTOR GENES

Genotyping of the serotonin receptor genes has been more recently introduced into clinical practice. Three serotonin receptor genes are currently available for clinical testing.

Serotonin 1A Receptor Gene

The serotonin 1A receptor gene (HTR1A) is a simple gene that consists of one exon and is located on chromosome 5. Although a number of variations have been identified, the focus of the majority of pharmacogenomic studies has been on a single variant, rs6295. This is a promoter variant at nucleotide location − 1019. In some European samples, the allele frequency of the guanine and cytosine alleles of rs2695 is nearly identical (26). In contrast, a Chinese sample reported an allele frequency of 78 percent for the guanine allele (27).

Patients who are homozygous for the cytosine allele of rs6295 have had a better response to antidepressant treatment with fluvoxamine, citalopram, and fluoxetine (26–28). A similar finding was reported in patients with schizophrenia. Those patients with schizophrenia who were treated with risperidone and olanzapine, and who were homozygous for the cytosine allele of rs6295, were found to have had more improvement in negative symptoms than patients with other genotypes (29).

Serotonin 2A Receptor Gene

The serotonin 2A receptor gene (HTR2A) is located on chromosome 13. rs6311, one of the primary variations of HTR2A that has been studied, is alternatively referred to as −1438G/A. rs6313, which is also referred to as 102T/C, is in strong linkage disequilibrium with rs6311. Another variant that has been more recently associated with antidepressant response is rs7997012.

The adenine allele of rs6311 has generally been associated with better response to antidepressants. However, in a Korean sample, the guanine allele of rs6311 was associated with a better response to citalopram (30). In a large North American sample, rs7997012 was reported to have a positive association with citalopram response. Specifically, patients who were homozygous for the adenine allele of rs7997012 had an approximately 80 percent response to citalopram treatment as opposed to only 64 percent of subjects who were homozygous for the guanine allele (31).

Studies focusing on the adverse effects of antidepressants have shown an association between being homozygous for the cytosine allele of rs6313 and having more difficulties continuing with paroxetine treatment. Specifically, approximately 40 percent of patients who were homozygous for the cytosine allele discontinued their treatment with paroxetine (32).

In samples of European ancestry, it has been repeatedly shown that the adenine allele of rs6311 is associated with a better response to clozapine (25, 33). There has also been some association between variation in 6311 and extrapyramidal effects in patients treated with antipsychotic medications. In a sample of Jewish ancestry, those patients who were homozygous for the guanine allele of rs6311 were more likely to have higher dyskinesia scores (34).

Serotonin 2C Receptor Gene

The serotonin 2C receptor gene (HTR2C) is located on the X chromosome. Variations in HTR2C have been reported to be associated with both a better response to antipsychotic medication and the occurrence of weight gain. The most widely studied variant of HTR2C is rs3813929. This variant is also referred to as −759C/T and is located in the promoter region of the gene.

The most consistent finding related to rs3813929 is that patients who are homozygous for the cytosine allele are at increased risk for weight gain when taking antipsychotic medications. This has been reported in patients of both Chinese (35) and European ancestry (36). As a consequence of this association, the thymine allele of rs3813929 is sometimes referred to as the "protective allele."

DOPAMINE RECEPTOR GENES

Dopamine 2 Receptor Gene

The dopamine 2 receptor gene (DRD2) is located on chromosome 11. Variations of DRD2 have been associated with the development of side effects when taking antipsychotic medication.

Although a number of DRD2 variations have been studied, an rs1801028 variant, which is also referred to as the Ser311Cys polymorphism, has been associated with better antipsychotic response. In addition, patients who are homozygous for the thymine allele of rs1800497, which is also referred to as Taq1A, have been shown to be at an increased risk for weight gain in comparison with patients with other genotypes (37, 38).

Dopamine 3 Receptor Gene

The dopamine 3 receptor gene (DRD3) is located on chromosome 3. The most extensively studied variant in DRD3 is rs6280. rs6280 is also referred to as the Ser9Gly polymorphism. The serine allele of rs6280 has been associated with a more positive response to typical antipsychotic medications (39), whereas the glycine allele has been associated with a more positive response to atypical antipsychotic medications (40). The glycine allele of rs6280 has also been associated with increased risk for the development of tardive dyskinesia (41–43).

Dopamine 4 Receptor Gene

The dopamine 4 receptor (DRD4) is located on chromosome 11. Variants of DRD4 have been associated with differential stimulant medication response. Although a number of variants have been described, the strongest association with medication response has been the variations of the 48-base pair VNTR. This variant is located in exon 3. Several repeat alleles have been described. The four-repeat allele is found in all populations and is considered to be the ancestral allele (44, 45). However, the evolution of the relatively common two-repeat allele and the seven-repeat allele is believed to have occurred before the global level dispersion from South Africa because these alleles are found in most geographic locations (46). Considerable research has examined variations in the VNTR alleles and methylphenidate response. Patients who are homozygous for the four-repeat allele of the VNTR in exon 3 have been more likely to respond positively to methylphenidate than those who are homozygous for the seven-repeat allele (48, 49).

ETHICAL CONSIDERATIONS

The clinical application of psychiatric pharmacogenomics has raised ethical issues that differ from the ethical issues that must be considered in conducting research designed to clarify the influence of gene variations on drug response. There are four broad ethical guidelines that should be considered as part of the clinical implementation of pharmacogenomic testing. The first principle is that clinical testing must be preceded by appropriate consent. This is true for the conduct of any medical procedure. The second principle is that clinical consent should be voluntary, unless there is an overriding public health concern. The third principle is that the confidentiality of medical information must be maintained. Finally, the fourth principle is that pharmacogenomic testing must yield accurate and reliable results. At this point in time, the primary barrier to the implementation to pharmacogenomic testing is related to the cost. As genotyping technology evolves and the cost decreases, it is widely anticipated that the utilization of clinical testing will increase.

APPLICATIONS

At this point in the implementation of psychiatric pharmacogenomic testing, the primary goal of testing has been to minimize the side effects and adverse events. This has been the focus of much of the research related to the genotyping of drug-metabolizing enzyme genes. The other primary consideration is the identification of patients who are unlikely to respond to certain medications at traditional doses. This is particularly well demonstrated by evaluating the outcomes of patients who are 2D6 ultrarapid metabolizers and have been treated with 2D6 substrate medications. A current objective of pharmacogenomic research is to make more accurate predictions of a positive medication response for an individual patient. As more research is conducted, the accuracy of individualizing molecular psychiatric predictions is anticipated to become increasingly precise.

REFERENCES

1. Mrazek DA. *Psychiatric Pharmacogenomics*. New York: Oxford University Press; 2010.
2. Kirchheiner J, Nickchen K, Bauer M, Wong M-L, Licinio J, Roots I, et al. Pharmacogenetics of antidepressants and antipsychotics: the contribution of allelic variations to the phenotype of drug response. *Mol Psychiatry*. 2004;**9**:442–73.
3. McAlpine DE, O'Kane DJ, Black JL, & Mrazek DA. Cytochrome P450 2D6 genotype variation and venlafaxine dosage. *Mayo Clin Proc*. 2007;**82**:1065–8.
4. Sallee FR, DeVane CL, & Ferrell RE. Fluoxetine-related death in a child with cytochrome P-450 2D6 genetic deficiency. *J Child Adolesc Psychopharmacol*. 2000;**10**:27–34.
5. Ingelman-Sundberg M. Genetic polymorphisms of cytochrome P450 2D6 (CYP2D6): clinical consequences, evolutionary aspects and functional diversity. *Pharmacogenomics J*. 2005;**5**:6–13.
6. Goetz MP, Rae JM, Suman VJ, Safgren SL, Ames MM, Visscher DW, et al. Pharmacogenetics of tamoxifen biotransformation is associated with clinical outcomes of efficacy and hot flashes. 2005;**23**:9312–18.
7. Sim SC, Risinger C, Dahl ML, Aklillu E, Christensen M, Bertilsson L, et al. A common novel CYP2C19 gene variant causes ultrarapid drug metabolism relevant for the drug response to proton pump inhibitors and antidepressants. *Clin Pharmacol Ther*. 2006;**79**:103–13.
8. Weinshilboum RM & Raymond FA. Inheritance of low erythrocyte catechol-O-methyltransferase activity in man. *Am J Hum Genet*. 1977;**29**:125–35.
9. Berrettini WH, Wileyto EP, Epstein L, Restine S, Hawk L, Shields P, et al. Catechol-O-methyltransferase (COMT) gene variants predict response to bupropion therapy for tobacco dependence. *Biol Psychiatry*. 2007;**61**:111–18.
10. Illi A, Kampmanc O, Anttilac S, Roivasa M, Mattilab K, Lehtimäkib T, et al. Interaction between angiotensin-converting enzyme and catechol-O-methyltransferase genotypes in schizophrenics with poor response to conventional neuroleptics. *Eur Neuropsychopharmacol*. 2003;**13**:147–51.
11. Bertolino A, Caforio G, Blasi G, De Candia M, Latorre V, Petruzzella V, et al. Interaction of COMT Val108/158 Met genotype and olanzapine treatment on prefrontal cortical function in patients with schizophrenia. 2004;**161**:1798–805.
12. Kereszturi E, Tarnok Z, Bognar E, Lakatos K, Farkas L, Gadoros J, et al. Catechol-O-methyltransferase Val158Met polymorphism is associated with methylphenidate response in ADHD children. *Am J Med Genet B Neuropsychiatr Genet*. 2008;**147B**:1431–5.
13. Kim H, Lim SW, Kim S, Kim JW, Chang YH, Carroll BJ, et al. Monoamine transporter gene polymorphisms and antidepressant response in Koreans with late-life depression. *JAMA*. 2006;**296**:1609–18.
14. Meary A, Brousse G, Jamain S, Schmitt A, Szoke A, Schurhoff F, et al. Pharmacogenetic study of atypical antipsychotic drug response: involvement of the norepinephrine transporter gene. *Am J Med Genet B Neuropsychiatr Genet*. 2008;**147B**:491–4.
15. Sano A, Kondoh K, Kakimoto Y, & Kondo I. A 40-nucleotide repeat polymorphism in the human dopamine transporter gene. *Hum Genet*. 1993;**91**:405–6.
16. Kirchheiner J, Nickchen K, Sasse J, Bauer M, Roots I, & Brockmoller J. A 40-basepair VNTR polymorphism in the dopamine transporter (DAT1) gene and the rapid response to antidepressant treatment. *Pharmacogenomics J*. 2007;**7**:48–55.
17. Winsberg BG & Comings DE. Association of the dopamine transporter gene (DAT1) with poor methylphenidate response. *J Am Acad Child Adolesc Psychiatry*. 1999;**38**:1474–7.
18. Roman T, Szobot C, Martins S, Biederman J, Rohde L, & Hutz M. Dopamine transporter gene and response to methylphenidate in attention-deficit/hyperactivity disorder. *Pharmacogenetics*. 2002;**12**:497–9.
19. Stein MA, Waldman ID, Sarampote CS, Seymour KE, Robb AS, Conlon C, et al. Dopamine transporter genotype and methylphenidate dose response in children with ADHD. *Neuropsychopharmacology*. 2005;**30**:1374–82.
20. Joober R, Grizenko N, Sengupta S, Amor LB, Schmitz N, Schwartz G, et al. Dopamine transporter 3′-UTR VNTR genotype and ADHD: a pharmaco-behavioural genetic study with methylphenidate. *Neuropsychopharmacology*. 2007;**32**:1370–6.

21. Nakamura M, Ueno S, Sano A, & Tanabe H. The human serotonin transporter gene linked polymorphism (5-HTTLPR) shows ten novel allelic variants. *Mol Psychiatry.* 2000;**5**:32–8.
22. Hu XZ, Rush AJ, Charney D, Wilson AF, Sorant AJ, Papanicolaou GJ, et al. Association between a functional serotonin transporter promoter polymorphism and citalopram treatment in adult outpatients with major depression. *Arch Gen Psychiatry.* 2007;**64**:783–92.
23. Serretti A, Kato M, De Ronchi D, & Kinoshita T. Meta-analysis of serotonin transporter gene promoter polymorphism (5-HTTLPR) association with selective serotonin reuptake inhibitor efficacy in depressed patients. *Mol Psychiatry.* 2007;**12**:247–57.
24. Mrazek DA, Rush AJ, Biernacka JM, O'Kane DJ, Cunningham JM, Wieben ED, et al. SLC6A4 variation and citalopram response. *Am J Med Genet Part B.* 2009;**150**:341–51.
25. Arranz MJ, Munro J, Birkett J, Bolonna A, Mancama DT, Sodhi M, et al. Pharmacogenetic prediction of clozapine response. *Lancet.* 2000;**355**:1615–16.
26. Serretti A, Artioli P, Lorenzi C, Pirovano A, Tubazio V, & Zanardi R. The C(-1019)G polymorphism of the 5-HT1A gene promoter and antidepressant response in mood disorders: preliminary findings. *Int J Neuropsychopharmcol.* 2004;**7**:453–60.
27. Yu YW, Tsai SJ, Liou YJ, Hong CJ, & Chen TJ. Association study of two serotonin 1A receptor gene polymorphisms and fluoxetine treatment response in Chinese major depressive disorders. *Eur Neuropsychopharmacol.* 2006;**16**:498–503.
28. Arias B, Catalan R, Gasto C, Gutierrez B, & Fananas L. Evidence for a combined genetic effect of the 5-HT1A receptor and serotonin transporter genes in the clinical outcome of major depressive patients treated with citalopram. *J Psychopharmacol.* 2005;**19**:166–72.
29. Reynolds GP, Arranz B, Templeman LA, Fertuzinhos S, & San L. Effect of 5-HT1A receptor gene polymorphism on negative and depressive symptom response to antipsychotic treatment of drug-naive psychotic patients. *Am J Psychiatry.* 2006;**163**:1826–9.
30. Choi MJ, Kang RH, Ham BJ, Jeong HY, & Lee MS. Serotonin receptor 2A gene polymorphism (-1438A/G) and short-term treatment response to citalopram. *Neuropsychobiology.* 2005;**52**:155–62.
31. McMahon FJ, Buervenich S, Charney D, Lipsky R, Rush AJ, Wilson AF, et al. Variation in the gene encoding the serotonin 2A receptor is associated with outcome of antidepressant treatment. *Am J Hum Genet.* 2006;**78**:804–14.
32. Murphy GM, Kremer C, Rodrigues HE, & Schatzberg AF. Pharmacogenetics of antidepressant medication intolerance. *Am J Psychiatry.* 2003;**160**:1830–5.
33. Arranz M, Collier D, Sodhi M, Ball D, Roberts G, Price J, et al. Association between clozapine response and allelic variation in 5-HT2A receptor gene. *Lancet.* 1995;**346**:281–2.
34. Segman RH, Heresco-Levy U, Finkel B, Goltser T, Shalem R, Schlafman M, et al. Association between the serotonin 2A receptor gene and tardive dyskinesia in chronic schizophrenia. *Mol Psychiatry.* 2001;**6**:225–9.
35. Reynolds GP, Zhang ZJ, & Zhang XB. Association of antipsychotic drug-induced weight gain with a 5-HT2C receptor gene polymorphism. *Lancet.* 2002;**359**:2086–7.
36. Templeman LA, Reynolds GP, Arranz B, & San L. Polymorphisms of the 5-HT2C receptor and leptin genes are associated with antipsychotic drug-induced weight gain in Caucasian subjects with a first-episode psychosis. *Pharmacogenet Genomics.* 2005;**15**:195–200.
37. Xing Q, Qian X, Li H, Wong S, Wu S, Feng G, et al. The relationship between the therapeutic response to risperidone and the dopamine D2 receptor polymorphism in Chinese schizophrenia patients. *Int J Neuropsychopharmcol.* 2007;**10**:631–7.
38. Stice E, Spoor S, Bohon C, & Small DM. Relation between obesity and blunted striatal response to food is moderated by TaqIA A1 allele. *Science.* 2008;**322**:449–52.
39. Scharfetter J. Pharmacogenetics of dopamine receptors and response to antipsychotic drugs in schizophrenia–an update. *Pharmacogenomics.* 2004;**5**:691–8.
40. Szekeres G, Keri S, Juhasz A, Rimanoczy A, Szendi I, Czimmer C, et al. Role of dopamine D3 receptor (DRD3) and dopamine transporter (DAT) polymorphism in cognitive dysfunctions and therapeutic response to atypical antipsychotics in patients with schizophrenia. *Am J Med Genet B Neuropsychiatr Genet.* 2004;**124B**:1–5.
41. Arranz MJ & de Leon J. Pharmacogenetics and pharmacogenomics of schizophrenia: a review of last decade of research. *Mol Psychiatry.* 2007;**12**:707–47.
42. Segman R, Neeman T, Heresco-Levy U, Finkel B, Karagichev L, Schlafman M, et al. Genotypic association between the dopamine D3 receptor and tardive dyskinesia in chronic schizophrenia. *Mol Psychiatry.* 1999;**4**:247–53.
43. Liao DL, Yeh YC, Chen HM, Chen H, Hong CJ, & Tsai SJ. Association between the Ser9Gly polymorphism of the dopamine D3 receptor gene and tardive dyskinesia in Chinese schizophrenic patients. *Neuropsychobiology.* 2001;**44**:95–8.
44. Chang FM, Kidd JR, Livak KJ, Pakstis AJ, & Kidd KK. The world-wide distribution of allele frequencies at the human dopamine D4 receptor locus. *Hum Genet.* 1996;**98**:91–101.
45. Ding Y. Evidence of positive selection acting at the human dopamine receptor D4 gene locus. *Proc Natl Acad Sci USA.* 2002;**99**:309–14.
46. Wang E, Ding Y, Flodman P, Kidd J, Kidd K, Grady D, et al. The genetic architecture of selection at the human dopamine receptor D4 (DRD4) gene locus. *Am J Hum Genet.* 2004;**74**:931–44.
47. Yang L, Wang YF, Li J, Faraone SV. Association of norepinephrine transporter gene with methylphenidate response. *J Am Acad Child Adolesc Psychiatry.* 2004;**43**:1154–8.
48. Kereszturi E, Tarnok Z, Bognar E, Lakatos K, Farkas L, Gadoros J, et al. Catechol-O-methyltransferase Val158Met polymorphism is associated with methylphenidate response in ADHD children. *Am J Med Genet B Neuropsychiatr Genet.* 2008;**147B**:1431–5.
49. Cheon K-A, Kim B-N, & Cho, S-C. Association of 4-repeat allele of the dopamine D4 receptor gene exon III polymorphism and response to methylphenidate treatment in Korean ADHD children. *Neuropsychopharmacology.* 2007;**32**:1377–83.

20 Pain and Anesthesia

Konrad Meissner and Evan D. Kharasch

Pain is a fundamental biological response to noxious stimuli, comprising both unpleasant physical perceptions and negative emotions. It signals actual or potential tissue damage, and elicits protective responses needed for survival, such as withdrawal (1). There is profound interindividual variability in the response to noxious stimuli, susceptibility to pain, and the response to analgesics. This chapter will review the genetic factors that influence pain and sensitivity, and the response to drugs for acute and chronic pain therapy. Issues associated with analgesics, such as tolerance, dependence, and withdrawal, will not be addressed. Significant advances in understanding the neurobiology of pain have been made using genetic models (1). This chapter will focus on human pain and analgesia.

General anesthesia, whose goal is to render a patient insensitive to pain and control physiologic responses in the perioperative period, has as its two most basic elements analgesia and hypnosis ("sleep"). Nevertheless, a typical anesthetic comprises drugs from numerous classes, including benzodiazepine anxiolytics, sedative-hypnotics, inhalation anesthetics, opioids, muscle relaxants and their antagonists, and cardiovascular/vasoactive drugs (so-called balanced anesthesia). This chapter will also review the genetic factors that influence the response to anesthetics, with a focus on drug classes not covered in other chapters. Because opioids are central to both anesthesia and pain therapy, these will be a central focus. It is instructive to note that the field of pharmacogenetics had its birth in anesthesiology, with the discovery by Werner Kalow of heritable responses to the muscle relaxant succinylcholine (2, 3).

PAIN

Responses to pain are complex and depend on genetic and environmental factors, and on the type of pain, as well. Information on genetic contributions to pain is derived from studies in twins, identification of genetically based pain syndromes, and gene association studies.

Twin studies of chronic pain typically report 30 percent to 60 percent heritability (the proportion of variance explained by genetic factors). For example, with the use of experimental acute-pain models, the heritability of heat and cold pain sensitivity was estimated at 10 percent to 20 percent (4), whereas more recently it was estimated at 26 percent for heat pain and 60 percent for cold pain (5). The heritability of chronic pelvic pain and fibromyalgia was estimated at 40 percent to 50 percent (4). By comparison, less than 10 percent of the variability was considered to be environmental (5).

One major category of genetically based pain syndromes are the monogenic disorders that cause pathological pain or congenital insensitivity to pain (1, 6). The former disorders include familial hemiplegic migraine and others. Patients with the latter, generally hereditary sensory and autonomic neuropathy types I to V, have a diffuse incapacity to perceive all or most pain modalities (4). Various genes have been implicated in these disorders, but the one of greatest interest is *SNC9A*, which encodes the voltage-gated $Na_v1.7$ channel. Impulse propagation in peripheral neurons depends on voltage-gated sodium channels, and a loss-of-function mutation in the $Na_v1.7$ α-subunit causes a channelopathy and a congenital inability to experience pain (7). $Na_v1.7$ mutants cause various neuropathic pain disorders (8). Several drugs used to treat neuropathic pain, including local anesthetics, nonselectively block sodium channels. Because humans with nonfunctional $Na_v1.7$ channels have no phenotype other than pain insensitivity, selective blockers of this target are the focus of much drug development activity, because it potentially portends analgesia without side effects.

Genetic association studies of pain and analgesia in humans have identified more than twenty genes that appear to alter clinical or experimental pain and analgesia (1). Three prominent examples, which also illustrate the murky nature of genetic association studies, include polymorphisms in the catechol-*O*-methyltransferase (*COMT*), melanocortin-1 receptor (*MC1R*), and guanosine triphosphate cyclohydrolase 1 (*GCH1*) genes (1). The *COMT* G472A val158met variant in humans was linked to altered pain sensitivity; individuals carrying the met/met genotype that codes for diminished COMT activity had lower pain tolerance (higher sensory and affective pain ratings) in response to an experimental pain stimulus. Follow-up studies showed that pain susceptibility was inversely related to *COMT* activity (9), but replication of the results suggested that different loci were responsible for distinct pain characteristics (10). Moreover, a postoperative pain study failed to show an association with the *COMT* genotype (11). Other inconsistencies have also been reported; thus, the role of *COMT* variants remains incompletely understood. *MC1R* variants coding for a loss of function were associated with reduced sensitivity to electrical pain (12). Conversely, women with red hair, which is also coded for by these *MC1R* variants, were reported to have increased sensitivity to thermal pain (6). These contradictory findings remain unresolved. Most recently, *GCH1*, which synthesizes tetrahydrobiopterin, was found to be a key modulator of peripheral neuropathic and inflammatory pain in animals, and a relatively common *GCH1* haplotype (15 percent allele frequency) in humans was associated with a lower degree of persistent back pain after spine surgery (13). Nevertheless, the changes in *GCH1* activity were considered to be modest (6), and follow-up studies were contradictory (1). In another example, genetic variations in genes coding for interleukin 1β (*IL-1β*) and its endogenous receptor antagonist (*IL-1Ra*) have been associated with differences in pain sensitivity and postoperative morphine consumption (14). Experimental data using thermal and cold stimuli in healthy volunteers suggested an influence of known single-nucleotide polymorphisms (SNPs) in genes coding for the *vanilloid receptor subtype 1 (TRVP1)* (A18753G, associated with a higher cold pressor pain tolerance in females but not males), and the *delta opioid receptor subtype 1 (OPRD1)* (T80G, associated with lower pain scores for cold pressor pain) (15). In general, although studies reporting gene associations with pain and analgesia have generated considerable interest, few have been replicated, those which were replicated were often contradictory, and the gene effects appear small (1).

ANALGESIA

Analgesics are generally classified as opioid or nonopioid, with the former used to treat moderate to severe pain, and the latter used to treat mild pain or as first-line drugs in the treatment of more severe pain to which opioids are added. Opioids, because they are a component of most anesthetic regimens, are discussed in "Anesthesia."

Nonopioid analgesics include nonsteroidal anti-inflammatory drugs (NSAIDs, such as ibuprofen, diclofenac, and naproxen) and acetaminophen. The majority of NSAIDs are metabolized by CYP2C9 (16). *CYP2C9* variant alleles code for enzymes with diminished catalytic activity and reduced NSAID metabolism, and individuals with *CYP2C9* polymorphisms have impaired clearance and increased plasma concentrations of the NSAIDs studied to date. Examination of the clinical implications of these polymorphisms has focused mainly on the side effects rather than clinical effectiveness. In general, the risk for upper gastrointestinal tract side effects (mainly gastric bleeding) is higher in *CYP2C9* poor metabolizers. For example, the odds ratio for bleeding is 2.5 for heterozygous and 3.7 for homozygous carriers of variant *CYP2C9* alleles (16).

ANESTHESIA

Anesthesia practice is unusual, in that (a) a practitioner is continually present and personally administering drugs, as opposed to writing a prescription for medications taken later as an outpatient, and (b) intraoperative drug administration occurs exclusively parenterally. Therefore, in clinical operating room practice, an anesthesiologist constantly titrates drugs to effect, and genetic variabilities in intestinal absorption and first-pass metabolism are not a clinical concern. Nevertheless, genetic variability does contribute to other types of pharmacokinetic (distribution, metabolism, and excretion) and pharmacodynamic (receptors, ion channels, enzymes) variability, and can be a major factor in the predictability of response. In clinical pain practice, where oral opioid use is common, genetic determinants of intestinal absorption and first-pass metabolism are clinically relevant.

Intravenous Anesthetics

The intravenous anesthetics (in general, sedative-hypnotics) are used to produce unconsciousness and "induce" anesthesia. Most commonly used are propofol, thiopental, etomidate, and ketamine. Propofol is often given by continuous infusion to maintain unconsciousness. The molecular target for ketamine is the *N*-methyl-D-aspartate (NMDA) receptor, whereas for the others it is the γ-aminobutyric acid A ($GABA_A$) receptor. There are no known clinically relevant genetic polymorphisms in these receptors that alter intravenous anesthetic pharmacodynamics or dose requirements (17, 18). In general, much is known about the metabolism of intravenous anesthetics, but there is little evidence for clinically relevant genetic polymorphisms (19). Both ketamine and propofol are metabolized by CYP2B6, which is highly polymorphic. Although interindividual differences in human liver microsomal propofol hydroxylation are correlated with CYP2B6 expression and activity, and certain CYP2B6 variants have altered activity, no CYP2B6 polymorphisms have to date been linked to altered propofol metabolism, and propofol clearance has not been altered in patients with *CYP2B6*5* and **6B* haplotypes, in comparison with wild-type patients (18). Propofol and its hydroxylated metabolite also undergo glucuronidation by UGT1A9, although no influence of UGT1A9 polymorphisms on propofol disposition have been identified.

Inhaled Anesthetics

Inhaled anesthetics, which include the halogenated volatile anesthetics (halothane, enflurane, isoflurane, sevoflurane, and desflurane) and nitrous oxide, are used to maintain unconsciousness and to produce some analgesia. The volatile anesthetics act at the $GABA_A$ receptor and are metabolized mainly by CYP2E1, whereas nitrous oxide acts at the NMDA receptor and is not metabolized. Other than the above-described report of a higher sevoflurane requirement in women with red hair, which is associated with mutations in the *MC1R* gene, in comparison with dark-haired women (6), there are no known clinically relevant genetic polymorphisms altering the pharmacodynamics, pharmacokinetics, or dose requirements for the volatile anesthetics. Volatile anesthetic biotransformation is well characterized, and these drugs undergo varying extents of metabolism, catalyzed mainly by CYP2E1, and to a lesser extent (for halothane) by CYPs 2A6 and 3A4 (19). Although genotypic variants in these CYPs are well described, there are no known clinically relevant implications of such variants for volatile anesthetic metabolism.

An early example of a pharmacogenetically based, albeit to this day still unsolved, adverse drug reaction is "halothane hepatitis" (20). This is a rare (1 in 6,000 to 35,000 patients), but often fatal, fulminant hepatic necrosis with severe jaundice. In a series of elegant investigations that were among the first to elucidate the immunologic and metabolic basis of drug hypersensitivities, halothane hepatitis was identified as a hypersensitivity phenomenon initiated by oxidative halothane metabolism. The metabolite, trifluoroacetyl chloride, can inconsequentially form trifluoroacetic acid or bind to several liver proteins. In susceptible (for reasons not yet understood) individuals, in the seminal toxification event, these trifluoroacetylated proteins act as neoantigens to stimulate the formation of antibodies. Upon subsequent exposure to halothane or other structurally related volatile anesthetics (enflurane, isoflurane, and desflurane), and a renewed burst of protein trifluoroacetylation, these antibodies mediate an immune response causing hepatic necrosis. Although all patients anesthetized with halothane form trifluoroacetylated proteins, only rarely does this stimulate the immune system. Although a genetically based immune etiology has long been suspected, the mechanism has never been elucidated.

The predominant reason to focus on volatile anesthetics with respect to genetic polymorphisms is malignant hyperthermia (MH), a rare (1 in 5,000 to 50,000 patients) skeletal muscle disorder that is inherited in an autosomal-dominant fashion (http://medical.mhaus.org) (21). MH patients are otherwise phenotypically normal, and the syndrome presents clinically as a hypermetabolic disorder in response to "triggering agents" (specifically volatile anesthetics and the muscle relaxant succinylcholine), and is usually fatal if untreated (although mortality today is less than 5 percent with proper therapy). Mechanistically, excessive calcium release from the sarcoplasmic reticulum and dysfunctional magnesium regulation eventually results in a generalized hypermetabolic syndrome, which leads to hyperthermia, tachycardia, acidosis, and rhabdomyolysis. MH is caused by a defect in a sarcoplasmic reticulum calcium channel, termed the ryanodine receptor (RYR). The estimated prevalence of the genetic abnormality is 1 in 3,000 to 8,500 patients. More than 100 mutations in the *RYR1* gene have been described to date, at least thirty are known to cause MH, and they account for approximately 50 percent of MH cases (21). At least six other genetic loci have been implicated in MH, including a mutation in the gene for the voltage-dependent calcium channel (*CACNA13*) that codes for the dihydropyridine receptor. Diagnosis of MH or MH-susceptibility currently uses a phenotypic assay, the "in vitro contracture test," also called the "caffeine halothane contracture test." An excised muscle biopsy is exposed to halothane and caffeine, and the test is positive when there is an abnormal hypercontractile response.

Identification of a genetic basis for MH stimulated the attempted development of a DNA test for clinically diagnosing MH or susceptibility to MH. Unfortunately, this has not been successful. DNA testing has been confounded by the genetic heterogeneity and metabolic complexity of MH, the multiplicity of mutations identified, the discordance between mutations and the contracture test, and the false positives and false negatives; at present, the predictive value of genetic testing has been suggested to be only about 50 percent (21). Therefore, despite the identification of causative mutations and advanced DNA testing, the in vitro contracture test remains the standard test to diagnose MH. For the identification of carriers within a family, DNA testing is helpful but does not rule out MH susceptibility, and current guidelines recommend contracture testing before genetic testing in a family. Practically, patients with a prior MH episode, or their relative, are anesthetized using "non-triggering" anesthetics.

Nitrous oxide is one of the most widely used inhalational agents worldwide. The only relevant pharmacogenetic issue pertains to the ability of nitrous oxide to irreversibly oxidize the cobalt atom of cobalamin (vitamin B_{12}), and inhibit enzymes by using B_{12} as a cofactor, most notably methionine synthase. In the folate cycle, the enzyme 5,10-methylenetetrahydrofolate reductase (*MTHFR*) converts 5,10-methylenetetrahydrofolate to 5-methyltetrahydrofolate, which in turn donates a methyl group for the methionine synthase-catalyzed conversion of homocysteine to methionine. Two common SNPs in the *MTHFR* gene, which together have a combined prevalence of approximately 20 percent in individuals of European ancestry, code for reduced enzyme activity and thus diminished formation of 5-methyltetrahydrofolate and, consequently, methionine. Exposure to nitrous oxide of individuals with reduced MTHFR activity further impairs methionine formation, and may lead to substantial neurologic deficits because of numerous impaired biochemical pathways including synthesis of the myelin sheath (similar to the effects of vitamin B_{12} deficiency) (22). Serum homocysteine concentrations are increased after nitrous oxide exposure, and even more so in those homozygous for the *MTHFR C677T* SNP (23). Clinical consequences of increased homocysteine concentrations are not yet known.

Neuromuscular Blocking Drugs

Anesthesiologists need muscle relaxation to perform endotracheal intubation and to enable surgery on the abdomen and sometimes the limbs. Indeed, the discovery of the muscle relaxant effects of curare in the first half of the last century, and later succinylcholine, in part, enabled modern anesthesia practice. The pharmacologic target is the neuromuscular junction, which consists of a complex system of presynaptic nerve endings that release acetylcholine, postsynaptic acetylcholine receptors that cause muscle contraction upon activation, and postsynaptic acetylcholinesterase enzymes that hydrolyze the acetylcholine and terminate its effect. Neuromuscular blocking drugs either mimic the effect of acetylcholine ("depolarizing blockers," causing profound depolarization and, hence, muscle relaxation), or noncompetitively block the acetylcholine receptor, exclude acetylcholine, and thereby prevent depolarization and muscle contraction ("nondepolarizing blockers"). Effects of the depolarizing blocker succinylcholine are terminated by hydrolysis of the drug. Effects of nondepolarizing blockers are terminated by administering an inhibitor of acetylcholinesterase, thereby causing the accumulation of acetylcholine at the neuromuscular junction, which "out-competes" and displaces the nondepolarizing blocker from the acetylcholine receptor.

Succinylcholine is the only depolarizing neuromuscular blocker. Succinylcholine is hydrolyzed by butyrylcholinesterase (also known as plasma cholinesterase, serum cholinesterase, pseudocholinesterase, and nonspecific cholinesterase), which is a different enzyme than acetylcholinesterase at the neuromuscular junction (24). Butyrylcholinesterase is widely distributed in blood, liver, and other tissues, has an unknown physiologic role, but great relevance to drug metabolism and pharmacogenetics. Succinylcholine hydrolysis is rapid, with 90 percent of a dose metabolized within one minute, and presystemic clearance is substantial (most of an intravenous dose is hydrolyzed before it reaches the tissues). Butyrylcholinesterase also metabolizes other esters, including the nondepolarizing muscle relaxant mivacurium, and the local anesthetics cocaine, procaine, chloroprocaine, and tetracaine.

Butyrylcholinesterase was one of the first enzymes for which pharmacogenetic variation was elucidated, based on heritable interindividual variability in the response to succinylcholine (24). Beginning in the 1950s, cases of prolonged succinylcholine paralysis, termed *suxamethonium (succinylcholine) apnea*, alarmed clinicians. By analyzing the heritability of postanesthetic apnea, Werner Kalow discovered a genetic mutation as the cause (2, 3), and also published the world's first textbook on pharmacogenetics, in 1962. Modern molecular genetics techniques have identified several butyrylcholinesterase polymorphisms with clinical consequence. Approximately 1 in 2,000 patients is homozygous for the "atypical" A209G variant with markedly reduced activity, approximately 1 percent of patients have the milder G1615A "K" variant with 66 percent reduced activity, and more than sixty rare mutations have been identified to date. Mivacurium is another neuromuscular blocker, which is hydrolyzed by butyrylcholinesterase at a rate 70 percent to 90 percent of that of succinylcholine, and patients with butyrylcholinesterase variants respond similarly to mivacurium and succinylcholine.

Unexpected prolonged apnea (inability to breathe) after succinylcholine or mivacurium, because of an otherwise unknown butyrylcholinesterase variant, is managed clinically by mechanically ventilating a patient until the drug ultimately wears off. Although phenotypic tests of butyrylcholinesterase activity have been available for four decades, and genotyping is also now available, it is generally considered that the rarity of butyrylcholinesterase deficiency (and increased use of other muscle relaxants) renders routine phenotypic or genotypic screening cost-ineffective (25). In contrast, retrospective genotyping of patients is recommended in cases of prolonged succinylcholine paralysis, and prospective testing is recommended for their relatives. Indeed, such testing was considered (in 2006) as one of only two tests for drug-metabolizing enzymes that were considered relevant or potentially clinically relevant (25).

Butyrylcholinesterase deficiency is one of the most widely recognized examples of a genetic defect in drug metabolism that affects clinical outcome.

Local Anesthetics

Local anesthetics exert their therapeutic and many side effects by blocking one or more of the nine isoforms of voltage-gated sodium channels, and preventing the generation and propagation of action potentials in nerves and other excitable tissues. Mutations in various sodium channels alter the binding of local anesthetics; however, the influence of polymorphisms on local anesthetic effects and any clinical significance are unknown (26). Loss-of-function mutations in the *SNC5A* cardiac sodium channel $Na_V1.5$ cause the Brugada syndrome, characterized by electrocardiographic abnormalities, malignant arrhythmias, and sudden death in patients with otherwise normal hearts. Rare case reports suggest that the local anesthetic bupivacaine can unmask a subclinical form of the Brugada syndrome, causing life-threatening arrhythmias. Overall, however, few clinically significant genetic polymorphisms that affect local anesthetics have been identified to date.

Opioids

Opioids constitute an essential component of anesthesia and pain therapy. The clinical response to opioids varies substantially among patients, because of numerous factors, only one of which is genetics. For example, morphine dose requirements for postoperative analgesia after hip replacement surgery in more than 3,000 patients varied more than fortyfold, and similar differences are observed for the treatment of cancer pain (15). Most opioids used clinically are μ-receptor agonists and act in the central nervous system (brain and spinal cord) to relieve pain. The therapeutic index of opioids is relatively narrow, with respiratory depression the most clinically significant unwanted side effect, which, if severe enough, can be fatal. The need to provide adequate pain relief, together with the narrow therapeutic index of opioids, has driven the intense interest in opioid pharmacogenetics. Polymorphisms in proteins mediating opioid metabolism, transport, pharmacokinetics, and pharmacodynamics have been identified. Evidence unambiguously supporting pharmacogenetic contributions to opioid variability derives from laboratory, patient, and human volunteer studies (15, 27, 28). However there are also a plethora of reports that are inconsistent if not contradictory, or based purely on speculation (extrapolating from a known pathway or process and an identified polymorphism, to suggest some actual biological effect and/or clinical significance of the polymorphism), and of unproven validity and application.

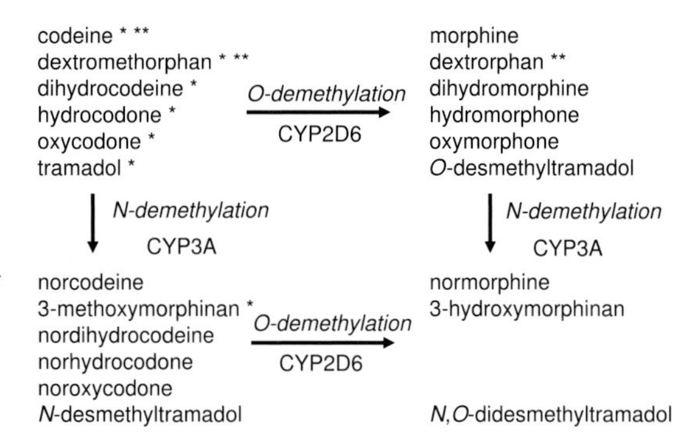

Figure 20.1. Role of CYP2D6 and CYP3A4 in the bioactivation and inactivation of oral opioids. Pathways known to be metabolized by CYP2D6 and CYP3A4 are indicated by * and **, respectively.

Phase I Metabolism

One of the most important genetic polymorphisms affecting opioid analgesia, and what is now a classic story in pharmacogenetics, is the cytochrome P4502D6 (CYP2D6) polymorphism affecting codeine metabolism and analgesia (29, 30). Codeine is now known to be a prodrug, requiring CYP2D6-catalyzed *O*-demethylation and bioactivation to morphine (Figure 20.1). CYP2D6-deficient individuals ("poor metabolizers") minimally demethylate codeine and have markedly diminished or absent morphine formation, lower plasma and cerebrospinal fluid morphine concentrations, and minimal if any analgesia. Patients who are unrecognized CYP2D6 poor metabolizers, given codeine-containing analgesic preparations for pain, will unfortunately derive no therapeutic analgesic benefit from such drugs. If attempting to obtain pain relief by increasing their codeine consumption, such patients may incorrectly be labeled "problematic" or "drug seekers." Conversely, individuals with CYP2D6 gene duplication ("ultrarapid metabolizers") have greater morphine formation, and may be predisposed to toxicity, such as respiratory depression. For example, unfortunate instances of fatal neonatal opioid toxicity have occurred when mothers who were unsuspected CYP2D6 ultrarapid metabolizers used codeine while breastfeeding their infant. In a typical European population, the frequency distribution of CYP2D6 activity is 80 percent to 90 percent extensive and intermediate metabolizers (normals), 7 percent to 8 percent poor metabolizers, and 2 percent to 10 percent ultrarapid metabolizers. The CYP2D6 genetic polymorphism exhibits ethnic and racial diversity. The frequency of poor metabolizers is 7 percent to 8 percent of whites, 2 percent of African Americans, and <1 percent of Asians, whereas that of ultrarapid metabolizers is 30 percent of Ethiopians, 10 percent of Spanish and Italians, 1 percent to 2 percent of Northern Europeans, and absent in Asians (31). Ethnic differences in codeine effects attributable to variable bioactivation have been observed. For example,

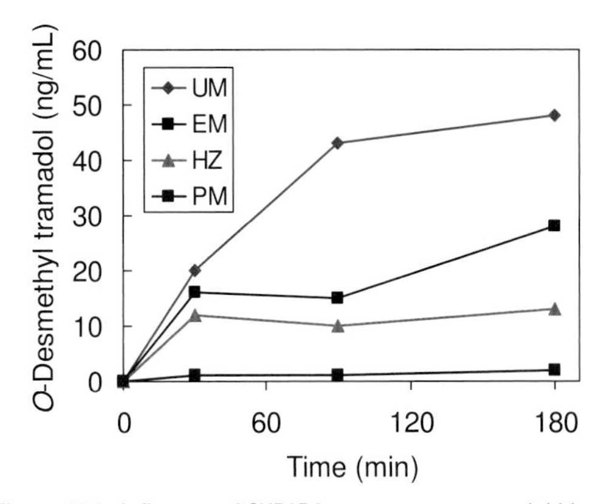

Figure 20.2. Influence of CYP2D6 genotype on tramadol bioactivation to *O*-desmethyltramadol. Shown are plasma concentrations of *O*-desmethyltramadol (mean ± SEM) in CYP2D6 ultrarapid metabolizers (UM), extensive metabolizers (EM), heterozygotes (HZ), and poor metabolizers (PM) after intravenous tramadol (3 mg/kg). Redrawn from Stamer and Stuber with permission (29).

Chinese produced less morphine from codeine than did whites and showed diminished opioid effects, and were therefore also expected to experience reduced analgesia.

From the perspective of influence on clinical pain treatment, the *CYP2D6* genetic polymorphism affecting codeine (and tramadol, below) is the most important example of opioid pharmacogenetics. Even in subjects with "wild-type" *CYP2D6* alleles, there is a range of CYP2D6 activity and considerable variability in codeine demethylation to morphine.

Several other opioids also undergo CYP2D6-mediated *O*-demethylation. The clinical significance of such metabolism and CYP2D6 polymorphisms depends on the extent to which metabolism constitutes a bioactivation pathway (i.e., rate of metabolism and μ-receptor affinity of the metabolite). Tramadol undergoes CYP2D6-catalyzed metabolism to the main product *O*-desmethyltramadol (Figure 20.1), which has μ-receptor affinity 200 times greater than that of the parent drug (29). CYP2D6 polymorphisms markedly influence the formation rate and plasma concentrations of *O*-desmethyltramadol (Figure 20.2), as well as clinical dose requirements and the quality of analgesia. CYP2D6 poor metabolizers have lower plasma *O*-desmethyltramadol concentrations, higher loading dose requirements, increased tramadol consumption, and a greater need for rescue analgesic medication. In contrast, other opioids undergoing CYP2D6-catalyzed metabolism, even to active metabolites, have not been found (to date) to have altered the clinical effects in patients with CYP2D6 polymorphisms, even though metabolism in vitro and/or in vivo has been affected (29). Such opioids include hydrocodone, oxycodone, and dihydrocodeine. This may result from smaller differences between parent and metabolite μ-receptor affinities, relative amounts of active metabolite formed, or other factors. *O*-demethylation of the opioid dextromethorphan to dextrorphan is a standard phenotypic probe for the activity of CYP2D6.

CYP3A4, which metabolizes about half of all therapeutic drugs, also metabolizes numerous opioids. The synthetic piperidine class of opioids, including fentanyl, sufentanil, and alfentanil, all undergo nearly complete metabolism, mainly by *N*-dealkylation, whereas the morphinoids (such as codeine) undergo CYP3A-catalyzed *N*-demethylation (in addition to metabolism by CYP2D6) (Figure 20.2). All CYP3A-dependent opioid metabolites are inactive. The polymorphically expressed CYP3A5 variant, which has similar (but not totally overlapping) substrate specificity with CYP3A4, metabolizes many of the CYP3A4 opioid substrates. For example, CYP3A5 metabolized alfentanil at rates similar to CYP3A4, and liver microsomes from individuals with at least one *CYP3A5*1* allele (thus expressing active CYP3A5 enzyme) metabolized alfentanil at significantly (about five- to tenfold) higher rates than did those from *CYP3A5**3 heterozygotes (expressing little if any active CYP3A5). Nevertheless, in a clinical study, *CYP3A5* genotype had no apparent effect on the clearance and pharmacokinetics of intravenous or oral alfentanil (Figure 20.3) (32). To date, genetic differences in CYP3A expression and activity have not translated into clinically relevant pharmacologic differences in opioid disposition or effects. The reason for such in vitro-in vivo discrepancies in *CYP3A5* pharmacogenetics remains unknown.

Methadone is a very long-lasting synthetic opioid that undergoes hepatic demethylation to inactive metabolites. An appreciation of methadone metabolism and the role of pharmacogenetics has evolved substantially over time, and yet these still remain incompletely elucidated and persistently misunderstood. Early reports suggested methadone metabolism by CYP2D6, and an influence of *CYP2D6* genotype on methadone plasma concentrations, but accumulated evidence for either is unconvincing (15, 33). There is very good evidence for methadone metabolism by CYP3A4 in vitro, although human studies have suggested no role for CYP3A in clinical methadone metabolism and clearance (34). Accumulated data now strongly support a major role for CYP2B6 in methadone metabolism and clearance (34, 35), and some clinical reports suggest an influence of *CYP2B6* polymorphisms on methadone plasma concentrations (36), although the role of *CYP2B6* genotype on clinical methadone clearance and metabolism is unknown.

Phase II Metabolism

Opioids of the morphinan family (codeines, codones, morphine, and morphones) contain a hydroxyl group at position 6 on the molecule, and some (morphine)

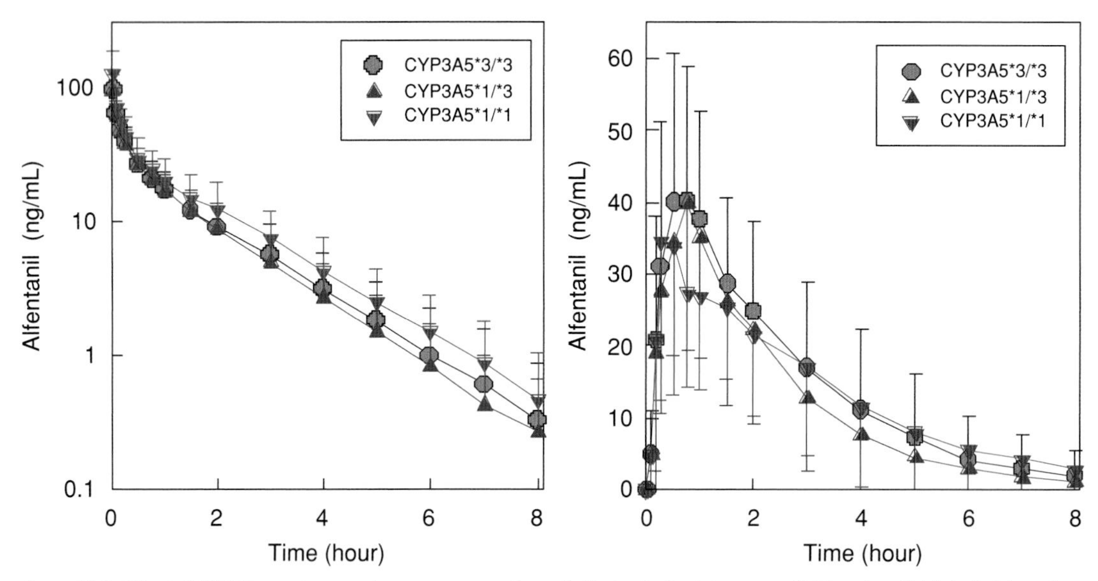

Figure 20.3. Effect of CYP3A5 genotype on plasma concentrations of alfentanil after intravenous (left) and oral (right) administration. Results are the mean ± SD. Redrawn from Kharasch et al. with permission (32).

also at position 3. Morphine-6-glucuronide is pharmacologically active, producing analgesia and opioid-related side effects, whereas morphine-3-glucuronide is considered inactive, as are other morphinan-3-glucuronides (15, 16). Morphine 3- and 6-glucuronides are formed by the uridine diphosphate glucuronosyltransferase enzyme UGT2B7, as are the 6-glucuronides of codeine, dihydrocodeine, dihydromorphine, hydromorphone, oxymorphone, and other opioids, whereas morphine-3-glucuronide is also formed by UGT1A1. Variants in the *UGT2B7* gene have been described, with varying results regarding any influence on morphine disposition and clinical effects. The *UGT2B7*2* allele had no effect on glucuronidation in vitro, nor on plasma glucuronide concentrations in vivo (16). In contrast, a promoter variant of *UGT2B7* did influence morphine plasma concentrations and the ratio of morphine-6-glucuronide to morphine in patients (37). A *UGT1A1* variant had no effect on morphine-glucuronide to morphine plasma concentration ratios (16). Although UGT genotypes may affect morphine disposition, information to date suggests that the influence of UGT polymorphisms on morphine analgesia is minor and clinically unimportant (15). Considerably less is understood about the influence of UGT polymorphisms on the disposition and analgesic effects of other opioids.

Drug Transport

Active transport across membranes may theoretically affect opioid and/or metabolite intestinal absorption, tissue distribution and penetration, metabolism, and hepatic and renal elimination. Both uptake and efflux transporters have been variously implicated for assorted opioids. The greatest interest has been focused on the intestinal transporters affecting absorption and bioavailability and on the blood-brain barrier (Figure 20.4) or blood-cerebrospinal fluid transporters affecting the pharmacodynamic, analgesic, and side effects of opioids (38). Information on the identification of opioids as transporter substrates, and the influence of genotype, derives primarily from in vitro studies using cell expression systems and transporter inhibitors and from gene knockout studies in animals; there is comparatively much less information on the pharmacogenetics of opioid transporters from actual clinical studies. Extrapolation of in vitro and animal studies to predict clinical scenarios, in the absence of human data, is not wise.

The most widely studied class of proteins potentially mediating opioid transport are members of the ATP-binding cassette (ABC)-transporter family, most notably the efflux transporter P-glycoprotein (P-gp, also called MDR1 or ABCB1). P-gp, which pumps substrates against a concentration gradient, and conceptually functions in defense against foreign compounds in the body, is expressed in classic defense sites such as the intestinal mucosa and the endothelial cells of organs with barrier function, such as the brain, the testes, and the placenta. Brain capillary endothelial cells are the major component of the blood-brain barrier, and P-gp is expressed on the luminal side of these cells, where it pumps drugs from the brain back into blood. P-gp is also expressed at the apical side of intestinal epithelial cells, where it pumps drugs back into the gut lumen. Opioids identified as P-gp substrates, on the basis of laboratory studies, include morphine, morphine-6-glucuronide,

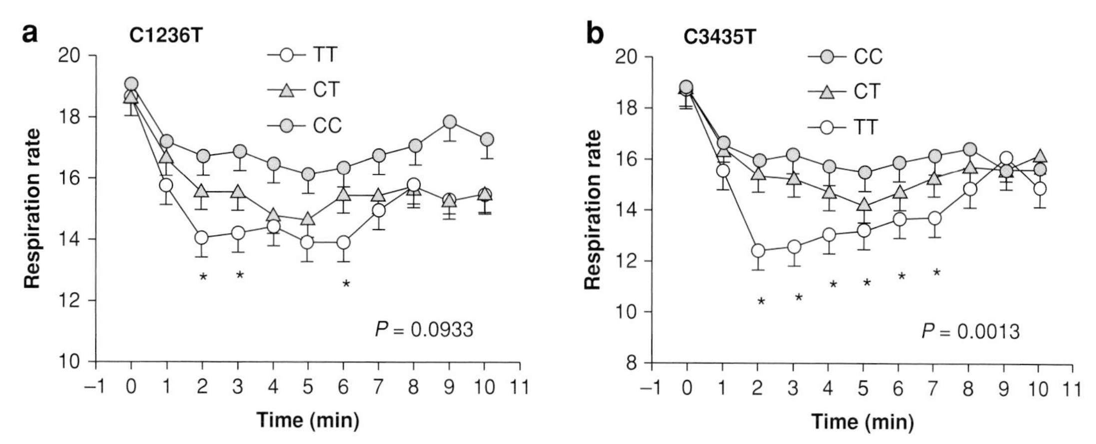

Figure 20.4. Influence of ABCB1 (P-glycoprotein) genotype on fentanyl respiratory depression. Respiratory rates during the initial 10 min after intravenous fentanyl administration are shown for the 1236C>T (**a**) and 3435C>T (**b**) alleles. Reproduced from Park et al. with permission (47).

fentanyl, sufentanil, methadone, meperidine, hydrocodone (but not oxycodone), tramadol, and loperamide (15, 28, 39). In humans, P-gp inhibition influenced the absorption, bioavailability, plasma concentrations, and clinical effects of oral morphine, methadone, and fentanyl, suggesting these opioids as intestinal P-gp substrates (40). In contrast, studies using the same P-gp inhibitors failed to show any increase in the clinical effects of these opioids, which did not provide evidence for P-gp as a major determinant of their brain access (40, 41).

P-gp is highly polymorphic, with numerous variants identified, but relatively few clinical studies to date have assessed the influence of P-gp genotype on opioid effects, and some have been contradictory (15, 16). The most commonly studied variant is the C3435T polymorphism, found in 50 percent to 60 percent of European Americans, 40 percent to 50 percent of Asians, but only 10 percent to 30 percent of Africans. Even though the C3435T polymorphism is synonymous, does not alter the coding sequence, and therefore is not expected to change protein structure, and studies on the expression level of P-gp resulting from the C3435T SNP are inconclusive, recent data suggest that it (alone and in combination with the G2677T/A SNP) can result in a P-gp with altered function (42).

To date, no in vitro studies have assessed the influence of the C3435T SNP on opioid transport, and clinical studies have been inconsistent and inconclusive. In neurosurgical patients, peak cerebrospinal fluid concentrations of morphine were higher in C3435T homozygotes compared with wild types (43). In cancer patients, the pain relief response to morphine was significantly greater in 3435 T/T homozygotes, compared with C/C wild types or C/T heterozygotes (44). These two studies could support a conclusion that blood-brain barrier P-gp is responsible for brain morphine efflux. However, because plasma morphine concentrations were not measured in the cancer patients, any mechanisms explaining the greater pain relief are unknown. Because morphine was given orally, the effect could be attributable to increased intestinal morphine absorption due to decreased efflux by P-gp. In contrast to the previous investigations, in patients treated postoperatively for pain, P-gp SNPs (C3435T and G2677T/A) were not associated with different morphine clinical effects, based on morphine dose requirements (45).

Like morphine, the role of P-gp variants in human methadone brain transport remains inconclusive (16). In healthy volunteers, with the use of pupil diameter change as a measure of opioid effect, neither the C3435T nor G2677T/A SNPs had any influence on maximum pupil constriction (46). Similarly, in patients on methadone maintenance therapy for opiate addiction, the *C3435T* SNP was associated with slightly lower trough plasma methadone concentrations (opposite to that expected for a functionally less active efflux transporter), but not peak concentrations, and no influence on clinical effects, together arguing against a clinically relevant genetic polymorphism (36). In contrast, methadone maintenance patients who were heterozygous or homozygous for any one of several P-gp SNPs had methadone doses about half that of homozygous wild types, and carriers with one or two copies of the wild-type haplotype required moderately higher doses (33). These results were postulated to reflect greater blood-brain barrier P-gp activity and decreased brain methadone exposure in wild types, requiring higher daily doses. Nevertheless, those results have been challenged.

Similar to morphine and methadone, the role of P-gp variants in human brain transport of other opioids remains inconclusive. In patients given fentanyl intravenously, which removes oral absorption as a potential

a Blood-to-brain influx transporters

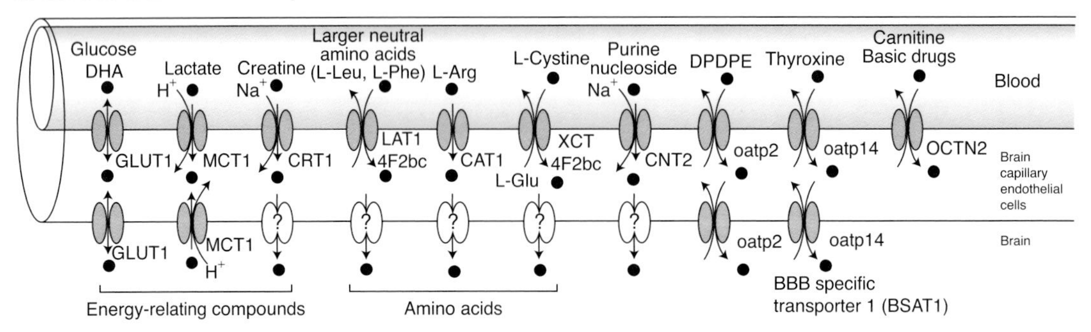

b ATP-binding cassette transporters

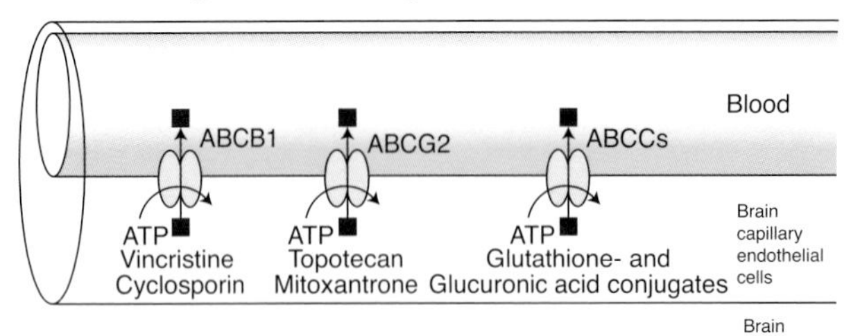

c Brain-to-blood efflux transporters

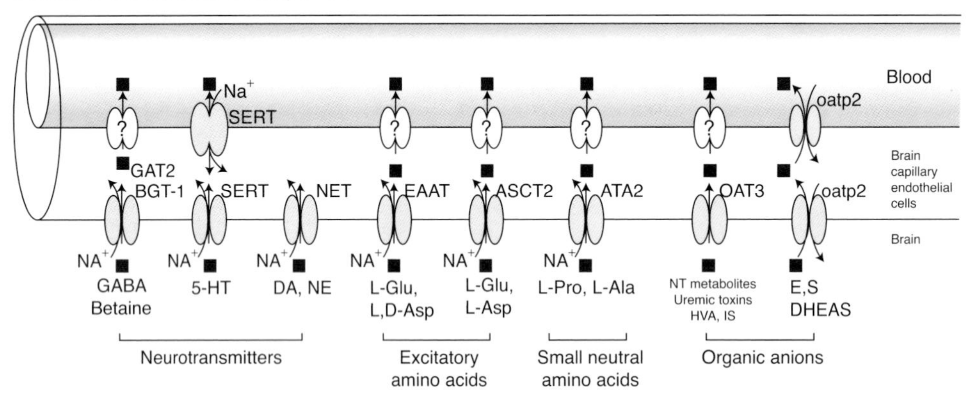

Figure 20.5. Schematic of drug transport proteins involved in the formation of the blood-brain barrier: (**a**) transporters mediating blood-to-brain influx transport; (**b**) ATP-binding cassette transporters as drug efflux pumps; (**c**) transporters mediating brain-to-blood efflux transport. Reproduced from Ohtsuki et al. with permission (59).

confounder, respiratory depression was greater in C3435T T/T homozygotes and C2677T T/T homozygotes (Figure 20.5), although analgesia was not assessed (47). The opioid agonist loperamide, a well-known substrate of P-gp in vitro, in humans acts primarily in the gastrointestinal tract because of very limited brain access, postulated to occur because of active brain efflux mediated by P-gp. Chemical inhibition of P-gp increases loperamide clinical effects. However, in healthy volunteers, neither the pupil nor respiratory effects of loperamide were associated with the C3435T polymorphism, suggesting no P-gp influence on brain loperamide access (48, 49). In summary, the role of P-gp in clinical opioid pharmacokinetics and pharmacodynamics remains to be clearly elucidated, the extent to which P-gp is clinically relevant at the blood-brain barrier is still controversial, and the clinical implications of P-gp genetic variants in opioid pharmacodynamics are not well understood.

Compared with P-gp, there is little known about the role of other efflux or uptake transporters in the disposition of opioids. On the basis of their expression in the intestine, liver, kidney, and/or brain, other ABC transporters such as breast cancer resistance protein (BCRP, ABCG2) and the multidrug resistance proteins (ABCC1, ABCC2, ABCC3), as well as organic anion-transporting polypeptides and organic cation transporters, have been implicated or speculated to be involved in the transport of opioids and/or opioid peptides (15). Nevertheless, there are scant in vitro data, no clinical data, and no studies to date on genetic polymorphisms in these transporters and

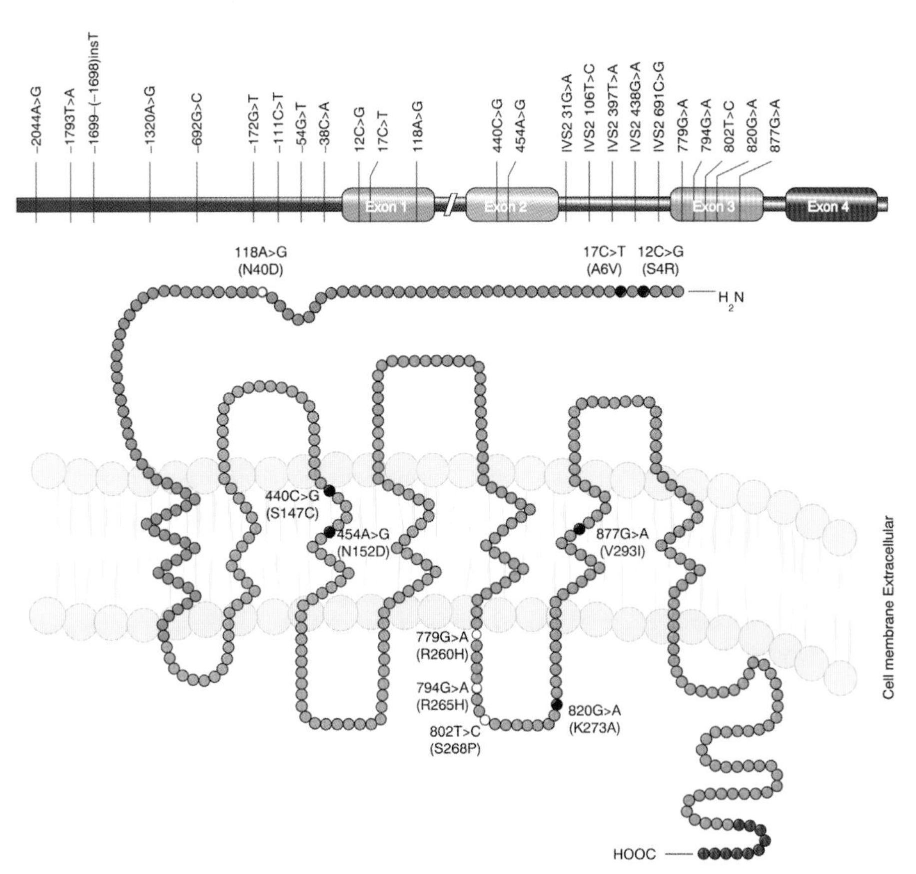

Figure 20.6. Reported mutations in the μ-opioid receptor related to the exonic organization of the *OPRM1* gene. Twenty-four mutations that produce an amino acid exchange and are frequently reportedly (>1 percent) or are proposed to have functional consequences are indicated in the gene. Amino acids are symbolized as circles, colored according to the exons by which they are coded. Black circles represent a naturally occurring mutation at the respective position, and red circles when functional consequences are shown at the molecular level. Mutations are indicated by the nucleotide exchange and the resulting amino acid exchange. Reproduced from Lötsch and Geisslinger with permission (51).

their influence on the transport of therapeutic opioids. Consequently, any genetic influence on clinical opioid transport and clinical effects remains unknown.

Opioid Receptors

The opioid receptor family consists of mu (μ), kappa (κ), and delta (δ) receptor subtypes. The μ-receptor, encoded by the *OPRM1* gene, is the primary target of clinically used opiates and opioids, mediating all of the major desirable (analgesia) and unwanted (respiratory depression, pupil constriction, constipation, sedation, itching, etc.) effects. Opioid analgesia is elicited predominantly in the central nervous system (brain and spinal cord) by the μ-opioid receptor. Genetically determined interindividual differences in opioid receptors, opioid analgesia, and addiction have been the focus of considerable interest (50).

Numerous SNPs have been identified in the *OPRM1* gene, more than twenty of which are nonsynonymous and lead to amino acid changes (Figure 20.6) (16, 28, 51). The most common and the most studied μ-receptor polymorphism is the A118G mutation on exon 1, with an allelic frequency of 10 percent to 30 percent, and which changes an asparagine to an aspartate (Asp40Asn) with the loss of a putative *N*-glycosylation site on the extracellular domain of the receptor. Initial studies in expressed cell lines reported that the A118G receptor had increased affinity for β-endorphin; however, later studies reported no significant receptor affinity for β-endorphin, morphine, or morphine-6-glucuronide. The functional significance of the A118G polymorphism is therefore unknown, and the clinical significance is presently inconclusive, if not confusing, and at most small.

Clinical studies in healthy volunteers generally, but not consistently, suggest moderately (two- to threefold) decreased opioid potency in A118G carriers (15, 16, 28, 46). By the use of pupil constriction as a measure of opioid effect, the dose of morphine-6-glucuronide required to decrease pupil size in half was increased two- and three- to fourfold in A118G hetero- and homozygotes, respectively, but there was no difference in response to morphine. A follow-up study showed an influence of the A118G polymorphism on the potency of both morphine and morphine-6-glucuronide based on pupil constriction, yet no influence on analgesia for either opioid. When the authors combined data from several studies, they concluded that the A118G polymorphism reduced the pupil-constricting potency of both morphine and morphine-6-glucuronide twofold (46). Other studies of morphine-6-glucuronide, by a different group of investigators, suggested the A118G SNP (a) reduced analgesic potency threefold, (b) reduced maximum pain tolerance slightly but had no effect on potency, and (c) had no effect on respiratory depression (response to hypoxia) (15). In contrast, similar investigations with alfentanil showed that A118G homozygotes required ten to twelve times higher alfentanil concentrations to elicit the same degree of respiratory depression and two to four times higher concentrations to achieve the same degree of analgesia (52). A study of single-dose *R*-methadone and pupil constriction found that opioid potency was 1.7-fold lower in A118G carriers (53). Human volunteer studies therefore suggest a small, at most, influence of the A118G polymorphism, which varies somewhat with the opioid and opioid response studied.

There are relatively few clinical studies of μ-receptor polymorphisms in patients, and they afford an even more inconsistent perspective on the A118G SNP (15, 16, 28). In the two days after total knee replacement surgery, morphine requirements were higher (1.6-fold) in A118G homozygotes, but not in heterozygotes (54). In women, during the first day after cesarean delivery, pain scores and average self-administered intravenous morphine doses were reported to be statistically higher in A118G heterozygotes (8 mg) and homozygotes (9 mg) than in wild types (6 mg) in one investigation (55), but were not different in another (56). In patients in the immediate postoperative period, no statistically significant association was found between the A118G SNP and morphine consumption or the average postoperative pain score (57). In patients with chronic cancer pain, those homozygous for the A118G polymorphism needed higher morphine doses for pain control, or were reported to have achieved lesser pain control with morphine (44, 46). In contrast to all of the preceding studies, suggesting *increased* opioid requirements, the *OPRM1* A118G polymorphism was recently reported to *decrease* the opioid (intrathecal fentanyl) dose requirement for analgesia in laboring women, with the authors suggesting greater opioid effects with this SNP (58).

Although the subject of considerable interest and investigations, genetic influences of *OPRM1* polymorphisms appear small in the studies to date. Investigations of the clinical relevance of the A118G polymorphism are continuing, as are larger gene association studies. For example, the A118G SNP has been associated with a greater impact on opioid effects in comparison with the impact of mutations in the *ABCB1* gene (44, 46). No other significant *OPRM1* polymorphisms have been identified to date. Compared with the highly clinically significant influence of *CYP2D6* polymorphisms on codeine and tramadol clinical effectiveness, the influence of *OPRM1* polymorphisms is minor.

Nonopioid Receptors

Nonopioid receptors have been shown to modulate pain and the response to analgesics. In addition to influencing the response to experimental pain, as described above, SNPs in these nonopioid pathways can influence the response to opioids (16). In patients with cancer pain, in two separate studies, the COMT G472A (val 158 met) SNP (homozygotes more than heterozygotes) significantly (by about 1.3- to 2-fold) decreased daily morphine dose requirements. In healthy volunteers, with an electrical pain model, subjects carrying at least two variants of the *MCR1* gene coding for reduced activity exhibited greater (1.6-fold) morphine potency and an increased (1.4-fold) analgesic response to morphine-6-glucuronide (12).

GENETIC TESTING

Discoveries of gene polymorphisms inevitably lead to discussions of pharmacogenetic testing, often with enthusiasts advocating for widespread implementation. The utility of such testing depends on multiple considerations, including the consequence and magnitude of the SNP effect on the biological pathway, the overall relevance of the pathway to disease and response to therapy, cost-effectiveness of testing, the therapeutic setting, and the available of alternative pharmacologic interventions. It is apparent from this chapter that genetic polymorphisms affecting anesthetic and analgesic pharmacodynamics (essentially receptor polymorphisms) have relatively small effects with little clinical significance. Conversely, SNPs modulating the pharmacokinetics of certain analgesics (codeine and tramadol, requiring CYP2D6-catalyzed bioactivation) markedly influence both therapeutic response and toxicity, and are clearly of clinical significance. One potential approach to improved therapy is the implementation of widespread pharmacogenetic testing for *CYP2D6* (whose advocates also suggest *OPRM1* testing), promoted as an implementation of "personalized medicine." Genetic testing for *CYP* variants might presently be expensive, but could become a standard of care in the future, thereby helping not only to better treat patients in pain, but also to diminish the probability of potentially dangerous side effects like respiratory depression. Nevertheless, the cost-effectiveness of such testing is unknown. Because numerous analgesics exist that are minimally affected by genetic polymorphisms, an alternative approach is to simply use these drugs and avoid those susceptible to the *CYP2D6* polymorphism. Because anesthesia practice occurs "at the bedside," where drugs are titrated to effect, and patient responses provide immediate feedback on drug effectiveness and dosing, implementation of genetic testing for anesthesia does not appear imminent.

SUMMARY

The field of pharmacogenetics has its roots in anesthesia, grounded in searching for the inherited basis of the abnormal response to succinylcholine, and also to volatile anesthetics (MH). The bioactivation of codeine, and its susceptibility to genetic polymorphisms, serves as a now classic example of genetically determined therapeutic response. Nevertheless, the evaluation of genetic influences on pain and analgesia is just beginning.

REFERENCES

1. Lacroix-Fralish ML & Mogil JS. Progress in genetic studies of pain and analgesia. *Annu Rev Pharmacol Toxicol.* 2009;**49**:97–121.
2. Kalow W. Hydrolysis of local anesthetics by human serum cholinesterase. *J Pharmacol Exp Ther.* 1952;**104**:122–34.
3. Kalow W. Pharmacogenetics and anesthesia. *Anesthesiology.* 1964;**25**:377–87.
4. Edwards RR. Genetic predictors of acute and chronic pain. *Curr Rheumatol Rep.* 2006;**8**:411–17.
5. Nielsen CS, Stubhaug A, Price DD, Vassend O, Czajkowski N, & Harris JR. Individual differences in pain sensitivity: genetic and environmental contributions. *Pain.* 2008;**136**:21–9.
6. Oertel B & Lötsch J. Genetic mutations that prevent pain: implications for future pain medication. *Pharmacogenomics.* 2008;**9**:179–94.
7. Cox JJ, Reimann F, Nicholas AK, Thornton G, Roberts E, Springell K, et al. An SCN9A channelopathy causes congenital inability to experience pain. *Nature.* 2006;**444**:894–8.
8. Drenth JP & Waxman SG. Mutations in sodium-channel gene SCN9A cause a spectrum of human genetic pain disorders. *J Clin Invest.* 2007;**117**:3603–9.
9. Diatchenko L, Slade GD, Nackley AG, Bhalang K, Sigurdsson A, Belfer I, et al. Genetic basis for individual variations in pain perception and the development of a chronic pain condition. *Hum Mol Genet.* 2005;**14**:135–43.
10. Diatchenko L, Nackley AG, Slade GD, Bhalang K, Belfer I, Max MB, et al. Catechol-O-methyltransferase gene polymorphisms are associated with multiple pain-evoking stimuli. *Pain.* 2006;**125**:216–24.
11. Kim H, Lee H, Rowan J, Brahim J, & Dionne RA. Genetic polymorphisms in monoamine neurotransmitter systems show only weak association with acute post-surgical pain in humans. *Mol Pain.* 2006;**2**:24.
12. Mogil JS, Ritchie J, Smith SB, Strasburg K, Kaplan L, Wallace MR, et al. Melanocortin-1 receptor gene variants affect pain and mu-opioid analgesia in mice and humans. *J Med Genet.* 2005;**42**:583–7.
13. Tegeder I, Costigan M, Griffin RS, Abele A, Belfer I, Schmidt H, et al. GTP cyclohydrolase and tetrahydrobiopterin regulate pain sensitivity and persistence. *Nat Med.* 2006;**12**:1269–77.
14. Bessler H, Shavit Y, Mayburd E, Smirnov G, & Beilin B. Postoperative pain, morphine consumption, and genetic polymorphism of IL-1beta and IL-1 receptor antagonist. *Neurosci Lett.* 2006;**404**:154–8.
15. Somogyi AA, Barratt DT, & Coller JK. Pharmacogenetics of opioids. *Clin Pharmacol Ther.* 2007;**81**:429–44.
16. Rollason V, Samer C, Piguet V, Dayer P, & Desmeules J. Pharmacogenetics of analgesics: toward the individualization of prescription. *Pharmacogenomics.* 2008;**9**:905–33.
17. Galley HF, Mahdy A, & Lowes DA. Pharmacogenetics and anesthesiologists. *Pharmacogenomics.* 2005;**6**:849–56.
18. Iohom G, Ni Chonghaile M, O'Brien JK, Cunningham AJ, Fitzgerald DF, & Shields DC. An investigation of potential genetic determinants of propofol requirements and recovery from anaesthesia. *Eur J Anaesthesiol.* 2007;**24**:912–19.
19. Restrepo JG, Garcia-Martin E, Martinez C, & Agundez JA. Polymorphic drug metabolism in anaesthesia. *Curr Drug Metab.* 2009;**10**:236–46.
20. Kharasch ED. Adverse drug reactions with halogenated anesthetics. *Clin Pharmacol Ther.* 2008;**84**:158–62.
21. Rosenberg H, Davis M, James D, Pollock N, & Stowell K. Malignant hyperthermia. *Orphanet J Rare Dis.* 2007;**2**:21.
22. Selzer RR, Rosenblatt DS, Laxova R, & Hogan K. Adverse effect of nitrous oxide in a child with 5,10-methylenetetrahydrofolate reductase deficiency. *N Engl J Med.* 2003;**349**:45–50.
23. Nagele P, Zeugswetter B, Wiener C, Burger H, Hupfl M, Mittlbock M, et al. Influence of methylenetetrahydrofolate reductase gene polymorphisms on homocysteine concentrations after nitrous oxide anesthesia. *Anesthesiology.* 2008;**109**:36–43.
24. Davis L, Britten JJ, & Morgan M. Cholinesterase. Its significance in anaesthetic practice. *Anaesthesia.* 1997;**52**:244–60.
25. Gardiner SJ & Begg EJ. Pharmacogenetics, drug-metabolizing enzymes, and clinical practice. *Pharmacol Rev.* 2006;**58**:521–90.
26. Wang GK, Calderon J, & Wang SY. State- and use-dependent block of muscle Nav1.4 and neuronal Nav1.7 voltage-gated Na+ channel isoforms by ranolazine. *Mol Pharmacol.* 2008;**73**:940–8.
27. Samer CF, Desmeules JA, Dayer P. Individualizing analgesic prescription. Part I: pharmacogenetics of opioid analgesics. *Pers Med.* 2006;**3(3)**:239–69.
28. Kadiev E, Patel V, Rad P, Thankachan L, Tram A, Weinlein M, et al. Role of pharmacogenetics in variable response to drugs: focus on opioids. *Expert Opin Drug Metab Toxicol.* 2008;**4**:77–91.
29. Stamer UM & Stuber F. The pharmacogenetics of analgesia. *Expert Opin Pharmacother.* 2007;**8**:2235–45.

30. Madadi P & Koren G. Pharmacogenetic insights into codeine analgesia: implications to pediatric codeine use. *Pharmacogenomics.* 2008;**9**:1267–84.
31. Ingelman-Sundberg M, Sim SC, Gomez A, & Rodriguez-Antona C. Influence of cytochrome P450 polymorphisms on drug therapies: pharmacogenetic, pharmacoepigenetic and clinical aspects. *Pharmacol Ther.* 2007;**116**:496–526.
32. Kharasch ED, Walker A, Isoherranen N, Hoffer C, Sheffels P, Thummel K, et al. Influence of CYP3A5 genotype on the pharmacokinetics and pharmacodynamics of the cytochrome P4503A probes alfentanil and midazolam. *Clin Pharmacol Ther.* 2007;**82**:410–26.
33. Coller JK, Barratt DT, Dahlen K, Loennechen MH, & Somogyi AA. ABCB1 genetic variability and methadone dosage requirements in opioid-dependent individuals. *Clin Pharmacol Ther.* 2006;**80**:682–90.
34. Kharasch ED, Hoffer C, Whittington D, & Sheffels P. Role of hepatic and intestinal cytochrome P450 3A and 2B6 in the metabolism, disposition, and miotic effects of methadone. *Clin Pharmacol Ther.* 2004;**76**:250–69.
35. Totah RA, Sheffels P, Roberts T, Whittington D, Thummel K, & Kharasch ED. Role of CYP2B6 in stereoselective human methadone metabolism. *Anesthesiology.* 2008;**108**:363–74.
36. Crettol S, Deglon JJ, Besson J, Croquette-Krokar M, Hammig R, Gothuey I, et al. ABCB1 and cytochrome P450 genotypes and phenotypes: influence on methadone plasma levels and response to treatment. *Clin Pharmacol Ther.* 2006;**80**:668–81.
37. Sawyer MB, Innocenti F, Das S, Cheng C, Ramirez J, Pantle-Fisher FH, et al. A pharmacogenetic study of uridine diphosphate-glucuronosyltransferase 2B7 in patients receiving morphine. *Clin Pharmacol Ther.* 2003;**73**:566–74.
38. Loscher W & Potschka H. Role of drug efflux transporters in the brain for drug disposition and treatment of brain diseases. *Prog Neurobiol.* 2005;**76**:22–76.
39. Liang DY, Liao G, Lighthall GK, Peltz G, & Clark DJ. Genetic variants of the P-glycoprotein gene Abcb1b modulate opioid-induced hyperalgesia, tolerance and dependence. *Pharmacogenet Genomics.* 2006;**16**:825–35.
40. Kharasch ED, Hoffer C, & Whittington D. The effect of quinidine, used as a probe for the involvement of P-glycoprotein, on the intestinal absorption and pharmacodynamics of methadone. *Br J Clin Pharmacol.* 2004;**57**:600–10.
41. Kurnik D, Sofowora GG, Donahue JP, Nair UB, Wilkinson GR, Wood AJ, et al. Tariquidar, a selective P-glycoprotein inhibitor, does not potentiate loperamide's opioid brain effects in humans despite full inhibition of lymphocyte P-glycoprotein. *Anesthesiology.* 2008;**109**:1092–9.
42. Kimchi-Sarfaty C, Oh JM, Kim IW, Sauna ZE, Calcagno AM, Ambudkar SV, et al. A "silent" polymorphism in the MDR1 gene changes substrate specificity. *Science.* 2007;**315**:525–8.
43. Meineke I, Freudenthaler S, Hofmann U, Schaeffeler E, Mikus G, Schwab M, et al. Pharmacokinetic modelling of morphine, morphine-3-glucuronide and morphine-6-glucuronide in plasma and cerebrospinal fluid of neurosurgical patients after short-term infusion of morphine. *Br J Clin Pharmacol.* 2002;**54**:592–603.
44. Campa D, Gioia A, Tomei A, Poli P, & Barale R. Association of ABCB1/MDR1 and OPRM1 gene polymorphisms with morphine pain relief. *Clin Pharmacol Ther.* 2007;**83**:559–66.
45. Coulbault L, Beaussier M, Verstuyft C, Weickmans H, Dubert L, Tregouet D, et al. Environmental and genetic factors associated with morphine response in the postoperative period. *Clin Pharmacol Ther.* 2006;**79**:316–24.
46. Lötsch J, & Geisslinger G. Relevance of frequent mu-opioid receptor polymorphisms for opioid activity in healthy volunteers. *Pharmacogenomics J.* 2006;**6**:200–10.
47. Park HJ, Shinn HK, Ryu SH, Lee HS, Park CS, & Kang JH. Genetic polymorphisms in the ABCB1 gene and the effects of fentanyl in Koreans. *Clin Pharmacol Ther.* 2007;**81**:539–46.
48. Pauli-Magnus C, Feiner J, Brett C, Lin E, & Kroetz DL. No effect of MDR1 C3435T variant on loperamide disposition and central nervous system effects. *Clin Pharmacol Ther.* 2003;**74**:487–98.
49. Skarke C, Jarrar M, Schmidt H, Kauert G, Langer M, Geisslinger G, et al. Effects of ABCB1 (multidrug resistance transporter) gene mutations on disposition and central nervous effects of loperamide in healthy volunteers. *Pharmacogenetics.* 2003;**13**:651–60.
50. Mayer P & Hollt V. Pharmacogenetics of opioid receptors and addiction. *Pharmacogenet Genomics.* 2006;**16**:1–7.
51. Lötsch J & Geisslinger G. Are mu-opioid receptor polymorphisms important for clinical opioid therapy? *Trends Mol Med.* 2005;**11**:82–9.
52. Oertel BG, Schmidt R, Schneider A, Geisslinger G, & Lötsch J. The mu-opioid receptor gene polymorphism 118A>G depletes alfentanil-induced analgesia and protects against respiratory depression in homozygous carriers. *Pharmacogenet Genomics.* 2006;**16**:625–36.
53. Lötsch J, Skarke C, Wieting J, Oertel BG, Schmidt H, Brockmoller J, et al. Modulation of the central nervous effects of levomethadone by genetic polymorphisms potentially affecting its metabolism, distribution, and drug action. *Clin Pharmacol Ther.* 2006;**79**:72–89.
54. Chou WY, Yang LC, Lu HF, Ko JY, Wang CH, Lin SH, et al. Association of mu-opioid receptor gene polymorphism (A118G) with variations in morphine consumption for analgesia after total knee arthroplasty. *Acta Anaesthesiol Scand.* 2006;**50**:787–92.
55. Sia AT, Lim Y, Lim EC, Goh RW, Law HY, Landau R, et al. A118G single nucleotide polymorphism of human mu-opioid receptor gene influences pain perception and patient-controlled intravenous morphine consumption after intrathecal morphine for postcesarean analgesia. *Anesthesiology.* 2008;**109**:520–6.
56. Chou WY, Wang CH, Liu PH, Liu CC, Tseng CC, & Jawan B. Human opioid receptor A118G polymorphism affects intravenous patient-controlled analgesia morphine consumption after total abdominal hysterectomy. *Anesthesiology.* 2006;**105**:334–7.

57. Janicki PK, Schuler G, Francis D, Bohr A, Gordin V, Jarzembowski T, et al. A genetic association study of the functional A118G polymorphism of the human mu-opioid receptor gene in patients with acute and chronic pain. *Anesth Analg.* 2006;**103**:1011–17.
58. Landau R, Kern C, Columb MO, Smiley RM, & Blouin JL. Genetic variability of the mu-opioid receptor influences intrathecal fentanyl analgesia requirements in laboring women. *Pain.* 2008;**139**:5–14.
59. Ohtsuki S & Terasaki T. Contribution of carrier-mediated transport systems to the blood-brain barrier as a supporting and protecting interface for the brain; importance for CNS drug discovery and development. *Pharm Res.* 2007;**24**:1745–58.

21 HIV and Antiretroviral Therapy

Amalio Telenti

BASIC CONCEPTS IN HIV THERAPEUTICS

Considerable progress has been made during the past twenty years in the development of effective treatment against HIV-1, the agent of AIDS. Today, there are more than twenty agents commercialized that target various steps in the viral life cycle (Figure 21.1). These drugs, used in combination, inhibit two or more steps of the viral cycle: viral entry (fusion inhibitors, coreceptor antagonists), reverse transcription (nucleoside analog and nonnucleoside inhibitors), integration into the host genome (anti-integrases), and viral polyprotein processing (protease inhibitors). The need for combination antiretroviral therapy (ART) is determined by the ease of viral escape from the selective pressure when confronted with single antiretroviral agents. The ability to escape from drug pressure is, in turn, the result of rapid mutation rates due to a viral polymerase (the reverse transcriptase) that lacks proofreading activity.

HIV treatment is highly standardized, with defined criteria for its initiation and detailed knowledge on comparative efficacy (1). Decisions to initiate ART are primarily based on the degree of immunosuppression (generally defined by the numbers of CD4+ T cells – the target of HIV-1 – in blood), with additional consideration given to the levels of viral replication (defined by the level of viremia – viral genome copy numbers per milliliter), and the presence of clinical symptoms. In the absence of treatment, most individuals infected with HIV-1 will have progression of immunosuppression, leading to AIDS-related opportunistic infections and cancer. Individuals progress to AIDS over a mean of eight years after infection; however, rates of progression are highly variable among individuals.

The initiation of therapy halts the ineluctable progression of disease in most individuals. After one year of treatment, greater than 80 percent of individuals will have undetectable levels of viremia, and will be greatly protected from opportunistic diseases. However, the eradication of the virus cannot be achieved with current medication because of the characteristics of the viral life cycle (e.g., viral integration) and the basic mechanisms of pathogenesis (e.g., failure of effective immune clearance). Treatment includes several drugs simultaneously that are given long term. This paradigm leads to one of the key issues in HIV medicine: how to ensure maximal efficacy and minimal short- and long-term toxicity.

Central to these practical issues is the observation of the significant interindividual differences in plasma drug levels (drug exposure) and in the susceptibility to adverse drug events that, for a number of drugs, is genetically determined.

This chapter will review current knowledge on genetic and genomic determinants of pharmacokinetics and pharmacodynamics of ART. The various technical and conceptual approaches for the identification and validation of pharmacogenetic markers will be illustrated with representative clinical vignettes.

CURRENT KNOWLEDGE IN PHARMACOGENETICS OF ANTIRETROVIRAL TREATMENT

Drug Pharmacokinetics

A practical approach to investigating the genetic determinants of the pharmacokinetics of a drug requires an understanding, for a given medication, of the drug absorption, distribution, metabolism, and excretion (ADME) and of drug-drug (and drug-environment) interactions and the observation of the distribution of drug levels (or other pharmacokinetic parameters) at the population level (Figure 21.2).

Current knowledge concerns primarily the contribution of variation in cytochrome P450 genes to metabolism of nonnucleoside reverse transcriptase

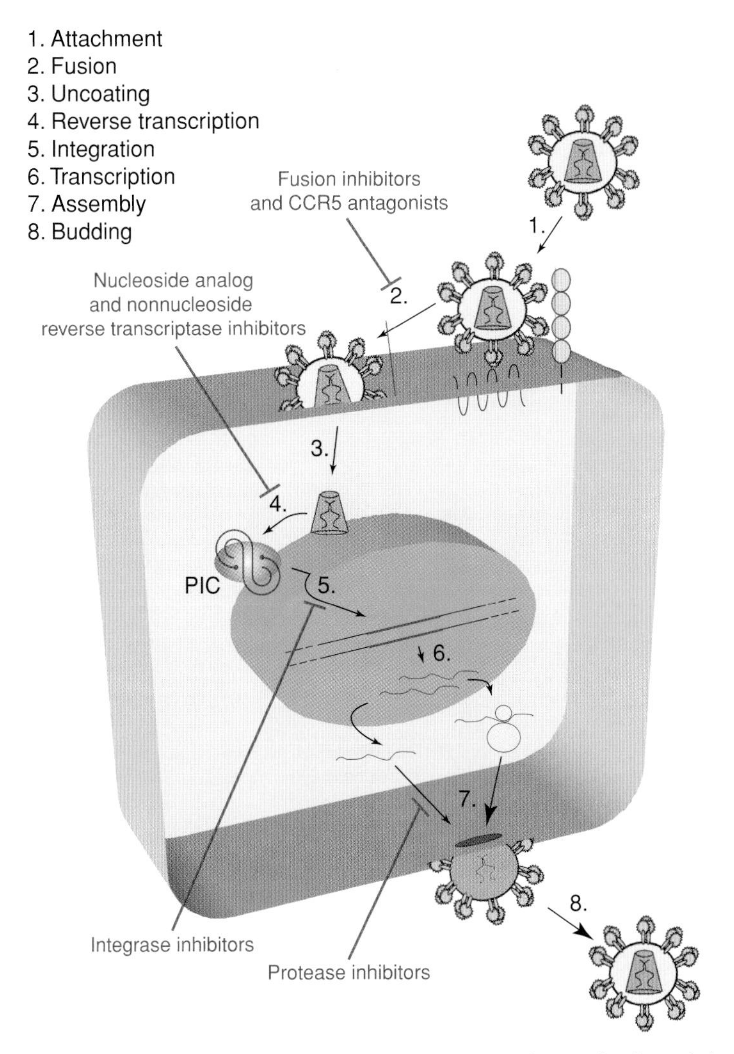

Figure 21.1. HIV life cycle. The site of action of the current classes of antiretroviral agents is indicated by red bars.

inhibitors (NNRTIs) and of protease inhibitors (Table 21.1). CYP450 isozymes are responsible for most phase I biotransformation of drugs and other xenobiotics. The expression and function of these proteins are highly variable both inter- and intraindividually, and thus are a major contributors to unpredictable drug/metabolite plasma concentrations. Genetic polymorphisms in some CYP450 genes have been studied intensely for more than twenty years, but more recently discovered examples are less completely studied, and novel variants are continuously being described.

In an initial study with HIV patients receiving different protease inhibitors (nelfinavir, saquinavir, indinavir) and NNRTIs (nevirapine, efavirenz), *CYP3A4*1B* and the *CYP3A5*3* alleles were weakly associated with plasma efavirenz exposure (2), in agreement with a limited contribution of CYP3A to efavirenz metabolism (3). However, no such relationships were identified in other studies (reviewed in reference 4). On the other hand, CYP3A5 expressor genotype (carriers of *CYP3A5*1*) was associated with moderate increases in the oral clearance of saquinavir and indinavir (5, 6). A fact that limits the ability of CYP3A pharmacogenetics to predict enzyme activity is that these enzymes are potently and irreversibly inhibited by almost all protease inhibitor drugs.

In vitro studies identified CYP2B6 as the enzyme responsible for the conversion of efavirenz into its major 8-hydroxylated and 8-, 14-dihydroxylated metabolites, with minor contributions by other isoforms, including CYP3A4/5 and CYP1A2 (3). The *CYP2B6* gene is highly polymorphic, with pronounced racial influences (7). The most frequent allele in all ethnic populations studied is *CYP2B6*6*, defined by two amino acid alterations,

Table 21.1. Inherited Differences in the Pharmacokinetics of Antiretroviral Drugs

Gene or Protein (*)	*Allele or Variant Evaluated*	*Reported Consequence for Antiretroviral Drugs*
Metabolism		
CYP3A5	*CYP3A5*3* and *CYP3A5*6* (alleles associated with severely reduced enzyme expression due to aberrant splicing)	Higher saquinavir area under the curve (AUC) and metabolite ratio. Reduction of oral indinavir clearance
CYP2C19	*CYP2C19*2* (null allele due to aberrant splicing; poor metabolizer phenotype results from homozygous condition)	Higher nelfinavir AUC and lower M8/nelfinavir AUC ratio
CYP2B6	CYP2B6*6, *11, *18, *27, *28, *29 (alleles with diminished or loss of function, associated with decreased expression, or decreased function, protein truncation or gene deletion)	Higher efavirenz and nevirapine plasma drug levels
Transport		
P-glycoprotein (*MDR1, ABCB1*)	3435C>T (synonymous I1145I, in linkage disequilibrium with *ABCB1*_1236, and 2677). Limited data on 61A>G (N21D), 1199G>A (S400N), other variants, or on haplotypes	Controversial data with several reports indicating an association of 3435T with decreased transport function resulting in increased protease inhibitor exposure. Recent data indicate that this synonymous SNP results in altered codon usage for isoleucine, leading to a change in timing of cotranslational folding of the P-glycoprotein, and results in changes in substrate specificity
MRP4 (ABCC4)	3724G>A (A1203A), 4131T>G, 669C>T (I223I)	Elevated zidovudine- and 3TC-triphosphate concentration
SCLO1B1	*SCLO1B1*4, SCLO1B1*5*	Changes in Lopinavir clearance
Protein Binding		
α-1-acid glycoprotein (*ORM1)*	F1 and S protein variants of ORM1, which result from amino acid changes at variant positions 20 and 156 (determined phenotypically)	Higher apparent clearance in F1F1 individuals compared with SS for indinavir and, weakly, for lopinavir

Only the most relevant associations are indicated. Compiled from www.hiv-pharmacogenomics.org.

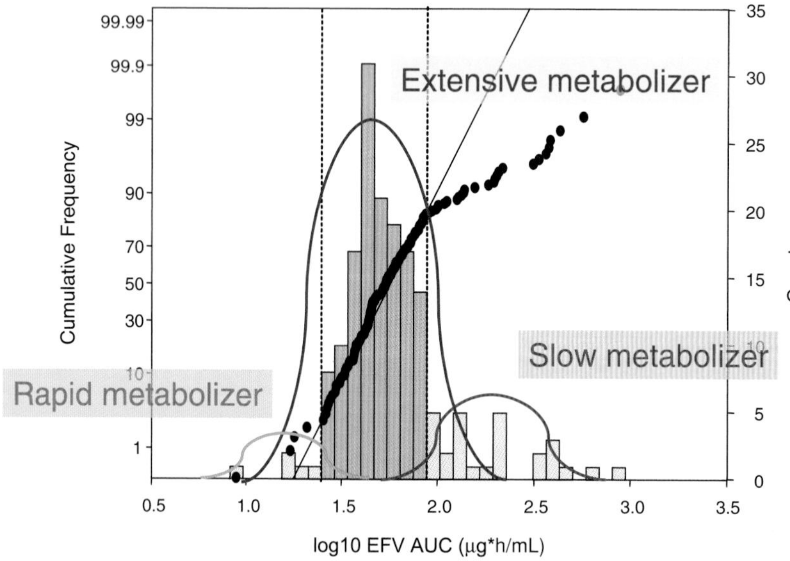

Figure 21.2. Distribution of efavirenz plasma concentration and identification of metabolizer profiles. Plasma concentrations from 169 individual values are expressed as log of the area under the curve (AUC) values. Displayed are the frequency histogram of values and the corresponding probits in the study population. Individuals with clearly higher (slow metabolizers) or lower AUC (rapid metabolizers) than predicted by normal distribution are shown with hatched bars.

Q172H and K262R. However, a comprehensive study that included testing for all known functional variants of CYP2B6 demonstrated that the ability of CYP2B6 genotyping to predict unexpectedly high efavirenz plasma levels is significantly enhanced by including the rarer loss-of-function alleles (8). Of particular relevance is *CYP2B6*18* (I328T), a variant that showed extremely low expression in recombinant systems and was found at frequencies of 4 percent to 10 percent in various populations of African descent and in Turks, but not in white Europeans or Asians (4). Several other rarer alleles show pronounced functional differences in vivo and/or in vitro with lower activity in comparison with the reference allele. In addition, a number of variants that increase enzyme expression or activity have been described. The *CYP2B6*4* allele (K262R) was associated with decreased efavirenz drug levels (7). The latter study also identified the variant –82T>C (*CYP2B6*22*), which increases expression via a complex rearrangement of the transcriptional machinery at the *CYP2B6* core promoter (9), in a patient with very low efavirenz concentration. The metabolism of efavirenz is also a model for the study of the role of primary and accessory pathways. In the absence of CYP2B6 activity due to the homozygous loss of function variants, the metabolism of efavirenz critically depends on CY3A4 and CYP2A6. Individuals with simultaneous loss-of-function alleles on all three pathways will have massive accumulation of efavirenz (10) (see case 1). The NNRTI nevirapine is metabolized by CYP3A4/5 and CYP2B6. Patients with the *CYP2B6*6* allele have significant increases in plasma drug levels (11).

HIV drugs are rarely substrates of CYP2C19. An exception is the formation of nelfinavir hydroxy-*t*-butylamide (M8), a major and pharmacologically active metabolite of nelfinavir whose concentration varies according to *CYP2C19* genotype and may influence treatment outcome (12, 13).

Raltegravir, the first commercially available member of the new drug class of inhibitors of the viral integrase undergoes UGT1A1-mediated glucuronidation (14). The nucleos(t)ide analog reverse transcriptase inhibitors (NRTIs) are not metabolized as extensively by cytochromes P450 as are other antiretroviral agents. Host cell-mediated sequential enzymatic phosphorylation steps are required for activating the NRTIs (15). There is, however, limited pharmacogenetic data on these pathways. Identification of novel genetic markers may require the detailed genotyping of the ADME pathway for a given drug (see case 2).

Drug Pharmacodynamics

Pharmacodynamics describes the effect of a drug on the organism. Whereas less information is available on the relationship between genetic polymorphism and treatment outcome in ART, more data are emerging on the pharmacogenetics of drug toxicity (Table 21.2). This is of particular interest because of the frequency of the short- and long-term adverse effects of ART (16, 17). Data are currently available on the genetic predisposition to neuropsychological toxicity of efavirenz, protease inhibitor dyslipidemia, hypersensitivity reactions to abacavir, hyperbilirubinemia of atazanavir, and lipodystrophy.

Although the association of *CYP2B6* alleles and the pharmacokinetics of efavirenz is well characterized, the data on the role of high plasma levels on the characteristic neuropsychological toxicity of efavirenz has remained controversial. However, a number of publications describe the association of acute (early) toxicity, as well as of chronic toxicity and efavirenz pharmacokinetics, and the reversal of neuropsychotoxicity upon dose reduction (2, 18) (see case 1).

Rapidly increasing data on the determinants of dyslipidemia (and metabolic syndrome) in the general population have been made possible by recent genome-wide studies. Dyslipidemia in HIV-infected patients can include increased non-high-density lipoprotein cholesterol and low-density lipoprotein cholesterol, hypertriglyceridemia, and decreased high-density lipoprotein cholesterol. Each of these plasma lipid abnormalities is considered an independent cardiovascular risk factor. Dyslipidemia is a common adverse effect of protease inhibitors. In particular, ritonavir-boosted (enhanced) protease inhibitor treatment acts as a trigger for a constitutional predisposition to dyslipidemia (19). This situation is similar to that described after administration of 13-*cis*-retinoic acid (isotretinoin) for acne (20). Here, the lipid response to isotretinoin was closely associated with the *apoE* gene, and identified individuals at increased risk for future hyperlipidemia and the metabolic syndrome.

The use of abacavir is associated with drug hypersensitivity reactions in 5 percent to 8 percent of abacavir recipients (21). Retrospective studies indicated a strong association of the abacavir hypersensitivity reaction with the presence of the major histocompatibility complex class I allele *HLA-B*5701* (22, 23) in chromosome 6. The usefulness of genetic screening for reducing the incidence of ABC-HSR has been demonstrated in white populations (24), and it has been confirmed in the large randomized PREDICT-1 trial (25). In the later study, the negative predictive value of *HLA-B*5701* was 96 percent for clinically suspected and 100 percent for the immunologically confirmed abacavir hypersensitivity reaction. Therefore, screening for *HLA-B*5701* before the initiation of abacavir therapy is currently recommended (see case 3). Similarly, hypersensitivity to nevirapine has been associated with certain HLA alleles (Table 21.2). These associations are not as secure as that of abacavir and *HLA B*5701*.

Table 21.2. Toxicogenetics of Antiretroviral Drugs

Gene or Protein (*)	*Allele or Variant Evaluated*	*Reported Consequence for Antiretroviral Drugs*
HLA-B	HLA-B*5701	Hypersensitivity reaction to abacavir
HLA-C	HLA-Cw8	Hypersensitivity reaction to nevirapine
HLA-DR	*HLA-DRB1*0101*	High negative predictive value of hypersensitivity reactions to nevirapine
TNF-α	-238G/A TNF-α promoter polymorphism	Earlier onset of lipoatrophy
UGT1A1	*UGT1A1*28*, promoter region (insertion at TATA box associated with reduction in bilirubin-conjugating activity)	Gilbert syndrome Hyperbilirubinemia, increased levels of bilirubin in presence of atazanavir or indinavir
APOC3, APOE, APOA5, CETP, and *ABCA1*	APOC3 -482 C>T, -455 T>C, 3238 C>G, APOE ε2 and ε3 haplotypes	Increased risk of hypertriglyceridemia associated with use of ritonavir; identification of individuals at risk for low HDL-cholesterol
SPINK1, CFTR	Multiple variants associated with cystic fibrosis and pancreatitis	Susceptibility to pancreatitis
Mitochondrial DNA (mtDNA)	Tissue-specific mtDNA depletion may represent toxic effect of NRTI therapy on mDNA synthesis. Possibility for accumulation of mutations in mtDNA due to gamma polymerase damage due to NRTIs.	Certain human mtDNA haplotypes (haplotype T) may increase susceptibility to peripheral neuropathy; depletion and mutation of mtDNA likely associated with lipodystrophy
CYP2B6	CYP2B6*6, *11, *18, *27, *28, *29 (alleles with diminished or loss of function, associated with decreased expression, or decreased function, protein truncation or gene deletion)	Higher efavirenz and nevirapine drug levels
MRP2 (ABCC2)	1249G>A (V417I)	Associated with tenofovir-induced renal tubular acidosis
MRP4 (ABCC4)	3724G>A (A1203A), 4131T>G, 669C>T (I223I)	Associated with tenofovir-induced renal tubular acidosis

Only the most relevant associations are indicated. Compiled from www.hiv-pharmacogenomics.org.

Atazanavir, a protease inhibitor that is well tolerated, is conjugated with glucuronic acid by the microsomal enzyme UDP-glucuronosyltransferase 1A1 (UGT1A1). A common adverse effect is the development of unconjugated hyperbilirubinemia. The genetic basis of this (mild) toxicity is well established, because it reflects the carriage of the diminished function *28 allele of *UGT1A1*. This allele decreases transcriptional activity, and it is responsible for the unconjugated hyperbilirubinemia observed in the context of the Gilbert syndrome. Atazanavir and the unconjugated bilirubin compete for the limited enzyme available in the setting of the Gilbert syndrome.

Lipodystrophy, and its two components lipoaccumulation and lipoatrophy, is common among individuals with HIV infection on treatment, especially among those who started therapy in the late 1990s. The choice and duration of NRTI therapy (stavudine > zidovudine) is the main risk factor for clinical lipoatrophy. Genetic prediction of lipodystrophy has been limited by the lack of a unifying hypothesis of the pathophysiological mechanisms underlying this adverse effect of drugs. A widely implicated pathogenic mechanism underlying lipoatrophy is mitochondrial dysfunction. On the one hand, the depletion of mitochondrial DNA is the result of toxicity, and, on the other hand, mitochondrial genome variation may predispose to this adverse effect (19). Other genetic variants have been proposed as associated with the more rapid development of lipodystrophy (Table 21.2).

Transporter Genes and Antiretroviral Treatment

As in other fields in therapeutics, there has been great interest in evaluating the influence of genetic variation in transporters in treatment outcome and toxicity. One of the most studied transporters is P-glycoprotein (MDR1), encoded by *ABCB1*. Various single-nucleotide polymorphisms (SNPs) and haplotypes have been associated with drug efficacy and pharmacokinetics (Table 21.1). However, a significant level of controversy remains because of conflicting results for the same drug and substrate-specific associations (4). For one particular variant (*MDR1 3435C>T*), a possible answer to the observed substrate variation has been reported. This synonymous

Table 21.3. Genetic Variants Modulating HIV-1 Susceptibility and Disease

Gene	*Function, Role in HIV-1 Disease*	*Variant*	*Influence on HIV-1 Disease*
HLA	Major histocompatibility complex Acquired immunity	*MHC I A, B, C* homozygosity, or selected *HLA B* alleles, and *HLA C* SNPs	Protection or progression depending on allele
CCR2-CCR5 Locus	Decrease in co-receptor expression altering viral entry; possible immunoregulatory effect	Coding (e.g., CCR2-64I, CCR5 Δ32) and promoter variants (e.g., CCR5 P1)	Protection from infection; better viral control and slower disease progression; development of entry antagonists for treatment
KIR	Innate immunity Regulation of NK call response	Specific KIR-HLA associations	Protection or progression depending on the specific association

Shown are genes and variants for which there is sufficient evidence (36), or emerged from sizeable studies. Updated in www.HIV-pharmacogenomics.org.

SNP represents a rare codon usage for isoleucine that leads to a change in timing of cotranslational folding of the P-glycoprotein and results in changes in substrate specificity (26). This novel mechanism explains, to a large extent, the controversial results or the lack of association of this variant, directly, or through linkage disequilibrium, with causal variants or with changes in mRNA expression. Various combinations of *MDR1/ABCB1* variants and SNPs in metabolic genes have also been associated with the toxicity/outcome of treatment with NNRTIs (27).

Other transporters have been investigated for their role in the renal tubular toxicity of the first-line agent tenofovir – an NRTI. Here, a restricted number of nonsynonymous variants of multidrug-resistant proteins MRP2 and MRP4 have been associated with tenofovir-induced proximal tubulopathy (Table 21.2). However, these are small studies that need to be confirmed.

WHEN TO START ANTIRETROVIRAL TREATMENT

In "Basic concepts in HIV therapeutics," I presented the current criteria for initiating ART. From this, it appears that complementary to the interest of identifying the best drug for each patient, the field of HIV would benefit from genetic markers that help identify those most in need of treatment. Some of these markers of disease progression are emerging from studies of the genetic determinants of HIV-1 pathogenesis (Table 21.3), and under the impulsion of genome-wide association analyses (28, 29). It remains speculative how and when these markers would be used in the clinical setting. One possibility is that the pharmacogenetic arrays would include genetic markers of pharmacokinetics (to assist in drug dosing), toxicogenetics (to assist in drug selection), and disease progression (to assist in timing of ART). Prediction of the rate of disease progression would be of particular interest for decisions to initiate treatment at the time of primary infection – when the viral set point has not been reached – and to guide treatment interruption by identifying patients likely to maintain prolonged spontaneous control of viral replication in the absence of treatment. In the setting of chronic disease, genetic markers could guide the optimal timing for the initiation of ART. Personalized medicine in HIV will, however, need a significant effort in the design and completion of dedicated randomized clinical trials testing the relative value of pharmacogenetics in routine clinical care.

CASE STUDIES

This section will illustrate the key principles in pharmacogenetics and genomics of ART. The chosen cases will illustrate the application of principles of candidate gene analysis (*case study 1*), of a comprehensive approach for the identification of genetic variants influencing the ADME of protease inhibitors (*case study 2*), and the genome-wide approach to the identification of genetic determinants of severe toxicity reactions (*case study 3*).

Gene Candidate Approach

Case study 1. Efavirenz pharmacokinetics, primary and accessory metabolic pathways.

A fifty-two-year-old, HIV-1-infected black woman initiates therapy with a CD4+ T-cell count of 327 cells/μl and a viral load of 57,000 copies/ml. First-line treatment includes tenofovir (245 mg/day), lamivudine (300 mg/day), and efavirenz (600 mg/day). Two weeks after initiation of treatment, the patient describes significantly disturbed sleep. Therapeutic drug monitoring identifies a plasma drug level of 9,270 ng/ml 19 hours after the last dose of efavirenz (suggested therapeutic range 1,000–4,000 ng/ml), Figure 21.3A shows that, despite efavirenz dose reduction to 400 mg/day, plasma drug levels remain well above the therapeutic

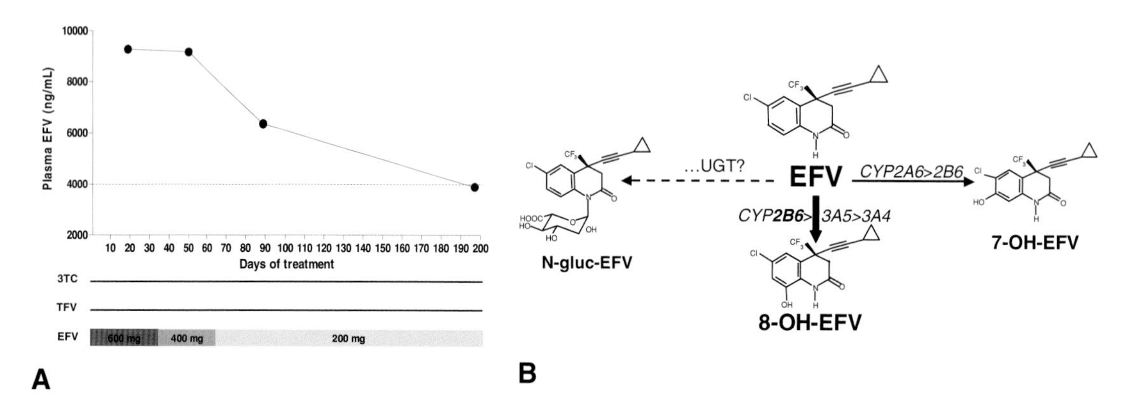

Figure 21.3. (**A**) Progressive correction of excessive plasma drug levels in an individual with loss-of-function alleles in CYP2B6 and 2A6. (**B**) Primary and accessory metabolic pathways of efavirenz.

range. An additional reduction of efavirenz dosing to 200 mg/day brings plasma drug levels into therapeutic range and coincides with improved sleep. No relevant drug interactions were identified.

Pharmacogenetic characterization of *CYP2B6* identifies the presence of allele *CYP2B6*6*, associated with reduced expression and function of the isoenzyme, and of *CYP2B6*18*, associated with markedly reduced protein expression. Both alleles are common in individuals of African descent. However, the difficulty of bringing the plasma drug levels into therapeutic range motivated the analysis of the accessory metabolic pathways of efavirenz, depicted in Figure 21.3B. Efavirenz metabolism can proceed via CYP2A6 and CYP3A. This patient was homozygous for *CYP2A6*9* (loss of function), *CYP3A5*6* (loss of function), and *CYP3A4*1B* (controversial, associated with diminished function). This complex genotype maximally limits metabolism of efavirenz, which can only proceed through glucuronidation (Figure 21.3B).

This case illustrates the interest of genotyping primary and accessory metabolic pathways for the understanding of unusually elevated plasma drug levels and pharmacokinetic profiles.

Pathway Approaches

Case study 2. ADME pathway of protease inhibitors.

A thirty-year-old, HIV-1-infected white man participates in a study of the pharmacokinetics of lopinavir in fixed combination with ritonavir. His plasma drug level is low (high clearance of 12 L/h). No relevant drug interactions are identified. To better understand this situation, the patient contributes DNA to the study of the pharmacogenetics of ADME pathways for this drug (Figure 21.4A). This includes the genotyping of known genetic variants and tagging SNPs in fifteen selected genes, These include genes coding for CYP450 isoenzymes (Locus *CYP3A4, 3A5, 3A7, 3A43*), transporters (*MDR1/ABCB1*), plasma-binding proteins (Albumin, *ORM1* and *ORM2*), and nuclear receptors (*VDR, NR3C1, NR1I3, NR1H4, NR1I2, HNF4A, TCF1*). Analysis of 117 individuals with low lopinavir clearance (cases) and 90 with high clearance (controls) identified candidate variants associated with the study phenotype. Figure 21.4B is a display, referred to as a "Manhattan plot," that places the various SNPs in their chromosomal context (*x*-axis), and the statistical test for the association with the study phenotype (elevated lopinavir high clearance) in the *y*-axis. The analysis identified an association of functional variants of *SCLO1B1*, and, through a second round of validation (genotyping in an independent cohort), variants of *ABCC2* and variants in the intergenic region of *CYP3A4, 3A7,* and *3A5.*

The patient described was himself homozygous for *SCLO1B1*4*, an allele that results in P155T, an amino acid change in the substrate recognition domain of the transporter. However, the study underscores the limits to genetic analyses because there is no certitude that the genotyping array includes all genes/SNPs that are relevant to the pharmacokinetics of a given medication. The study may also lack the statistical power to identify variants that contribute little to the final phenotype. The latter pattern is commonly defined as a "complex genetic trait."

This case illustrates the interest of large-scale genotyping of ADME pathway genes in situations where the genetic basis of a pharmacokinetic phenotype is not well understood.

Genome-Wide Approaches

Case study 3. Genome analysis of abacavir hypersensitivity.

A thirty-nine-year-old man initiates ART at a CD4+ T-cell count of 190/μl and a viral load of 4,500 copies/ml. Treatment includes the fixed-dose combination of zidovudine, lamivudine, and abacavir. Ten days

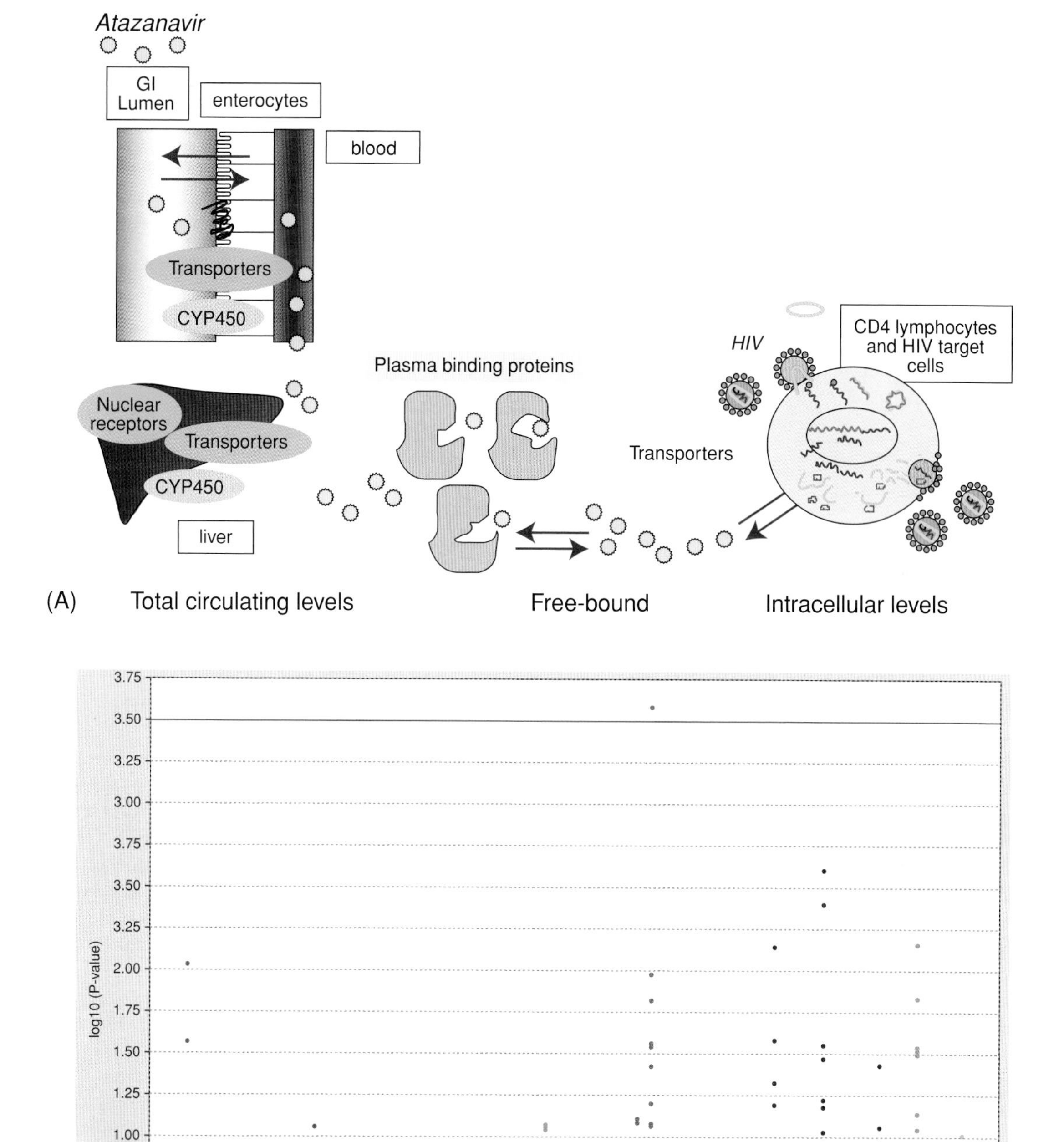

Figure 21.4. (**A**) Graphical representation of the steps and genes participating in the ADME (absorption, distribution, metabolism, and excretion) of lopinavir. (**B**) Manhattan plot of the genetic association study of ADME variants and lopinavir clearance.

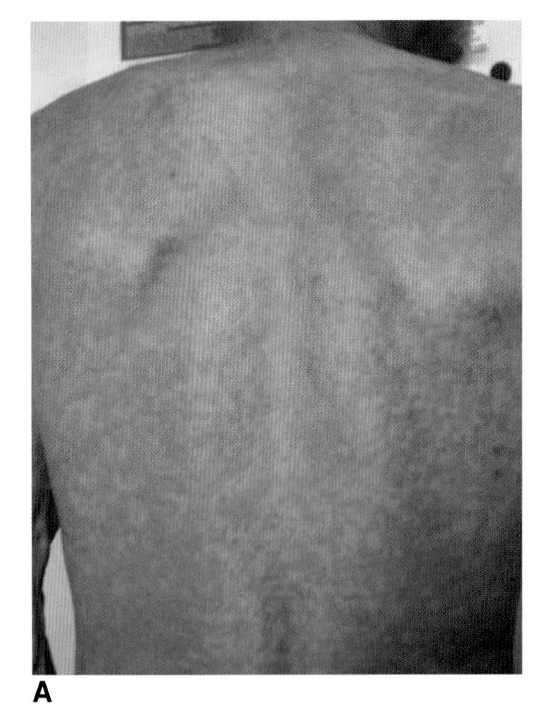

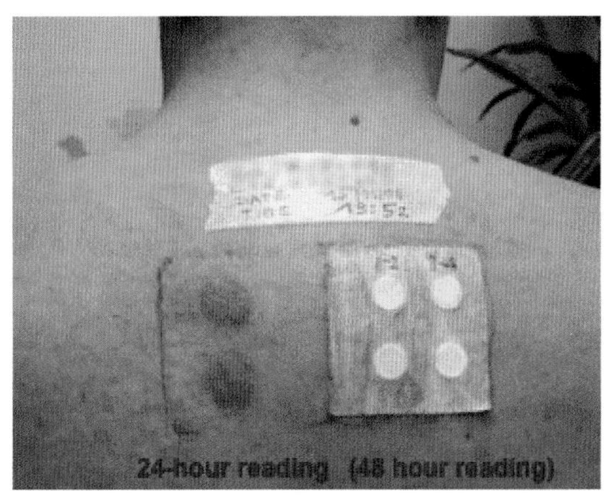

A B

Figure 21.5. (**A**) Macular skin rash in an individual receiving abacavir. (**B**) A positive skin "patch test" that confirms that the hypersensitivity reaction is due to abacavir.

later he develops nausea, fever, and myalgia. A few days later, he presents to the consultation with a skin rash (Figure 21.5A). He is included in a case-control study that aims at assessing the power of genome-wide association analysis to confirm the role of the *HLA-B*5701* allele. For this, cases proven by a positive skin "patch test" to have a hypersensitivity reaction to abacavir (Figure 21.5B) were compared with individuals that received the medication and did not develop toxicity (abacavir-tolerant). The patch test allowed the ruling out of other cases of hypersensitivity unrelated to abacavir. The study allowed the unequivocal identification of the determinant role of the *HLA-B*5701* after genome-wide analysis of only 15 cases and 200 abacavir tolerant controls.

This proof-of-concept study indicates that genome-wide studies should be considered in the investigation of severe toxicity to ART or other medications, as long as a precise, definitive diagnosis can be established.

DISCUSSION PROBLEMS

The problems presented in this section are intended as advanced exercises for independent or group work.

1. Increasingly, ART is provided to different human populations. It is conceivable that the allelic frequency of genetic variants of specific CYP450 genes varies in the target population. Discuss the steps that you would complete to assess this possibility. Background reference: Rotger et al. (8).
2. It is proposed that lipoatrophy may have its genetic basis in the diversity of the mitochondrial genome. Identify the optimal tissue for research, and describe the basis of the human mitochondrial haplogroups. Background reference: McRae et al. (30).
3. Recent genome-wide association analyses of genetic variants associated with diabetes mellitus has identified multiple SNPs that confer a small incremental risk of this complication. The risk of diabetes is also associated with the use of ART. Discuss the construction of a predictive genetic score that would help identify the patients at increased risk for this complication. Background reference: Pharoah et al. (31).
4. Use of nevirapine is occasionally associated with Stevens-Johnson syndrome (32). Describe an approach that makes use of the power of genome-wide association analysis. Background reference: Nelson et al. (33).
5. You wish to evaluate the usefulness of pharmacogenetic markers to guide the dose of efavirenz. Discuss the design of a theoretical trial that assesses this possibility. Background references: Nyakutira et al. (34) and Rotger and Telenti (35).

REFERENCES

1. Hammer SM, Saag MS, Schechter M, et al. Treatment for adult HIV infection: 2006 recommendations of the International AIDS Society-USA panel. *JAMA*. 2006;**296**: 827–43.
2. Haas DW, Ribaudo HJ, Kim RB, et al. Pharmacogenetics of efavirenz and central nervous system side effects: an Adult AIDS Clinical Trials Group study. *AIDS*. 2004;**18**: 2391–400.
3. Ward BA, Gorski JC, Jones DR, Hall SD, Flockhart DA, & Desta Z. The cytochrome P450 2B6 (CYP2B6) is the main catalyst of efavirenz primary and secondary metabolism: implication for HIV/AIDS therapy and utility of efavirenz as a substrate marker of CYP2B6 catalytic activity. *J Pharmacol Exp Ther*. 2003;**306**:287–300.
4. Telenti A & Zanger UM. Pharmacogenetics of anti-HIV drugs. *Annu Rev Pharmacol Toxicol*. 2008;**48**:227–56.
5. Frohlich M, Hoffmann MM, Burhenne J, Mikus G, Weiss J, & Haefeli WE. Association of the CYP3A5 A6986G (CYP3A5*3) polymorphism with saquinavir pharmacokinetics. *Br J Clin Pharmacol*. 2004;**58**:443–4.
6. Anderson PL, Lamba J, Aquilante CL, Schuetz E, & Fletcher CV. Pharmacogenetic characteristics of indinavir, zidovudine, and lamivudine therapy in HIV-infected adults: a pilot study. *J Acquir Immune Defic Syndr*. 2006;**42**:441–9.
7. Rotger M, Csajka C, & Telenti A. Genetic, ethnic, and gender differences in the pharmacokinetics of antiretroviral agents. *Curr HIV/AIDS Rep*. 2006;**3**:118–25.
8. Rotger M, Tegude H, Colombo S, et al. Predictive value of known and novel alleles of CYP2B6 for efavirenz plasma concentrations in HIV-infected individuals. *Clin Pharmacol Ther*. 2007;**81**:557–66.
9. Zukunft J, Lang T, Richter T, et al. A natural CYP2B6 TATA box polymorphism (-82T–> C) leading to enhanced transcription and relocation of the transcriptional start site. *Mol Pharmacol*. 2005;**67**:1772–82.
10. di Iulio J, Fayet A, Arab-Alameddine M, et al. In vivo analysis of efavirenz metabolism in individuals with impaired CYP2A6 function. *Pharmacogenet Genomics*. 2009;**19**(4): 300–9.
11. Rotger M, Colombo S, Furrer H, et al. Influence of CYP2B6 polymorphism on plasma and intracellular concentrations and toxicity of efavirenz and nevirapine in HIV-infected patients. *Pharmacogenetics Genomics*. 2005;**15**:1–5.
12. Baede-van Dijk PA, Hugen PW, Verweij-van Wissen CP, Koopmans PP, Burger DM, & Hekster YA. Analysis of variation in plasma concentrations of nelfinavir and its active metabolite M8 in HIV-positive patients. *AIDS*. 2001;**15**:991–8.
13. Haas DW, Smeaton LM, Shafer RW, et al. Pharmacogenetics of long-term responses to antiretroviral regimens containing efavirenz and/or nelfinavir: an Adult Aids Clinical Trials Group Study. *J Infect Dis*. 2005;**192**:1931–42.
14. Kassahun K, McIntosh I, Cui D, et al. Metabolism and disposition in humans of raltegravir (MK-0518), an anti-AIDS drug targeting the human immunodeficiency virus 1 integrase enzyme. *Drug Metab Dispos*. 2007;**35**: 1657–63.
15. Anderson PL, Kakuda TN, & Lichtenstein KA. The cellular pharmacology of nucleoside- and nucleotide-analogue reverse-transcriptase inhibitors and its relationship to clinical toxicities. *Clin Infect Dis*. 2004;**38**:743–53.
16. Fellay J, Boubaker K, Ledergerber B, et al. Prevalence of adverse events associated with potent antiretroviral treatment: Swiss HIV Cohort Study. *Lancet*. 2001;**358**:1322–7.
17. Keiser O, Fellay J, Opravil M, et al. Adverse events to antiretrovirals in the Swiss HIV Cohort Study: impact on mortality and treatment modification. *Antivir Ther*. 2007;**12**:1157–64.
18. Rotger M, Colombo S, Furrer H, et al. Influence of CYP2B6 polymorphism on plasma and intracellular concentrations and toxicity of efavirenz and nevirapine in HIV-infected patients. *Pharmacogenet Genomics*. 2005;**15**:1–5.
19. Tarr PE & Telenti A. Toxicogenetics of antiretroviral therapy: genetic factors that contribute to metabolic complications. *Antivir Ther*. 2007;**12**:999–1013.
20. Rodondi N, Darioli R, Ramelet AA, et al. High risk for hyperlipidemia and the metabolic syndrome after an episode of hypertriglyceridemia during 13-cis retinoic acid therapy for acne: a pharmacogenetic study. *Ann Intern Med*. 2002;**136**:582–9.
21. Cutrell AG, Hernandez JE, Fleming JW, et al. Updated clinical risk factor analysis of suspected hypersensitivity reactions to abacavir. *Ann Pharmacother*. 2004;**38**:2171–2.
22. Hetherington S, Hughes AR, Mosteller M, et al. Genetic variations in HLA-B region and hypersensitivity reactions to abacavir. *Lancet*. 2002;**359**:1121–2.
23. Mallal S, Nolan D, Witt C, et al. Association between presence of HLA-B*5701, HLA- DR7, and HLA-DQ3 and hypersensitivity to HIV-1 reverse-transcriptase inhibitor abacavir. *Lancet*. 2002;**359**:727–32.
24. Rauch A, Nolan D, Martin A, McKinnon E, Almeida C, & Mallal S. Prospective genetic screening decreases the incidence of abacavir hypersensitivity reactions in the Western Australian HIV cohort study. *Clin Infect Dis*. 2006;**43**:99–102.
25. Hughes S, Hughes A, Brothers C, Spreen W, & Thorborn D. PREDICT-1 (CNA106030): the first powered, prospective trial of pharmacogenetic screening to reduce drug adverse events. *Pharm Stat*. 2008;**7**:121–9.
26. Kimchi-Sarfaty C, Oh JM, Kim IW, et al. A "silent" polymorphism in the MDR1 gene changes substrate specificity. *Science*. 2007;**315**:525–8.
27. Haas DW, Bartlett JA, Andersen JW, et al. Pharmacogenetics of nevirapine-associated hepatotoxicity: an Adult AIDS Clinical Trials Group collaboration. *Clin Infect Dis*. 2006;**43**:783–6.
28. Fellay J, Shianna KV, Ge D, et al. A whole-genome association study of major determinants for host control of HIV-1. *Science*. 2007;**317**:944–7.
29. Loeuillet C, Deutsch S, Ciuffi A, et al. In vitro whole-genome analysis identifies a susceptibility locus for HIV-1. *PLoS Biol*. 2008;**6**:e32.
30. McRae AF, Byrne EM, Zhao ZZ, Montgomery GW, & Visscher PM. Power and SNP tagging in whole mitochondrial genome association studies. *Genome Res*. 2008;**18**:911–17.
31. Pharoah PD, Antoniou AC, Easton DF, & Ponder BA. Polygenes, risk prediction, and targeted prevention of breast cancer. *N Engl J Med*. 2008;**358**:2796–803.
32. Roujeau JC & Stern RS. Severe adverse cutaneous reactions to drugs. *N Engl J Med*. 1994;**331**:1272–85.

33. Nelson MR, Bacanu SA, Mosteller M, et al. Genome-wide approaches to identify pharmacogenetic contributions to adverse drug reactions. *Pharmacogenomics J.* 2009;**9**: 23–33.
34. Nyakutira C, Roshammar D, Chigutsa E, et al. High prevalence of the CYP2B6 516G–>T(*6) variant and effect on the population pharmacokinetics of efavirenz in HIV/AIDS outpatients in Zimbabwe. *Eur J Clin Pharmacol.* 2008;**64**: 357–65.
35. Rotger M & Telenti A. Optimizing efavirenz treatment: CYP2B6 genotyping or therapeutic drug monitoring? *Eur J Clin Pharmacol.* 2008;**64**:335–6.
36. Telenti A & Goldstein DB. Genomics meets HIV. *Nat Rev Microbiol.* 2006;**4**:9–18.

Application of Pharmacogenetics and Pharmacogenomics in Pediatrics: What Makes Children Different?

22

Jennifer A. Lowry and J. Steven Leeder

Historically, submissions to the Food and Drug Administration (FDA) have been based on safety and efficacy data obtained from clinical trials conducted in adults with limited or no data from children. As a result, pediatricians and other health care professionals have relied on empiric therapeutic strategies, largely the consequence of treatment on a trial-and-error basis. In essence, the absence of information in the product label forces pediatricians to choose between avoiding the use of a medication that may be beneficial and using a medication "off-label" in the absence of evidence-based safety and efficacy data with the accompanying potential for ineffective and harmful outcomes.

During the past fifteen years, new federal laws and regulations have increased the level of scientific and clinical rigor of investigations aimed at ensuring that the use of medications by children is, indeed, safe and effective. Interested readers are referred to a detailed chronology of events occurring between 1994 and 2002 (1), and a contemporary discussion of the issues surrounding more recent legislative activities, such as the Pediatric Research Equity Act (PREA) of January 2003 (2). In general, it is recognized that growth and development are accompanied by changes in the physiological and biochemical processes determining drug disposition and response, for example, drug absorption, distribution, metabolism, excretion, and targets of drug response (3). However, the acquisition of information that can inform safe and efficacious use of medications in children of different ages or developmental stages has been a relatively recent development, increasing dramatically in recent years as a consequence of the various legislative initiatives. It is now apparent that extrapolation of adult data to pediatric populations is quite inappropriate because drug clearance may be greater than or, in some cases, less than that observed in adults (4). Thus, even though weight-based dosing strategies are becoming more sophisticated and have improved abilities to use adult data to infer drug clearance in children (5), they are unlikely to provide consistent dosing guidelines across all pediatric age groups or chemical classes. This is largely due to the variability in the developmental patterns of expression of the various drug-metabolizing enzymes and transporters involved in the disposition of individual compounds. Furthermore, evidence that the response to some medications may be different in children relative to adults despite comparable drug exposure is beginning to accrue (e.g., buspirone [4]), implying that age-related differences in drug targets and downstream signal transduction pathways may also be present.

As in adults, the dose → exposure → response relationship is at the heart of pediatric pharmacotherapy. Our understanding of the contribution of genetic variation to the variability in drug disposition and response in adults has increased substantially during the past twenty-five years (6). However, the application of pharmacogenetic and pharmacogenomic principles to pediatric drug therapy has lagged well behind (7). In comparison with adults, *pediatric pharmacogenetics* involves an added measure of complexity because the variability due to developmental processes, or ontogeny, is superimposed on genetic variation. In other words, a specific phenotype may change over time between conception and senescence. In the context of drug biotransformation, fetuses and newborns may be phenotypically "slow" or "poor" metabolizers for certain drug-metabolizing pathways, acquiring a phenotype consistent with their genotype at some point later in the developmental process as those pathways mature (e.g., glucuronidation, some cytochrome P450 (CYP) activities [8]). It is apparent that not all infants will acquire drug metabolizing enzyme activity at the same rate because of the interaction between genetics and environmental factors, such as formula versus breastfeeding, for example. Therefore, interindividual variability in the trajectory (rate and

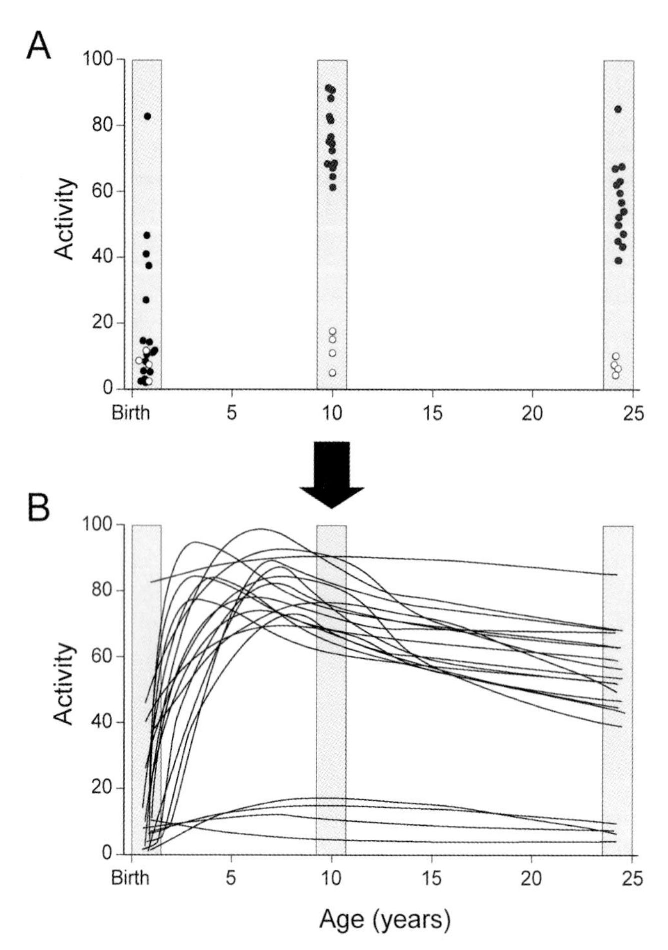

Figure 22.1. Concept of "developmental trajectory." Ontogeny and genetic variation both contribute to observed variability in drug disposition and response in pediatric populations with developmental changes in gene expression and functional enzyme activity superimposed on pharmacogenetic determinants. (**A**) Phenotyping of an adult population ("age 25 years") for a particular drug biotransformation enzyme or drug response may reveal allelic variation within the coding region of the gene that gives rise to two distinct phenotypes, high activity in 92 percent of the population ("extensive metabolizers"; solid circles) and low activity in 8 percent of the population ("poor metabolizers"; open circles). A similar phenotyping strategy in newborns ("Birth") may reveal that the two phenotypes are not readily distinguishable because of the relative absence of that pathway at birth. Furthermore, there may be discrete periods during childhood ("age 10 years") in which the genotype-phenotype relationship may differ from that observed in adults (e.g., developmental stages where enzyme activity appears to be greater in children compared with adults). (**B**) The rate at which functional activity in a drug biotransformation enzyme or pathway, or drug target and downstream signal transduction cascade, is acquired after birth can be referred to as a "developmental trajectory." Interindividual variability in the developmental trajectory may give rise to different distributions of phenotype measurements ("developmental phenotypes") at discrete ages or developmental stages. Modified from Leeder JS. *Drug Discov Today* 2004;**9**:567–73.

extent) of acquired drug biotransformation capacity may be considered a *developmental phenotype* (Figure 22.1) and helps to explain the considerable variability in some CYP activities observed immediately after birth (8).

Fortunately, examples of how pharmacogenetic and pharmacogenomic approaches can improve pediatric pharmacotherapy are now appearing in the literature on a more frequent basis (9–11). At the level of individual medications, however, the relative contributions of ontogeny and genetic variation to observed variability in drug disposition and response may not be readily apparent from the published literature. Therefore, the goals of this chapter are to (1) describe some fundamental differences between children and adults that have an impact on the application of pharmacogenetics and pharmacogenomics in a pediatric context, and (2) present a general paradigm that can be used as an aid to proactively assess the contribution of ontogeny and genetic variation for drug-related issues in children.

IMPORTANT CONSIDERATIONS IN PEDIATRIC PHARMACOGENETICS AND DEVELOPMENTAL PHARMACOGENOMICS

1. The primary function of "drug" biotransformation and transport pathways is the biosynthesis and catabolism of endogenous molecules responsible for normal growth and development. The biotransformation of exogenous molecules, xenobiotics, is superimposed on these essential developmental functions.

A fundamental difference between children and adults is that children are undergoing myriad molecular, biochemical, and physiological changes in the course of achieving full reproductive potential. In contrast, adults, whether they like it or not, experience slowly progressive changes in function associated with senescence. To illustrate this point, it is quite evident that more dramatic physiological change occurs in the five-year interval between birth and five years of age, or between five years and ten years of age, relative to any five-year interval beyond twenty years of age. In fact, anatomic differences in body proportions and age-related differences in body composition readily distinguish newborns, infants, and children from adults. Furthermore, the biological cadence in the developing human is characterized by a more rapid pulse and respiratory rates and markedly increased nutritional requirements that are essential to support a rapid rate of somatic and brain growth.

To carry out the developmental program from conception to reproductive capability, small-molecular-weight endogenous compounds must be maintained within tightly regulated, physiological ranges with appropriate temporal and spatial patterns of expression. The CYPs catalyze steps in the biosynthesis and catabolism of a wide variety of endogenous compounds, including androgens, arachidonic acid and eicosanoids, bile acids, adrenal corticoids, estrogens, progestins, retinoic acid, and sterols, among others. It has been estimated that at least fourteen CYPs are involved in the arachidonic acid cascade, seven CYPs in cholesterol metabolism and bile acid biosynthesis, six CYPs in steroidogenesis, and four CYPs in vitamin D_3 synthesis and metabolism (12). Likewise, three CYPs in the CYP26 family are involved in retinoic acid hydroxylation, and the critical importance of maintaining concentrations of endogenous molecules within tightly regulated physiological ranges to ensure proper control of growth and development is exemplified by retinoic acid teratogenicity, which is considered to be the consequence of an imbalance between synthesis and degradation.

The involvement of "drug-metabolizing" enzymes during development is not restricted to CYPs. Multiple sulfotransferases (SULTs) have been detected in various organs of the developing fetus, in particular, liver, kidney, lung, and adrenal (13). SULTs modulate the activity of several hormones and neurotransmitters involved in fetal growth and development, including triiodothyronine (SULT1A1), dopamine and other catecholamines (SULT1A3/1A4), estradiol (SULT1E1), and dehydroepiandrosterone (SULT2A1). In general, expression of UDP-glucuronosyltransferases (UGTs) is limited in the fetus, and functional expression of various UGT1 and UGT2 family members is acquired postnatally in isoform-specific patterns (8). Analogous to the situation with CYPs and SULTs, several endogenous compounds are substrates for UGTs, including bilirubin (UGT1A1), serotonin (UGT1A6), dopamine (UGT1A10), bile acids (UGT2B4), and androgens (UGT2B7, UGT2B17), among others. Likewise, endogenous compounds feature prominently among the substrates of xenobiotic transporters, such as P-glycoprotein (ABCB1), multidrug resistance proteins (ABCC1–5), and organic anion-transporting polypeptides (SLCO1A1, 1B1, 2B) (14, 15).

It is important to recognize that CYPs and other drug-metabolizing enzymes do not function in isolation but protect the organism from foreign chemicals through a coordinated system that involves cellular uptake, oxidative biotransformation, conjugation, and cellular export of the hydrophilic metabolites of those foreign compounds. Likewise, CYPs and other drug-metabolizing enzymes function during development within pathways or networks necessary to generate and degrade endogenous molecules critical for maturation. For example, CYP3A7 is considered to be a fetal liver-specific member of the human CYP3A subfamily, and most drug metabolism researchers think of CYP3A7 in the context of its ability to metabolize drugs. In reality, CYP3A7's ability to metabolize drugs is rather limited compared with that of CYP3A4 or CYP3A5 (16), but it plays a critical role in fetal development, specifically in the pathway responsible for estriol biosynthesis. Humans and great apes are unique in that estriol is the primary estrogen of pregnancy, and it is produced by the placenta at levels that are severalfold higher than estradiol and estrone (17). The human fetus is actively involved in estriol biosynthesis through a process that involves multiple steps. The first step is production of dehydroepiandrosterone 3-sulfate (DHEA-S) by SULT2A1 in the fetal adrenal cortex. DHEA-S is then taken up by the fetal liver (presumably by an OATP transporter), and a third hydroxyl group is added by CYP3A7 to form 16α-hydroxy-DHEA-S (18). After deconjugation by sulfatase in placental syncytiotrophoblasts, 16α-hydroxy-DHEA undergoes a multistep process culminating in aromatization by CYP19A1 (aromatase) to form estriol. In fact, it has been estimated that $>$90 percent of estriol originates from fetal DHEA-S. Thus, the estriol biosynthetic pathway during pregnancy illustrates the interrelationships between gene products that provide important endogenous functions during development, but that may also play an important protective function in protecting the developing human from xenobiotic exposure as well. A similar situation exists for detoxifying bilirubin after birth (19).

It is important to note that many currently marketed drugs are designed to treat diseases that are the consequence of disrupted or dysregulated physiological pathways. Therefore, drug exposure during prenatal and postnatal development, whether unintended or intended for therapeutic purposes, must take into consideration the potential to disrupt or dysregulate normal developmental processes. Thus, the presence of drug during critical periods of development may lead to an increased risk of unintended consequences – teratogenesis during fetal life or an age-related increased risk of adverse drug reactions in susceptible individuals.

2. Genotype-phenotype relationships change during development.

Whereas genetic constitution or "genotype" is invariant throughout life (acquired genomic changes occurring in tumor cells, especially leukemias [20], being an exception), the process of growth and development involves multiple phenotypic changes. One needs only to view a picture album containing photos of him- or herself as a baby, at each birthday, and at each school grade, to be reminded of the discrete phenotypes (and profound changes in phenotype) that occur between birth and high school graduation. These obvious, superficial variations in phenotype are accompanied by myriad systemic changes that affect drug disposition and response. For example, drug absorption in neonates is highly variable because of the developmental changes in intragastric pH (relatively higher in neonates in comparison with older ages), and maturational changes in intestinal motor activity and mucosal surface area, both of which increase with age. Drug disposition is also influenced by age-dependent changes in body composition (e.g., total body water, adipose tissue) and protein binding, which ultimately affect volume of distribution (3). Finally, maturation of renal function occurs during the first year of life with the functional components of glomerular filtration, tubular secretion, and tubular reabsorption processes, each exhibiting independent rates and patterns of development (21). As a consequence, ontogeny contributes substantially to interindividual variability in pharmacokinetic parameters, particularly early in life, as these various physiological processes mature (21, 22).

Maturation of functional drug biotransformation capacity is probably the best understood of all the processes contributing to variability in drug disposition and response in children. However, available in vitro and in vivo data must be interpreted in the context of the dramatic anatomic and physiological changes that occur during prenatal and postnatal hepatic development. For example, the fetal liver is an important hematopoietic organ, and apparent changes in the developmental pattern of expression of some enzyme activities reflect the decline in hematopoietic stem cells and the increase in parenchymal hepatocytes that occurs as fetal life progresses. Several factors may contribute to developmental differences in drug biotransformation between children and adults: decreased numbers and smaller hepatocytes in infant livers in comparison with adults, the reduced amount of endoplasmic reticulum as measured by the ratio of microsomal to total cellular protein, and changes in the ratio of liver mass to total body mass throughout development. More comprehensive discussions of these factors can be found in reviews by Ring et al. (23), Alcorn and McNamara (21), and Hines (8).

A key concept for the application of pharmacogenetics in a pediatric context is that *genotype-phenotype relationships may not be apparent until the gene product is fully expressed*. Thus, knowledge of the ontogeny of genes involved in drug biotransformation and transport, specifically *when* the gene products are fully expressed, is essential for determining the contribution of genetic variation to observed variability in drug clearance in children. Available data reveal that functional drug biotransformation capacity tends to be acquired in gene-specific patterns. A comprehensive review of the current state of knowledge, derived from in vitro investigations and insights gained from in vivo pharmacokinetic studies of drugs considered to be prototypical substrates of specific drug-metabolizing enzymes, has been published recently (8).

In brief, the ontogeny of hepatic drug-metabolizing enzymes appears to follow three general patterns of expression. The first pattern is characterized by relatively high expression during fetal life with expression maintained at low levels or silenced during postnatal life. CYP3A7 and FMO1 are the best known examples of this group, which also includes SULT1A3. Some genes, such as CYP3A5, SULT1A1, and TPMT, are expressed at relatively constant levels throughout gestation and postnatal life. Most hepatic drug-metabolizing enzymes either are not expressed in fetal liver or begin to appear in the late second or third trimester of pregnancy, but their expression increases dramatically after birth. Within this group, which includes CYPs 1A2, 2C9, 2C19, 2D6, 2E1, 3A4, FMO3, and most UGTs, the onset of functional activity occurs within hours to days after birth, but complete maturation may not be achieved until much later in life – adolescence, or later in some cases (8). In fact, it is difficult to determine definitively at what age a particular pathway is fully developed because in vitro and in vivo data may not lead to consistent conclusions because of the developmental changes in microsomal to total cellular protein or liver to total body mass.

The time course for acquisition of fully functional drug biotransformation activity can be considered as a "developmental trajectory" (Figure 22.1). A major challenge for the application of pharmacogenetics in a pediatric context is that variability due to developmental changes in gene expression and functional enzyme activity is superimposed on pharmacogenetic determinants. For example, the considerable variability in some CYP activities observed immediately after birth, particularly pronounced for CYP2C9 and CYP2C19 (24), probably involves interindividual variability in the trajectory (rate and extent) of the acquisition of drug biotransformation capacity, and genetic variation, as well. To add further complexity, the developmental trajectories described above may also be influenced by environmental factors as illustrated by the accelerated acquisition of CYP1A2 activity in formula-fed infants compared with breastfed infants (25).

3. Drug response in adults does not necessarily predict patterns of drug response in pediatric populations.

Children provide unique challenges with respect to the safety and efficacy of medications. For example, some pediatric diseases have no adult correlate or are more prevalent in children than in adults. Given the role of prematurity and immaturity in disorders such as patent ductus arteriosus and persistent pulmonary hypertension of the newborn, it is not difficult to accept that there are no adult treatment paradigms that can be simply extrapolated to patients in the newborn intensive care unit. Similarly, pediatric diseases like acute lymphoblastic anemia, Kawasaki's disease, neuroblastoma, and Wilms' tumor are encountered in children and have no close correlate in adults. Conditions such as attention deficit hyperactivity disorder, autism, and other pervasive developmental disorders have their origins early in life and are more prevalent in children, at least at the time of diagnosis. Finally, it is less clear whether, or to what extent, the mechanisms of diseases that are nominally similar and prevalent across all age groups, such as asthma, epilepsy, type 2 diabetes, depressive disorders and other psychiatric illnesses, hypertension, inflammatory bowel disease, and arthritis, among others, are shared by the pediatric form of the disease and adult forms of that disease. This latter issue is extremely important because the anticipated benefits of therapies designed and evaluated in adults may, or may not, be realized in children.

Although pediatric and adult diseases with similar names share some common characteristics, there are also important differences in disease presentation, treatment, and consequences of medication use. Perhaps the most obvious difference shared by pediatric forms of disease relative to their adult counterparts is an earlier age of onset. Less apparent are subtle differences in disease pathogenesis due, in part, to the limited appreciation by adult-oriented practitioners and researchers that pediatric disease may be a distinct entity as well as the relatively smaller numbers of affected children in comparison with adults. For example, the clinical characteristics of the various subtypes of juvenile idiopathic arthritis (JIA) are quite different than those found in adult rheumatoid arthritis patients. Only approximately 5 percent of JIA patients will be positive for the rheumatoid factor, and it is rarely found in children less than ten years of age. Furthermore, uveitis is much more frequently a complication in JIA than in adults (26). Likewise, family history, exposure to environmental tobacco smoke, and atopic sensitization are common features of child- and adult-onset asthma. In contrast, however, symptoms are intermittent in children and persistent in adults, and there are differences in sex predilection between children (more boys than girls) and adults (more women than men) (27). In addition, hypertension in children is most frequently secondary to chronic renal disease, rather than the essential hypertension prevalent in adults. As a consequence, patterns of drug use are different; a high percentage of children (>80 percent) receive angiotensin-converting enzyme inhibitors or angiotensin receptor blockers relative to the use of diuretics, β-blockers, and other antihypertensive agents (28).

These two factors, limited appreciation that pediatric disease may be a distinct entity and the relatively smaller numbers of affected children compared to adults, contribute to the paucity of studies describing the application of genomic tools, especially genome-wide association (GWA) studies, to pediatric disease and its treatment. Although modest sample sizes limit the power to detect associations, such studies are critical as a preliminary step to providing more targeted treatment for children. These would provide an improved understanding of disease risk and the identification of the gene pathways and networks involved in disease pathogenesis. Several significant advances have been made in this area (29–31). A GWA study of Kawasaki disease, an inflammatory vasculitic disease in children with a propensity for coronary artery damage that may have serious cardiovascular sequelae later in life, identified a set of biologically plausible, functionally interrelated genes associated with disease susceptibility (32). Because an infectious trigger of Kawasaki disease has been implicated but remains elusive, identification of *LNX1*, encoding a protein ligand of the coxsackievirus and adenovirus receptor, provides a compelling hypothesis for disease pathogenesis given that the development of myocarditis is associated with coxsackievirus. Furthermore, conducting GWA studies in patients with early-onset disease has been implemented as a new strategy to identify novel genes in asthma (31) and pediatric inflammatory bowel disease (30). In the context of asthma, genetic variation

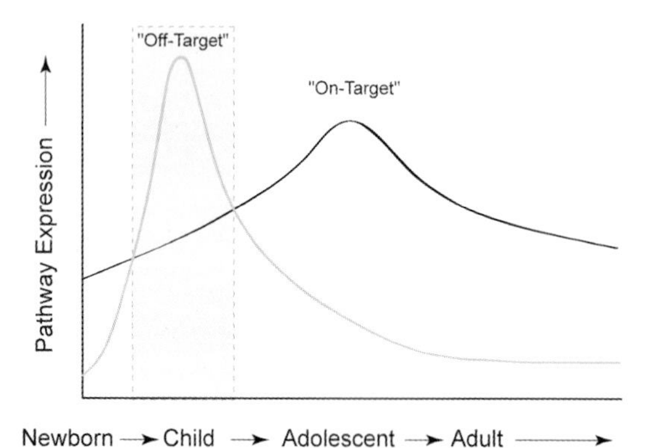

Figure 22.2. Paradigm for age- or developmental stage-specific differences in drug response. Different developmental patterns of expression of the intended drug target ("on-target"; black line) and unintended ("off-target"; gray line) pathways may lead to an increased risk of drug toxicity at a specific age or developmental stage, especially in situations where the unintended off-target pathway is expressed to a much greater extent at one developmental stage relative other ages (shaded rectangle) or when it is expressed to a greater extent compared with the target pathway.

in *ORMDL3*, a gene of unknown function, is associated with the onset of asthma before four years of age (33). Because almost all children with asthma, epilepsy, type 2 diabetes, depressive disorders and other psychiatric illnesses, hypertension, inflammatory bowel disease, and arthritis will grow up to be adults with those diseases, early identification of new disease pathways may lead to effective pediatric disease-appropriate therapies that have the potential to fundamentally alter the pattern and progression of disease later in life.

Most drugs are developed to treat disease processes in adults. Whether they can also be used to treat similar or unrelated diseases in children will depend on whether the drug target and the physiological context in which it operates are operative in the pathogenesis of the pediatric disease. Furthermore, the balance between therapeutic benefit and the risk of adverse effects will be a function of the ontogeny of the intended drug target and its associated pathway/network ("on-target" effect) relative to any unintended pathways with which the drug may interact ("off-target" effect). Specifically, different developmental patterns of expression of on-target and off-target pathways may lead to therapeutic windows associated with profound therapeutic benefit (target pathway expressed, off-target pathway not expressed) or increased risk of drug toxicity at a specific age or developmental stage (off-target pathway expressed to a much greater extent than the target pathway; Figure 22.2). This paradigm will be key to understanding why children appear to be uniquely susceptible or at higher risk for certain adverse drug reactions in comparison with adults. To some degree, knowledge of the ontogeny of drug biotransformation pathways is particularly informative. Delayed maturation of glucuronosyltransferases and the cardiovascular collapse associated with "gray baby syndrome" in newborns treated with chloramphenicol is an often cited example. Cutaneous toxicity associated with lamotrigine, which occurs more frequently in children than in adults (34), also may be related to developmental expression of glucuronosyltransferases that are involved in its elimination from the body, such as UGT1A4 and UGT2B7. In other situations, such as the development of Reye's syndrome in children treated with aspirin and the hepatotoxicity associated with valproic acid, drug biotransformation alone can not explain the increased risk observed in children. Even more complex are the factors determining susceptibility to drug toxicities related to drug exposures or chronic drug therapy initiated early in life but that do not become apparent until later in life. Examples include the development of secondary malignancies following treatment for various childhood cancers (35) and the effect of steroid therapy on linear growth (26). Accumulating evidence indicates that the application of genomic approaches holds considerable potential to identify pediatric patients most at risk for long-term toxicities (36, 37).

GENERAL PARADIGM TO SYSTEMATICALLY ASSESS THE ROLE OF ONTOGENY AND GENETIC VARIATION IN PEDIATRIC DRUG DISPOSITION AND RESPONSE

A series of legislative initiatives, including the FDA Modernization Act (38), the Best Pharmaceuticals for Children Act (39), and PREA (40) have resulted in the submission of new information related to the use of medications in pediatric patients. Approximately half of these studies resulted in unanticipated differences in dosing, safety, or efficacy in children compared with adults (4), which may be due, in some cases, to the lack of an appropriate drug formulation, clinical trial design issues, or data analysis considerations (41, 42). However, the extent to which developmental or pharmacogenetic factors may contribute to the dosing requirements, drug toxicity, or inability to demonstrate efficacy or statistically significant dose-response relationships in children remains to be investigated on a routine basis. As indicated at the beginning of this chapter, elucidation of the dose → exposure → response relationship in children of different ages and developmental stages requires deconvolution of developmental processes and genetic variation during growth and development.

The failure to observe efficacy in children may result from the inability to achieve adequate systemic drug exposure due to higher than anticipated drug clearance in children (4). In the presence of adequate exposure, however, treatment failure may occur if there are fundamental differences in the expression of the drug target throughout development. To facilitate the systematic

Table 22.1. General Paradigm for the Systematic Evaluation of Drug-Related Safety and Efficacy Problems in Children

1. What gene products are quantitatively important in the disposition (absorption, distribution, metabolism, and excretion) of the drug?
2. What is the developmental trajectory for the acquisition of functional activity for each gene/gene product?
3. Is allelic variation in the gene of interest associated with any functional consequences in vivo?
4. Is there any evidence that allelic variation affects the developmental trajectory of the drug disposition phenotype?
5. What gene products are primarily involved in drug response?
6. What is the developmental trajectory for the acquisition of functional activity of receptors, ion channels, and downstream signal transduction pathways involved in drug response?
7. Are there any known in vivo consequences of allelic variation in any of the genes of interest?
8. Is there any evidence that allelic variation affects the developmental trajectory of the drug target or downstream signal transduction pathway?

evaluation of drug-related safety and efficacy concerns in children, in particular, the relative contribution of ontogeny and genetic variation, the paradigm presented in Table 22.1 is proposed. This paradigm was originally developed to address variability in drug disposition and response in neonates (70), and subsequently applied to the recent controversy regarding the safety of over-the-counter cough and cold remedies in young children (71). The basic principles of this approach are applied to a case of dextromethorphan toxicity.

CASE SCENARIO (ADAPTED FROM CASES IN THE LITERATURE AND CLINICAL EXPERIENCE OF THE AUTHORS)

A 3-year-old female presents to an Urgent Care setting with altered mental status, disorientation, slurred speech, and nystagmus. By history, the child had developed symptoms of an upper respiratory tract infection and cough during the past 24 hours for which the mother had been treating the symptoms with over-the-counter cough and cold preparations. Specifically, the child had been receiving therapeutic doses of a combination product of phenylephrine, chlorpheniramine, and dextromethorphan, as well as a product containing only dextromethorphan. The total dextromethorphan in the past 24 hours was 30 mg (or 2 mg/kg), which was the maximum daily dose recommended for children 2 to 6 years of age at the time of presentation. The last dose was 2 hours before presentation.

On examination, the child is lethargic and disoriented. Vital signs reveal a heart rate of 140 beats per minute, respirations of 28 breaths per minute, blood pressure of 115/65 mmHg, and a temperature of 38°C. The child is noted to have horizontal nystagmus. Her pupils measure 5 mm and are sluggish to react to light. Mucous membranes are moist. No lymphadenopathy is present. A cardiovascular examination reveals tachycardia. No murmurs or gallops are present. Capillary refill is brisk. She is breathing without difficulty and has clear breath sounds. The abdominal examination results are benign. No rashes are present. Her neurological examination is significant for decreased tone, slurred speech, and ataxia. In addition, the physician notes that the child does answer questions appropriately, but she is slow to respond.

The treating physician believes the mother to be a good historian and does not believe the child received an overdose of the medication. However, central nervous system symptoms, including drowsiness, dizziness, and fatigue, are unusual at doses less than 7.5 mg/kg. Symptoms that the child is exhibiting are generally observed at doses that are much higher.

What do we know about dextromethorphan that can help determine the cause of these symptoms? What questions should we ask regarding this child that not only can help determine the cause of these symptoms, but also can prevent future adverse events?

1. What gene products are quantitatively important in the disposition (absorption, distribution, metabolism, and excretion) of dextromethorphan?

Dextromethorphan is a nonprescription antitussive agent found in many pediatric and adult over-the-counter cough and cold formulations. At recommended doses, it acts centrally to elevate the threshold for coughing without the addictive, analgesic, or sedative properties or respiratory depression associated with the use of structurally related opiate analgesics. After it has been absorbed, dextromethorphan undergoes extensive hepatic oxidative biotransformation to its primary metabolites dextrorphan, the product of *O*-demethylation, and the *N*-demethylated metabolite, 3-methoxymorphinan. A secondary metabolite, 3-hydroxymorphinan, is the product of sequential *O*- and *N*-demethylation (Figure 22.3). In vivo, dextrorphan is the most abundant metabolite of dextromethorphan present in urine, representing 70 percent to 80 percent of the total molar recovery of drug and metabolites in adults (43, 44).

Multiple CYPs have been identified as dextromethorphan *O*- and *N*-demethylases in vitro (45, 46). Therefore, it is important to differentiate between those that are *capable* of metabolizing dextromethorphan and those that are functionally involved or *quantitatively important* in its clearance from the body. In vitro investigations conducted with individual heterologously expressed CYPs provide information on which CYPs are capable of metabolizing a given substrate as no other competing pathways are present. Studies conducted with human

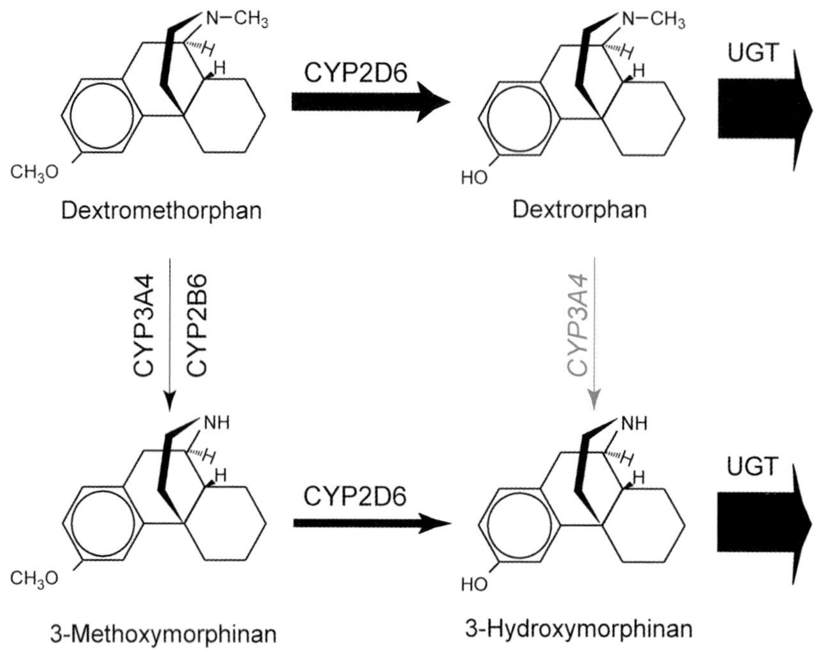

Figure 22.3. Major pathways of dextromethorphan biotransformation. CYP2D6-mediated *O*-demethylation of dextromethorphan results in the formation of dextrorphan, the major metabolite found in urine. The *N*-demethylation pathway leading to the formation of 3-hydroxymorphinan is a relatively minor pathway in human liver, and is catalyzed by CYP3A4 and CYP2B6. A secondary metabolite, 3-hydroxymorphinan, is the product of both *O*- and *N*-demethylation reactions, but enzyme kinetic considerations imply a pathway that involves initial *N*-demethylation followed by rapid CYP2D6-mediated *O*-demethylation. Both dextrorphan and 3-hydroxymorphinan are recovered in urine as glucuronide conjugates. UGT, glucuronosyltransferase.

liver microsomal fractions (corresponding to the endoplasmic reticulum of hepatocytes, the subcellular localization of CYPs) provide a more realistic context because of the presence of other CYPs. In this system, variability in the formation of a metabolite by a panel of different microsomal preparations is a function of interindividual variability in the complement of CYPs that is present in each preparation, and the relative abundance of each CYP. Cumulatively, data from in vitro studies indicate that multiple CYPs can catalyze dextrorphan formation, including CYP2D6, CYP2C9, CYP2C19, and, to a lesser extent, CYP3A4 and CYP2B6, whereas CYP2B6 and CYP3A4 are capable of dextromethorphan *N*-demethylation with small contributions from CYPs 2D6, 2C9, and 2C19. When corrected for relative abundance, CYP2D6 accounts for at least 83 percent of dextrorphan formation in human liver microsomes, although the contribution of CYP2C19 increases at higher substrate concentrations (46). Similarly, CYP3A4 is the dominant enzyme catalyzing *N*-demethylation with an important contribution from CYP2B6 (47).

In vivo, the quantitative importance of CYP2D6 for dextromethorphan *O*-demethylation and overall dextromethorphan clearance is confirmed by pharmacokinetic studies in which coadministration of the CYP2D6 inhibitor quinidine resulted in a 40-fold increase in dextromethorphan systemic exposure as defined by the area under the plasma concentration-time curve (48, 49). Similarly, a 150-fold increase in systemic exposure was observed in individuals in whom CYP2D6 activity was absent because of genetic polymorphism in comparison with subjects in which CYP2D6 was present (48). Thus, because neither dextromethorphan nor dextrorphan appear to be substrates for the drug transporter, P-glucoprotein (50), the primary determinant of dextromethorphan disposition is CYP2D6 activity. In fact, dextromethorphan is a preferred phenotyping probe for assessing CYP2D6 activity in vivo (51).

2. What is the developmental trajectory for the acquisition of functional CYP2D6 activity?

Early in vitro studies of CYP2D6 ontogeny revealed that expression of CYP2D6 immunoreactive protein was minimal in fetal liver microsomes but could be consistently demonstrated in neonates by seven days of age. Correspondingly, CYP2D6 activity assessed by dextromethorphan *O*-demethylation was absent or only detected at levels approximately 1 percent to 5 percent of adult values in fetal liver microsomes, and did not exceed 20 percent of that observed in adult samples during the first month

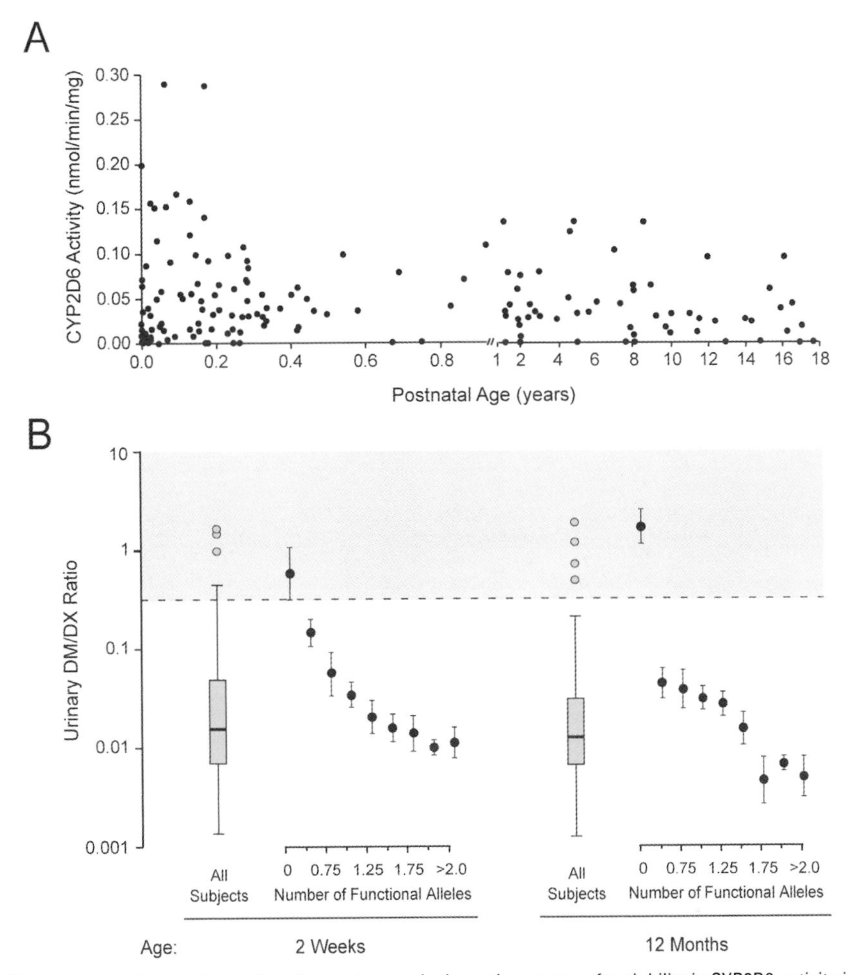

Figure 22.4. Pharmacogenetic variation, rather than ontogeny, is the major source of variability in CYP2D6 activity in children. (**A**) In vitro studies with human hepatic microsomal fractions indicate that there is minimal change in CYP2D6 activity beyond the first week or two of postnatal life (adapted from reference 53). (**B**) A similar conclusion can be drawn from in vivo data by the use of dextromethorphan as a phenotyping probe. Data from two time points, 2 weeks postnatal age ($n = 147$) and 12 months postnatal age ($n = 112$) are shown. The population distribution for all subjects at each time point is presented as a box plot of the logarithm of the ratio of dextromethorphan to dextrorphan (DM/DX ratio) in urine. Boxes represent interquartile range (25th to 75th percentile); bars are medians. Whiskers represent the 10th to the 90th percentile, and • indicates outlying values between 1.5 and 3 box lengths from the interquartile range. Adjacent to each box plot is the same dataset (log(DM/DX) $\pm$ s.e.m.) stratified according to the genotype of each subject as indicated by the number of functional alleles in each genotype. This was accomplished by assigning a value of "0" for loss-of-function" or "null" alleles, a value "1" for the fully functional reference sequence or "wild-type" allele, and intermediate values for "partial function" alleles as described in reference 65. The dashed line corresponds to a DM/DX ratio of 0.3, the cutoff for assigning poor-metabolizer status. Poor metabolizers are found in the shaded area above the dashed line (adapted from reference 25).

of life (52). In contrast, more recent in vitro analyses in a larger number of samples revealed that CYP2D6 protein and activity were similar between third trimester samples and those from infants in the first week of life, and both protein and activity remained relatively constant after one week of age up to eighteen years (53) (Figure 22.4A).

In vivo results derived from a longitudinal phenotyping study in more than 100 infants over the first year of life with the use of dextromethorphan as a probe compound demonstrated that there was considerable interindividual variability in CYP2D6 activity as measured by the ratio of dextromethorphan to dextrorphan in urine. However, no relationship between CYP2D6 activity and postnatal age was apparent between two weeks and twelve months of age (25) (Figure 22.4B). An earlier cross-sectional study involving 586 children (480 whites and 106 African Americans) indicated that the distribution of CYP2D6 phenotypes in children, based on dextromethorphan/dextrorphan ratios, was comparable to that observed in adults by at least ten years of age, and probably much earlier (54).

3. Is allelic variation in the gene of interest associated with any functional consequences in vivo?

Several comprehensive reviews of CYP2D6 pharmacogenetics have been published in recent years and may be referred to for more detailed information (55–57). The *CYP2D6* gene locus is highly polymorphic with more than seventy allelic variants (and additional subvariants) identified (http://www.imm.ki.se/CYPalleles/cyp2d6.htm). Among those are fully functional, reduced function, and nonfunctional alleles, which convey a wide range of activity from none to ultrarapid metabolism. Consequently, poor (PM), intermediate (IM), extensive (EM), and ultrarapid (UM) metabolizer phenotypes are observed when a population is challenged with a probe substrate. Allelic variants are the consequence of point mutations, single base pair deletions or additions, gene rearrangements, or deletion of the entire gene that results in a reduction or complete loss of activity. Inheritance of two recessive loss-of-function alleles results in the "poor-metabolizer phenotype," which is found in approximately 5 percent to 10 percent of whites and approximately 1 percent to 2 percent of Asian subjects.

Individual alleles are designated by the gene name (*CYP2D6*) followed by an asterisk and an arabic number, and *CYP2D6*1* designates, by convention, the fully functional reference or "wild-type" allele. In whites, the *3, *4, *5, and *6 alleles are the most common loss-of-function alleles and account for approximately 98 percent of poor metabolizer phenotypes (58). In contrast, CYP2D6 activity on a population basis tends to be lower in Asian and African American populations because of a lower frequency of nonfunctional alleles (*3, *4, *5, and *6) *and* a relatively high frequency of population-selective alleles that are associated with decreased activity relative to the wild-type *CYP2D6*1* allele. For example, the *CYP2D6*10* allele occurs at a frequency of approximately 50 percent in Asians, whereas *CYP2D6*17* and *CYP2D6*29* occur at relatively high frequencies (20 percent to 25 percent) in subjects of black African origin (55, 56).

In general, CYP2D6 genotypic poor metabolizers are at increased risk of drug accumulation and resulting concentration-dependent toxicities because of the reduced ability to eliminate medications that are primarily dependent on CYP2D6 for their clearance from the body. On the other hand, some medications, like codeine, tramadol, and tamoxifen, require functional CYP2D6 activity to be converted to pharmacologically active forms. In these situations, CYP2D6 poor metabolizers are at increased risk for treatment failure. At the other end of the spectrum are the CYP2D6 ultrarapid metabolizers, who have three or more functional copies of the gene. Enhanced clearance of CYP2D6 substrates is associated with lower drug concentrations and, thus, may result in lack of therapeutic efficacy in patients prescribed "normal" doses of medications (55). For prodrugs like codeine, however, increased formation of pharmacologically active metabolite may lead to an increased incidence of side effects (59) or, in extreme cases, may be overtly toxic (60).

Pharmacogenetic variation in *CYP2D6* may also lead to clinically important consequences in children. Sallee et al. (61) reported a case of fluoxetine-related death of a nine-year-old child with encephalopathy secondary to fetal alcohol syndrome, attention-deficit hyperactivity disorder, Tourette's syndrome, and obsessive-compulsive disorder. Autopsy results revealed blood and liver concentrations of fluoxetine that were severalfold higher than expected, consistent with subsequent genotype analysis on postmortem tissue that revealed that the child was a CYP2D6 poor metabolizer. A case of apnea was attributed to excessive morphine formation in a twenty-nine-month-old child receiving codeine (62). Although morphine concentrations were not determined, the authors noted that symptoms appeared following a relatively low dose in a child with a *CYP2D6*1/*2* genotype – two functional alleles.

Consequences of *CYP2D6* genetic variation may affect children in unexpected ways. Madadi et al. reported a fatal case of a breastfed infant whose mother was a CYP2D6 ultrarapid metabolizer (*CYP2D6*2A/*2x2* genotype) and was taking codeine for postpartum pain. Maternal milk saved during the last days of the baby's life contained 87 ng/mL morphine, ten- to twentyfold higher than values reported in the literature, consistent with increased maternal metabolism of codeine to morphine. In response to this case report, both the FDA (63) and Health Canada (64) issued public health advisories highlighting a risk for opioid toxicity in breastfed infants whose codeine-prescribed mothers are CYP2D6 ultrarapid metabolizers.

4. Is there any evidence that allelic variation affects the developmental trajectory of the drug disposition phenotype?

Although more than seventy allelic variants of *CYP2D6* have been annotated and some estimate of functional consequence is inferred from broad classification as "null" alleles with no functional activity, "partial" activity alleles, and fully functional alleles (65), a relationship between allelic variation and the rate at which functional CYP2D6 activity is acquired after birth has not been addressed. However, a longitudinal phenotyping study conducted in infants who were phenotyped with dextromethorphan at time points corresponding to well-baby visits in the first year of life (2 weeks and 1, 2, 4, 6, and 12 months of age) revealed that variation due to

CYP2D6 genotype was greater than that due to postnatal age (25). Cumulatively, these in vivo data (25), together with in vitro studies (53) (Figure. 22.4) and studies of a CYP2D6 substrate, tramadol, in infants (66), imply that the *CYP2D6* genotype, and not developmental considerations, is probably the most important factor contributing to the symptoms observed in the case scenario.

CASE SCENARIO (CONTINUED)

Given the significant symptoms at such a low dose, the physician assumes that this child may be a CYP2D6 poor metabolizer. A comprehensive urine drug screen is obtained and is positive for the medications found in the cough and cold preparations. In addition, the urine is sent to the Clinical Pharmacology Laboratory for assessment of the child's phenotype. The child improves with symptomatic and supportive care and is able to be discharged to the care of his family after observation for 8 hours in the Urgent Care facility.

One week later, the physician receives a report that states that the subject's urinary dextromethorphan to dextrorphan ratio was 0.0001, a value consistent with the phenotype of an extensive or ultrarapid CYP2D6 metabolizer. This finding rules out CYP2D6 *poor metabolizer status as an explanation for the clinical presentation and suggests the alternative that unexpectedly high conversion of dextromethorphan to dextrorphan may be responsible for the patient's symptoms. The child was, subsequently, determined to have a CYP2D6*1/*2xN genotype (i.e., carrying at least three fully functional gene copies), which is in accordance with an ultrarapid metabolizer (UM) phenotype. Reviewing the patient's symptoms in the context of the pharmacogenetic results, the physician realizes that they are more consistent with those seen with a phencyclidine (PCP) ingestion. Animal studies (67, 68) have shown that, whereas dextromethorphan exposures largely result in sedation, PCP-like symptoms are more commonly seen with exposure to dextrorphan. Similar PCP-like effects can be seen with high doses of dextromethorphan and/or in extensive metabolizers. The physician contacts the family to discuss the results and offers information on other medications, such as codeine and serotonin reuptake inhibitors, that, if exposed, may cause toxicity or decreased effect, respectively.*

This case scenario illustrates how a systematic analysis of ontogeny and pharmacogenetic information related to drug disposition can aid physicians and other health care providers in making sound clinical decisions in the care of a child presenting with an unusual response to a medication. Our knowledge of ontogeny and pharmacogenetics related to the processes involved in drug disposition generally exceeds that related to drug response. However, a similar systematic approach can be applied to drug targets (e.g., receptors, ion channels) and downstream signal transduction pathways. Increased understanding of the relative roles of ontogeny and genetic variation during development of these pathways may provide some important clues in situations where children display responses to medications that are unexpected on the basis of experience in adults. In this context, dextrorphan is a noncompetitive antagonist at the *N*-methyl-D-aspartate (NMDA) receptor, binding to which results in the PCP-like symptoms seen after high exposures to dextromethorphan (67, 68). Analogous to the ontogeny of drug biotransformation pathways, similar events are occurring with neurotransmitter receptors and reuptake pumps in the developing brain of fetuses, infants, and children. As an example, NMDA and serotonin both are important for the formation and maintenance of synapses during brain development. Although extremely limited, available data indicate that NDMA receptor subunits are differentially expressed during postnatal development of the human hippocampus. Specifically, NR1 subunit messenger RNA levels are lower in neonates compared with older groups, whereas the opposite situation is the case for the NR2B subunit (69). It is interesting to speculate that changes in subunit expression and assembly of the heteromeric receptor complex at different stages of development may lead to age-related differences in the response to ligands, such as dextrorphan, and possibly also contribute to the behavioral effects experienced by the child in the case scenario. However, much more work related to ontogeny and the role of genetic variation in developmental patterns of expression of drug targets will be needed to increase our understanding of why children appear to be uniquely susceptible to or at higher risk for certain adverse drug reactions in comparison with adults.

SUMMARY

As clinicians and researchers are, finally, beginning to learn, children are not "little adults" with regard to their diseases or their treatment. Vast developmental changes occur between the fetal, neonate, infant, child, and adolescent stages. These include not only the physiological changes that are commonly known, but also the pathways primarily responsible for the biosynthesis and catabolism of endogenous compounds, as well as, secondarily, the biotransformation and transport of exogenous molecules, such as drugs, plant-derived compounds, and environmental contaminants. Superimposed on the ontogeny of drug-metabolizing enzymes is the large variability contributed by genetic variation. It is not until the enzyme system has matured that the genotype-phenotype relationship can be fully appreciated. Only by asking the right questions can those who treat children fully understand the role that pharmacogenetics and pharmacogenomics may have in the interindividual variability in drug disposition and response observed in children of different ages and developmental stages.

REFERENCES

1. Steinbrook R. Testing medications in children. *N Engl J Med.* 2002;**347**:1462–70.
2. Ward RM & Kaufman R. Future of pediatric therapeutics: reauthorization of BPCA and PREA. *Clin Pharmacol Ther.* 2008;**81**:477–9.
3. Kearns GL, Abdel-Rahman SM, Alander SW, Blowey DL, Leeder JS, & Kauffman RE. The impact of ontogeny on drug disposition and action. *N Engl J Med.* 2003;**349**:1157–67.
4. Rodriguez W, Selen A, Avant D, et al. Improving pediatric dosing through pediatric initiatives: what have we learned. *Pediatrics.* 2008;**121**:530–9.
5. Anderson BJ & Holford NHG. Mechanism-based concepts of size and maturity in pharmacokinetics. *Annu Rev Pharmacol Toxicol.* 2008;**48**:303–32.
6. Weinshilboum RM & Wang L. Pharmacogenetics and pharmacogenomics: development, science and translation. *Annu Rev Genomics Hum Genet.* 2006;**7**:223–45.
7. Leeder JS. Developmental and pediatric pharmacogenomics. *Pharmacogenomics.* 2003;**4**:331–41.
8. Hines RN. The ontogeny of drug metabolism enzymes and implications for adverse drug events. *Pharmacol Ther.* 2008;**118**:250–67.
9. Cheok M & Evans WE. Acute lymphoblastic leukaemia: a model for the pharmacogenomics of cancer therapy. *Nat Rev Cancer.* 2006;**6**:117–29.
10. Husain A, Loehle JA, & Hein DW. Clinical pharmacogenetics in pediatric patients. *Pharmacogenomics.* 2007;**8**:1403–11.
11. Kronenberg S, Frisch A, Rotberg B, Carmel M, Apter A, & Weizman A. Pharmacogenetics of selective serotonin reuptake inhibitors in pediatric depression and anxiety. *Pharmacogenomics.* 2008;**9**:1725–36.
12. Nebert DW & Russell DW. Clinical importance of the cytochromes P450. *Lancet.* 2002;**360**:1155–62.
13. Coughtrie MWH. Sulfation through the looking glass – recent advances in sulfotransferase research for the curious. *Pharmacogenomics J.* 2002;**2**:297–308.
14. Choudhuri S & Klaassen CD. Structure, function, expression, genomic organization, and single nucleotide polymorphisms of human ABCB1 (MDR1), ABCC (MRP), ABCG2 (BCRP) efflux transporters. *Int J Toxicol.* 2006;**25**:231–59.
15. Hagenbuch B & Gui C. Xenobiotic transporters of the human organic anion transporting polypeptides (OATP) family. *Xenobiotica.* 2008;**38**:778–801.
16. Williams JA, Ring BJ, Cantrell VE, et al. Comparative metabolic capabilities of CYP3A4, CYP3A5 and CYP3A7. *Drug Metab Disp.* 2002;**30**:883–91.
17. Witorsch RJ. Low-dose in utero effects of xenoestrogens in mice and their relevance to humans: as analytical review of the literature. *Food Chem Toxicol.* 2002;**40**:905–12.
18. Kitada M, Kamataki T, Itahashi K, Rikihisa T, & Kanakubo Y. P-450 HFLa, a form of cytochrome P-450 purified from human fetal livers, is the 16α-hydroxylase of dehydroepiandrosterone 3-sulfate. *J Biol Chem.* 1987;**262**:13534–7.
19. Lin ZL, Fontaine J, & Watchko JF. Coexpression of gene polymorphisms involved in bilirubin production and metabolism. *Pediatrics.* 2008;**122**:e156–62.
20. French D, Wang W, Cheng C, et al. Acquired variation outweighs inherited variation in whole genome analysis of methotrexate polyglutamate accumulation in leukemia. *Blood.* 2009;doi:10.1182/blood-2008-07-172106.
21. Alcorn J & McNamara PJ. Ontogeny of hepatic and renal systemic clearance pathways in infants. Part I. *Clin Pharmacokinet.* 2002;**41**:959–98.
22. Alcorn J & McNamara PJ. Ontogeny of hepatic and renal systemic clearance pathways in infants. Part II. *Clin Pharmacokinet.* 2002;**41**:1077–94.
23. Ring JA, Ghabrial H, Ching MS, Smallwood RA, & Morgan DJ. Fetal hepatic drug elimination. *Pharmacol Ther.* 1999;**84**:429–45.
24. Koukouritaki SB, Manro JR, Marsh SA, et al. Developmental expression of human hepatic CYP2C9 and CYP2C19. *J Pharmacol Exp Ther.* 2004;**308**:965–74.
25. Blake MJ, Gaedigk A, Pearce RE, et al. Ontogeny of dextromethorphan O- and N-demethylation in the first year of life. *Clin Pharmacol Ther.* 2007;**81**:510–16.
26. Wagner-Weiner L. Pediatric rheumatology for the adult rheumatologist. *J Clin Rheumatol.* 2008;**14**:109–19.
27. Gelfand EW. Is asthma in childhood different from asthma in adults? Why do we need special approaches to asthma in children? *Allergy Asthma Proc.* 2008;**29**:99–102.
28. Flynn JT, Mitsnefes M, Pierce C, et al. Blood pressure in children with chronic kidney disease. A report from the Chronic Kidney Disease in Children Study. *Hypertension.* 2008;**52**:1–7.
29. Hinks A, Barton A, Shephard N, et al. Identification of a novel susceptibility locus for juvenile idiopathic arthritis by genome-wide association analysis. *Arthritis Rheum.* 2009;**60**:258–63.
30. Kugathasan S, Baldassano RN, Bradfield JP, et al. Loci on 20q13 and 21q22 are associated with pediatric-onset inflammatory bowel disease. *Nat Genet.* 2008;**40**:1211–15.
31. Moffatt MF, Kabesch M, Liang L, et al. Genetic variants regulating ORMDL3 expression contribute to the risk of childhood asthma. *Nature.* 2007;**448**:470–3.
32. Burgner D, Davila S, Breunis WB, et al. A genome-wide association study identifies novel and functionally related susceptibility loci for Kawasaki disease. *PLoS Genet.* 2009;**5**:e1000319.
33. Bouzigon E, Corda E, Aschard H, et al. Effect of 17q21 variants and smoking exposure in early-onset asthma. *N Engl J Med.* 2008;**359**:1985–94.
34. Guberman AH, Besag FMC, Brodie MJ, et al. Lamotrigine-induced rash: risk/benefit considerations in adults and children. *Epilepsia.* 1999;**40**:985–91.
35. Relling MV, Boyett JM, Blanco JG, et al. Granulocyte colony-stimulating factor and the risk of secondary myeloid malignancy after etoposide treatment. *Blood.* 2003;**101**:3862–7.

36. Edick MJ, Cheng C, Yang W, et al. Lymphoid gene expression as a predictor of risk of secondary brain tumors. *Genes Chromosomes Cancer.* 2005;**42**:107–16.
37. Hartford C, Yang W, Cheng C, et al. Genome scan implicates adhesion biological pathways in secondary leukemia. *Leukemia.* 2007;**21**:2128–36.
38. Food and Drug Modernization Act of 1997: Publ L No. 105–115 (Nov. 21, 1997).
39. Best Pharmaceuticals for Children Act: Publ L No. 107–109 (Jan. 4, 2002).
40. Pediatric Research Equity Act of 2003: S 650 (Jan. 7, 2003).
41. Abdel-Rahman SM, Reed MD, Wells TG, & Kearns GL. Considerations in the rational design and conduct of phase I/II pediatric clinical trials: avoiding the problems and pitfalls. *Clin Pharmacol Ther.* 2007;**81**:483–94.
42. Benjamin DKJ, Smith PB, Jadhav P, et al. Pediatric antihypertensive trial failures: analysis of end points and dose range. *Hypertension.* 2008;**51**:834–40.
43. Chládek J, Zimová G, Martínková J, & Tuma I. Intraindividual variability and influence of urine collection period on dextromethorphan metabolic ratios in healthy subjects. *Fundam Clin Pharmacol.* 1999;**13**:508–11.
44. Jones DR, Gorski JC, Hamman MA, & Hall SD. Quantification of dextromethorphan and metabolites: a dual phenotypic marker for cytochrome P450 3A4/5 and 2D6 activity. *J Chromatogr B.* 1996;**678**:105–11.
45. Kerry NL, Somogyi AA, Bochner F, & Mikus G. The role of CYP2D6 in primary and secondary oxidative metabolism of dextromethorphan: *in vitro* studies using human liver microsomes. *Br J Clin Pharmacol.* 1994;**38**:243–8.
46. Von Moltke LL, Greenblatt DJ, Grassi JM, et al. Multiple human cytochromes contribute to biotransformation of dextromethorphan in-vitro: role of CYP2C9, CYP2C19, CYP2D6, and CYP3A. *J Pharm Pharmacol.* 1998;**50**:997–1004.
47. Wang Y & Unadkat JD. Enzymes in addition to CYP3A4 and 3A5 mediate *N*-demethylation of dextromethorphan in human liver microsomes. *Biopharm Drug Dispos.* 1999;**20**:341–6.
48. Capon DA, Bochner F, Kerry N, Mikus G, Danz C, & Somogyi AA. The influence of CYP2D6 polymorphism and quinidine on the disposition and antitussive effect of dextromethorphan in humans. *Clin Pharmacol Ther.* 1996;**60**:295–307.
49. Pope LE, Khalil MH, Berg JE, Stiles M, Yakatan GJ, & Sellers EM. Pharmacokinetics of dextromethorphan after single or multiple dosing in combination with quinidine in extensive and poor metabolizers. *J Clin Pharmacol.* 2004;**44**:1132–42.
50. Kanaan M, Daali Y, Dayer P, & Desmeules J. Lack of interaction of the NMDA receptor antagonists dextromethorphan and dextrorphan with P-glycoprotein. *Curr Drug Metab.* 2008;**9**:144–51.
51. Frank D, Jaehde U, & Fuhr U. Evaluation of probe drugs and pharmacokinetic metrics for CYP2D6 phenotyping. *Eur J Clin Pharmacol.* 2007;**63**:321–33.
52. Treluyer J-M, Jacqz-Aigrain E, Alvarez F, & Cresteil T. Expression of *CYP2D6* in developing human liver. *Eur J Biochem.* 1991;**202**:583–8.
53. Stevens JC, Marsh SA, Zaya MJ, et al. Developmental changes in human liver CYP2D6 expression. *Drug Metab Dispos.* 2008;**36**:1587–93.
54. Relling MV, Cherrie J, Schell MJ, Petros WP, Meyer WH, & Evans WE. Lower prevalence of the debrisoquin oxidative poor metabolizer phenotype in American black versus white subjects. *Clin Pharmacol Ther.* 1991;**50**:308–13.
55. Zanger UM, Raimundo S, & Eichelbaum M. Cytochrome P450 2D6: overview and update on pharmacology, genetics and biochemistry. *Naunyn-Schmiedeberg's Arch Pharmacol.* 2004;**369**:23–37.
56. Ingelman-Sundberg M. Genetic polymorphisms of cytochrome P450 2D6 (CYP2D6): clinical consequences, evolutionary aspects and functional diversity. *Pharmacogenomics J.* 2005;**5**:6–13.
57. Sistonen J, Sajantila A, Lao O, Corander J, Barbujani G, & Fuselli S. *CYP2D6* worldwide genetic variation shows high frequency of altered activity variants and no continental structure. *Pharmacogenet Genomics.* 2007;**17**:93–101.
58. Gaedigk A, Gotschall RR, Forbes NS, Simon SD, Kearns GL, & Leeder JS. Optimization of cytochrome P450 2D6 (CYP2D6) phenotype assignment using a genotyping algorithm based on allele frequency data. *Pharmacogenetics.* 1999;**9**:669–82.
59. Kirchheiner J, Schmidt H, Tzvetkov M, et al. Pharmacokinetics of codeine and its metabolite morphine in ultrarapid metabolizers due to *CYP2D6* duplication. *Pharmacogenomics J.* 2007;**7**:257–65.
60. Gasche Y, Daali Y, Fathi M, et al. Codeine intoxication associated with ultrarapid CYP2D6 metabolism. *N Engl J Med.* 2004;**351**:2827–31.
61. Sallee FR, DeVane CL, & Ferrell RE. Fluoxetine-related death in a child with cytochrome P-450 2D6 genetic deficiency. *J Child Adol Psychopharmacol.* 2000;**10**:27–34.
62. Voronov P, Przybylo HJ, & Jagannathan N. Apnea in a child after oral codeine: a genetic variant – an ultra-rapid metabolizer. *Ped Anesth.* 2007;**17**:684–7.
63. U.S. Food and Drug Administration. Public Health Advisory: Use of Codeine by Some Breastfeeding Mothers May Lead to Life-threatening Side Effects in Nursing Babies (August 17, 2007). http://www.fda.gov/cder/drug/advisory/codeine.htm (accessed February 12, 2009).
64. Health Canada. Advisory: Use of Codeine Products by Nursing Mothers (October 8, 2008). http://www.hc-sc.gc.ca/ahc-asc/media/advisories-avis/_2008/2008_164-eng.php (accessed February 12, 2009).
65. Gaedigk A, Simon SD, Pearce RE, Bradford LD, Kennedy MJ, Leeder JS. The CYP2D6 activity score: translating genotype information into a quantitative measure of phenotype. *Clin Pharmacol Ther.* 2007;**83**:234–242.
66. Allegaert K, Van Den Anker JN, Verbesselt R, et al. *O*-Demethylation of tramadol in the first months of life. *Eur J Clin Pharmacol.* 2005;**61**:837–42.
67. Dematteis M, Lallement G, & Mallaret M. Dextromethorphan and dextrorphan in rats: common antitussives – different behavioural profiles. *Fundam Clin Pharmacol.* 1998;**12**:526–37.
68. Nicholson KL, Hayes BA, & Balster RL. Evaluation of the reinforcing properties and phencyclidine-like discriminative stimulus effects of dextromethorphan and dextrorphan

in rats and rhesus monkeys. *Psychopharmacology.* 1999;**146**:49–59.

69. Law AJ, Weickert CS, Webster MJ, Herman MM, Kleinman JE, & Harrison PJ. Expression of NMDA receptor NR1, NR2A and NR2B subunit mRNAs during development of the human hippocampal formation. *Eur J Neurosci.* 2003;**18**:1197–205.

70. Leeder JS. Developmental pharmacogenetics: a general paradigm for application to neonatal pharmacology and toxicology. *Clin. Pharmacol. Ther.* 2009;86:687–2.

71. Leeder JS, Kearns GL, Spielberg SP, van den Anker J. Understanding the relative roles of pharmacogenetics and ontogeny in pediatric drug development and regulatory science. *J. Clin. Pharmacol.* 2010;50:1377–87.

23

Fetal and Neonatal Pharmacogenomics

Yair Blumenfeld

No area within the field of pharmacogenomics generates greater excitement and potential for altering the way we practice medicine than fetal and neonatal pharmacogenomics. Though only in its infancy (no pun intended), this area that lies at the crossroads of perinatology, neonatology, pharmacology, and genetics will undoubtedly not only shape the way we diagnose and treat diseases of expectant mothers and their infants, but also improve our overall understanding of fetal and neonatal development. By focusing our attention on variations in genetic code and gene expression between the pre- and postnatal life, pharmacogenomics will likely contribute greatly to our understanding of cellular, tissue, and organ differentiation, physiologic and pathologic fetal development, and important variation in fetal and neonatal drug metabolism.

Fetal exposure to maternal medications may occur as a result of deliberate maternal drug therapy or inadvertent drug exposure. Since the discovery linking thalidomide with birth defects, obstetricians and their expectant patients have had to weigh the balance between maternal drug therapy and potential fetal exposure. As we gain insight into specific medications and their adverse fetal effects, it is clear that not all fetuses exposed to a given medication are at risk for developing its associated malformation. Moreover, if we can identify which fetuses may be at risk during the course of pregnancy, it would greatly benefit expectant mothers taking medications, especially in cases where exposure cannot be avoided as with antiepileptic and antithrombotic medications (1).

This chapter will attempt to review and summarize the great progress that has already been made in the field, and explore a few potential future directions. Although it is titled "Fetal and Neonatal Pharmacogenomics," it is difficult to exclude some discussion of maternal (adult) pharmacogenomics. Thus, although not separately categorized, we will incorporate this topic where clinically relevant.

FETAL AND PLACENTAL DEVELOPMENT

Before delving into fetal and neonatal pharmacogenomics, it is important to review some basic in utero fetal development. The *preimplantation period*, the two-week period from fertilization to implantation, has often been called the "all or none" period (Table 23.1). During this time, the zygote undergoes cleavage, and cells divide into an outer and inner cell mass. A large insult at this stage of embryogenesis will usually result in death of the embryo (2). The period between the second week through the first eight weeks of gestation, the *embryonic period*, is the period when each of the three germ layers – ectoderm, mesoderm, and endoderm – gives rise to a number of specific tissues and organs (3). By the end of this stage, the main organ systems, including the central nervous system (CNS), cardiovascular system, and gastrointestinal systems, have been established. Thus, the embryonic period is the most crucial in terms of structural malformations, because this is the period of major organogenesis. The period from the ninth week to birth is known as the *fetal period* and is characterized by maturation of tissues and organs and rapid growth of the body (3). Fetal drug exposure during this period may result in either congenital malformations or, more often, growth abnormalities.

Even though a complete embryology review and discussion is beyond the scope of this chapter, it is important to recognize that differences in gene expression occur as the embryo develops, and sequence variations as well as other factors play a role in the fetal response to small molecules such as drugs (pharmacogenomics) or environmental toxins (toxicogenomics) (1). Thus, variations at the genomic level and variations in gene expression may only be identified and relevant at specific critical points in the developmental continuum (1).

This process obviously continues well into the neonatal period and early childhood, and thus advances in pharmacogenomic variations discovered in adults may

Table 23.1. Fetal Development Time Line

Gestational Age	*Development*
Preimplantation period	The first two weeks of embryo development from fertilization to implantation
Embryonic period	The second through the eighth week of embryo development Cardiac development: 3rd to 9th week CNS: 3rd week to term Limbs: 4th to 9th week Lips: 5th to 6th week Teeth: 6th week to 3rd trimester Palate: 6th to 9th week External genitalia: 7th week to term
Fetal period	From 9 weeks until term

not be relevant for neonates and children. Moreover, variability in drug disposition (pharmacokinetics) and action (pharmacodynamics) not only changes during fetal, neonatal, and pediatric development, but also ultimately has an impact on drug response. Thus, drug exposure in early gestations may lead to fetal loss or severe structural abnormalities, whereas exposure in later gestations may lead to fetal growth restriction or developmental delay not identified until several years of life. For example, if fetal exposure to warfarin (Coumadin) occurs between the sixth and ninth week, the fetus is at increased risk for warfarin embryopathy, which is characterized by nasal hypoplasia and tippled vertebral and femoral epiphyses. It is doubtful that these defects result from fetal hemorrhage because vitamin K clotting factors are not demonstrable in the embryo at this age. Instead, it is thought that the teratogenic effect occurs as a result of inhibition of posttranslational carboxylation of coagulation proteins (4). Fetal defects due to exposure to warfarin in the second and third trimesters likely result from hemorrhage leading to scarring. Such abnormalities include dorsal CNS dysplasia and developmental and mental retardation.

Several layers of defense exist to protect the developing embryo from drug exposure and toxicity. The first of these is maternal drug metabolism and biotransformation. In addition to terminating pharmacologic activity, biotransformation enhances drug elimination from the body, thereby effectively reducing the amount of parent compound available to cross the placenta (1). As such, one must consider maternal liver and/or renal disease when using potentially toxic medications because these may alter drug metabolism and excretion. For example, pre-eclampsia, a pregnancy-associated disorder characterized by generalized vasoconstriction, may lead to both maternal hepatic dysfunction and renal arteriolar vasoconstriction and thus decreased urinary excretion.

The second line of defense is the placenta. Substances that pass from maternal blood to fetal blood must traverse syncytiotrophoblasts, cytotrophoblasts, stroma of the intravillous space, and fetal capillary wall (2). Several variables are important in the effectiveness of the human placenta as an organ of transfer, including the concentration of substance in maternal plasma, the rate of maternal blood flow through the intervillous space, the area available for exchange, the amount of substance metabolized by the placenta during transfer, the specific binding or carrier proteins in the fetal or maternal circulation, and the rate of fetal blood flow through the villus capillaries (2). Moreover, for substances transferred by diffusion or active transport, the physical properties of the tissue barrier, or capacity of the biochemical machinery of the placenta for effective active transfer, respectively, are important. In general, most substances with a molecular mass less than 500 daltons diffuse readily through placental tissue, and lipophilic molecules tend to cross more readily (2).

Besides acting as a mechanical barrier, and because of its origin in fetal trophectoderm, various enzymes involved in drug metabolism have been shown to be expressed by the placenta, including cytochrome P450 enzymes, superoxide dismutase, glutathione peroxidase, glutathione reductase, and glutathione *S*-transferases (GSTs) (1). Allelic variations in these enzymes have the potential to affect fetal exposure to exogenous compounds and thus, individual susceptibility to adverse consequences in the fetus and newborn (1). Examples of placental cytochrome P450 gene expression include expression of CYP2J2, CYP4B1, CYP19, CYP2C, CYP2D6, and CYP3A7 (5, 6). Upregulation of placental CYP1A1 and CYP2E1 have also been described in pathological conditions such as nicotine exposure (7).

FETAL PHARMACOGENOMICS

Approximately 3 percent of all infants are born with one or more major birth defects, and although the risk of exposure to "high-risk" teratogens such as thalidomide or isotretinoin is low, it is currently unclear what predisposes certain fetuses to the harmful effects of the more commonly used "moderate-risk" medications such as anticonvulsants and anticoagulants (8). In addition, 7 percent of children in the United States have developmental disorders by age 1, 12 percent to 14 percent by the time they enter school, and 17 percent before age 18 (9). Several examples of known teratogens are listed in Table 23.2. This differential effect whereby some infants exposed manifest the untoward outcome and others do not may be explained, at least to some degree, by variations in the fetal genotype, specifically variations in the enzymes involved in drug metabolism.

It is important to note that the same enzymes currently hypothesized to be involved in adverse pregnancy outcomes play critical roles in the synthesis and catabolism of endogenous compounds important for fetal growth and development, such as androgens,

Table 23.2. Common Teratogens and Their Effects

Medication	*Teratogenic Effects*
Phenytoin	7–10% fetal hydantoin syndrome (craniofacial defects, limb abnormalities, and mental retardation)
Valproic acid	1–2% spina bifida with first trimester exposure
Warfarin	9% permanent deformity and 17% death (31)
Isotretinoin	33% spontaneous abortion and 33% with at least one major malformation with first trimester exposure (32)
Diethylstilbestrol (DES)	Clear cell adenocarcinoma of vagina and cervix
Tetracyclines	Yellow-brown discoloration of teeth and deposition in fetal long bones
Thalidomide	Malformations in 20% of exposed individuals including limbs, ears, cardiovascular, and gastrointestinal

cholesterol, eicosanoids, estrogens, progestins, retinoic acid, thyroxine, and vitamin D (8). Thus, compounds administered with therapeutic intent have the potential to disrupt normal cellular function through competition with endogenous substrates and ligands (8).

When evaluating the role of fetal pharmacogenomics in adverse birth defects, it is important to also analyze the variations in gene expression in a variety of fetal tissues and to distinguish those from expression in neonates and adults. For example, recognizing differences in pre- and postnatal gene expression, several studies have explored specific fetal pharmacogenomic enzyme polymorphisms and their activity levels in various fetal tissues. In their study of CYP2J2, an enzyme that metabolizes arachidonic acid, Gaedigk et al. (10) found that CYP2J2 mRNA is expressed in human fetal liver as early as eleven weeks gestation and that expression levels are comparable to those observed in the fetal heart. Their group also showed that interindividual variability is relatively low and that there are differences in pre- and postnatal protein expression patterns (10). Another investigation by Bieche et al. (5) studied CYP 1, 2, and 3 mRNA from twenty-two different human tissues, including the uterus, fetal liver, and placenta, and showed that, contrary to the adult liver, in which several CYPs were expressed, in the fetal liver the main and almost the lone CYP expressed was CYP3A7.

Despite these data, little is known regarding the presence of different isoforms of these enzymes in different populations, let alone how these variations may be involved in actual disease processes. In 2005, recognizing the importance of CYP3A7 in fetal liver, Rodriguez-Antona et al. (11) reported a CYP3A7 variant that leads to increased catalytic activity (CUpsilonP3A7*2). Interestingly, the frequency of CUpsilonP3A7*2 was 8 percent, 17 percent, 28 percent, and 62 percent in white, Saudi Arabian, Chinese, and Tanzanian individuals, respectively. It is still unclear whether this mutation plays a role in any human birth defects, but future studies evaluating the prevalence of specific genotypes in a variety of patient populations may one day clarify differences in disease prevalence within distinct populations.

NEONATAL PHARMACOGENOMICS

Although fetal pharmacogenomics focuses primarily on variations in adverse fetal effects of inadvertent in utero drug exposure, neonatal pharmacogenomics may play a role in variations of both inadvertent drug exposure via breast milk and intentional therapeutic administration of medications. Similarly to fetal growth, neonatal growth may also involve variations in gene expression based on age, and genes that may be involved in drug response at a given time point may not be relevant in another. The following sections present an overview of recently published data linking neonatal pharmacogenomics with specific clinical outcomes (see Table 23.3).

Opioids

One of the first clinical depictions of the importance of neonatal pharmacogenomics followed the description of adverse effects seen in neonates breastfed by mothers receiving codeine. Codeine is normally converted to morphine by CYP2D6. The conversion of codeine into norcodeine by CYP3A4 and into codeine-6-glucuronide by glucuronidation usually represents 80 percent of codeine clearance, and conversion of codeine into morphine by CYP2D6 represents only 10 percent of codeine clearance (12). Morphine is further metabolized into morphine-6-glucuronide and morphine-3-glucuronide. Morphine and morphine-6-glucuronide have opioid activity. Glucuronides are eliminated by the kidney and are thus susceptible to accumulation in cases of acute renal failure.

CYP2D6 is one of the most studied cytochrome P450 enzymes, because it metabolizes approximately 25 percent of all medications in the human liver (13). About 7 percent to 10 percent of whites lack any CYP2D6 activity because of deletions and frame-shift or splice-site mutations of the gene (13). Approximately 1 percent to 3 percent of Middle-Europeans, and up to 29 percent of Ethiopians display gene duplications, leading to elevated so-called ultrarapid metabolization rates (13).

In 2004, Gasche et al. (12) described an episode of codeine intoxication in a male patient receiving codeine because of chronic pain related to chronic lymphocytic leukemia (12). A duplication of the CYP2D6 gene indicated that the patient was an "ultrametabolizer" of codeine (12). In that case, the codeine toxicity was further

Table 23.3. Pharmacogenomics in Maternal-Fetal-Neonatal Medicine

Maternal Enzymes Linked with Neonatal Effects	*Fetal/Placental Enzymes*	*Neonatal Enzymes*
CYP2D6, CYP3A4: codeine metabolism	CYP1A1, CYP2E1: upregulated by placenta with nicotine exposure	CYP2D6: opioid metabolism
CYP2D6 and CYP2C19: antidepressant metabolism	CYP2J2, CYP4B1, CYP19, CYP2C, CYP2D6 and CYP3A7: expressed by the placenta	SLC6A4 Serotonin transporter promoter: adverse neonatal effects of selective serotonin reuptake inhibitors (SSRIs)
	CYP3A7: the primary CYP gene expressed by fetal liver	NAT1, GSTM1: phase II drug metabolism enzymes influence the outcomes of childhood neuroblastoma
		CYP1A1: susceptibility to childhood acute lymphoblastic leukemia

compounded by the fact that the patient received clarithromycin and voriconazole antibiotics that are known inhibitors of CYP3A4, thus reducing codeine clearance (12).

In 2006, Koren et al. (14) reported a case of neonatal death secondary to maternal codeine use following a vaginal delivery. Maternal genotyping of CYP2D6, the enzyme catalyzing the *O*-demethylation of codeine to morphine revealed a CYP2D6*2x2 duplication, classifying the mother as an ultrarapid metabolizer. Elevated morphine concentrations were also found in both the mother's serum and neonatal postmortem blood.

Similarly, tramadol hydrochloride is an analog of codeine, and is a racemic mixture of two enantiomers. CYP2D6 and to a lesser extent CYP2B6 catalyze the *O*-demethylation of Tramadol. In 2005, Allegaert et al. suggested that CYP2D6 polymorphisms may explain the observed variability in *O*-demethylation activity of tramadol in the first months of life, and in 2008 the same group reported their data on the effects of CYP2D6 polymorphisms on Tramadol *O*-demethylation in critically ill neonates and infants (15, 16).

Selective Serotonin Reuptake Inhibitors

A recent analysis of depression in pregnancy found the prevalence of major depression to be as high as 5 percent, and that of major and minor depression to be as high as 10 percent (17). Selective serotonin reuptake inhibitors (SSRIs) are commonly used antidepressants that act by inhibiting serotonin reuptake in the synaptic cleft. Medications in this group include fluoxetine (Prozac), paroxetine (Paxil), sertraline (Zoloft), fluvoxamine (Luvox), citalopram (Celexa), and escitalopram (Lexapro). At higher doses, paroxetine and sertraline also block dopamine reuptake, which may contribute to their antidepressant action (18). Venlafaxine (Effexor) is a combined serotonin-norepinephrine reuptake inhibitor. Common side effects of these medications include insomnia and agitation, nausea and other gastrointestinal effects, sexual dysfunction, and weight gain (18).

Recently, multiple studies on depression in pregnancy have correlated SSRI use with adverse neonatal effects including neonatal respiratory distress, persistent pulmonary hypertension, jaundice, feeding problems, abnormal movements and tonus abnormalities, birth weight below the tenth percentile, and even congenital cardiac disease (Paxil only) (19–21).

In 2008, a study by Oberlander et al. (22) investigated whether neonatal effects of SSRIs are related to genotypes for the serotonin transporter (SLC6A4) promoter; specifically whether affected neonates were carriers of the short (s) or long (l) SLC6A4 alleles. The study suggested that prenatal SSRI exposure was associated with adverse neonatal outcomes, and these effects were moderated by infant SLC6A4 genotypes. Moreover, the relationships between polymorphisms and specific outcomes varied during the neonatal period, suggesting that beyond the apparent gene-medication interactions, multiple mechanisms contribute to the adverse neonatal outcomes following prenatal SSRI exposure.

Other studies exploring the role of maternal cytochrome P450 enzymes in SSRI metabolism have not found a similar link. In their study of twenty-five breastfeeding mothers treated with citalopram, sertraline, paroxetine, fluoxetine, and venlafaxine, Berlo et al. (23) evaluated multiple variations in maternal CYP2D6 and CYP2C19 genotypes, including genotypes associated with "poor metabolizing." The authors found that no specific genotypes correlated with variations in excreted medications in breast milk and all of the infants' serum levels were either undetectable or low.

Sepsis

A major area that is currently under investigation is that of genetic susceptibility to neonatal sepsis. In

2003, Carcillo et al. described reduced cytochrome P450-mediated drug metabolism in children with sepsis-induced multiple organ failure (24). Because cytokines such as interleukin-6 and nitric oxide production increase in children with sepsis, and because such cytokines have been shown to reduce CYP450 activity both in vivo and in vitro, it was hypothesized that the degree of CYP450 inhibition is related in part to the degree of inflammation and organ failure. Children ages one day to eighteen years old with or without sepsis were enrolled and administered antipyrine via nasogastric tube, and antipyrine elimination rate, elimination half-life, and apparent oral clearance were measured and calculated. Since antipyrine metabolism was lower in children with sepsis, especially those with multiple organ failure, the authors concluded that cytochrome P450-mediated drug metabolizing enzyme (DME) activity is markedly decreased in such circumstances.

Although a promising area, no studies have further correlated specific cytochrome P450 enzymes with either sepsis susceptibility or antibiotic response. In their review of this topic, Del Vecchio et al. suggest that, in light of the growth continuum of neonates on a molecular, cellular, and tissue level, further research in drug transport, drug biotransformation, receptors, and signal transduction processes is needed before the promise of pharmacogenetics and pharmacogenomics for rational therapeutics can be realized in such cases (25).

Antineoplastic Agents

Various DMEs are responsible for the metabolic activation and inactivation of potential carcinogenic agents and drugs used in the treatment of childhood malignancies (26). This complex interplay of specific enzymes and agents involves multiple phases. Phase I metabolism converts many compounds to reactive, electrophilic, water-soluble intermediates, some of which can damage DNA (27). This metabolic step is carried out by CYP enzymes, and allele variants in phase I genes such as CYP1A1 have been shown to be associated with higher enzymatic induction and have been implicated in susceptibility to childhood acute lymphoblastic leukemia (28). Phase II DMEs such as *N*-acetyltransferases (NATs) and GSTs are important modifiers that selectively target specific chemotherapeutic drugs and potential carcinogenic compounds. These enzymes also detoxify various reactive species by catalyzing the conjugation of potentially mutagenic electrophilic substrates to glutathione. A recent study of 209 children with neuroblastoma, 64 percent of whom were younger than two years of age, detailed specific NAT1 and GSTM1 genotypes that result in a more favorable outcome in patients treated with standard chemotherapeutic agents targeting this malignancy (29).

Table 23.4. Drug Classification in Pregnancy

Drug Category	*Ramifications*
A	Controlled studies in humans have demonstrated no fetal risks. There are few category A drugs, and examples include prenatal vitamins.
B	Animal studies indicate no fetal risks, but there are no human studies; or adverse effects have been demonstrated in animals, but not in well-controlled human studies. Several classes of commonly used drugs, for example, penicillins and metronidazole, are in this category.
C	There are either no adequate studies, either animal or human, or there are adverse fetal effects in animal studies but no available human data. Many drugs commonly taken during pregnancy are in this category including β-blockers, heparin, and indomethacin used to treat preterm labor.
D	There is evidence of fetal risk, but benefits are thought to outweigh these risks. Tetracyclines, carbamazepine, and phenytoin are examples.
X	Proven fetal risks clearly outweigh any benefits. An example is the acne medication isotretinoin, which may cause multiple CNS, facial, and cardiovascular anomalies.

FUTURE DIRECTIONS

Drug Classification

Fetal and neonatal pharmacogenomics may one day completely alter the way in which patients are counseled regarding antenatal drug exposure. The current drug classification system devised by the Food and Drug Administration is divided into five categories ranging from A to X (see Table 23.4). Patients and their caretakers must often make decisions based on limited human data and crude animal studies. For example, anticonvulsant medications are associated with major malformations including congenital heart defects, neural tube defects, oral-facial clefts, and other minor anomalies. However, only a relatively small percentage (4 to 12 percent) of patients exposed to anticonvulsants will manifest these symptoms. Pharmacogenomics may one day identify a subgroup of patients at high risk for developing these untoward effects, thereby individualizing treatment. In other words, a subset of patients may harbor 100 percent of the risk, but the large majority of patients do not have any risk. For example, after monitoring families with phenytoin (antiepileptic) exposure, Van Dyke et al. (30) describe certain families with multiple exposed but normal offspring, whereas other families had multiple affected children.

Antenatal Drug Efficacy

Other than preexisting maternal medical conditions necessitating pharmacological intervention, obstetricians use a variety of medications during the course of pregnancy to treat pregnancy-related conditions. Examples of this include tocolytics for preterm labor, antihypertensives for preeclampsia, and antibiotics for chorioamnionitis. Unlike known teratogens, most of these medications are relatively safe, but pharmacogenomics may assist clinicians in identifying subsets of the population that may benefit from one drug rather than another. For example, knowing that a patient's contractions will improve with nifedipine rather than indocin, or that labetalol is preferred over hydralazine in the case of hypertension may one day result in individualized care and improved drug efficacy.

Recognizing the limited data, research, and resources allocated toward the study of pharmacology in pregnant women and their infants, the National Institute of Child Health and Human Development recently developed the Obstetric-Fetal Pharmacology Research Unit intended to provide the "expert infrastructure needed to test therapeutic drugs during pregnancy." Several studies are already underway including ones evaluating glyburide for gestational diabetes and 17-α-hydroxyprogesterone caproate for preterm birth. Undoubtedly, pharmacogenomics will aid this research consortium and others to achieve their goals of conducting "a whole new generation of safe, technically sophisticated, and complex studies that will help clinicians protect the health of women, while improving birth outcomes and reducing infant mortality."

NONINVASIVE MATERNAL AND FETAL PHARMACOGENOMIC TESTING

Since the discovery of fetal cells in maternal blood several decades ago, investigators have used both intact fetal cells and cell-free fetal DNA in the maternal circulation as potential targets for noninvasive prenatal diagnosis. Although much of the attention has focused on fetal aneuploidy detection, the development of advanced molecular tools including digital polymerase chain reaction, microarray comparative genomic hybridization, and mass sequencers have opened the door to the possibility of complete noninvasive fetal genotyping while still in utero. Because fewer than 1 percent of all neonates born to women younger than age thirty-five are at risk for chromosomal abnormalities, fetal pharmacogenomic profiling may one day be as important a target as aneuploidy, if not even more so, for noninvasive prenatal diagnosis.

CONCLUSION

As our pregnant population ages, practitioners can expect to see a greater number of patients with medical conditions necessitating chronic drug use. Hypertension, diabetes, thyroid disorders, and chronic infections are only some examples, and daily medication use by expectant mothers is an area of high anxiety because of the fear of potential adverse fetal and neonatal effects. As such, expectant mothers and their caretakers are constantly forced to weigh the balance between risk and benefit of antepartum drug exposure (8). Although research in the area of drug metabolism in pregnancy is rapidly evolving, physicians are often hindered by limited knowledge of the pathophysiologic mechanisms linking a specific drug and adverse outcomes, as well as by a current drug classification system that rates a drug's safety in pregnancy based on limited animal data and weak associative studies. In today's world of rapid genetic testing, this approach seems increasingly archaic. Rather than stratifying patients based on individual genetic risk profiles, physicians overly generalize both the potential benefits and even harms of a given medication. Research in maternal-fetal-neonatal pharmacogenomics will undoubtedly not only improve our overall understanding of fetal and neonatal physiology and development, but also help us to stratify patients based on individual drug risk, susceptibility, and response. More importantly, this relatively new field will help clinicians improve maternal, fetal, and neonatal care in the years to come.

REFERENCES

1. Leeder JS. Developmental and pediatric pharmacogenomics. *Pharmacogenomics.* 2003;**4**:331–41.
2. Cunningham FG & Williams JW. *Williams Obstetrics.* 21st ed. New York: McGraw-Hill; 2001.
3. Sadler TW & Langman J. *Langman's Medical Embryology.* 10th ed. Philadelphia: Lippincott Williams & Wilkins; 2006.
4. Hall JG, Pauli RM, & Wilson KM. Maternal and fetal sequelae of anticoagulation during pregnancy. *Am J Med.* 1980;**68**:122–40.
5. Bieche I, Narjoz C, Asselah T, et al. Reverse transcriptase-PCR quantification of mRNA levels from cytochrome (CYP)1, CYP2 and CYP3 families in 22 different human tissues. *Pharmacogenet Genomics.* 2007;**17**:731–42.
6. Hakkola J, Pelkonen O, Pasanen M, & Raunio H. Xenobiotic-metabolizing cytochrome P450 enzymes in the human feto-placental unit: role in intrauterine toxicity. *Crit Rev Toxicol.* 1998;**28**:35–72.
7. Wang T, Chen M, Yan YE, Xiao FQ, Pan XL, & Wang H. Growth retardation of fetal rats exposed to nicotine in utero: possible involvement of CYP1A1, CYP2E1, and P-glycoprotein. *Environ Toxicol.* 2009;**24**:33–42.

8. Leeder JS & Mitchell AA. Application of pharmacogenomic strategies to the study of drug-induced birth defects. *Clin Pharmacol Ther.* 2007;**81**:595–9.
9. Boyle CA, Decoufle P, & Yeargin-Allsopp M. Prevalence and health impact of developmental disabilities in US children. *Pediatrics.* 1994;**93**:399–403.
10. Gaedigk A, Baker DW, Totah RA, et al. Variability of CYP2J2 expression in human fetal tissues. *J Pharmacol Exp Ther.* 2006;**319**:523–32.
11. Rodriguez-Antona C, Jande M, Rane A, & Ingelman-Sundberg M. Identification and phenotype characterization of two CYP3A haplotypes causing different enzymatic capacity in fetal livers. *Clin Pharmacol Ther.* 2005;**77**:259–70.
12. Gasche Y, Daali Y, Fathi M, et al. Codeine intoxication associated with ultrarapid CYP2D6 metabolism. *N Engl J Med.* 2004;**351**:2827–31.
13. Cascorbi I. Pharmacogenetics of cytochrome p4502D6: genetic background and clinical implication. *Eur J Clin Invest.* 2003;**33**(suppl 2):17–22.
14. Koren G, Cairns J, Chitayat D, Gaedigk A, & Leeder SJ. Pharmacogenetics of morphine poisoning in a breastfed neonate of a codeine-prescribed mother. *Lancet.* 2006;**368**: 704.
15. Allegaert K, Van den Anker JN, Verbesselt R, et al. O-demethylation of tramadol in the first months of life. *Eur J Clin Pharmacol.* 2005;**61**:837–42.
16. Allegaert K, van Schaik RH, Vermeersch S, et al. Postmenstrual age and CYP2D6 polymorphisms determine tramadol o-demethylation in critically ill neonates and infants. *Pediatr Res.* 2008;**63**:674–9.
17. Gavin NI, Gaynes BN, Lohr KN, Meltzer-Brody S, Gartlehner G, & Swinson T. Perinatal depression: a systematic review of prevalence and incidence. *Obstet Gynecol.* 2005;**106**:1071–83.
18. Mann JJ. The medical management of depression. *N Engl J Med.* 2005;**353**:1819–34.
19. Ferreira E, Carceller AM, Agogue C, et al. Effects of selective serotonin reuptake inhibitors and venlafaxine during pregnancy in term and preterm neonates. *Pediatrics.* 2007;**119**:52–9.
20. Chambers CD, Hernandez-Diaz S, Van Marter LJ, et al. Selective serotonin-reuptake inhibitors and risk of persistent pulmonary hypertension of the newborn. *N Engl J Med.* 2006;**354**:579–87.
21. Oberlander TF, Warburton W, Misri S, Aghajanian J, & Hertzman C. Neonatal outcomes after prenatal exposure to selective serotonin reuptake inhibitor antidepressants and maternal depression using population-based linked health data. *Arch Gen Psychiatry.* 2006;**63**:898–906.
22. Oberlander TF, Bonaguro RJ, Misri S, Papsdorf M, Ross CJ, & Simpson EM. Infant serotonin transporter (SLC6A4) promoter genotype is associated with adverse neonatal outcomes after prenatal exposure to serotonin reuptake inhibitor medications. *Mol Psychiatry.* 2008;**13**:65–73.
23. Berle JO, Steen VM, Aamo TO, Breilid H, Zahlsen K, & Spigset O. Breastfeeding during maternal antidepressant treatment with serotonin reuptake inhibitors: infant exposure, clinical symptoms, and cytochrome p450 genotypes. *J Clin Psychiatry.* 2004;**65**:1228–34.
24. Carcillo JA, Doughty L, Kofos D, et al. Cytochrome P450 mediated-drug metabolism is reduced in children with sepsis-induced multiple organ failure. *Intensive Care Med.* 2003;**29**:980–4.
25. Del Vecchio A, Laforgia N, Capasso M, Iolascon A, & Latini G. The role of molecular genetics in the pathogenesis and diagnosis of neonatal sepsis. *Clin Perinatol.* 2004;**31**:53–67.
26. Nebert DW, McKinnon RA, & Puga A. Human drug-metabolizing enzyme polymorphisms: effects on risk of toxicity and cancer. *DNA Cell Biol.* 1996;**15**:273–80.
27. Daly AK. Pharmacogenetics of the major polymorphic metabolizing enzymes. *Fundam Clin Pharmacol.* 2003; **17**:27–41.
28. Joseph T, Kusumakumary P, Chacko P, Abraham A, & Radhakrishna Pillai M. Genetic polymorphism of CYP1A1, CYP2D6, GSTM1 and GSTT1 and susceptibility to acute lymphoblastic leukaemia in Indian children. *Pediatr Blood Cancer.* 2004;**43**:560–7.
29. Ashton LJ, Murray JE, Haber M, Marshall GM, Ashley DM, & Norris MD. Polymorphisms in genes encoding drug metabolizing enzymes and their influence on the outcome of children with neuroblastoma. *Pharmacogenet Genomics.* 2007;**17**:709–17.
30. Van Dyke DC, Hodge SE, Heide F, & Hill LR. Family studies in fetal phenytoin exposure. *J Pediatr.* 1988;**113**:301–6.
31. Ginsberg JS & Hirsh J. Anticoagulants during pregnancy. *Annu Rev Med.* 1989;**40**:79–86.
32. Dai WS, LaBraico JM, & Stern RS. Epidemiology of isotretinoin exposure during pregnancy. *J Am Acad Dermatol.* 1992;**26**:599–606.